Atomic Masses and Fundamental Constants 5

Previously Published Conferences on Atomic Masses

Nuclear Masses and Their Determinations, H. Hintenberger (ed.), Pergamon Press, London (1957).

Proceedings of the International Conference on Nuclidic Masses, H. E. Duckworth (ed.), University of Toronto Press, Toronto (1960).

Nuclidic Masses, Proceedings of the Second International Conference on Nuclidic Masses, W. H. Johnson, Jr. (ed.), Springer-Verlag, Vienna and New York (1964).

Proceedings of the Third International Conference on Atomic Masses, R. C. Barber (ed.), University of Manitoba Press, Winnipeg (1968).

Atomic Masses and Fundamental Constants 4, J. H. Sanders and A. H. Wapstra (eds.), Plenum Press, London and New York (1972)

Atomic Masses and Fundamental Constants 5

Edited by

J. H. Sanders

Clarendon Laboratory
University of Oxford

and

A. H. Wapstra

University of Technology
Delft
and
Instituut voor Kernphysisch Onderzoek
Amsterdam

PLENUM PRESS · New York - London

Library of Congress Cataloging in Publication Data

International Conference on Atomic Masses and Fundamental Constants, 5th, Paris, 1975.
  Atomic masses and fundamental constants, 5.

  Includes index.
  1. Physical measurements—Congresses. 2. Atomic mass—Measurement—Congresses.
I. Sanders, John Howard. II. Wapstra, Aaldert Hendrik. III. Title.
QC39.I48 1975   539.7'2   75-44356
ISBN 0-306-35085-8

Proceedings of the Fifth International
Conference on Atomic Masses and Fundamental
Constants held in Paris, June 1975

© 1976 Plenum Press, New York
A Division of Plenum Publishing Corporation
227 West 17th Street, New York, N.Y. 10011

United Kingdom edition published by Plenum Press, London
A Division of Plenum Publishing Company, Ltd.
Davis House (4th Floor), 8 Scrubs Lane, Harlesden, London, NW10 6SE, England

Printed in the United States of America

The present fifth conference in this series - counting the 1956 conference in Mainz the same way as the "zeroth" symphonies of Beethoven and Bruckner - was held in conjunction with the centennial of the "Convention du Mètre" which was signed May 20, 1875 in Paris. Because of this occasion, an extensive honorary committee was appointed:

Honorary Committee; Chairman A. Kastler

R.C. Barber(M)	W.H. Johnson(M)
U. Bonse	K. Ogata(M)
E.R. Cohen(M)	Y. Sakurai
A.H. Cook	J.H. Sanders(M)
P. Dean	H.H. Staub(M)
R.D. Deslattes	U. Stille(M)
J.V. Dunworth	J. Terrien(M)
V.I. Goldanskii(M)	M.A. Thompson
P. Grivet	A.H. Wapstra(M)

The members (marked M) of the Commission on Atomic Masses and Fundamental Constants of the International Union of Pure and Applied Physics belonged ex officio to this Committee.

The Organizing Committee; chairman A. Horsfield

H. Capptuller
J.E. Faller
J.L. Hall
B.W. Petley
B.N. Taylor

was helped by the following two committees:

Program Committee; chairman A.H. Wapstra

C. Audoin   W.H. Johnson
W. Benenson   R. Klapisch
E.R. Cohen   A. Rytz
P. Giacomo   J.H. Sanders

Local Committee; chairman P. Grivet

C. Audoin   R. Klapisch
N. Elnekave   Y. Le Gallic
S. Gerstenkorn   P. Petit
F. Hartmann   M. Thue

Organizing this conference at the present time, now that it is more difficult to obtain money for science, was made possible by the generous help of the following sponsors:

International Union of Pure and Applied Physics
    (Commission on Atomic Masses and Fundamental Constants)
    (IUPAP)
International Union of Pure and Applied Chemistry (IUPAC)
Bureau International des Poids et Mesures (BIPM, Sèvres,
    France)
Committee on Data for Science and Technolgoy (CODATA)
Délégation Générale à la Recherche Scientifique et
    Technique (DGRST, France)
Centre National de la Recherche Scientifique (CNRS, France)
Comité Français de Physique
Bureau National de Métrologie (BNM, France)
Direction des Recherches et Moyens d'Essais (DRME, France)
National Bureau of Standards (NBS, U.S.A.)
National Physical Laboratory (NPL, G.B.)
Physikalisch-Technische Bundesanstalt (PTB, Braunschweig)
Rockwell International (U.S.A.)
International Business Machines (IBM, France) et (IBM,
    Europe)
International Telegraph and Telephone (ITT, Europe)
Commission des Communautés Européennes (Bruxelles)
Centre National d'Etudes des Télécommunications (CNET,
    France)
Association des Ouvriers en Instruments de Précision (AOIP,
    France)

A summary of the conference is given in the last paper.   The
progress since the last conference, AMCO-4, has been vigorous.
In the field of nuclear masses there have been some surprises;
they can be summarized in the statement that it appears that the
"Magic numbers" (for which enhanced stability occurs) are not at
all the same near the line of stability and near the nucleon drip
line.

In the field of atomic constants the big advance since AMCO-4
has been the new determination of the speed of light, but it
happened too long ago to make it 'hot news' at AMCO-5.   The
relevant techniques are, however, described in detail and their
achievements are impressive.   New determinations of the Avagadro
Constant and the Faraday have proved to be of considerable
interest, and in other fields the emphasis has been on consolida-
tion.

The Editors thank the authors of the contributions to this
volume for their collaboration in conforming with the stringent
requirements for brevity and promptness.   On this occasion they
feel confident in thanking the publishers in advance for their
efforts to ensure speedy publication.

A.H. WAPSTRA, Amsterdam

J.H. SANDERS, Oxford

LIST OF PARTICIPANTS

ALBURGER D.E.

ALECKLETT K.

ALLISY A.

ANDERSON G.

ANDREONE D.

ARDITI M.

AUDI G.

AUDOIN C.

BARBER R.C.

BARGER R.L.

BAUER M.

BAYER HELMS F.

BEHRENS H.

BEINER M.

BENENSON W.

BERTINETTO F.

BESSON J.

BESSON R.

BLEULER K.

BORCHERT G.

BOS K.

BOUCHAREINE P.

BOWER V.E.

BRACK M.

BRILLET A.

BROWNE C.P.

BRUN H.

BURCHAM W.E.

CAGNAC B.

CAMPI X.

CAPPTULLER H.

CEREZ P.

CHARTIER J.M.

CHRISTMAS P.

COHEN E.R.

COMMUNEAU F.

CURIEN H.

DAVIS R.S.

DEAN P.

DEHMELT H.G.

DELAHAYE F.

DENEGRE G.

DESLATTES R.D.

DETRAZ C.

DORENWENDT K.

DRATH P.

DUCKWORTH H.E.

DUNN A.F.

EBEL G.

ELNEKAVE N.

EPHERRE M.

ESKOLA K.

EZEKIEL S.

FALLER J.E.

FARLEY F.J.M.

FAU A.

FERRO MILONE A.

FINNEGAN T.F.

FREEMAN J.M.

GAGNEPAIN J.J.

GALLMANN A.

GARTNER G.

GERMAN S.

GERSTENKORN S.

GIACOMO P.

GIRARD G.

GISQUET M.F.

GOLDANSKII V.I.

GOVE N.B.

GOZZINI A.

GRIVET P.

GUELACHVILI G.

GUINIER A.

GUINOT B.

GYGAX F.N.

HAJDUKOVIC S.

HALL J.L.

HAND J.W.

HARDY J.C.	LE HONG L.	REYMANN D.
HARTMANN F.	LUND E.	RICH A.
HELLWIG H.	LUNDEEN S.R.	RICHARD A.C.
HELMCKE J.	LUTHER G.	RITTER R.C.
HELMER R.G.	MASUI T.	ROBERTS D.E.
HENRY L.	MATSUDA H.	ROBERTSON R.G.H.
HILF E.R.H.	MERCEREAU J.E.	ROECKL E.
HILL H.A.	MESNAGE P.L.	ROLFS C.
HOATH S.D.	MUIJLWIJK R.	ROUSSEL A.
HORSFIELD A.	MÜLLER J.W.	ROWLEY W.R.C.
HUANG DING L.	MYERS W.M.	RUTHMANN J.
HUARD S.	NAKABUSHI H.	RYTZ A.
HUBER G.	NAKAMURA A.	SACCONI A.
HUENGES E.	NGUYEN DUC V.	SAKURAI Y.
HUSSON J.P.	NGUYEN TUONG V.	SANDERS J.H.
ISLAM S.	NGUYEN VAN GIAI	SAUDER W.C.
ISRAEL G.	NOLEN J.A.	SCHELLEKENS P.
JANECKE J.	NYMAN G.	SCHULT O.
JENNINGS J.C.E.	OGATA K.	SEN S.K.
JOHNSON W.H.	OLSEN P.T.	SIEMSEN K.J.
JONSON B.N.G.	OPRAN R.	SPIEWECK F.
KASHY E.	PANAGIOTOU A.	SQUIER G.T.A.
KASTLER A.	PEOVER M.	STAUB H.H.
KELSON I.	PETIT P.	STILLE U.
KERN B.D.	PETTLEY B.W.	TAKAHASHI K.
KIBBLE B.P.	PHILLIPS W.D.	TANAKA K.
KLAPISCH R.	PICQUE J.L.	TAYLOR B.N.
KLEPPNER D.	PRESTON THOMAS H.	TERRIEN J.
KOETS K.	QUENTIN P.	THEOBALD J.G.
KOSE V.E.	RAMSEY N.F.	THIBAULT C.
LARINE S.	RAUCH F.	TORGERSON D.F.
LECLERC G.	REASENBERG R.D.	TRAHA M.
LEHANY F.J.	REMIDDI E.	TURNEAURE J.P.

TURNER R.	VOORHOF H.	WHITE R.E.
VAN ASSCHE P.H.M.	WALLARD A.J.	WILLIAMS E.R.
VAN BAAK D.	WAPSTRA A.H.	WITT T.J.
VAN DERLEUN C.	WAY K.	WOLLNIK H.
VASS D.G.	WEISS C.	ZELDES N.
VERMA A.R.	WERTH G.	ZOSI G.
VESSOT R.F.C.	WERTHEIMER R.	ZYGAN H.
VONACH H.	WESTGAARD L.	ZYLICZ J.
VON GROOTE H.		

CONTENTS

Part 1   The Fundamental Constants and Metrology

Part 2 Gamma rays

CONTENTS

Part 12   Miscellaneous Constants

ALLOCUTION DU PROFESSEUR Alfred KASTLER

Monsieur le Délégué Général,
Messieurs les Présidents,
Mesdames, Messieurs,

C'est pour moi un grand honneur de m'associer à mes collègues français pour vous souhaiter à tous la bienvenue à ce congrès international destiné à faire le point de nos connaissances sur les Masses Atomiques et les Constantes Fondamentales de la Physique. C'était certes une heureuse idée de jumeler ce congrès avec la réunion de la Commission Internationale des Poids et Mesures à l'occasion du Centenaire de la Convention Internationale du Mètre.

Vous attendiez de moi un exposé introductif, et je dois tout de suite vous dire que mon état de santé ne m'a pas permis, à mon grand regret de le préparer. Mais mon collègue et ami Pierre Grivet qui a déjà assumé de concert avec Monsieur Audoin, la lourde tâche de l'organisation de ce congrès, a bien voulu se substituer à moi. Il m'a fait l'amitié de me montrer son texte, et je peux vous assurer que vous gagnez au change. Mais avant de lui céder la parole, permettez-moi quelques remarques ayant trait à la Métrologie générale plutôt qu'à l'objet de votre congrès. Nous sommes certainement tous convaincus de l'importance, pour le progrès des sciences et des techniques et pour la promotion de notre civilisation, de la nécessité de savoir faire des mesures précises. Ceux d'entre nous qui ont eu le plaisir d'écouter samedi dernier, l'exposé de M. Terrien au Palais de la Découverte, se sont rendus compte du progrès accompli dans ce domaine depuis deux siècles.

Lorsque, comme nous l'apprend la bible, la malédiction de la Tour de Babel a frappé les hommes, elle a non seulement créé la multiplicité des langues, mais aussi la diversité et la confusion des unités de mesure, même et surtout de celles dont nous nous servons pour nos besoins quotidiens.

Si en ce qui concerne les langues, nous vivons toujours sous le signe de la malédiction, en ce qui concerne les unités de mesure, cette malédiction a été conjurée et l'universalité se trouve aujourd'hui rétablie.

ADDRESS BY PROFESSOR Alfred KASTLER

Monsieur le Délégué Général,
Messieurs les Présidents,
Mesdames, Messieurs,

It is a great honour for me to join my French colleagues in welcoming you to this international conference devoted to our knowledge of Atomic Masses and the Fundamental Constants of Physics. It was certainly a happy idea to link this conference with the meeting of the Commission Internationale des Poids et Mesures on the occasion of the Centenary of the Convention Internationale du Metre.

You were expecting me to give an introductory talk, but I have to tell you straight away that, to my great regret, my state of health has prevented me from preparing one. But my colleague and friend Pierre Grivet, who has already undertaken, together with Monsieur Audoin, the heavy task of organizing this conference, has kindly agreed to take my place. He has been good enough to let me see the text of his talk, and I can assure you that you will benefit from the change. But, before I hand over to him, I should like to make some remarks on Metrology in general terms, rather than on the topics with which this Conference is mainly concerned. We are all without doubt convinced of the importance to the progress of science and technology of being able to make precise measurements, as well as for the advance of our civilization. Those of us who had the pleasure of hearing M. Terrien's address at the Palais de la Découverte last Saturday are aware of the progress that has been made in this field over the past two centuries.

When, as the Bible teaches us, the curse of the Tower of Babel fell upon mankind, it was responsible not only for a multiplicity of tongues but also for the diversity and confusion of units of measurement, not least of those in daily use.

As far as language is concerned we still live under the sign of the curse, but when it comes to units of measurement the curse has been charmed away, and a universal system has now been established.

Personnellement, je dois à la métrologie le succès qui marque le début de ma carrière universitaire. Lorsque, jeune lycéen, il y a plus de cinquante ans, je me présentai au concours de l'Ecole Polytechnique, je dus aborder à l'écrit une épreuve redoutable, celle de la composition française. Le candidat avait le choix entre trois sujets. Les deux premiers m'étaient interdits car ils faisaient appel à des connaissances en littérature française qui, vu ma formation d'Alsacien passé par l'école allemande, me faisaient défaut. Le troisième sujet me sauva. En voici le texte : "Commentez cette pensée du philosophe Le Dantec : Il n'y a de science que de mesurable".

L'ayant choisi, je fus reçu glorieux 228e. Ce succès me délivra d'un complexe d'infériorité et me permit d'entrer en toute quiétude à l'Ecole Normale par la petite porte.

Bien plus tard, lorsque j'eus l'occasion d'enseigner la Physique, je ne manquais pas de proposer à mes étudiants des méditations et des exercices pratiques sur les unités de mesure. Ainsi lorsque j'eus à commenter les célèbres expériences par lesquelles James Prescott Joule détermina l'équivalent mécanique de la calorie, je leur fis comprendre que le moindre potache a aujourd'hui sur Joule une supériorité certaine, car Joule ignorait une unité de travail que nous manipulons fréquemment et que nous appelons le "Joule". Alors comment faisait-il ? Il mesurait le travail en "foot-pounds", produit de 1 foot (pied anglais) = 30,5 centimètres par 1 pound (livre anglaise) = 453 grammes.

Et il exprimait les quantités de chaleur en B.T.U. ou British Thermal Units, cette unité représentant la chaleur nécessaire pour élever de 1° Fahrenheit la température de 1 livre d'eau. Et Joule trouva que pour produire 1 B.T.U. de chaleur, il lui fallait dépenser 778 foot-pounds. Je vous laisse le soin de convertir en calories et en joules !

Je voudrais raconter à ce propos une petite anecdote relative aux années 1946-47 où tous les pays d'Europe connaissaient encore un rationnement alimentaire sévère. Quelques députés ayant en effet osé reprocher au Gouvernement lors d'une séance à la chambre de tolérer l'exportation en Angleterre de notre beurre breton, le Ministre du ravitaillement, indigné, rétorqua que c'était là pure calomnie. Alors, pourquoi donc, lui demanda insidieusement un député, fabrique-t-on dans certaines laiteries de Bretagne des paquets de beurre de 453 grammes ?

For my own part, I owe the success which marked the beginning of my university career to metrology.   When, as a young high school student more than fifty years ago, I had to sit for the entrance examination of the Ecole Polytechnique I was faced with the awe-inspiring task of writing a French essay.   Candidates had the choice of three subjects.   The first two were beyond me, as they required a knowledge of French literature which I lacked, due to my education as an Alsatian in a German school.   The third subject was my salvation.   Here is text of it : "Discuss this observation of the philosopher Le Dantec : The only knowledge is that which can be measured".

Having chosen this, I was placed a glorious 228th.   This success freed me from an inferiority complex, and enabled me to slip quietly into the Ecole Normale by the back door.

Long after, when I came to teach physics, I took care to suggest to my pupils the study of units of measurements and exercises in their use.   Again, when I discussed the celebrated experiments in which James Prescott Joule measured the mechanical equivalent of heat I told them that the humblest schoolboy has an advantage over Joule, since Joule knew nothing of the unit of work which we often use and which we call the "joule".   Then how did he manage ?   He measured work in "foot - pounds", the product of 1 foot (the length of an Englishman's foot *) = 30.5 centimetres, and 1 pound = 453 grammes.

Again, he expressed quantities of heat in B.T.U. or British Thermal Units, this unit being the amount of heat required to raise the temperature of one pound of water through 1° Fahrenheit. Joule also found that to produce 1 B.T.U. of heat one had to use 778 foot - pounds.   I leave it to you to convert to calories and to joules !

I should like to tell you a story in this connexion about the years 1946-47 when all European countries were subjected to severe food rationing.   Several deputies having dared, in effect, to accuse the Government of allowing the export to England of our Brittany butter, the Food Minister, outraged, retorted that this was sheer slander.   Why then, insisted one deputy, were some dairies in Brittany making up packets of butter each weighing 453 grammes ?

---

* Footnote :   I apologise to Prof. Kastler for any liberty I have taken in translating this phrase.
I am not sure that this was the meaning he intended. — J.H.S.

Devons-nous tirer de cette histoire la conclusion qu'il est beacoup plus facile de faire l'Europe dans l'illégalité, que de la réaliser légalement ? Espérons que mes amis britanniques, gardant quelque reconnaissance aux laiteries bretonnes, voudront bien dans peu de jours, confirmer leur association à l'Europe[x].
N'ont-ils pas déjà accompli une véritable révolution lorsqu'il y a quelques années, ils ont adopté une monnaie à subdivision centesimale !

Ne nous est-il pas permis dans ces conditions de rêver à un avenir où nous aurons en Europe une monnaie unique et, pourquoi pas, une langue commune ?

Ayant eu, nous autres Francais une grande satisfaction d'amour-propre dans le domaine des unites de mesure, sommes-nous prêts à faire, et ceci s'adresse à nous tous, une concession en vue du choix d'une langue commune ?

Pouvons-nous demander en attendant à chacun de ceux qui prennent la parole à une tribune internationale de bien vouloir parler lentement, distinctement et en mettant les décibles nécessaires ? Tout en nous rappelant l'heureux temps où dans nos Universités Européennes du Moyen Age, professeurs et étudiants, de Bologna à Oxford et de Coimbra à Uppsala parlaient la même langue universelle !

---

[x] Ce qui s'est fort heureusement réalisé.

Should we draw from this the conclusion that it is much easier to unite Europe by underhand means rather than openly ? Let us hope that my British friends, remembering the dairies of Brittany, will in a few days' time confirm their membership of Europe.*   Have they not already achieved a real revolution by adopting a few years ago a centesimal monetary system ?

Can we not be allowed, in these circumstances, to dream of a single European coinage, and, why not, of a common language ?

Since we Frenchmen have derived a great deal of self-satisfaction and pride in the field of the units of measurement would we, and this concerns us all, be ready to make a concession over the choice of a common language ?

In the meantime may we ask all those who speak at an international gathering to talk slowly, clearly and with the necessary decibels ?   All of us recall that happy time when in the Universities of Europe in the Middle Ages both students and teachers, from Bologna to Oxford, and from Coimbra to Uppsala, spoke the same universal language!

---

* This, very fortunately, has now been done   [by the Referendum of 5 June 1975].

OPENING REMARKS

E. Richard Cohen

Science Center, Rockwell International

Thousand Oaks, California 91360

I wish to welcome you, in these pleasant surroundings, to the
Fifth International Conference on Atomic Masses and Fundamental
Constants. The capital I in "International" implies sponsorship
by one or more of the ICSU family of unions, in this case primarily
IUPAP, but also IUPAC, more recently, CODATA. However there are
three other conferences of importance in this series. There was
the so-called "zero conference" on atomic masses which was held in
1956 in Mainz as a portion of the celebration of Prof. Mattauch's
60th birthday, and a smaller, more informal conference, also in
Mainz, ten years later to celebrate his 70th. This mini-conference
is important, among other reasons because there DuMond, almost immediately after the publication of the details of the 1963 adjustment of the fundamental constants was already questioning the
accuracy of the value of the fine structure constant, α, and
emphasizing the importance of this constant in the determination
of the entire set. Furthermore, one should not ignore the independent International Conference on Precision Measurement and Fundamental
Constants held at the U.S National Bureau of Standards in 1970 and
well-known simply as the "Gaithersburg Conference".

The conjunction of atomic masses and fundamental constants may
seem remote to some, and indeed the connection, apart from a common
interest in accuracy and precision and the physics that is to be
learned from the next decimal place - rather than merely the numer-
ology of being able to quote a physical magnitude to 9 or 10 or
even 12 digits - is a narrow one. However, if it is a narrow
channel, it is one which connects two large and important bodies of
knowledge, and is important in defining the level of each.

In the late 50's when the question of a redefinition of the
scale of atomic weights was under heated discussion, Mattauch and
Nier strongly opposed the suggestion that $m(^{19}F) \equiv 19$ be used as
the definition of a unified scale of atomic masses (replacing the
chemical scale, $m(0) = 16$, and the physical scale $m(^{16}0) = 16$.)
Mattauch pointed out that using fluorine and fluorine compounds in
a mass-spectrometer was almost an impossibility and hence one would
have a standard which was unuseable. He also pointed out that
defining $^{12}C = 12$ would be an excellent choice, both because it
involved only a small change in the numerical values assigned to
the masses, (65 ppm to the chemical scale, 340 ppm to the physical
scale) and because C was on many points a fundamentally better
standard than 0. Because hydrocarbon fragments, C_nH_m, in several
ionization states are easy to introduce into mass spectrometers
one could use these fragments, as well as $^{12}C_n{}^{13}C_mH_rD_s$ fragments
to provide calibration points throughout the mass range from 4 to
at least 200.

The Commission on Atomic Masses and Related Constants was
established to provide liaison between the physics community and
the IUPAC Commission on Atomic Weights during the transition
period of the introduction of the unified mass scale. Mattach and
Wapstra would collaborate on the least-squares analysis of all of
the mass doublet and reaction energy data to provide an extensive
and consistent set of atomic masses. To combine mass doublet data
with β-decay and nuclear reaction energetics data one needed a
conversion constant between atomic mass units and electron volts.
This conversion factor would come from DuMond.

The statistics on the conferences from 1960 to 1975 are interesting and show the growth and development:

	Attendance	Papers	Atomic Masses	Fund. Const.
1960	65	34	34	0
1963	65	38	32	6
1967	75	40	33	7
1971	90	62	30	32
1975	>150	90	40	50

The zero papers on fundamental constants in 1960 is a little mis-
leading; there were in some sense two or three: two papers on the
general area of the mathematics of least-squares adjustments and a
paper on the isotopic abundances and atomic weight of Ag. In 1963
one of the six sessions was specifically devoted to fundamental
constants. The jump in papers on fundamental constants in 1971 is
a reflection of the increased emphasis on this area within the
Commission and the strongly correlated fact that the location of
the conference was NPL.

The fact that this week's conference is part of the celebration of the centennial of the "Convention du Mètre" and the establishment of BIPM is partially responsible for the strong emphasis here on fundamental constants. But there is a widening of the scope of the conference on the mass side as well: in 1971 there was some discussion of the masses (or energy levels) of excited states of nuclei and increased discussion of the influence of excited electronic states of the products in nuclear reactions on Q-value measurements. In this conference we have a strong emphasis in this area, and on the masses of nuclei far from the stability line. There are also several papers on the mass-spectroscopy of radioactive nuclides. There has been, and I hope will continue to be, an increasing number of papers on the theory of nuclear masses and nuclear mass formulae. I pointed out in my summary of AMCO-4 that there are two distinct uses for atomic mass formulae: one for interpolation to evaluate masses of unmeasured or unmeasurable nuclides and to resolve such experimental questions as whether a measured mass refers to the ground state or an excited nuclear state; two, to gain a theoretical understanding of nuclear matter, shell structure and deformed nuclei. Until we have a complete theory, rather than simply a model, of nuclear structure there is no a priori reason why the "best" formulas for these two very different purposes should be identical. I hope in my summary remarks on Friday to be able to report on this aspect in more detail.

With this brief sketch of the history of these conferences and some incomplete analysis of the scope of the present one, I shall end these remarks and allow the conference to speak for itself.

Tycho Brahe

THE MEASUREMENT OF FUNDAMENTAL CONSTANTS (METROLOGY)

AND ITS EFFECT ON SCIENTIFIC AND TECHNICAL PROGRESS

A. Kastler, Nobel Prize, Member of the Academy of Science
P. Grivet, President of the National Bureau of Metrology,
    Member of the Academy of Science

Académie des Sciences, 23 quai Conti, 75006 Paris, France

I. INTRODUCTION

It is perhaps appropriate that in this year 1975, which marks
the centenary of the *Système Métrique* and that of the *Bureau International des Poids et Mesures*, scientific activity in the field of
Metrology appears to be flourishing. This is witnessed on the one
hand by the number, range and quality of the contributions to this
Symposium and, on the other, by the importance which most of the 44
Governments affiliated to the BIPM attach both to their Basic Laboratories and to the network of technical centers engaged in developing measuring methods and in disseminating their results to all levels of the various branches of industry. Should this state of af-
fairs be considered a happy but ephemeral stage in the march of
science, or does it represent a tendency of fundamental importance
in the evolution of science and technology ? Enough is now known
about the history of civilization for us to affirm that, even since
the Renaissance, Metrology has payed an ever greater part in both
scientific and civic activities.

II. THE OBSERVATORY AT URANIBOURG, THE FIRST LABORATORY FOR INVESTIGATING THE METROLOGY† OF ANGLES

The 16th century offers a very good example of the trend. At
first, it is worth noting that there are many analogies between the
historical evolution that followed the invention of the printing
press and the discovery of the New World, and what we see in our
own 20th century: in both periods civilized man was subjected to the
impact of great discoveries which, for two or more generations,
had a profound influence on scientists, explorers, inventors

---

† A popular science known as *Gnomonic* at the time.

Table 1

Items	16th century	20th century
	N.B. The rate of evolution seems rather slower than in our own time	
Application	Firearms (15th-16th century)	Atomic bomb (1945)
Religious and social doctrines	Luther (1525) - Calvin (1535)	Lenin (1918) - Mao Tsé Tung (1950)
Applied mathematics	Use of arabic numerals, algebra, trigonometry	Electronic computers
Great discoveries	Voyages to America: C. Columbus (1492) - J. Cartier (1534) Global circumnavigation: Vasco de Gama (1497) - Magellan (1520)	Transcontinental flights for the public Space exploration Gagarine (1957) Conquest of the moon (1969)
Physical astronomy	Accurate description of the solar system. Copernicus' theory (1543) Tycho-Brahe's observations (1576-1601) Kepler's laws (Astronomia Nova): 1st-1606, 2nd-1616, 3rd-1618 (Harmonies Mundi) Galileo's telescope (phases of Mars and Venus, Jupiter's satellites, sunspots)	Exploration of Venus, Mars and Saturn by automatic space probes.

	17th and 18th centuries	20th century
Scientific achievements	Newtonian dynamics (Principia) Refractive and reflective telescopes J. Watt's steam engine	Einsteinian dynamics Quantum mechanics Nuclear energy

and engineers; and this influence rapidly extended to the man in
the street. Table I gives some very significant bases of comparison. The dates given either identify an event directly or desi-
gnate the time of its impact on the public.

This very succint table of 16th century achievements reveals
a great proliferation of ideas, which, in spite of causing painful
upheavals in the life of all Europeans, resulted in the birth of a
sober and meticulous science based on the measurement of angles -
that of Situation Astronomy. Thus, 500 years ago, the time became
ripe for Copernicus. After travelling widely in Europe and visit-
ing the leading scientists and erudites, he returned to his uncle's
home at Frauenberg and concentrated on a penetrating criticism of
Ptolemee's system. An Arabic translation of the celebrated Alexan-
drian astronomer, in particular that of his treatise on astronomy,
the *Almagest*, had reached the Middle Ages intact and exerted a
profound influence on the astronomical concepts of the age. His
theory was that all the celestial bodies revolved round the earth.
The Sun and thousands of other stars described a simple circle in
24 hours, but the Moon and the planets had zig-zag trajectories.
To explain this anomaly there were various theories involving *cycles*
and *epicycles*. Copernicus had at his disposal not only data collected by the Greeks but also later and more accurate observations
made by Arab astronomers. These had never ceased to maintain and
use their observations, especially that at Samarkand. Their observations were assembled during an astronomical conference held
at Toledo under the patronage of Alfonso X, King of Castile, who
brought and kept together for four years the best 13th century
Christian, Jewish and Arab astronomers. The results of this, the
first-ever astronomical symposium, were published at Venice in
1483. The results consisted of the celebrated *Alphonsine Tables*
which contain observations with a mean error of about 10' and from
which were expurgated most of the erroneous information introduced
over the centuries to justify the Ptolemaic theory. In these tables
Copernicus found all the data required to support his heliocentric
theory of the solar system. His brilliant calculations of the trajectories of Mars and Venus, which are not only the nearest to the
Earth but are also the brightest and the best observed, reveal the
astonishing simplicity of their circular and heliocentric trajectories within the ecliptic (the plane of the Earth's orbit). The
very harmony and simplicity of this view of the universe gave
Copernicus inexhaustible faith in the reality of his discovery.
His personal conviction was, however, difficult to pass on to his
contemporaries because it was based on mathematical calculations
which only a hand ful of astronomers could understand, let alone
believe. There was some excuse for the unbelievers in the fact
that the inaccuracy of the Alphonsine Tables (about 10' of arc)
helped the Copernican theory to gain acceptance (Copernicus himself was not much of an observer and only made about thirty ob-

servations) as it did not reveal orbital ellipticity and thus simplified the theory and the computations by reducing the elliptic
orbits of Mars and the Earth to circles.

During the 16th century, only Kepler and Galileo accepted the
Copernican theory[†] though Galileo was careful not to own his conviction for some considerable time in deference to the state of public
and clerical opinion, thus revealing much about the initial fragi-
lity of this great discovery.

Copernicus had indeed confronted European scientists with a
new and major problem, which was only to be solved after an earnest
quest for accuracy, lasting more than 30 years, by two men of genius,
Tycho-Brahé the observer, and Kepler the mathematician. These two
were only able to collaborate fully for a period of eighteen months,
Tycho-Brahé dying in 1601 without ever being fully convinced of the
validity of the Copernican theory. It was only after Kepler died in
1630 that his persistence was crowned and the scientific world
yielded Copernicus his due.

Tycho-Brahé was a noble man and a nephew of the high Admiral of
Frederick the Second, King of Denmark. He was 13 years old when he
began classical studies at the University of Copenhagen, but became
an enthusiastic astronomer after witnessing an eclipse of the sun
occuring at the time officially predicted for it. When he was 16, he
began to travel throughout Europe in search of knowledge, a practice
which he continued profitably for more than 10 years, nourishing his
passion for astronomy by inventing precision instruments for measuring angles and dreaming of reducing the standard minute of arc to
smaller sub-divisions. He thus built a collection of instruments
(alidades, astrolabes, etc.), drawing and projects, all with the
characteristic of operating with the *naked eye*. He also set himself
to master their use; to such good effect that on being 17 years old
in August 1563, he was able to observe from his room a conjunction
between Jupiter and Saturn - and was outraged because the predicted
date was several days out.

When being 26, he felt that his education was complete and returned to Denmark, where he devoted all his time to studying astronomy and astrology, abetted in this by an uncle, Steen, who was rich
and something of an alchemist. His renown as an astronomer soon
crossed the Danish border and fame came when, on November 11, 1572,
using a large sextant he had made himself, he established to within
a minute the position of the first nova identified in modern times.

This feat considerably impressed his astronomer peers but also
brought him to the notice of the princes and peoples of Europe all
believers in astrology, who found the nova disquieting. He then
published his first book *De Stella Nova* , which was a judicious
mixture of astrology and real science. It was a resounding success
and three years later, he was so well known that the Landgraf of

---

† in *De Orbium Celestium Revolutionibus,* Nuremberg 1543, see the
   beautiful reedition by the French CNRS.

Kassel (an enlightened man) and King Frederick of Denmark fought
to secure his services. In 1576, Frederick won offering Tycho -
Brahé the island of Ven[†], a flat chalky rock some five kilometers
long near Elsinore in the Oresund Strait. Here Tycho built the
first observatory in Europe and called it Uranibourg, designing
and constructing it both as a Palace of Science and as a baronial
hall capable of welcoming noblemen. Many were the lordlings and
princes who subsequently went there for their first taste of what
they considered merely a revolutionary form of astrology. As his
fame grew and his talents ripened, Tycho extended Uranibourg, adding to it a few years later an annex which he named Stellabourg.
This incorporated the prototype of the modern observatory, which
is semi-buried in order to protect its instruments from the effects
of wind and changes in temperature. For twenty years Tycho worked
incessantly at Stellabourg listing the positions of nearly 800 stars
and compiling an impressive number of observations on the planets -
all accurate to within *one minute* of arc.

Arthur Koestler calls this *Tycho's Treasure* in his book *The
Watershed*[††] devoted to the lives of Tycho-Brahé and Kepler and
written in 1960 with all the spirit and talent of the human and
imaginative author that he is.

Johannes Kepler was born in 1571 and was a convert to
Copernicus before he graduated at the age of 20 from the Protestant
University at Tübingen. All his life he was attracted by the in-
trisic harmony displayed by numbers and figures such as polygons
and regular polyhedrons. Hence the inspired and profoundly mystic
student that he was could not fail to be immediately drawn by the
symmetry of Copernicus' circles round the sun.

In 1593, two years after graduating, he became Professor of
Astronomy and Mathematics in the very Roman Catholic city of Gratz
in Austria, where his passion for the Copernican theory and astronomy in general became much more rigorous and scientific without
losing its initial warmth. This passion was to guide his actions
for the rest of his life. Koestler's *Watershed* paints a vivid
picture of his incident-full maneuverings and his repeated and
often heroic efforts to have the Copernican theory accepted scientifically. Here we can only highlight the fact that the *Mysterium*

[†] Ven, from Venus, the goddess to whom the island was dedicated
in Roman times. The spelling often adopted in English or Danish
documents somewhat masks this derivation.

[††] A fragment from his book *The Sleepwalkers* or *A history of man's
changing view of the Universe* published in 1959.
The Watershed or *A Biography of Johannes Kepler* was published
by Heinemann in 1961 (Science Study, No 14).

Cosmographicum, his first book published in 1597, only reveals the mystical side of his thinking. It does not bring out the hours of thought spent in seeking if not an explanation at least a mathematical expression of the Copernican theory. From his correspondence it appears that the mass of his observations and calculations led him to criticize the means of measurement at his disposal. Dissatisfaction with his best figure of 8' spurred him to seek contact with Tycho, for he knew of the latter's treasury of observations and fully appreciated its worth.

A first exchange of correspondence between the two proved unfruitful, but both were to fall on evil days and thus be drawn irresistibly and providentially together. In 1598, the closure of his Protestant school by the Archduke of Habsburg, an intolerant Roman Catholic, forced Kepler to leave Gratz. A year previously Tycho had irreparably fallen out with his new young King Christian IV and had left his island and Uranibourg[†] to seek protection at Prague from Emperor Rudolph II, another prince attracted by the sciences, astronomers and astrologers.

Tycho had not forgotten to take with him from Uranibourg his instruments, which were all as disassemblable (one of his first inventions) as they were accurate. He used them to equip the new observatory he created in 1599 at Benatek Castle. Here Kepler joined him in February 1600 after experiencing various vicissitudes. Kepler's first law dates from that year. Tycho, eighteen months before his death in 1601, had Kepler officially appointed as his assistant, opened his *Treasury* of observations to him and gave him the task of studying Mars, the nearest to the Earth of the planets farther from the sun and the one with the easiest amount of ellipticity to determine (although only 1/60). Two days after Tycho's death in November, Kepler has the unexpected good fortune to be appointed his successor as Imperial Mathematician. From then on, working unceasingly, Kepler used the accuracy of Tycho's *Treasury* to establish his second and third laws. His basic advance was to calculate the shape of Mars' orbit to the nearest 1' and to recognize that it formed an ellipse of which the shorter axis is

---

† Uranibourg has left no historical traces, for after Tycho's departure, King Christian VII gave the island of Ven and its serfdom, to one of his favourites, who soon had the two observatories levelled to the ground. So well, that in 1671, when the Abbé Picard, a noted and enterprising astronomer, planning an expedition to Lapland in order to measure an arc of the meridian there, and, wishing to use Tycho's observations, needed to verify the latitude and longitude of the observations from which they were made, he had to dig to find Uranibourg foundations.

only 4,29 10^{-3} less than the longer. From this discovery he formulated in 1606 his second law relative to the constant areolar
speed. His third law was formulated much later and was not published till 1618 in one of his last works: the *Earth's Harmonies*.
More admirable still, after this sum of exhausting labour, the now
ailing Kepler spent six years of his waning energy in making
Tycho's Treasure available to all astronomers, sorting and completing the information they contained before publishing, largely at
his own expense, the celebrated *Rudolphine Tables* printed at Ulm
in 1627. In addition to Tycho's 777 stars, the Tables contain another 226 added by Kepler, accurate refraction[†] tables and a table
of logarithms, over the use of which Kepler had enthused ever since
their discovery in 1617; he had indeed written an explanatory
booklet to popularize their use.

   Newton was the first to take advantage of Kepler and Tycho's
common legacy, making it the basis of his first investigation into
universal gravity when he left Cambridge for his Woolsthorpe estate
in 1666. He was also able to make good use of Galileo's knowledge
of optics. Galileo's telescope had, in Kepler's lifetime, given
the public qualitative but striking evidence of the truth of the
Copernican theory by making it both possible and easy to make new
observations. It enabled not only professionals but also numerous
amateurs to discover :

- Jupiter's four satellites. This reduced the Moon from the rank
  of a Ptolemean celestial body to the more modest role of a satellite.

- Sunspots and their movement. By analogy, this gave credibility
  to the theory that the Earth rotated.

- Finally, the phases of Mars and Venus. The differences between
  the evolution of their phases show clearly that, in the heliocentric system, one is further from the Sun than the earth, while
  the other is closer.

   Fifty years after these discoveries, Galileo's telescope,
equipped with a convergent ocular (added by Kepler) and later with
spider-lines, enabled Halley and Flamsteed to improve positional
accuracy by a factor of nearly 10, reducing the error to less than
10" . Flamsteed's Tables, which proved this, were published in

---

† The correction required is small (< 2') unless the body is close
  to the horizon. Tycho's empirical tables contained a scholastic
  reminiscence: no correction was required in the region of 45°
  above the horizon. In fact, the error at 45° is 1'. Kepler, one
  of the founders of optical science (he published *Dioptics* in 1610),
  did not indulge in the same error.

1675, but Newton[†] had knowledge of the information they contained long before that time.

Here at last was the time when Copernicus'ideas would sweep all before them Europe-wide. Two of the greatest states in Europe were to build observatories. In 1667, France's Colbert inaugurated the Paris Observatory headed by Cassini, and in 1675, England's Charles II yielded to Flamsteed's urging and decided to build Greenwich Observatory.

Table II summaries the main stages of this revolutionary period in the history of astronomy.

Admittedly the creation of these observations was also a triumph for the precise measurement of angles which Tycho-Brahé first undertook at Uranibourg. In our time we can appreciate the importance of this innovation by comparing it to what lay ahead of Newton's inverse square law: Coulomb used it as a basis for establishing the first and fundamental quantitative law in electricity and later a similar law for magnetism. Thus, after more than two centuries, the work of Tycho-Brahé and Kepler indirectly provided a basic contribution to the increasing importance of electromagnetism, one of the main springs of our 20th century civilization. So, too, Tycho-Brahé's search for a minute of accuracy in the 16th century was not an expensive astrologer's folly but a bona-fide investment whose only disadvantage lay in the fact that it fructified at long term .

III. THE VELOCITY OF LIGHT IN VACUUM c

The social value of Metrology is well brought out by considering the part played in the evolution of science and technology by precise knowledge of the velocity of light c . Its influence extends from more than three hundred years ago to the 5th AMCO conference at which will be announced and discussed an improvement of two significant figures in the known velocity of light. This was achieved by using a laser, an invention scarcely 15 years old.

----

† Newton also owes a debt to the length metrology. His theory of gravitation was not published till 1682, sometime after he had learnt the results of the Abbé Picard's first ever measurement in 1659 of the length of an arc of a meridian (that between Sourdon near Amiens and Mallevoisine to the South of Paris).The resulting new figure for the Earth's radius fitted in Newton's calculation of the value of g, and the value as measured by pendulum.

Table 2

Contributor	Published work	Instruments	Accuracy	Discoveries
Ptolemy (130) (Alexandria)	Almagest			
Arab Astronomers		Naked eye & alidade		
Alfonso of Castile (1252-1284) (4-year conference at Toledo)	Alphonsine Tables (Printed at Venice in 1483)		10'	
Copernicus (1543)	De orbium celestium Revolutionibus	30 observations		Heliocentric system
Tycho Brahé (1576)		Naked eye: improved alidades Uranibourg Observatory	1'	Ellipticity of orbits of Mars and the Earth
Kepler	Rudolphine Tables (Ulm, 1627) Booklet on Logarithms	Naked eye and Tycho's alidades at Prague	1'	Kepler's 3 laws
Galileo (1609)		Refracting telescope		- 4 Jupiterian satellites - Phases of Venus & Mars - Sunspots
I. Newton (1666-1686)	Principia			Newton's laws
J. Flamstead (1675)	Flamstead's Tables	Refracting and reflecting telescopes	10"	Gravitation and dynamics of solar system
Modern observatories	Paris (1667) J.D. Cassini Greenwich (1676) J. Flamstead			

Here again, however, the contribution made by the past is as interesting as that of the present, for it brings out clearly, like a film run in slow motion, that the apparently pure research of ever greater levels of accuracy has had an impressive impact not only on the way a few European states evolved but nowadays on the direction in which the whole of humanity is to go. Knowledge of c , already fundamental at the earth scale in maritime and air navigation, is also necessary for controlling satellites and rockets on their interplanetary trajectories; finally, through Einsteinian mass-energy equivalence, knowledge of c contributes to the design and evaluation of tomorrow's sources of energy, atomic reactors, and the day after tomorrow's nuclear fusion.

Galileo's dream of measuring the velocity of light was first realized in 1675, 33 years after his death by the Danish astronomer Olaf Römer, using Galileo's telescope and thanks to the Jupiterian satellites he had discovered. Since Uranibourg Laboratory no longer existed, Römer had come to work in the Paris Observatory where his brilliant reputation gave him the coveted post of mathematics teacher to the Dauphin. At the observatory he noticed a small[†] biannual variation in the period (about 1,8 days) of Jupiter's first satellite Io, and correctly attributed this to the difference in time light took to travel the varying distance between Jupiter's satellite and the Earth. Although this difference is barely perceptible between two of the satellite's successive appearances either side of Jupiter's shadow, it is fortunately cumulative[††] from the time of the maximum range to the time of the minimum range[†††].

Römer obtained a value for c $= 214\ 000$ km s^{-1}, which he finally expressed in the formula $\tau = 2a/c$ where τ is expressed in seconds and is the resultant of all the small discrepancies in period observed. If 2a is the length of the greater axis of the Earth's elliptic orbit (i.e. the present unit of astronomic length) then τ is the time light takes to travel that distance, $\tau = 11' = 660''$.

[†] There is evidence to support the claim that the original idea was Cassini's but he decided not to pursue it when he found it impossible to use Jupiter's 2nd satellite Europa (period: 3.5 days) or the 3rd, Ganymede (period: 7.1 days) for the purpose. This objection was not to be proved invalid till much later. Cassini did, however, authorize Römer to proceed with both observations and the resulting theory.

[††] See pages 60-62 of *Optics* 1954 for an elegant and original presentation of the theory by A. Sommerfeld.

[†††] Jupiter, which revolves round the sun once every 11 years, can be considered stationary during the six months separating the 2 times presently considered.

Römer's contemporary astronomers considered this discovery so important that they called his equation *light's equation*, thus for the first time in astronomical history spectacularly honouring a method of measuring time. It must be admitted, however, that Römer's exploit owed much to Huyghens' years of work on clocks culminating in the publication of his treatise *Horlogiorum Oscillationum*, as, the introduction of the Galileo's pendulum in clockmaking technology, had brought the time measurement accuracy to the level of real metrology. Römer's method continued to be used up to the end of the 19th century by Delambre ($\tau = 493''$ in 1817), Glasenapp ($\tau = 500''$ in 1875) and finally Sampson at Harvard in 1909 ($\tau = 498.8''$). Sampson was the last astronomer to compete with the physicists, with Michelson in fact. Michelson's figure for c gives a $\tau = 500''$. Thereafter discussions over the errors caused by Jupiter's penumbra showed that there was little hope of achieving greater accuracy by the use of Römer's method.

Fizeau was the first to measure c on the earth itself in 1849, using a toothed wheel to chop light in his laboratory. His result of 310 000 km s^{-1} was achieved just at the time that two German physicists, Kohlrauch and Weber, were starting to determine the charge in a condenser by two different methods based, on one hand, on the measurement of the electrostatic force between its 2 plates and, on the other hand, on the measurement of the magnetic force produced by the discharge current. Each method used a different unit of charge, one magnetic the other electric, thus making it possible to establish a ratio between the two. The *Annalen der Physik* of 1856, recounting ten years of memorable experiments, published c' = q_m/q_e as being the equivalent of c , giving c' = 310 000 km s^{-1}. At about the same time Maxwell had established on a theoretical basis his 2 fundamentally important equations, which he announced for the first time in a memoir dated 1855[†]. One cannot do better than quote one of his own letters at the time to reveal the profound impact the measures of c and c' and the verification of their equality had on him. He said *we can scarcely avoid the inference that light consists in the transverse undulation of the same medium which is the cause of electric and magnetic phenomena.*

Theory and experiment then proceeded at the same pace. In 1862 Foucault with his *rotating mirror* achieved at his first try a remarkable degree of accuracy[††] finding c = 298 000 km s^{-1}. In 1864

[†] The reader will find a deep and detailed analysis of the thought of the main protagonists of this scientific revolution in the book entitled *History of the Theories of Aether and Electricity*, t.I, § VIII, Maxwell, by Sir Ed. Whittaker .

[††] A degree of accuracy that the successors of Fizeau, such as Cornu in 1875 and Perrotin in 1902, were unable to achieve.

Maxwell disclosed the details of his theory to the Royal Society[†].
H. Hertz's discovery[††] in 1888 of the electromagnetic waves which
bear his name was a direct result of an effort to provide experimental proof of Maxwell's theory. Very soon afterwards in 1890,
E. Lecher, the Austrian physicist, measured on a two-wire line the
velocity at which Hertzian waves travelled and found that it was
within 2% of c .

    This chain of events shows clearly the importance of the con-
tribution to our current knowledge of electromagnetism made by
attempts to measure c . With hindsight they can be seen to have
been of fundamental importance: at a critical time they provided
crucial support to the thinking of founders of electrical theory.
This theory, as yet unchallenged, has greatly hastened the widespread use of electrical power, made possible radio communications,
and provided a quantitative basis for Einstein's Restricted Relativity. If there is accordingly little point in expatiating further on the major role electrical theory has played in the social
and industrial revolutions of the last hundred years, the recent
history of c is worth summarizing, for it throws light on noteworthy advances in Metrology:

    - c continued to be measured over wires until 1923, the year
in which Mercier working in Professor H. Abraham's laboratory at
the Ecole Normale Supérieure in Paris, obtained a figure for
c = 299 782 ± 30 km s^{-1} and demonstrated that further progress
with this method was highly problematical, due to the important
corrections made necessary by the radiation of the open-ended type
of line used by Lecher. It was not until 1947 that this method
was again to be used very successfully by replacing the resonant
line with a closed electromagnetic cavity.

    - between 1879 and 1926 Michelson developed Foucault's method
to an extraordinary degree, and the accuracy he obtained in 1927,
in an evacuated tube one mile long, with c = 299 796 ± 4 km s^{-1}
was never to be bettered by Michelson's successors.

    - the static method, or ratio of units method, was to attain
its apogee when E.B. Rosa and N.E. Dorsey refined the measuring
processes and conducted experiments in what is a real laboratory
of metrology, the National Bureau of Standards at Washington,D.C.
Their 1907 results[†††] well deserved being updated by R.T. Birge in

---

† Page 256 of Scientific Paper I entitled *A Dynamical theory of
  the Electromagnetic Field.*

†† Hertz had first tried, in vain, to find closer and clearer reasons on which to base what Maxwell's equations expressed.

††† See Bureau of Standards Bulletin No3, pp.433-604 and 605-622,
   E.B.Rosa and N.E. Dorsey, 1907.

1941. Substituting *absolute ohm* for the *international ohm* used in 1907, Birge obtained : c = 299 790 ± 10 km s^{-1}[†], a remarkable accuracy for the resultant of two static measures conducted before the first World War.

- Fizeau's toothed wheel was from 1927 onwards replaced by a Kerr cell. This enabled Karolus and Mittelstedt to chop light very much faster than when using even the very latest toothed wheel. Next, Huttel successfully replaced optical occultation by a photoelectric method of determining the phase of the modulated return ray. Huttel in Germany and Anderson in the USA managed to measure c just before World War II, but the publication of their results in 1940 marked only a progress in the technique of measurement, as their results were no better than Michelson's. Only R.T. Birge, quite unexpectedly, contributed to improved accuracy when he noted that in all the experiments conducted in the free atmosphere, including Michelson's, the velocity measured was group velocity instead of phase velocity. Phase velocity for visible light is about 2 km s^{-1} greater than group velocity.

The war, with the advent of radar and its extensions to long-distance telemetry systems such as Oboe, Shoran and Loran, made the accurate knowledge of c mandatory. This knowledge became so important for air navigation that it was justified to devote numerous teams of engineers and researchers to measuring c . Thus, in 1949, Aslakson was able to publish the results of several campaigns of measurement extending over about five years and using Shoran measurements of distances of 67 to 357 miles made in the Canadian plains. The measurements were made at all seasons of the year, and especially in winter because the air then contains the least water vapour, thus diminishing the correction for air refraction and dispersion which have to be applied. The extreme care taken and the years of applying prudent metrological methods paid off, for in 1949 Aslakson obtained c = 299 792 ± 2.4 km s^{-1}, and in 1952 c = 299 794 ± 2 km s^{-1}. Today's value is inside the brackets defined by the margin error of each of these brilliant results.

At the same time, in England, the National Physical Laboratory's professional metrologists were also successful, L. Essen obtaining first class results with microwaves (λ = 9 to 10 cm). He used a resonant cavity and took many precautions, such as using an adjustable cavity and varying frequencies, modes and temperatures to achieve c = 299 792.5 ± 1 km s^{-1}. His evaluation of the margin of error is on the high side and today it seems likely that Essen was accurate to within a few hundreds of meters per second.

---

† See R.T. Birge's article of 1941, page 10.

Working between 1954 and 1958, the N P L's K.D. Froome did even
better with a very different method using free waves in the mil-
limeter band (24 to 70 GHz). The main corrective needed is for
diffraction effects. These, Froome calculated accurately and,
relying heavily on studies in depth of index and dispersion corrections, in 1958 he again reduced the margin of error with
c = 299 792 ± 0.1 km s^{-1} .

- Finally in Sweden the manufacture on an industrial scale of
an optical radar, with a transmitter using a Kerr cell and a re-
ceiver using photomultipliers, ensured a place in the vanguard of
progress for this technique. Bergstrand, its inventor, designed
his *geodimeter* for topographical telemetry, but it also provides
very good results for c .

In that case, innovation consists in the use of an original
double modulation method which ensures that the angle of scanning
around a minimum is accurately locked to $\pi/2$[†]. In 1950, at night
over ten or so kilometers of calm sea, the corrections required
were small and accurate enough to enable Bergstrand's *geodimeter* to
provide a value for c = 299 793 ± 0.3 km s^{-1}.

Such are the advances produced by a more objective understanding of the difficulties to be overcome and of the means required to
do so. The wanderings of 1939 to 1940 were corrected and Michelson's
1927 results greatly improved by several different methods. These
successes were not achieved easily and in spite of a laudable degree of inventiveness and treasures of skill in measurement, the
road was a long one and these talented teams were to take at least
half a decade before they could make any significant advance in our
knowledge of c .

AMCO-5 will record a quicker and more striking success, for
the results to be announced were published on a preliminary basis
in 1972, a year after AMCO-4, and then accepted as early as 1974
after a very thorough discussion by the *Comité Consultatif du Mètre*[††], which found itself able at the end of 1973 to recommend
acceptance of the value :

$$c = 299\ 792\ 458\ \text{m s}^{-1}$$

---

† See P. Grivet's *Mesures physiques basées sur des techniques
radioélectriques*, Part I, *Célérité de la lumière dans le vide*,
RGE, 67, 285-300, 1958, and E. Bergstrand, *Handbuch der Physik*,
t.24, p.1.

†† J. Terrien, *International Agreement on the Value of the Velocity
of Light*, Metrologia, 10, 9, 1974.

This fact victory was due to the introduction of a remarkable invention, the laser, into two of the most powerful tools of Metrology: the electronic measurement of high frequencies by arithme-
tical multiplication and beating on the one hand, and the determination of wavelength by multiple optical wave interferometry on
the other. This leads to a leap forward in accuracy in the shape
of an addition to the value of c_0 of two significant figures -
and all in the six years which elapsed between the first clear
description of the technique[†] by J.L. Hall in 1968 and official
recognition of his and others' success in 1974. J.L. Hall is himself to describe these memorable experiments in this conference.
Since several recent variants will then be discussed in detail, we
will only briefly mention here the three techniques which form the
foundations of the method:

     - C.H. Townes', A.L. Schawlow's and A. Javan's helium-neon
laser is still much as it was designed: a mixture of He-Ne contained in a Fabry-Perot interferometer and ionized by an electric
current. Taking into account the reaction effect resulting from
the stimulated emission of light, the equipment constitutes an oscillator whose naturally high frequency stability is still too low
for the purpose. It is thus further stabilized by a filter similar in operation to the quartz filter in the classic radio-electric
clock, but here consisting of a tube containing, for example, vaporized iodine. One of the molecular vibratory spectrum lines of
the vaporized iodine has a frequency approximating that of the
laser, $\lambda = 6328$ Å (Fig. 1). This spectrum line is very narrow since
the iodine is a passive agent. It is not subjected to any disturbance other than the gentle and stable excitation produced by the
laser light. Additionally, the line is easily and selectively saturable[††]; in other words luminous excitation results in opening a
very narrow window of transparency in the tank's absorption band.

[†] J.L. Hall, *The Laser Absolute Wavelength Standard Problem*,
   Quantum Electronics, 4, p.638, 1968.

[††] A simple form of the phenomenon of saturation of the absorption
   in nuclear magnetic resonance spectrum lines was identified in
   1945 (F. Bloch's equations). This was that which occurs under
   homogenous conditions i.e. a state where all atoms interact in
   the same way with the exciter field. The case in point is,
   however, *non-homogenous* and only those atoms having no longitudinal velocity become saturated, since for them there is no
   Doppler effect when they interact with the two plane waves in
   the resonator which both assist in reaching saturation, digging
   a *hole* in the absorption band similar to the *Lamb-dip* of ordinary gas lasers. See Haroche, S. and Hartman, F., *Theory of
   saturated-absorption line shapes*, Phys. Rev. A 5, 1280-1300,
   1972.

This remarkably simple device is now used as an *optical clock* with an extraordinary stable frequency of better than 10^{-13} (relative value).

This frequency stability was easily demonstrated by using two separate but identical lasers to illuminate a single photocell. At these very high frequencies the cell acts as a rectifying diode and reveals the low frequency beat $\nu_1 - \nu_2$ of the two independent sources (which are never truly identical). This technique is, however, unsuitable for constructing a frequency multiplying chain between the cesium microwave clock and the visible spectrum, since such a chain has meshes situated in the centimetric, millimetric and infrared bands, for which photoelectrically sensitive layers are not available. Other means had accordingly to be found and the final solution involved the development of three types of contact diode. One method was to miniaturize the metal-to-metal or metal-to-semiconductor rectifying point contacts, another was to construct semiconducting very small surface (e.g. $1\mu m^2$) Schottky-effect diodes[†] (Fig.2-a, b). The third solution involved research into the at-the-time unexplored field of superconducting Josephson-effect diodes. Here surprising success was achieved, for although the Cooper-pairs of electrons which carry the current by tunnel-effect in these diodes are easily disassociated by the photoelectric effect of infrared radiation, the diodes continue to detect efficiently up to $\lambda = 20$ μm. Hence what was announced in 1973[††] was the realisation of a complete, stable multiplier chain between a cesium clock and the frequency of the infrared line $\lambda = 3.39$ μm of an He-Ne laser. The source of accuracy is the rigidity of the multiplier link. The only thing which impairs this rigidity is the presence of background noise[†††] and no real corrections are required to the integer multiplication model.

Once the stabilized frequency ν, of the laser is known, the value of c is expressed by the simple basic relationship :

$$c = \lambda_o \nu$$

λ_o being the wavelength in vacuum of the laser's light whose interferential measurement is necessary; this measurement is not easy, but it can be done by using a very well-tried technique in which the wavelength of the orange line from a krypton lamp is used as the basic unit of length . The parallel, coherent and relatively powerful laser beam is much better suited to being measured than the low-level and incoherent orange light from a krypton lamp. A comparison is still necessary today within the framework of international metrological conventions. But this comparison took

---

[†] H.R.Fetterman, B.J.Clifton, P.E.Tannenwald, and C.D. Parker, *Appl. Phys. Lett.*, t.24, No2, 70-71, 1974.

[††] K.M.Evenson, J.S.Wells, F.R.Petersen, B.L.Danielson, G.W.Day, in *Appl.Phys.Lett.* 22, 192, 1973.

[†††] The effect of noise through the multiplying chain is accurary analyzed in H. Hellwig's communication.

only two years 1972-1974 to lead to a first well-deserved official recognition due to the fact that the relative value error in c_0 could be reduced to less than $3 \cdot 10^{-9}$.

The very rapidity of this authentificating process highlights the interest of the proposal formulated and discussed in detail by Z. Bay at AMCO-4, namely *can a stabilized laser be legitimately used as the basis of a single primary standard of measurement for both frequency and length ?*

Current metrological practice qualifies the answer, for if the laser provides an accurate, practical and very popular secondary standard of length measurement, it is still too difficult to measure frequency with it; or, in other words, physicists have confidence in the stability and accurary of this excellent time keeper but still find it very difficult to use it for telling the time. The quality of the results already obtained, however, shows that several Primary Laboratories can reasonably expect to be able in the near future to merge into one those so far separate branches of Metrology - Space measurement and Time measurement.

IV. THE ALTERNATIVE JOSEPHSON EFFECT AND THE NOTION OF PURITY OR PERFECTION

Over the last few years the Josephson diode has played a much more important part in Metrology than the above-mentioned frequency multiplier application would lead one to believe. This is well shown in the AMCO-4 Proceedings, both in Chapter IX (pages 387-410) and in Chapter 13, page 547, where E.R. Cohen and B.N. Taylor stress the large increase in accuracy obtained by basing the measurement of 2e/h on the alternative Josephson effect. The experiment is essentially very simple, consisting merely of plotting the I-V characteristics of any type of Josephson diode when subjected to fixed frequency radiation in the microwave band. Fig.3 shows the equipment normally used and Fig.4[†] the stepped curve obtained for the I to V relationship. The abscisse V_n in the middle of the nth order step is linked to the fixed irradiation frequency, by the Josephson's equation:

$$n = 2(e/h)V_n \qquad \text{being} \quad 483{,}593718 \text{ MHz} \quad \text{for} \quad V_n/n = 1 \ \mu V_{NBS69}$$

The NBS standard volt S69, is used. There is little point in converting it to the international volt, since doing so entails incurring errors much too great with respect to the probable error

---

[†] The figures are taken from *The Fundamental Physical Constants*, by B.N. Taylor, D.N. Langenberg, W.H. Parker, in the American Scientist, 62-78, October 1970.

of 0.12 10^{-6} to be incurred in the measurements described above[†].

  This contribution to progress dates from 1966[††] and, rapidly, improvements in accuracy were obtained by several laboratories bet-ween 1966 and 1971[†††]. This is an impressive aspect of the question and is no doubt explained by the fact that the results interest two different categories of worker: the physicist in order to have a really precise value for e/h and the metrologist for the advent of the new primary standard of voltage measurement provided by the diode's steps with scarcely any need for correction. The main source of errors is the disparity between the low value of the voltage de-fined by the steps (a few mV) and that of the voltage of standard cell batteries (about 1 V). These errors have, however, been pro-gressively reduced by improving diode and potentiometer technolo-gies. Accordingly, Finnegan and others, operating at 12.4 GHz, were able as early as 1971 to use the 450th step of their diode, thus raising the Josephson voltage to 10 mV. The ν to V_n relationship has, however, been found to be insensitive to diode structure; it may be made of two Pb or Sn thin films separated by a 20 Å thick slice of copper; or it can consist of Nb point contacts. Very varied and careful experiments[††††] have shown that the coefficient of the I-V relationship remained invariable. This is a very sur-prising aspect of the way these diodes behave, since to all atten-tive observers they seem complex devices on whose operation many technological advances have very little effect. The X-ray tube pre-viously used to measure e/h was visibly of a better quality and the Weston cell was more carefully constructed; yet both required nume-rous and difficult corrections. We are accordingly led to the con-clusion that both theoretical and practical studies of the Josephson diode have shown convincingly that, at the very low temperatures at

---

  [†] *A.C.Josephson-effect of e/h - A Standard of Electrochemical Potential Based on Macroscopic Quantum Phase Coherence in Su-perconductors.* F.F.Finnegan, A.Denenstein, D.N.Langenberg, in the *Phys. Rev.*, 3rd series, B 4, 1487-1522, 1971.

 [††] D.N.Langenberg, W.H.Parker, and B.N.Taylor, in the *Phys. Rev.* 150, 186, 1966, and *Phys. Rev. Lett.*, 22, 259, 1966.

[†††] as shown in the communication of Dr B.W.Petley.

[††††] The initial verification campaigns undertaken by J. Clarke alone are astonishingly complete and very convincing. Many others have, however followed these (J. Clarke : *Phys. Rev. Lett.*, 21, 1566-69, 1968 ; *Proc. Roy. Soc. A,* 308, 1969 ; *Phys. Rev.*, B 4, 2963, 1971).

which it is used, this diode is an excellent superconductor system
which always obeys a simple and inflexible ν to V law, even though
in practice it is complex and unrefined in manufacture. This is a
typical characteristic of superconductors' physics which drove a
celebrated pioneer in the field, Professor H.B.G.Casimir, to express
his scepticism with regard to the chances of a quantum theory[†] suc-
ceeding in supporting London's intuitive ideas expressed as follows
*Can one really imagine how the phase of a wave function can remain
coherent over a mile of dirty lead wire ?* As for the diode, it
should be said that the efforts of the theorists, as presented by
D.B. Josephson as early as 1969[††], are also very convincing. They
say that whatever the type of diode, whatever their physical appearance and whatever theory is used to describe them, all diodes exhibit the same I-V ideal relationship. This perfection is accordingly beyond the scope of direct observation and can only result
from a complete system of comparative measures and from different
theories. The Josephson diode is, however, sensitive to magnetic
disturbances or the presence of magnetic impurities. The measurements already described must accordingly be made within a screen
which reduces the mean level of magnetic influence to less than a
milligauss.

This sensitivity to magnetic fields can on the other hand, be
useful in measuring very weak fields, a very interesting by product
of the 2e/h metrology. The SQUID first developed by J.E.Mercereau
from 1964 onwards an then by J.E. Zimmermann has today reached extraordinary levels of sensitivity of the order of 10^{-9} oersteds in
its latest form. Bearing in mind that the fluctuations of the
earth's magnetic field on even the magnetically calmest days scarcely fall below 10^{-6} oersteds and that the magnetic field in interplanetary space is seldom more than ten times less, it will be appreciated that the latest magnetometers, and their *optical pumping*
and *Hanle effect*[†††] competitors, can only operate efficiently in

---

† See *N. Bohr and the Development of Physics,* p.118 (New York
   1955, Pergamon).

†† D.B. Josephson, for example, takes pains to show that in each
   weakly coupled sub-system, the electron pairs and unattached
   electrons are so strongly coupled by Pauli's principle, and
   B.C.S. theory, that it is a matter of indifference whether one
   or the other type of particle is considered to be the current
   carrier in the weak link. See Ch.9, *Weakly Coupled Superconductors,* t.I, 423-448, of R.D. Parks, *Treatise on Superconductivity.*

††† *Detection of Very Weak Magnetic Field (10^{-9} G) by ^{87}Rb Zero
   Field Level Crossing Reasonances,* J. Dupont-Roc, S. Haroche,
   C. Cohen-Tannoudji, *Phys. Letters,* 28 A, 638-9, 1969.

spaces artificially protected by a superconductor, or more easily
ferromagnetic, screening (Fig. 5). This constraint has not prevented the spread of a number of very interesting applications,
such as that demonstrated by Dr D. Cohen, a physicist interested
in biology, at MIT in Boston. He has constructed screening of an
original type which can contain a human being and permits measuring
the magnetic fields produced by the circulation of the blood at,
for example, heart level. See Figs. 6 and 7[†].

Finally, we have seen that the applications of a discovery do
not lag far behind it, even when the initial advantages seem to far
exceed expressed requirements at the time. Hence physicists are
hoping that this particular discovery will, by the greater degree
of precision it brings to the value of e/h, improve our knowledge
of the so-called *fine structure* constant and thus contribute to an
advance in quantum theory of the electromagnetic fields, much as
Fizeau's, Foucault's and Lecher's measurements contributed to
Maxwell's and Hertz's discoveries at the end of the 19th century.
Pure and applied science will thus continue forward hand-in-hand -
or such would, I am sure, be the fervent hope of the physicists and
metrologists assembled here at this Symposium.

[†] D. Cohen, E.A. Edelsack, J.E. Zimmermann, *Magnetocardiograms
Taken Inside a Shielded Room with a Superconducting Point-contact
Magnetometer* . Appl. Phys. Lett., t.16, No7, 278-80, 1970

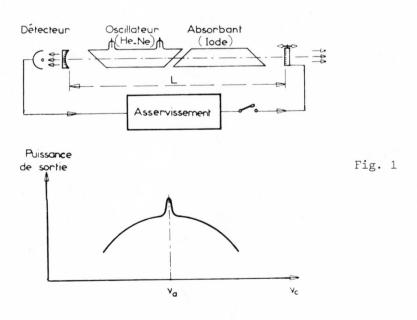

Fig. 1

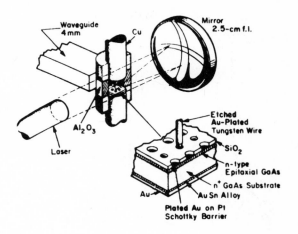

Fig. 2-a

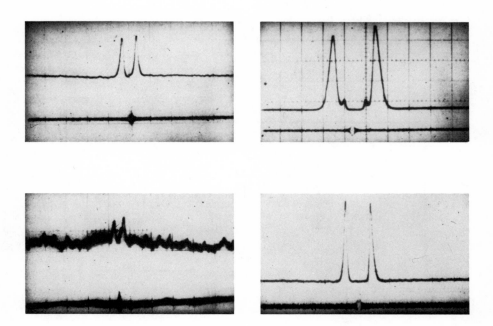

Fig. 2-b

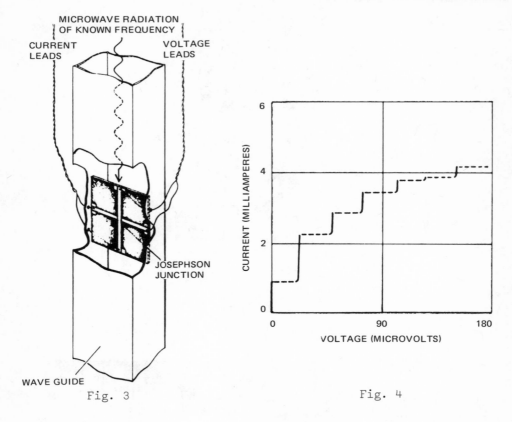

Fig. 3

Fig. 4

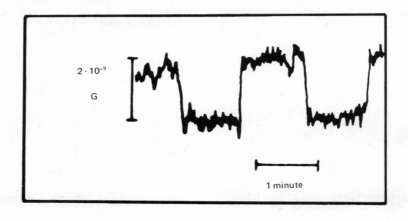

Fig. 5

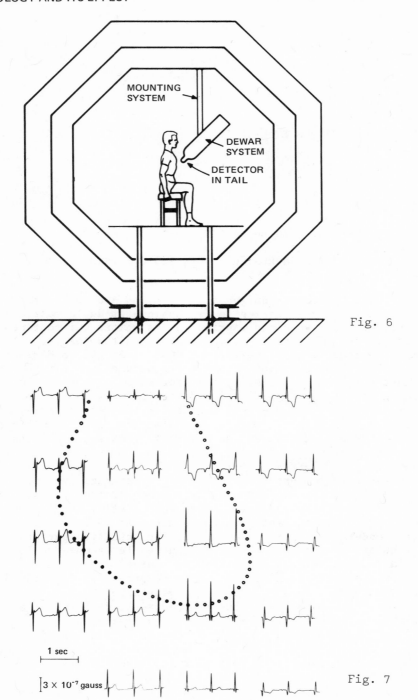

Fig. 6

Fig. 7

CONSTANTES PHYSIQUES ET MÉTROLOGIE

J. TERRIEN

Directeur du Bureau International des Poids et
    Mesures
Pavillon de Breteuil   F - 92310 SEVRES

A cette Conférence qui va s'achever, on a beaucoup parlé des constantes fondamentales de la physique et de la mesure de leur valeur. Leur valeur s'exprime en des unités qui sont des unités de base du Système International d'Unités (mètre, kilogramme, seconde, ampère, kelvin) ou des unités qui en dérivent. Il y a là un lien nécessaire, qui a toujours existé, entre les constantes fondamentales et la métrologie. Je prends ce mot métrologie avec une signification restreinte au domaine des unités de base et des quelques unités dérivées nécessaires, c'est-à-dire à peu près la métrologie dont s'occupe le Bureau International des Poids et Mesures, au niveau le plus élevé des étalons de départ.

Cette relation entre métrologie et constantes physiques est devenue encore plus évidente depuis que deux unités de base, le mètre et la seconde, ont reçu une nouvelle définition faisant appel à des constantes atomiques, le mètre en 1960 avec la longueur d'onde dans le vide d'une radiation spécifiée par deux états quantiques de l'atome de krypton 86, la seconde en 1967 avec la fréquence d'une transition de l'atome de césium 133. Ces deux grandeurs atomiques n'ont en théorie rien de remarquable, mais on a constaté que les moyens expérimentaux à notre disposition permettaient de les reproduire et de les utiliser avec plus de précision et de commodité que les centaines de milliers d'autres connues ; on continue de perfectionner avec un soin extrême ces méthodes expérimentales parce que cette longueur d'onde et cette fréquence particulières ont pris une grande importance depuis qu'elles ont été choisies par accord international, comme référence conventionnelle en vue de définir le mètre et la seconde. Le recours aux constantes atomiques confère

24

à ces deux unités une garantie d'invariabilité mieux assurée
qu'avec leurs définitions antérieures, car on peut dire que les
grandeurs atomiques sont invariables par nature, selon les théo-
ries les mieux admises ; au contraire, le platine iridié du prototype du mètre, présumé stable parce que cet alliage ne se
liquéfie qu'à haute température, a dû subir l'épreuve de l'expérience avant d'être choisi par Sainte-Claire Deville, et que la
rotation de la Terre, on le sait maintenant, fournissait une
échelle de temps non uniforme, c'est-à-dire incompatible avec
des lois simples de la dynamique.

Un autre lien entre constantes physiques et unités se trouve
dissimulé dans la définition de l'unité de base l'ampère ; en
effet cette définition implique que l'on attribue la valeur
$4 \pi \times 10^{-7}$ henrys par mètre à la constante magnétique μ_o, la
perméabilité magnétique du vide. On peut dire que la valeur de μ_o
est une mesure de la liaison quantitative entre les grandeurs
électromagnétiques (courant, champ magnétique par exemple) et les
forces qu'elles engendrent (loi de Coulomb entre deux charges
électriques, loi des forces entre deux conducteurs parcourus par
un courant). En conséquence, si l'on adopte par convention
l'unité de force, le newton, et si l'on fixe par convention μ_o,
l'unité de courant électrique se trouve fixée ; les autres unités
électriques, coulomb, ohm, volt, henry, farad en dérivent, grâce
aux lois élémentaires de l'électricité, selon un schéma bien
connu.

Si au contraire on avait choisi par convention une définition
indépendante de l'ampère, par exemple en adoptant une vitesse de
décomposition d'un électrolyte, il faudrait faire une mesure
expérimentale de μ_o. Pour l'expérimentateur dans son laboratoire
de métrologie, les opérations physiques seraient inchangées :
la balance de courant, qui sert à déterminer l'ampère avec la
convention actuelle qui fixe μ_o arbitrairement, servirait alors
à la mesure de μ_o en utilisant l'ampère défini par ailleurs. Cet
exemple montre que nos conceptions de constantes et d'unités sont
en quelque sorte interchangeables selon les conventions admises.
La distinction entre constantes physiques et unités n'est pas
imposée par la physique, elle résulte de choix, et ces choix
cherchent la commodité.

J'ai parlé d'une définition de l'ampère par l'électrolyse ;
cet exemple a une réalité historique, il est simple, mais il
est mauvais à cause de l'insuffisance de la précision accessible ;
un autre exemple, tout à fait d'actualité, est que l'effet
Josephson pourrait servir à définir le volt, par l'adoption
conventionnelle d'une fréquence ; partant du volt, on dériverait
l'ampère par le chemin inverse de celui qui conduit actuellement
de l'ampère au volt. Le résultat serait encore le même, c'est-à-

dire que μ_o deviendrait une constante dont la valeur doit être
déterminée par l'expérience, comme le sont actuellement les
constantes de Planck, de Faraday, de Rydberg, le coefficient gyromagnétique du proton, etc.

A propos du coefficient gyromagnétique du proton γ_p, on
pourrait se servir de cette constante pour définir l'ampère, en
fixant la valeur de γ_p par convention. Là encore, l'appareillage
expérimental qui sert actuellement à la mesure de γ_p resterait
inchangé, mais il servirait à déterminer l'ampère. Ce nouvel
exemple montre encore que les travaux expérimentaux du physicien
qui mesure la valeur d'une constante physique, et ceux du métrologiste qui matérialise la valeur d'une unité dans un étalon, sont
interchangeables, selon les conventions admises, et qu'elles ne
se distinguent pas par des propriétés naturelles.

En passant, je rappelle que si l'on voulait à la fois définir
le volt par une fréquence conventionnelle de l'effet Josephson,
et définir l'ampère par une fréquence conventionnelle de la
résonance gyromagnétique du proton, on renoncerait par là-même
à l'identité du watt mécanique et du watt électrique. Il deviendrait alors nécessaire de mesurer expérimentalement le rapport
entre le newton-mètre par seconde et le volt-ampère, et ce
rapport serait une nouvelle constante fondamentale de la physique.
Or tous les systèmes d'unités passés et actuels observent le
principe que ces deux façons de réaliser le watt doivent donner
un résultat identique par la seule vertu du choix des définitions
des unités, mais ce principe est purement conventionnel. Si l'on
ajoute une unité de base indépendante au système d'unités, il
en résulte une nouvelle constante physique dont la valeur doit
être déterminée par l'expérience. Inversement, si l'on retranche
une unité de base, on supprime une constante ; par exemple,
on pourrait supprimer le kelvin des unités de base en posant la
constante de Boltzmann égale à 1 par convention.

Revenons dans le cadre du Système International d'Unités
en vigueur actuellement. Le métrologiste n'a pas achevé son
travail lorsqu'il a déterminé l'ampère et le volt conformes à
leur définition. Il doit encore conserver le résultat de ces
déterminations absolues au moyen d'étalons qu'il voudrait stables,
et qui ne le sont qu'imparfaitement. Les constantes physiques
$2e/h$ et γ_p, c'est-à-dire les constantes de l'effet Josephson et
du coefficient gyromagnétique du proton, peuvent lui servir, et
lui servent effectivement, à vérifier la constance de ses étalons,
ou à mesurer leur dérive. Donc, sans qu'il soit nécessaire de
changer la définition de ces unités pour les rattacher directe-
ment à des constantes physiques, ces constantes et leur invariabilité sont mises à profit pour se rapprocher de l'idéal
d'étalons métrologiques invariables.

On trouvera peut-être à l'avenir des procédés analogues
permettant au métrologiste de relier d'autres étalons à des
constantes physiques, et de contrôler leur invariabilité par leur
comparaison à une propriété atomique. Les constantes physiques
dont il est question à cette conférence expriment en effet, pour
la plupart, des relations entre les grandeurs macroscopiques des
étalons matériels du métrologiste, et les grandeurs atomiques
invariables par nature.

Cette tendance vers l'utilisation métrologique des constantes physiques a toujours existé, mais elle est devenue pratiquement utile depuis que la précision expérimentale concernant
ces constantes rejoint la précision des étalons métrologiques
qui sont la matérialisation des unités. J'ai déjà dit que les
unités de base sont choisies en vue de la commodité d'emploi du
système d'unités, mais une autre condition impérative est la
précision avec laquelle leurs étalons peuvent être réalisés ; si
cette précision est moins bonne que la précision d'une expérience
de mesure d'une certaine grandeur, on ne pourra pas exprimer le
résultat de cette mesure d'une façon complète. Pour éviter cette
perte d'information, il y a un remède simple : c'est de renoncer
à la définition devenue trop imprécise de l'unité, et d'adopter
une nouvelle définition fondée sur cette expérience plus précise.

Nous sommes peut-être arrivés à cette situation en ce qui
concerne la définition du mètre. En effet, la mesure de la
vitesse de propagation des ondes électromagnétiques dans le
vide, appelons-la c, a pu être déterminée par la mesure de la
fréquence et de la longueur d'onde dans le vide d'une même
radiation électromagnétique que j'appellerai X. La fréquence X
est mesurée par comparaison à la fréquence d'une transition de
l'atome de césium 133, conformément à la définition actuelle
de la seconde, avec une précision surabondante. La longueur
d'onde X est mesurée par comparaison à la longueur d'onde de la
radiation du krypton 86 conformément à la définition du mètre.
Mais la longueur d'onde X est mieux définie que celle du
krypton ; ceci veut dire que la radiation X est mieux reproduc-
tible et possède un profil spectral plus étroit ; ou encore que
le rapport entre la longueur d'onde de cette radiation X et la
longueur d'onde d'une autre radiation qui aurait les mêmes
qualités est mesurable avec plus de précision que le rapport
à la longueur d'onde du krypton. Il me semble donc certain que
la définition du mètre sera changée. Ce n'est pas urgent, car le
bénéfice de ce changement n'intéresse que très peu d'expériences,
et de plus, en attendant, le Comité International des Poids et
Mesures a recommandé la valeur 299 792 458 m·s^{-1} pour c, ce
qui assure à la fois la continuité avec les mesures passées, et
l'uniformité des rares mesures qui nécessitent un niveau de
précision plus élevé que ne le permet la définition actuelle du
mètre.

Il me semble à peu près certain que la future définition
du mètre sera rédigée de façon à impliquer que la constante
physique c soit fixée par convention, de même que la définition
actuelle de l'ampère implique une valeur conventionnelle de μ_o.
Les expériences que nous appelons maintenant "mesure de la
vitesse de la lumière" resteront nécessaires et inchangées, mais
elles s'appelleront alors "détermination du mètre".

Les théoriciens de la relativité trouvent commode de poser
c = 1 ; alors masse et énergie ont la même dimension et la même
unité ; longueur et temps ont aussi la même unité qui peut être
soit le mètre, soit la seconde, mais l'une de ces deux unités
disparaît. Ils suppriment ainsi à la fois une unité de base et
une constante. La définition future du mètre, si elle implique une
valeur de c conventionnelle, contiendra la meilleure valeur expérimentale actuelle ; du fait que c sera différent de 1, le mètre
et la seconde seront tous deux conservés en qualité d'unités de
base du système d'unités, et c restera une constante physique.
Cet exemple illustre encore le fait que le nombre des unités de
base et le nombre des constantes physiques ne sont pas imposés
par la nature, mais résultent de nos conventions.

Les conventions qui définissent le système d'unités actuel
et l'ensemble actuel des constantes physiques sont-elles les
meilleures ? On peut en discuter sans fin ; en effet, ces conventions sont bonnes si ces systèmes sont commodes, et si elles
permettent en même temps des mesures précises. Or ce qui est le
plus commode pour l'un ne l'est pas nécessairement pour l'autre ;
si l'on change les unités pour rendre plus précises certaines mesures, on peut rendre plus difficiles et moins précises d'autres
mesures. Le compromis actuel est incontestablement bon, il ne
soulève aucune objection sérieuse, et le Système International
d'Unités est devenu seul légal dans presque tous les pays.
C'est la première fois, dans l'histoire de l'humanité, qu'un
seul et même système d'unités est approuvé par le monde entier.
Pour qu'il soit commode, il ne faut pas le changer, car il est
malcommode de changer d'habitudes. Ce qu'il ne faut pas changer,
c'est la liste des noms des unités de base et l'ordonnance du
système, mais on peut sans inconvénient changer leur définition.
Cela veut dire, par exemple, que la décision de fixer par convention la valeur de c ne devra pas substituer à l'unité de base
pour la longueur une unité de base qui serait une unité de vitesse ; il faut conserver les unités de base actuelles mètre et
seconde, et simplement formuler leur définition d'une façon qui
implique la valeur convenue de c.

Je pense avoir fait comprendre l'importance des constantes
physiques en métrologie. Le Bureau International des Poids et
Mesures, qui depuis 100 ans s'occupe uniquement de métrologie,

s'intéresse évidemment aux constantes physiques, qu'elles soient impliquées dans la définition des unités, ou qu'elles offrent un contrôle de la stabilité des étalons. Mais, en plus des unités que je viens de citer, le kilogramme joue un rôle essentiel, car une mesure de masse est nécessaire pour réaliser les étalons d'un grand nombre de grandeurs ; il s'agit de toutes les grandeurs dont le symbole dimensionnel contient la masse ; voici par exemple une liste de quelques grandeurs où la masse intervient : force, pression, énergie, chaleur, puissance, potentiel électrique, capacité, résistance, induction, inductance, viscosité, moment d'une force, tension superficielle, capacité thermique, entropie, conductivité thermique, champ électrique, permittivité, etc. Dans mon esprit d'expérimentateur métrologiste, la précision des mesures de masse et la stabilité des étalons de masse sont aussi importants, pour les mesures pratiques, que la connaissance des constantes physiques. Même avec le Prototype international du kilogramme en platine iridié, et malgré qu'il soit presque centenaire, il doit être possible d'améliorer encore les mesures de masse. Le B.I.P.M. y travaille, il vient même de proposer l'étude d'un programme international de recherche sur ce sujet, car il me semble utopique d'espérer trouver une constante atomique qui servirait à définir l'unité de masse, et qui ne ferait pas perdre la précision que l'on atteint en utilisant le Prototype international.

Je n'ose pas aller jusqu'à dire que ce Prototype fait partie des constantes physiques fondamentales ; mais il intéresse en premier lieu le B.I.P.M., car ce Prototype unique lui est confié. Mais je puis dire que le B.I.P.M. consacre ses efforts à la bonne utilisation de ce Prototype, en même temps qu'il s'intéresse de plus en plus aux constantes physiques parce que c'est exactement une conséquence de la mission qui lui est confiée.

GAMMA-RAY ENERGIES FOR CALIBRATION OF Ge(Li) SPECTROMETERS*

R. G. Helmer, R. C. Greenwood and R. J. Gehrke

Aerojet Nuclear Company, Idaho National Engineering

Laboratory, Idaho Falls, Idaho USA

INTRODUCTION

At the last Conference in this series, we reported (1) on our work in the comparison and measurement of γ-ray energies with multi-channel spectrometer systems using Ge(Li) detectors. Since that time two things have occurred that are pertinent to the definition of precise calibration energies. First, the 1973 adjustment of the values of the fundamental constants by Cohen and Taylor (2) can be used in place of the values of Taylor et al. (3) of 1969. Second, a number of new precise energy measurements have been made with bent crystal spectrometers. In this paper, we review the existing measurements and our work in this area.

ENERGY SCALES

In γ-ray spectrometry, energies have been measured commonly on two different scales. The first group of γ-ray energies has been measured with respect to the energies, or wavelengths, of the atomic x rays. These data have been (or can be) referred to the x-ray energies compiled and correlated by Bearden (4) where the energy scale is referenced to the $WK\alpha_1$ x ray. As shown in Table I the newest value of $E(WK\alpha_1)$ is down only 2.9 ppm from the 1969 value and up 13 ppm from the Bearden value (4). The reference error in this energy scale is taken to be the 5.6 ppm quoted in ref. 2 for the $WK\alpha_1$ energy.

* Work performed under the auspices of the U.S. Energy Research and Development Administration.

TABLE I

REFERENCE ENERGIES

	Value (keV)			Change (ppm)	
Quantity	1967 Bearden	1969 Taylor et al.	1973 Cohen & Taylor	67-73	69-73
$WK\alpha_1$	59.31824	59.31918(35)	59.31901(33)	+ 16	- 2.9
m_0c^2		511.0041(16)	511.0034(14)		- 1.4
^{198}Au-411		411.794(8)	511.794(7)		

The second group of γ-ray energies have been measured with respect to a scale based on the energy equivalent of the mass of the electron, m_0c^2. As indicated in Table I, the latest value of m_0c^2 is reduced by only 1.4 ppm from the previous value. The basic measurement on this energy scale is the single determination (5,6) of the energy of the 411-keV line from the decay of ^{198}Au relative to the m_0c^2 value as observed in positronium annihilation. For this line, the current error is 7 eV or 17 ppm which is taken to be the reference error in this scale.

Several measurements have been made to determine the consistency of the two energy scales, with varying results. This question will be discussed in another paper at this Conference (7). We have chosen not to make any adjustments between the energies measured on the two scales.

In all cases a distinction has been made between two components of the energy uncertainty: (1) the contribution from the reference energy defining the energy scale, σ_r, and (2) that from the specific measurements σ_m, and $\sigma_t^2 = \sigma_r^2 + \sigma_m^2$.

GAMMA-RAY ENERGIES

Comparison of Recent Bent Crystal Measurements

For primary calibration for our γ-ray energy measurements on Ge(Li) spectrometers, a set of precisely known γ-ray energies, measured by other techniques, was compiled. Besides having the necessary precision, only those data were included which were available with sufficient detail to allow reevaluation to correspond to the current set of fundamental constants.

Of special interest are two groups of recent precise measurements (8-11) relative to the Au-411 line made with bent-crystal (diffraction) spectrometers. Some of these new values are quoted with very high precision; for example, σ_m = 0.06 and 0.4 eV at 100 and 200 keV. For the energies of [182]Ta and [192]Ir a comparison is given in Table II. Of the 20 entries, the agreement is not quite within the quoted errors since 12 differences are over 1σ, five are over 2σ and three are over 4σ. At this level of precision, this agreement may be quite reasonable. However, it should be noted that for many lines the uncertainty in the discrepancy comes primarily from the work of Kern et al. (8-9) and as a result, does not indicate whether the Borchert et al. (10,11) errors are realistic.

These data also indicate that for [182]Ta any systematic difference is quite small while for ^{192}Ir the Borchert et al. values are larger by 5-10 ppm.

TABLE II

COMPARISON OF RECENT BENT-CRYSTAL MEASUREMENTS

Parent Isotope	Energy (keV)	Uncertainty Piller (8)	Borchert (10)	Difference (eV)
[182]Ta	65	0.2	0.15	1.1 ± 0.3
	67	0.2	0.13	1.3 ± 0.2
	100	0.3	0.06	2.7 ± 0.3
	113	1.2	0.26	-2.3 ± 1.2
	152	1.4	0.27	1.7 ± 1.4
	156	1.2	0.35	-1.6 ± 1.3
	179	1.8	0.53	0.5 ± 1.9
	198	2.2	0.44	0.6 ± 2.2
	222	2.4	0.45	-2.6 ± 2.4
	229	3.0	0.63	0.0 ± 3.1
	264	5.2	0.61	2.0 ± 5.2
		Beer (9)	Borchert (11)	
[192]Ir	136	0.7	0.7	-0.5 ± 1.0
	295	1.5	0.5	-2.4 ± 1.6
	308	1.7	2.2	-0.6 ± 2.8
	316	1.6	0.3	-2.3 ± 1.6
	416	5.	3.0	-8.1 ± 5.8
	468	4.	0.6	-8.2 ± 4.0
	588	8.	4.0	-22.0 ± 8.9
	604	8.	2.2	-2.1 ± 8.3
	612	9.	2.6	-5.3 ± 9.4

Energy from Other Authors

Our treatment of data in the literature is illustrated for the case of ^{192}Ir. The data used here for these γ-ray energies are indicated at the bottom of Table III along with the type of instrument used and the reference line. The various energies for one line were averaged (1/σ weighting). These average values were then used as input to a least-squares fitting program which determines the "best" level energies for the decay scheme. The "best" γ-ray energies and uncertainties were then computed from these level energies.

In addition to these γ-ray energy values, the input to this particular least-squares determination of the level energies included two γ-ray energy differences. These were our measured 588-604 and 588-612 differences. The inclusion of these differences is significant in that it increases directly the energy of the 588-keV line by 3 eV and indirectly the 416- and 884-keV energies. The average energies are quite consistent since the reduced-χ^2 of the least-squares fit is 0.60. This is in contrast to the various experiments which are somewhat inconsistent as indicated by the reduced-χ^2 values for the averages.

TABLE III

^{192}Ir GAMMA ENERGIES

| Average | | Least-Squares | Uncertainty (eV) | |
Error (eV)	Reduced-χ^2	Value (keV)	Measurement	Total (Au)
0.5	0.3	136.3398	0.5	2.4
0.8	1.8	205.7903	0.8	3.6
1.3	4.4	295.9496	1.0	5.1
2.4	3.4	308.4491	1.1	5.3
1.0	4.9	316.4994	0.9	5.5
3.7	1.9	416.4602	1.4	7.2
1.3	2.2	468.0588	1.0	8.1
6.6	3.2	588.5692	1.1	10.1
2.6	0.4	604.3982	1.0	10.3
3.0	0.7	612.4485	1.0	10.5
24.0	---	884.5179	1.8	15.1

Author	Date	Ref. Line	Spectrometer	No. Energies
Murray et al. (6)	1965	Au-411	Electron	7
Daniel et al. (12)	1966	Au-411	Electron	1
Reidy (13)	1970	Au-411	Bent-crystal	5
Helmer et al. (14)	1971	--	Ge(Li)	2
Inoue et al. (15)	1973	--	Ge(Li)	1
Beer & Kern (9)	1974	Au-411	Bent-crystal	10
Borchert et al. (11)	1975	Au-411	Bent-crystal	10

In this manner we have evaluated the γ-ray energies for a number of isotopes including only bent-crystal and electron spectrometer measurements. The recent measurements (8-11) for ^{192}Ir and ^{183}Ta and the low-energy ^{182}Ta lines have been combined with earlier data and evaluated by the above methods including a least-squares fit to the level scheme. In addition to these three isotopes, we have included energies from 12 other isotopes and a total of 21 lines.

Our Measurements

The goal of our measurements of γ-ray energies has been to extend the above set of calibration energies, especially above 400 keV. All of our measurements have been made on Ge(Li) spectrometers. In a few cases, our energies have been determined simply by determining the energy of a peak from spectra calibrated with several known lines. More generally our measurements have been of the difference in energy between closely spaced lines, one of which has a "known" energy. This method is used for two reasons. First, highly accurate differences can be obtained on Ge(Li) spectrometers in spite of their inherent nonlinearity and ill-defined "zero". Second, the values of the energy differences are independent of small changes in the calibration energies and, therefore, can be updated as the primary calibration energies change.

In many cases, cascade-crossover relationships have been used to compute the energy of one of the transitions involved. In particular, two chains of energy combinations have been developed separately up to above 1 MeV as shown in Fig. 1. One chain consists of the lines from ^{192}Ir and ^{160}Tb and the other consists of ^{198}Au, ^{59}Fe and ^{182}Ta. As indicated in the figure, precise comparisons between the two chains can be made to indicate the magnitude of potential systematic errors. In computing the uncertainties in the discrepancy, the 884-(^{192}Ir) and 1087-keV(^{198}Au) energies have first been assumed to be exact. Thus, the first error includes only the measurement uncertainties above these points in the chains. The average discrepancy for two comparisons is (11±4) eV. On the basis of this result, we might conclude that there is a small, real discrepancy between the energies of the two chains at above 1 MeV. However, if one includes the measurement errors in the 884(^{192}Ir) and 1087(^{198}Au) energies, the average discrepancy becomes (11±12) eV. This result indicates that this discrepancy is not unreasonable compared to the magnitude of the measurement errors. The major contribution to this 12 eV measurement error is the 11 eV term from the 675-keV line of ^{198}Au.

At 1 MeV typical errors are $\sigma_m = 4$, $\sigma_r = 17$ and $\sigma_t = 17$ eV for γ rays based on the ^{192}Ir-884 line and $\sigma_m = 11$, $\sigma_r = 17$ and $\sigma_t = 20$ eV for those based on the ^{198}Au-1087 line.

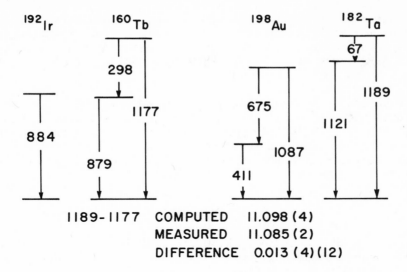

Fig. 1. Gamma-ray energy differences and sums involved in going
         from ^{192}Ir-884 to ^{160}Tb-1177 and from ^{198}Au-411 to
         ^{182}Ta-1189; and the results of comparing the ^{160}Tb-1177
         and ^{182}Ta-1189 energies.

COMMENTS

In summary, based on the primary calibration energies from
other types of spectrometers, our energy differences and energy
combinations have been used to determine the energies of 150 γ rays
below 3450 keV from 35 isotopes. The energies of 79 lines below
1300 keV have been published (14,16), but these have recently been
reevaluated based on the new fundamental constants and the new
bent-crystal measurements.

For the above treatment of γ-ray energies, the single measurement that would be the most effective in improving the existing
energies would be a more precise comparison of the energy of the
675-keV line from ^{198}Au to that of the 411-keV line of ^{198}Au.

Another significant improvement would result from a more
accurate measurement of the 411-keV energy from ^{198}Au. For example,
in Table III the 884-keV line from ^{198}Ir has a total uncertainty of

15 eV, but a measurement error of only 2 eV. So a very precise (error of $\sim$ 1 eV) 411-keV energy would make a vast improvement. If this were combined with a comparable measurement of the 675-keV line of ^{198}Au, one would have both energy chains with errors at 1 MeV of a few eV.

In contrast to the last Conference, we have not discussed the techniques used in Ge(Li) spectrometry. It should only be noted that on Ge(Li) detectors the apparent energy does depend on the source position. We have reported (17) measurements on the shift of peak positions with the source-detector distance, with observed shifts as large as 100 eV for a 2000 keV γ ray.

REFERENCES

1. HELMER, R. G., GREENWOOD, R. C., and GEHRKE, R. J., Atomic Masses and Fundamental Constants 4, eds. J. H. Sanders and A. H. Wapstra, Plenum Press (London) 1972.
2. COHEN, E. R. and TAYLOR, B. N., J. Phys. Chem. Ref. Data $\underline{2}$ (1973) 663.
3. TAYLOR, B. N., PARKER, W. H., and LANGENBERG, D. N., Rev. Mod. Phys. $\underline{41}$ (1969) 375.
4. BEARDEN, J. A., Rev. Mod. Phys. $\underline{39}$ (1967) 78.
5. MURRAY, G., GRAHAM, R. L. and GEIGER, J. S., Nucl. Phys. $\underline{45}$ (1963) 177.
6. MURRAY, G., GRAHAM, R. L. and GEIGER, J. S., Nucl. Phys. $\underline{63}$ (1965) 353.
7. VAN ASSCHE, P. H. M., BORNER, H., DAVIDSON, W., KOCH, H. R., and PINSTON, J. A., a paper presented at this Conference.
8. PILLER, O., BEER, W. and KERN, J., Nucl. Instr. and Meth. $\underline{107}$ (1973) 61.
9. BEER, W. and KERN, J., Nucl. Instr. and Meth. $\underline{117}$ (1974) 183.
10. BORCHERT, G. L., SCHECK, W. and SCHULT, O. W. B., Nucl. Instr. and Meth. $\underline{124}$ (1975) 107.
11. BORCHERT G. L., SCHECK, W. and WIEDER, K. P., Zeits. für Naturforsch. $\underline{30a}$ (1975) 274.
12. DANIEL, H., JAHN, P. and TODT, W., Bull. Akad. Sci. USSR, Phys. Ser. $\underline{30}$ (1966) 2107.
13. REIDY, J. J., private communication.
14. HELMER, R. G., GREENWOOD, R. C., and GEHRKE, R. J., Nucl. Instr. and Meth. $\underline{96}$ (1971) 1973.
15. INOUE, H. and YOSHIZAWA, Y., Nucl. Instr. and Meth. $\underline{108}$ (1973) 385.
16. GREENWOOD, R. C., HELMER, R. G., and GEHRKE, R. J., Nucl. Instr. and Meth. $\underline{77}$ (1970) 141.
17. HELMER, R. G., GEHRKE, R. J., and GREENWOOD, R. C., Nucl. Instr. and Meth. $\underline{123}$ (1975) 51.

PRIMARY STANDARDS FOR GAMMA ENERGY DETERMINATIONS

P.H.M. Van Assche (Nuclear Energy Centre, S.C.K./C.E.N., B-2400 Mol, Belgium and Leuven University)

H. Börner,W.F.Davidson and H.R. Koch (Institut Laue-Langevin, F-38042 Grenoble, France)

1. INTRODUCTION

It was intended to contribute by two different - but related - ways to the determination of gamma energy standards. A first approach consisted in the intercomparison of the two primary wavelength standards directly related to other Fundamental Constants, viz. the observation of a positon annihilation quantum, and the Röntgen ray wavelength standard. The aim is to observe in the same experiment both the 511 keV annihilation line and the W $K\alpha_1$ Röntgen line.

Once the intercomparison of these primary standards has been made, they can be used for measuring gamma ray energies. The most frenquently used transition for energy calibration purpose is the 411.8 keV gamma ray in ^{198}Hg following the 2.7 d. half-life β-decay of ^{198}Au.

Due to unexpected delays and the premature interruption of a successful experimental run, only preliminary results on the second point can be reported at the moment.

2. EXPERIMENTAL METHOD

The wavelength measurements of the different quantum energies have been measured with the bent-crystal diffractometers at the French-German-British Institut Laue-Langevin. The first spectrometer (GAMS 1) has the same crystal and curvature radius as used in the Risø-spectrometer[1]; however the angular measurement of the

crystal turning table has been improved through use of an interferometric system. The second spectrometer is made up of two identical diffractometers (GAMS 2 and 3) with a curvature radius of
24 m. They are mounted independently one above the other and observe simultaneously a given reflection on both sides of the crystal planes. This original spectrometer has been conceived by
Dr. O. Schult. Performances of the GAMS-spectrometers have been
reported recently[2]).

    All diffractometers use the (110) crystal planes in quartz
(d $\sim$ 250 pm). The routine precision is about 0.02 and 0.05 sec. of
arc for GAMS 2-3 and GAMS 1 respectively. Repeated measurements
in different diffraction orders enable a 1 ppm ultimate precision.

    In the previously reported experiment[1]) the exact observation
of the centroid of the positon annihilation peek was slighlty hampered by the presence of some (n,γ) lines superposed upon it. In
order to avoid this the radioactive β+ emitter ^{64}Cu was used
instead as a positon source. As it was intended to measure the
^{198}Au gamma decay line and the W Kα_1 Röntgen lines simultaneously,
a sandwich source was activated in a thermal neutron flux of
8.10^{14} cm^{-2} sec^{-1} for 5 days. The counting was initiated several
hours after the reactor has been stopped. This source consisted
of the following elements: Cu, Ta, Au, Ta and Cu with thicknesses
of 50, 7.5, 1, 7.5 and 50 μm resp.

3. RESULTS

    After some trials a source was successfully made such that it
did not show too apparent an asymmetry in its Cu-constituents
during the (n,γ) activation. A peak-fitting with Gaussian line
shapes and symmetric exponential tails yielded the best results
for both the annihilation and the gamma lines. Two unrelated corrections had to be applied to the observed annihilation spectrum
before it could be fitted. Firstly it happened that the measuring
time over the total reflected peak was not negligible compared to
the ^{64}Cu half-line; for some peaks this correction amounts to approximately one standard deviation in the line position. A second
correction had to be introduced for the change of total efficiency
over the energetically broadened line. The efficiency of the gamma diffractometers changes approximately as E^{-2} at this energy.
This correction was of the same order as the previous one.

    By misfortune the first successful experiment was interrupted
before the highest diffraction orders (4th and 5th) were even
measured.

In these spectrometers the width of the Doppler- broadened annihilation line could easily deduced from its observation in the different diffraction orders : the observed angular width is a *linear*
addition of the spectrometer resolution (1.0 keV for GAMS 1 and
0.3 keV for GAMS 2 and 3, all in fifth order) and the energetically
broadened 511 keV line. This yields a value of ΔE(FWHM)= 2.5+0.1
keV.

In a direct measure of the momentum of the annihilating electron
positon pair in copper, Varlashkin and Zganjar[3]) obtained ΔE(FWHM)=
2.56 keV. This indicates that with a good approximation their conclusions can be applied to this Cu, Ta and Au sandwich, Cu being the
main constituent anyway: these authors concluded that the net photon
energy shift of the Doppler-broadened annihilation peak is 1.1 eV
below mc^2 . With the data from Cohen and Taylor[4]) one abtains E(m)=
511.003 5 keV ± 3.1 eV (6ppm) and consequently 511.002 4 keV ± 3.1 eV
for the observed annihilation peak.

Relative to this energy, following results have been obtained for
the 411.8 keV gamma line of ^{198}Hg :

GAMS 1   : Eγ = 411.835 ± 0.013 keV
GAMS 2-3 : Eγ = 411.800 0 ± 0.005 2 keV.

It is now essential to mention that both spectrometers observe
the source from opposite directions; this means that any asymmetry
of the Cu with respect to the Au-layer will cause a decrease in
diffranction angle (and an increase in energy) of the 411.8 keV for
one spectrometer but an increase in the diffraction angle for the
other. Geometrically one expects an energy difference of 12.7 eV
between the two results of the 411.8 keV line per μm of asymmetry
in the source. From this, it may be deduced that the actual asy-
metry is about 2.8 μm. This allows us also to obtain a final result
from the two separately measured energies, viz.

Eγ = 411.805 ± 0.005 keV (12ppm)

for the gamma ray energy *of ^{198}Hg.

---

* As gamma quanta result commonly from transitions between nuclear
  levels, it is a good policy to make a clear distinction between gamma *ray* (or gamma *line*) and gamma *transition* energy , having in mind
  that the transition energy equals the energy of the observed gamma ray plus the kinetic energy of the recoiling nucleus with mass M
  and atomic A. This *recoil* energy Er is given by the equation Er =
  $E\gamma^2/2Mc^2$ and written more practically as follows :

  Er (in eV) = $\dfrac{533}{A}$ Eγ^2 (in MeV)

  In case of the 411.8 keV transition in ^{198}Hg the recoil energy is
  Er = 0.46 eV.

4. CONCLUSIONS

The most precise previous determination of the [198]Hg gamma line energy has been made by G. Murray et al.[5]. Their result of E_{tr} = 411 795 ± 7 eV was related to the 1963 evaluated value for the elec- tron rest mass of 511 006 ± 5 eV. The slight decrease of the 1973 evaluation brings the [198]Hg gamma ray of ref.[5] down to E_{tr} = 411 793 ± 7 eV, resulting is an energy difference with the present result of 12 ± 8 eV, sufficiently large to be mentioned.

As stated previously, the authors did not succeed in obtaining results on the intended intercomparison of primary gamma energy stand- ards. However, from G.L. Borchert's contribution to this Confer- ence[6], the energy of the W $K\alpha_1$ Röntgen line can be calculated pro- portionally to the actual 411.805 keV energy; this yields an energy of E = 59.319 53 keV ± 0.70 eV (12 ppm) and, from the energy-to- wavelength conversion factor of ref.[4]

E.λ = 1239.852 0 ± 0.003 2 (keV.pm)   (2.6 ppm)

one obtains a wavelength of

λ = 20.901 29 ± 0.000 25 pm   (12 ppm)

This wavelength is in agreement with Bearden's value quoted in ref. [4], when it is converted to pm. Therefore one needs the radio Λ of a wavelength expressed in Å (1Å = 100 pm) to the same wavelength expressed in kXU (1XU is slightly larger than 0.1 pm). With the adopted value [4] Λ = 1.002 077 2 (5.3 ppm) Bearden's $WK\alpha_1$ wave- length becomes

λ = 20.901 43 ± 0.000 12 pm   (5.6 ppm)

in agreement with the combined information of Borchert and the present contribution, as Λλ = 190 ± 270 am or 9 ± 13 ppm. It may therefore be concluded that the actual two primary wavelength stand- ards (the Compton wavelength of the electron and the Röntgen ray wavelengths) both given independently by the fundamental constants, are consistent.

When using the preliminary value of Λ = 1.002 084 1 (2.4 ppm) obtained independently by Deslattes and Sauder[7] from Röntgen ray interferometry on crystal lattice spancings, Bearden's value becomes :

λ = 20.901 57 ± 0.000 06 pm   (3.0 ppm)

yielding a larger difference of Δλ = 330 ± 270 am or 16 ± 13 ppm.

As a general conclusion, it may be stated that with the described experiment, an intercomparison of the two important standards for gamma transition energy determination can be performed;
the expected ppm precision is still three times better than the
actual precision on the two mentioned standrd wavelengths.

ACKNOWLEDGMENTS

The authors are indebted to Dr O.W.B. Schult who suggested
that the influence of the reflectivity on the position of the
observed annihilation line should be checked. They had also useful discussions with Drs R.G. Helmer and R.D. Deslattes.

REFERENCES

1) P.H.M. VAN ASSCHE, J.M. VAN DEN CRUYCE, G. VANDENPUT, H.A.
    BAADER, H.R. KOCH, D.BREITIG and O.W.B. SCHULT in *Precision
    Measurement and Fundamental Constants*, Ed.by D.N. Langenberg
    and B.N. Taylor, Nat. Bur. Stand. (U.S.) Spec. Publ. 343
    (Aug. 1971), p.271.

2) H. BÖRNER, P. GÖTTEL, H.R. KOCH, J. PINSTON and R. ROUSSILLE
    in *Neutron Capture Gamma-Ray Spectroscopy* (Petten, 1975), to
    be published.

3) P.G. VARLASHKIN and E.F. ZGANJAR, Nuclear Physics A 130 (1969)
    182.

4) E.R. COHEN and B.N. TAYLOR, J. Phys. Chem. Ref. Data 2 (1973)
    663.

5) G. MURRAY, R.L. GRAHAM and J.S. GEIGER, Nuclear Physics 45
    (1963) 177 and ibidem 63 (1965) 353.

6) G.L. BORCHERT, these Proceedings, p.

7) R.D. DESLATTES and W.C. SAUDER in *Atomic Masses and Fundamen-
    tal Constants 4*, Ed. by J.H. Sanders and A.H. Wapstra (Plenum
    Publishing Corp., New York, 1972), p. 337.

PRECISION MEASUREMENT OF RELATIVE γ-RAY ENERGIES WITH A

CURVED CRYSTAL DIFFRACTOMETER

G.L. Borchert, W. Scheck and O.W.B. Schult

Institut für Kernphysik, Kernforschungsanlage

517 Jülich, Box 1913, West Germany

1. Introduction

Very accurate energies of γ- and X-rays are needed as calibration lines for Ge(Li) work (1,2), for μ-onic and mesonic X- and γ-ray studies (3) and for theoretical work on heavy and superheavy atoms (4). With the increasing performance of Ge(Li) spectrometers the errors of several calibration lines became about as large as those of the measurement. However one would wish that in practically all cases the uncertainties of the standards were much smaller. A sufficiently precise set of γ-ray standards must therefore be measured with the use of another method. We have chosen a curved crystal spectrometer for the measurement of a set of relative γ-ray standards with 30 keV < E_r < 1.5 MeV and dE_r/E_r > $5 \cdot 10^{-7}$ in relative energies based on the 411.794 $\pm$ 0 keV gold standard.

2. The Spectrometer

As outlined in detail elsewhere (5) at crystal spectrometers E is obtained through the measurement of the Bragg angle ϑ:

$$E = nhc/(2d \sin \vartheta)$$

with n being the order of reflection. At low γ-ray energies a crystal spectrometer has better resolution $\Delta E/E$, better linearity and a smoother background than a Ge(Li) spectrometer.

The FWHM ΔE achieved at our spectrometer is related to E:

$$\Delta E \text{ (keV)} > 4.5\left(\frac{E}{MeV}\right)^2/n,$$

which for n = 5 yields ΔE = 150 eV for E = 411 keV
and ΔE = 900 eV for E = 1 MeV.

The exact value of the proportionality factor(> 4.5) depends on
the actual source width. The value 4.5 is verified with very thin
sources. The energy E of a reflection is determined with an error
dE which is composed of the following contributions:

a) the statistical error $dE_s = \Delta E/(a\sqrt{N})$
   with a ≈ 2 and N = total number of counts in the reflection. This error is small if ΔE is small and if N
   is large which implies narrow and very strong reflections (very good curved crystal) and high specific activity of the source.
b) the instrumental error of the angle measuring system
   which instead of a mechanical device (6-9) consists of
   a Michelson interferometer. The basic principle of this
   system which is described in detail elsewhere (5) is
   shown in fig. 1. The rotating unit, a pair of parallel
   mirrors S_1 and S_2, is rigidly connected with the diffrac-
   tometer crystal and does not respond on a translational
   movement (vibrations). The optical path is changed in
   proportion to sin ϑ if the spectrometer crystal is rotated through an angle of ϑ. Provided proper adjustments,
   the number of interference fringes recorded during such
   a rotation is directly proportional to the wavelength of
   the γ-ray being reflected. The sensivity of the interferometer is proportional to A_1 which was chosen so that 1
   fringe corresponds with 0.5" for ϑ << 1. A series of
   tests indicated that the uncertainty of the angle
   measurement is $d\vartheta$ < 0.01".
c) the errors of the mechanical parts, which lead to uncertainties in the Bragg angles. These errors are given
   by the ratios of the lateral displacements of either
   the spectrometer crystal or the source and the focal
   length. For the reduction of the influence of these
   errors we have chosen a rather large focal length (464 cm)
   and put additional load on the main spectrometer bearing

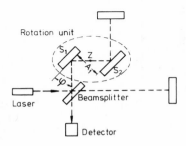

Fig. 1. Interferometric angle measuring system (principle)

so that its radial displacements were significantly reduced. The resulting uncertainties in angle measurement are estimated to be < 0.005".

Furthermore, a change of the temperature can easily produce uncertainties of this kind, but they should change gradually. In order to reduce this influence the whole spectrometer with its source was enclosed in a thermally stabilized housing. Fig. 3 shows that small drifts still remain – caused by temperature or air pressure changes etc. However, these drifts affect the different reflections in the same way. Therefore, during all of our measurements the γ-ray lines of interest were measured together with the line of the standard in different orders of reflections through the use of a twin source (10). The figure shows that although the drift amounts to about 0.2" during a measurement run of 3 days, the data points scatter around the smooth curves by only 0.01" or even less.

The linearity of a crystal spectrometer is in principle given through Bragg's law in the exact Laue transmission. Because of misalignments small deviations from linearity can be expected. These deviations can be determined by measuring many reflections in several orders of γ-rays with different energies. In this way we have found at our spectrometer a small systematic deviation from linearity by about $5 \cdot 10^{-6}$ which can be corrected for with an uncertainty of better than $2 \cdot 10^{-6}$ even for a single reflection.

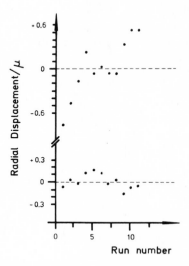

Fig. 2 Radial displacement of the spectrometer bearing (upper points without, lower points with additional load)

While it is quite cumbersome to correct with high accuracy for the non-linear background produced by Ge(Li) spectrometers because of the Compton edges and the Compton continuum, such difficulties are essentially absent at crystal spectrometers. Disregarding the immediate vincinity of the direct beam, the region where reflections of high energy γ-rays occur in low orders (n = 1,2), and where the background rapidly increases, the background is very smooth and its slope is extremely small. Therefore, it is very easy to correct for eventual small line energy shifts due to a change in background beneath the line.

3. Measuring Procedure

In each run, a total of eight reflections are recorded both at $\vartheta >$ 0 and at $\vartheta <$ 0. At least two of these pairs of reflections are used for the measurement of the standard line. Each run is repeated 10 – 20 times in order to reduce the statistical errors and to correct for systematic drifts. Often the same γ-ray reflections are measured in different programs so that the statistical errors can be further reduced (10). All energies are measured relative to the 411.794 ± 0.007 keV line from the decay of ^{198}Au. Since this line is taken as standard we used as basis for our relative γ-ray energies E_r and their errors dE_r the value 411.794 ± 0.000 keV.

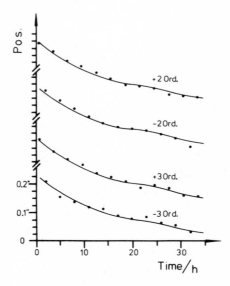

Fig. 3 Drifts of the reflections of the second and third order of the 411 keV line from the decay of ^{198}Au

4. Results

In table 1 are shown our results from the measurement of the
γ-ray energies from the decay of [183]Ta. The first column contains
the experimental data E_r. The next column shows the experimental
errors dE_r. All of them are smaller than 0.7 eV. In the third
column we have listed the absolute errors dE_a which include the
error of the gold standard of 7 eV corresponding to 1.7×10^{-5}.
We have also fitted our data into the level scheme of [183]W. The
results E_c and their corresponding errors dE_c are given in columns
4 and 5. These relative energies are the values which are recommended
as energy standards. The errors dE_c do not include the quality of
fit number which is about 0.7. But as the system is highly correlated (13 equations for 7 levels) we regard these errors as realistic
estimates for the standard deviation. In the last column there are
tabulated the weighted differences between the experimental and
the computed energy values. They show the very good consistency of
our data. Other precision energies for the γ-rays from the decay
of [182]Ta, and [199]Au were already published in ref. 5, those of
[169]Yb, [192]Jr, ^{51}Cr, [170]Tm and [203]Hg are given in ref. 11.

In table 2 are listed our results for the γ-lines from the
decay of [110]Ag. In this measurement we have even covered an energy
range from 440 to 1505 keV. Again the table shows very good overall consistency regarding our recommended values $E_c \pm dE_c$ and the
weighted differences in the last column.

Table 1 Energies of γ-rays from the decay of [183]Ta based on the
411.794 ± 0 keV gold standard

$\dfrac{E_r}{keV}$	$\dfrac{dE_r}{eV}$	$\dfrac{dE_a}{eV}$	$\dfrac{E_c}{keV}$	$\dfrac{dE_c}{eV}$	$\dfrac{\left|E_r - E_c\right|}{dE_r}$
46,48384	0,16	0,8	46,48402	0,12	1,12
52,59515	0,12	0,9	52,59524	0,11	0,75
99,07932	0,09	1,7	99,07925	0,09	0,78
107,93096	0,11	1,8	107,93088	0,12	0,73
144,12172	0,28	2,4	144,12187	0,27	0,54
160,52599	0,48	2,7	160,52608	0,16	0,19
161,34391	0,07	2,7	161,34392	0,08	0,14
162,32112	0,27	2,7	162,32117	0,28	0,19
209,86690	0,44	3,6	209,86727	0,30	0,84
244,26296	0,54	4,1	244,26315	0,33	0,35
246,05851	0,21	4,2	246,05823	0,19	1,33
291,72359	0,67	5,0	291,72436	0,22	1,15
353,98872	0,34	6,0	353,98895	0,19	0,68

Besides we have recently measured the 675 keV line from the decay of ^{198}Au which is also frequently used as calibration line. Our preliminary result is $E_r \pm dE_r$ = 675.871 $\pm$ 0.004 keV. All of our recommended secondary γ-ray standard energies are much more precise – up to one order of magnitude – than their relative energies known so far.

Table 2 Energies of γ-rays from the decay of ^{110}Ag based on the 411.794 $\pm$ 0 keV gold standard. The relative energies E_c result from a fit of the energies E_r into the level scheme of ^{110}Cd

$\dfrac{E_r}{keV}$	$\dfrac{dE_r}{eV}$	$\dfrac{dE_a}{eV}$	$\dfrac{E_c}{keV}$	$\dfrac{dE_c}{eV}$	$\dfrac{E_r - E_c}{dE_r}$
446.799	5	9	446.799	5	0.74
620.350	17	20	620.336	9	0.84
657.743	4	12	657.743	4	0.00
677.594	14	18	677.605	9	0.81
686.990	14	18	686.984	10	0.45
706.663	5	13	706.665	5	0.39
744.245	20	24	744.253	11	0.41
763.936	4	13	763.935	4	0.29
818.025	15	20	818.022	12	0.18
884.665	17	23	884.670	13	0.31
937.465	19	25	937.470	10	0.28
1384.300	80	83	1384.265	9	0.43
1505.051	90	94	1505.000	14	0.56

References

1) R.C. GREENWOOD, R.G. HELMER and R.J. GEHRKE, Nucl. Instr. and Meth. 77 (1970) 141
2) R.G. HELMER, R.C. GREENWOOD and R.J. GEHRKE, Nucl. Instr. and Meth. 96 (1971) 173
3) R. ENGFER, H.K. WALTER and H. SCHNEUWLY to be published in Physics of elementary particles and atomic nuclei (Joint Institute for Nuclear Research, Dubna)
4) J. RAFELSKI, W. GREINER and L.P. FULCHER, Nuovo Cimento 13 (1973) 135
5) G.L. BORCHERT, W. SCHECK and O.W.B. SCHULT, Nucl. Instr. and Meth. 124 (1975) 107
6) O. BECKMANN, P. BERGVALL and B. AXELSON, Arkiv Fysik 14 (1958) 419
7) A.I. SMIRNOW, V.A. SHABUROV, V.L. ALEXEYEV, D.M. KAMINKER, A.S. RYLNIKOV and O.I. SUMBAEV, Nucl. Instr. and Meth. 60 (1968) 103
8) D. BREITIG, Z. Naturf. 26a (1971) 371
9) O. PILLER, W. BEER and J. KERN, Nucl. Instr. and Meth. 107 (1973) 61
10) G.L. BORCHERT, Dissertation (Technische Hochschule München 1974)
11) G.L. BORCHERT, Z. Naturf. 30a (1975) 274

VISIBLE TO GAMMA-RAY WAVELENGTH RATIO

R.D. Deslattes, E.G. Kessler, W.C. Sauder, and A. Henins

National Bureau of Standards

Washington, D.C. 20234, USA

Over the past decade we have attempted to develop a measurement capability for connecting visible and γ-ray reference wavelengths (1). This communication summarizes our procedures and gives a preliminary result from an initial measurement of ^{198}Au (412 KeV).

In the past, the connection between visible and γ-ray wavelengths has been established using the X-ray wavelength scale to bridge the gap between nanometer and picometer regions. Such a route is limited in accuracy by the several ppm uncertainty arising from inherent imprecision of X-ray lines and the accumulation of errors in wavelength transfers (2). The required X-ray to visible comparison was traditionally by means of ruled grating measurements which reached the 10 ppm level only with difficulty (3).

Recently, the X-ray to visible comparison was improved by use of methods similar to those reported here (4). We emphasize that the direct approach to γ-ray sources avoids limitations due to the large widths of X-ray lines and the accumulation of uncertainty due to the several transfers required.

Our work begins with an iodine stabilized ^{3}He-^{22}Ne laser operating in the visible near 633 nm. This is stabilized with respect to an absorption transition in ^{129}I$_2$ conventionally labelled "B" (5). This ^{129}I$_2$ wavelength is known with respect to ^{86}Kr 606 nm to an accuracy limited principally by the definition. The value, namely 632 990.0742 ($\pm$ .0009) pm is also consistent with the provisional definition, c = 299 792 458 ms^{-1} (6). Perhaps of more direct importance in γ-ray measurements is that the reference point for the most recent Rydberg determination (7) is connected with this line by a heterodyne measurement (5).

This light source was used to establish the 220 repeat distance in a silicon crystal sample by a process of simultaneous optical and X-ray interferometry of a common baseline. The current result of this measurement corrected to 25C is d(220) = 192.01715 pm (.15 ppm). (Samples of this crystal were used in the determination of the Avogadro constant (8) and in evaluation of X-ray reference wavelengths (4)).

Next, a quasi-non-dispersive transfer is in progress between a sample of the above Si crystal and a sample of the Ge crystal used in the γ-ray measurements. This procedure follows that described by Hart (9) except that angle measuring interferometry permits acceptance of a larger difference in lattice parameter.

Pending completion of this exercise, we have made a preliminary estimate for the Ge crystals' lattice parameter by a more traditional two crystal measurement. In this, we had previously evaluated the wavelength of the peak of the $CuK\alpha_1$ profile from a conventional X-ray tube by means of diffraction from a Si crystal calibrated by simultaneous X-ray and optical interferometry (4). Radiation from the same source under similar operating conditions was in the present case diffracted (Bragg case) from Ge samples in 440 and 220 reflection. Estimates of the index of refraction were taken from the review by Hart (10). The resulting value for the cell edge dimension is a_o = 565.7787 pm (1 ppm) at the temperature of measurement, 21.66C. (The contribution of the Si-Ge transfer measurement to the Ge lattice constant uncertainty will be made small compared to the uncertainty of the Si repeat distance when the non-dispersive measurements are carried through.)

For comparison, we note that the previous measurements in which both Ge and Si lattice constants are reported as measured in the same apparatus also yield Ge to Si lattice constant ratios. Smakula and Kalnajs (11) give 1.041766 (5 ppm) for the ratio while Hom, Kiszenich and Post (12) report 1.041783 (4 ppm), whereas the present result is 1.041765 (1.2 ppm). These ratios are all referred to 25C. These discrepancies have not been analyzed in detail.

Slabs of this germanium sample were oriented for 400 diffraction in symmetric Laue transmission. These were used in a two-crystal instrument which was designed especially for the small angle measurements required in γ-ray wavelength determinations (13). Angle measurements were carried out by a modification of a Michelson (Marzolf) angle interferometer (14). In this modification, crystal rotation is encoded as polarization rotation with a 90° polarization change corresponding to 70 milli arc sec. Faraday modulators and step driven Glan-Thompson analysers permit convenient servo control to about 0.3 milli arc sec or better. The interfer-

ometers were calibrated by summing to closure measurements of the
interfacial angles of a 72-sided optical polygon. The provisional
result of this calibration is that angular departures from the
"zero" interferometer angle are given by $\phi = \sin^{-1} n/k$ where n is
the fringe count (including the fractional part) and k = 5 774 451.0
(0.4 ppm). Measurements taken for the two reflection conditions
give values for n_+ and n_- whence we obtain ϕ_+ and ϕ_- such that
$2\phi = \phi_+ + \phi_-$. In turn, $\lambda = \dfrac{2a_o \sin \theta}{\sqrt{k^2 + k^2 + \ell^2}}$. For the 400 reflection
this becomes $\lambda = \dfrac{1}{2} a_o \sin \theta$.

Quite recently, about mid-May, we emplaced a ^{198}Au source having
an initial activity of about 50 Ci. This was prepared by irradiation
of a natural Au foil ($\sim 6\,\mu m \times 4$ cm dia) for four days in a flux of
10^{14} n/cm^2 sec in the NBS research reactor (15). Due to a variety
of problems, useful data were not obtained until after about two
half-lives so that the activity was about 5 Ci during most of the
observation period of about 40 hours. Counting rates of about
1 sec^{-1} were then obtained in the presence of an equal background,
part of which was source associated. The detector was NaI-Tℓ,
7.5 cm dia x 7.5 cm long operated with a 20% pulse height window at
the 412 keV photopeak.

The full widths at half maxima of the two diffraction curves
were about 0.35 arc sec. Rough theoretical estimates indicate that
this is mostly due to intrinsic diffraction width. From this and
from other studies, the excess width which might be ascribed to
residual imperfection in these crystals is of the order of 0.1 arc
sec or less.(16).

Because of the low activity at the time of measurement, we had
to allow a larger than desirable vertical divergence through the
instrument. This angular spread was about 5 milli radians which
required a vertical divergence correction of 4.8 ppm, considerably
higher than should be tolerated in normal circumstances. There is
also apparent in the data a small but significant asymmetry in
both the plus position and minus diffraction profiles. Expressed
as a fractional displacement of the mid-chord at half height from
the peak relative to the width at half height, the asymmetry is
0.027. This effect is not yet understood but may arise from the
residual imperfection noted above; further measurements and crystal
characterization are required. Pending resolution of this question,
our initial results should be treated with caution somewhat greater
than that normally applied to first run data.

With these reservations, Table I indicates the results of this
preliminary measurement. The voltage-wavelength conversion factor
is taken from Cohen and Taylor (17) who associate with it a 1σ

estimate of 2.6 ppm.   (When comparing γ-ray energies derived from
wavelength measurements this is not a problem as long as the same
conversion factor is applied.)

Table I

Observed Bragg Angle	0.60980822° (0.7 ppm)
Bragg Angle Corrected for Vertical Divergence	0.60980528° (1.2 ppm)
Wavelength	3.010766 pm (2 ppm)
Energy	411.8062 KeV (3.3 ppm)

This table summarizes our measurements on 412 keV Au line.   The
uncertainty for the observed Bragg angle includes only statistical
uncertainty while that for the Bragg angle corrected includes
statistical and vertical divergence uncertainty.   The energy
uncertainty includes the total uncertainty in our measurement and
the uncertainty in the voltage-wavelength conversion factor.

To establish an uncertainty for this measurement, we have
considered the following sources of error – statistical uncertainty,
uncertainty of the vertical divergence correction, uncertainty due
to the line shape asymmetry, uncertainty in the calibration of the
angle measuring interferometers, and uncertainty of the lattice
spacing of germanium.   The statistical uncertainty of the three
measurements which were used to obtain the observed Bragg angle
in Table I was 0.7 ppm (one standard deviation).   The vertical
divergence correction was estimated to be correct to 1 ppm.   The
observed Bragg angles were determined using the peak of the line
as reference.   If the mid-chord at half height were used as
reference, the angles would increase by 4 ppm.   The uncertainty due
to line shape asymmetry is estimated to be 1 ppm.   The calibration
of the angle measuring interferometers is uncertain to 0.4 ppm.
The uncertainty in the lattice spacing including a contribution
from the expansion coefficient is 1.2 ppm.   We combine these errors
in the root-sum-square manner to obtain a total uncertainty in
wavelength of 2 ppm.

The reported value of 411.8062 keV (3.3 ppm) differs by more
than a standard deviation from the widely used 411.794 keV (17 ppm)
as well as the result 411.797 keV (17 ppm) given by Beer and Kern
(18).   In Ref. 18 and in the work of Bergvall (19) the basic
experimental measurement is that of the wavelength ratio between
WKα$_1$ (produced in a ^{187}W source) and ^{198}Au 412 keV.   Beer and Kern
report 6.94222 (20 ppm) for this ratio, while Bergvall gives
6.94156 (80 ppm).   (The suggested "chemical shift" of WKα$_1$ from a

[182]Ta K capture source (20) is not of importance in this comparison.)

The above situation is somewhat clarified by the following observations: Ref. 4 gives an optical scale value for the wavelength of MoKα_1, viz 70.93187 pm (0.6 ppm). This can be combined with the ratio $\dfrac{\lambda \text{ (MoK}\alpha_1)}{\lambda \text{ (WK}\alpha_1)} = 3.393620$ (1.3 ppm) (2) to obtain a value of

20.90154 pm (1.4 ppm) for WKα_1 produced by electron bombardment of natural W. If one neglects any possible shift between WKα_1 from electron bombardment of natural W and that from a [187]W source, then the above value for the wavelength of WKα_1 can be combined with the WKα_1 to [198]Au (412 keV) ratios to obtain alternate estimates of the [198]Au (412) keV energy. The result of applying this procedure with the ratio from Beer and Kern is 411.803 keV (20 ppm) while use of Bergvall's ratio gives 411.764 keV (80 ppm).

If further refinement of our work sustains the initial result reported here then the agreement between our direct approach yielding 411.8062 keV (3.3 ppm) and the route through the X-ray scale and the Beer and Kern ratio, viz 411.803 keV (20 ppm) is satisfying. It is premature to dwell on this for two reasons: Firstly, we wish to emphasize the provisional character of our initial result. Secondly, it is strongly felt that more thorough exploitation of our measurement capability will yield results significantly better than 1 ppm free of the reservations noted above.

References

1. R.D. Deslattes and W.C. Sauder, Atomic Masses and Fundamental Constants, J.H. Sanders and A.H. Wapstra, eds. (Plenum Press, N.Y., 1972) p. 337.

2. J.A. Bearden, A. Henins, J. Marzolf, W.C. Sauder and J.S. Thomsen, Phys. Rev. 135, A899 (1964).

3. A. Henins, Precision Measurement and Fundamental Constants, edited by D.N. Langenberg and B.N. Taylor, Nat. Bur. Stds. Spec. Pub. No. 343, (U.S. GPO, Washington, D.C., 1971) p. 255.

4. R.D. Deslattes and A. Henins, Phys. Rev. Lett. 31, 972 (1973).

5. W.G. Schweitzer, E.G. Kessler, R.D. Deslattes, H.P. Layer and J.R. Whetstone, Appl. Opt. 12, 2927 (1973).

6. J. Terrien, Nouv. Rev. Optique 4, 215 (1973).

7.   T.W. Hänsch, M.H. Nayfeh, S.A. Lee, S.M. Curry and I.S. Shakin, Phys. Rev. Lett. 32, 1336 (1974).

8.   R.D. Deslattes, A. Henins, H.A. Bowman, R.M. Schoonover, C.L. Carroll, I.L. Barnes, L.A. Machlan, L.J. Moore and W.R. Shields, Phys. Rev. Lett. 33, 463 (1974).

9.   M. Hart, Proc. Roy. Soc. A309, 281 (1969).

10.   M. Hart, Reports Progress in Physics 34, 435 (1971).

11.   A. Smakula and J. Kalnajs, Phys. Rev. 99, 1737 (1955).

12.   T. Hom, W. Kiszenich and B. Post, J. Appl. Cryst. (1975).

13.   Some features of this instrument were described by W.C. Sauder in Precision Measurement and Fundamental Constants, op cit., p. 275, and in Ref. 1.

14.   J.G. Marzolf, Rev. Sci. Inst. 35, 1212 (1964).

15.   We are indebted to R.J. Carter, T. Rabi, N. Bickford, H. Despain and P. Cassidy for their cooperation in the reactor irradiation.

16.   These excellent crystals were kindly furnished by Robert N. Hall of the General Electric Research Laboratory.

17.   E.R. Cohen and B.N. Taylor, Jour. Phys. Chem. Ref. Data 2, 663 (1973).

18.   W. Beer and J. Kern, Nuc. Inst. and Meth. 117, 183 (1974).

19.   P. Bergvall, Arkiv Fys. 17, 125 (1960).

20.   W. Beer and J. Kern, Phys. Lett. 47B, 345 (1973).

DETERMINATION OF PROTON BINDING ENERGIES FOR ^{89}Y, ^{90}Zr, ^{91}Nb AND ^{93}Tc FROM (p,γ) REACTION Q-VALUES.

U. Bertsche[+], F. Rauch and K. Stelzer

Institut für Kernphysik

Universität Frankfurt/Main

In the course of spectroscopic work with (p,γ) reactions in the A = 90 mass region, we had found discrepancies between the reaction Q-values, i.e., the proton binding energies, derived from our measurements and those given in the 1964 mass table /1/ and also in the 1971 mass table /2/. Therefore, we made a systematic investigation in order to obtain more accurate values. The reactions studied were ^{88}Sr(p,γ)^{89}Y, ^{89}Y(p,γ)^{90}Zr, ^{90}Zr(p,γ)^{91}Nb and ^{92}Mo(p,γ)^{93}Tc.

The experiments were performed at the 7 MeV Van de Graaff accelerator of the Institut für Kernphysik at Frankfurt. The proton beam energy was calibrated by measuring the ^{13}C(p,n) threshold, which is at E_p = 3235.7 $\pm$ 0.7 keV /6/. Including the relative error of our threshold determination of about 0.8 keV, the error for the proton beam energy in our (p,γ) measurements was about 1 keV.

The targets used had thicknesses of 0.30 mg/cm^2, 0.10 mg/cm^2, 0.50 mg/cm^2 and 0.73 mg/cm^2 for ^{88}Sr, ^{89}Y, ^{90}Zr and ^{92}Mo, respectively. The ^{90}Zr and ^{92}Mo targets were foils of enriched isotopes. The other targets were made by evaporating natural Sr resp. Y on Ta backings. A liquid nitrogen trap prevented carbon build-up during the measurements.

The γ spectra were taken at 90° with a 30 cm^3 Ge(Li) detector which had a resolution of about 9 keV for E_γ = 9.0 MeV. An accurate energy calibration for the high-energy region of the spectra was effected by measuring simultaneously γ-rays from the Ni(n_{th},γ) reaction. The principle of this method has been described by van der Leun and de Wit /3/.

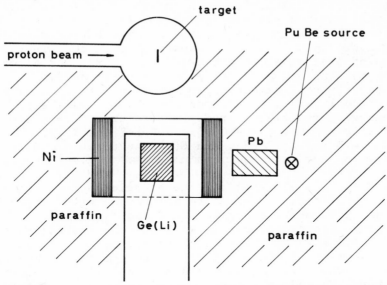

Fig.1. Sketch of the set-up for the simultaneous measurement of a (p,γ) spectrum and a $Ni(n_{th},\gamma)$ spectrum .

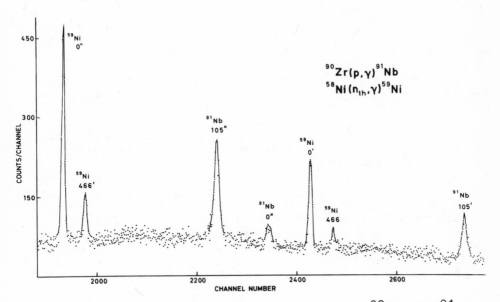

Fig.2. Part of a spectrum from the reactions $^{90}Zr(p,\gamma)^{91}Nb$ and $Ni(n_{th},\gamma)$. Each peak is labelled by the energy of the final state. (') denotes single-escape peaks, (") denotes double-escape peaks.

A sketch of our set-up is shown in fig. 1. A favourable geometry was obtained by surrounding the detector by a nickel cylinder. The neutrons were supplied by a PuBe source with a strength of 1.5×10^6 n/sec and were slowed down in paraffin. A piece of lead shielded the detector against radiation from the source. The energies of the three Ni transitions, which were used for calibration, were determined by measuring simultaneously a $Fe(n_{th}, \gamma)$ spectrum and taking the precise Fe γ energies reported in ref. 3 as reference energies. Our results are $E(\gamma_0) = 9000.6 \pm 0.7$ keV and $E(\gamma_1) = 8534.7 \pm 0.5$ keV for ^{59}Ni, and $\overline{E}(\gamma_0) = 7821.6 \pm 0.5$ keV for ^{61}Ni. An example of a spectrum is shown in fig.2.

For each (p, γ) reaction, the energies $E(\gamma_i)$ of three primary transitions γ_i could be determined with errors between 3 keV and 5 keV. The energies E_i of the final states i were known from literature or from work at our laboratory to better than 1.0 keV. An error-weighted average of the sums $E(\gamma_i) + E_i$ was calculated to yield the excitation energy E^+ of the compound system. E^+ was determined with an error of 5 keV for ^{93}Tc and an error of 3 keV for each of the three other compound nuclei. Recoil corrections were neglected since they were smaller than 0.5 keV in all cases. The proton binding energy was obtained as $B_p = E^+ - (E_p - \Delta E/2)$, where E_p is the proton energy in the center of mass system and ΔE is the proton energy loss in the target. The error in B_p includes an error of 2 keV to 3 keV in $\Delta E/2$ resulting from the estimated uncertainty in the target thicknesses.

The results for the proton binding energies are: $B_p(^{89}Y) = 7078 \pm 4$ keV, $B_p(^{90}Zr) = 8351 \pm 4$ keV, $B_p(^{91}Nb) = 5167 \pm 4$ keV, $B_p(^{93}Tc) = 4081 \pm 6$ keV.

As a check on the internal consistency of these results, we performed additional experiments, in which we measured relative reaction Q-values. Pairs of targets, one target placed behind the other, were bombarded simultaneously, and in the spectra the energy differences between primary transitions of the two (p, γ) reactions were determined. Three pairs of targets were used: $^{88}Sr/^{89}Y$, $^{88}Sr/^{90}Zr$ and $^{90}Zr/^{92}Mo$. The differences between B_p values obtained from these measurements, which had errors between 3 keV and 4 keV, agreed for each pair within 2 keV with the corresponding differences between the absolute Q-values listed above.

A comparison of the proton binding energies with values determined from other measurements is shown in table 1. The B_p values calculated from mass spectroscopic data and reaction Q-values given in part III of the 1971 mass tables are listed first. The references for these B_p values are abbreviated as

Table 1
Comparison of Proton Binding Energies obtained by Different
Methods

Method		B_p (keV)	Reference
^{89}Y	mass doublets	7069.0 ± 5.5	63RIO7
	(d,p) and ß$^-$	7063.0 ± 6.4	67SP09, 70W005
	(p,γ)	7052 ± 7	69IR01
	adjusted value	7067.2 ± 3.2	1971 mass tables /2/
	(p,γ)	7078 ± 4	this work
^{90}Zr	mass doublets	8382.7 ± 6.6	63RIO7
	(n,γ)/(d,p) and ß$^-$	8349.6 ± 4.2	67RA27,64WA14,
			66HA15,64LA13
	adjusted value	8366.2 ± 3.3	1971 mass tables /2/
	ß$^+$/(p,n) and (p,d)	8335 ± 11	60J05, 63OK01,/4/
	(p,γ)	8351 ± 4	this work
^{91}Nb	(d,p) and (p,n)	5152 ± 10	67SP09, 70KI01
	mass doublets	5159 ± 6	63RIO7, 67RA27
	and (n,γ) and (p,n)		70KI01, /4/
	adjusted value	5158 ± 7	1971 mass tables /2/
	(d,p) and (p,n)	5157.8 ± 8.5	67SP09 /5/
	(p,d) and (p,n)	5163 ± 10	/4/, /5/
	(p,γ)	5167 ± 4	this work
^{93}Tc	(n,γ) and ß$^+$	4106 ± 13	70MU07, 62VI4,
			51B48
	adjusted value	4104 ± 13	1971 mass tables /2/
	(p,γ)	4081 ± 5	this work

Table 2
Mass Excess of ^{89}Y

C7H5-89Y mass doublet (ref.2)	- 87671 ± 5 keV
mass doublets and reaction Q-values for lower A-values (ref.2)	- 87670 ± 10 keV
mass doublets and reaction Q-values for higher A-values (ref.2)	- 87715 ± 5 keV
C4H802-88SR mass doublet (ref.2) and 88SR(P,G) Q-value (this work)	- 87683 ± 6 keV
C4H1002-90ZR mass doublet (ref.2) and 89Y(P,G) Q-value (this work)	- 87706 ± 7 keV

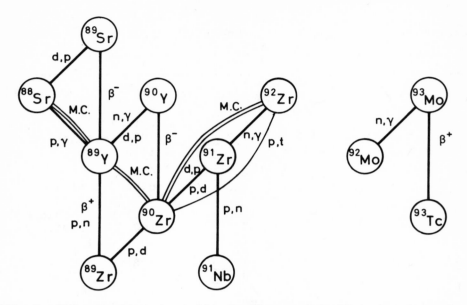

Fig.3. Illustration to table 1. The explanation is given in
the text.

in the tables of part III. We have also included the adjusted
values given in those tables. For ^{90}Zr and for ^{91}Nb also a B_p
value is listed which was calculated using Q-values given in
recent publications. In fig.3, which serves as illustration
to table 1, the reactions used in calculating the B_p values
are indicated, exept for our (p,γ) reactions. M.C. denotes
mass connections between nuclides whose masses have been determined from mass doublets.

As the data of the table shows, there is in general reasonable or good agreement of our new B_p values with previous
results, although in single cases there are deviations outside the quoted errors. There is no systematic deviation of
our values compared to those obtained by other methods. Since
also the internal consictency of our results is very satisfactory, we believe that the proton binding energies determined
in the present investigation should be valuable input data
for an improved mass adjustment in the A = 90 mass region.

Of special interest is the question of the ^{89}Y mass excess. As discussed in part III of the 1971 mass tables, there
are discrepancies between the mass excess derived from mass
doublets and Q-values for lower A-values and the mass excess
derived from mass doublets and Q-values for higher A-values.

In table 2 we compare these values with the values of the ^{89}Y
mass excess obtained from the C4H8O2-88SR mass doublet plus
our 88SR(P,G) Q-value and the C4H10O2-90ZR mass doublet plus
our 89Y(P,G) Q-value. Also listed is the 89Y mass excess derived from the C7H5-89Y mass doublet. The comparison shows
that with our new (p,γ) Q-values the discrepancy is reduced
but is still not removed. It seems that the uncertainty in the
mass doublets of ^{88}Sr, ^{89}Y, and ^{90}Zr is larger than the quoted
errors, as indicated also by the fact that the B$_p$ value of
^{90}Zr derived from mass doublets is much higher than that derived from Q-values.

+) Present address: Deutsche Bundeswehr, Bonn.

/1/ J.H.E.Mattauch, W.Thiele and A.H.Wapstra,Nucl.Phys. 67
    (1965) 1 .
/2/ A.H.Wapstra and N.B.Gove, Nuclear Data Tables 9 (1971) 265.
/3/ C.van der Leun and P. de Wit, Proceedings of the 4th Int.
    Conf.on Atomic Masses and Fundamental Constants,Plenum
    Press, London-New York, 1972.
/4/ J.B.Ball, R.L.Auble and P.G.Roos, Phys.Rev.C 4 (1971) 196.
/5/ S.Matsuki et al., Nucl.Phys. A 174 (1971) 343.
/6/ J.B.Marion and F.C.Young, Nuclear Reaction Analysis,
    North-Holland, Amsterdam , 1968.

A NEW METHOD FOR MEASUREMENT OF PROTON BEAM ENERGIES

C. Rolfs, W. S. Rodney and H. Winkler

Institut für Kernphysik, Münster, National Science
Foundation, Washington D. C. and California State
University, Los Angeles

The absolute energy of an ion beam can be measured with the
help of an electrostatic, a $180°$ magnetic or a velocity analyzer.
Most commonly a $90°$ magnetic analyzer is used as a relative
instrument that must be calibrated using the energies of certain
known resonances or threshold energies. The deduced calibration parameters relate the proton beam energies to the field
strength of the $90°$ magnetic analyzer. However, these parameters can change significantly with time and are critically
dependent on the settings of the energy-defining input and output slits and the stabilizer of the accelerator. At any given time,
a precise beam energy measurement requires one therefore to
check the calibration parameters, a lengthy and inconvenient
procedure. The present report describes a method which avoids
these inconveniences and which determines the proton beam
energy in a relatively short running period directly at any given
value of interest to a precision of the order of 0.4 keV, using a
Ge (Li) detector. This precision is well comparable with other
techniques mentioned above.

Since the advent of the Ge(Li) detector, it is possible to
determine γ-ray energies with an accuracy of better than one
part in 10^4. Consequently, a measurement of the energy of a
primary γ-ray following radiative capture gives the energy of
the captured particle to the same accuracy, if the Q-value of
the reaction is precisely known. The procedure is usually not

convenient because of the high Q-value of most capture reactions
($\sim$8 MeV) and consequently of the high energy of most primary
capture γ-rays. However, due to the exceptionally low Q-value
of 600 keV for the $^{16}O(p,\gamma)^{17}F$ capture reaction, the two observed
primary direct capture (=DC) γ-rays, DC$\rightarrow$O and DC$\rightarrow$495 keV
(fig. 1), have energies close to the energy of the incident proton
beam. Since the energies of these γ-rays lie in a region where
comparison with precisely known energies of γ-rays from
radioactive sources is possible, a precise determination of the
$^{16}O(p,\gamma)^{17}F$ γ-ray energies can be achieved. In addition, since
these γ-rays result from a non-resonant direct capture reaction[1],
they are observable over a continuous range of beam energies
(fig. 1) and thus the beam energy measurement is not restricted
to resonant energies as it is the case in the usual relative methods. The measurement of the DC γ-ray energies in conjunction
with the known Q-value for this reaction allow one to determine
proton beam energies above 0.8 MeV (fig. 1) directly at any
given value of interest in a relatively short running time without
interpolation procedures as it is the case in the relative methods.
The 90° magnetic analyzer serves in this method only as a beam
energy stabilizing device and no precise knowledge of its calibration parameters is required.

Since the Q-value of the $^{16}O(p,\gamma)^{17}F$ reaction is important in
the application of this method, an independent determination of
this value was desirable. This determination was carried out by
turning the method around. The proton beam energy was set to
precisely known values, namely at the E_p = 991.88+ 0.04 keV
resonance in $^{27}Al(p,\gamma)^{28}Si$ as well as the $^7\overline{Li}(p,n)^7Be$
threshold at E_p = 1880.59+0.08 keV. At these beam energies the
DC transitions of the $^{16}O(\overline{p},\gamma)^{17}F$ reaction were observed at θ_γ =
0° concurrently with well known γ-calibration lines from radioactive sources. A schematic diagram of the experimental set-up
is shown in fig. 2. An individual run for the Q-value measurement consisted of the following measurements: First the
excitation function of the 992 keV resonance was measured on
a fresh ^{27}Al target at target location 1 and midpoint (NMR
frequency) and width determined. The beam energy was then
set at the midpoint and a ^{16}O target irradiated at target location
2. A sample γ-ray spectrum obtained at location 2 with a
72 cm^3 Ge(Li) detector is shown in fig. 3. No beam or stabilizer control of the accelerator was changed from the beginning
of the excitation function measurement. For runs over two
hours on the ^{16}O target, the yield from the ^{27}Al target at
location 1 was checked at least once in

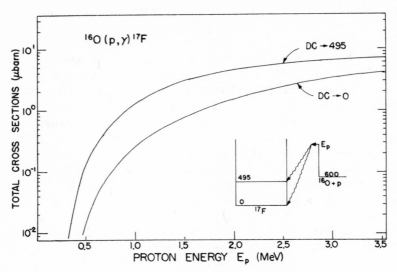

Fig. 1. Observed excitation functions for the direct capture transitions DC→ 495 keV and DC→O of the ^{16}O (p, γ) ^{17}F reaction. The inset shows the level diagram of ^{17}F. For details, see ref. [1].

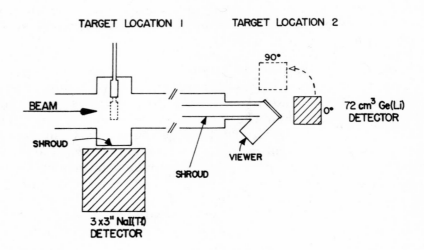

Fig. 2. Schematic diagram of the experimental set-up. The two target locations were separated by 3 m.

order to test the stability of the midpoint of the beam energy distribution. Finally another excitation function of the 992 keV resonance was taken after the end of the ^{16}O irradiation. The energy of the DC→O γ-ray transition was extracted from the midpoint of the highenergy slopes (fig. 3). A similar procedure also was applied in the Q-value determination via the ^{7}Li (p, n) ^{7}Be threshold. From several runs, an average value of Q = 600.35±0.28 keV was obtained[+], in fair agreement with the value of Q = 600.7±0.4 keV from the Wapstra tables[2].

With the Q-value for the ^{16}O (p, γ)^{17}F reaction known to an accuracy of ± 280 eV, one can now find the energy of the proton beam at any value of interest to a precision of the same order with a single measurement. The overall accuracy in such measurements depends on the precision to which the energies of the DC transitions are determined.

The present method can be applied to determine Q-values of proton induced reactions to a higher precision than those given in the tables[2], where errors in Q as large as 8 keV are not uncommon. Two examples of such Q-value measurements of (p, γ) reactions are descirbed below, which are of astrophysical interest.

The ^{17}O(p, γ) ^{18}F reaction: The Q-value of this reaction (Q = + 5.61 MeV) is of interest in the CNO-cycle for hydrogen burning of ^{17}O in stars, since it determines the position of excited states in ^{18}F at $E_x \sim 5.6$ MeV with regard to the proton threshold. For this Q-value measurement, the 1099 keV resonance in the reaction was chosen. An excitation function at this resonance was first measured at target location 1 (fig. 2). For a proton energy at the midpoint of this yield curve, the DC transitions of the ^{16}O(p, γ) ^{17}F reaction were observed at target location 2. The techniques applied are identical to those described above. The resulting proton energy is E_p = 1098.9 ±0.4 keV. The excitation energy of the resonance state was determined

---

[+] Further details on experimental and analyses procedures (angular position of the detector, Dopplershift-reduction due to the finite size of the detector, Lewis effect, carbon buildup, assymetry of high energy slope of DC γ-ray peaks) are described in ref. [3]

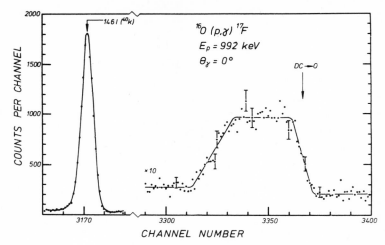

Fig. 3.

Relevant part of the γ-ray spectrum for the DC→ O transition of
^{16}O (p, γ) ^{17}F. Also shown is the 1461 keV γ-ray line of ^{40}K recorded
concurrently during the run.

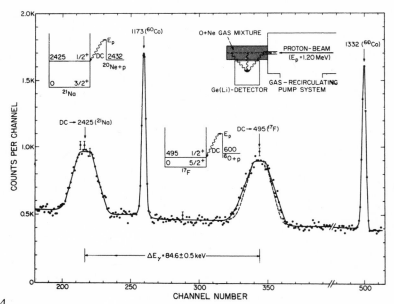

Fig. 4.

Relevant part of a γ-ray spectrum obtained by proton bombarde-
ment (E_p = 1.20 MeV) of a gas mixture of neon and oxygen. Also
shown are the γ-rays from a ^{60}Co source observed concurrently
in the experiment 4).

in an additional experiment at the flat plateau of the resonance yield curve by adding the γ-ray energies of the $\gamma\gamma$ cascades: $E_x = 6643.6 \pm 0.4$ keV. From these two results one derives a Q-value of $\bar{Q} = 5606.2 \pm 0.6$ keV for the $^{17}O(p, \gamma)$ ^{18}F reaction, which has to be compared with the 1972 Q-value table[2] of Q = 5609.0 ± 1.2 keV.

The $^{20}Ne(p, \gamma)$ ^{21}Na reaction: During the course of detailed studies of this reaction using a differentially pumped gas target, a DC transition to the 2425.2 ± 0.4 keV threshold state was observed[4]. The knowledge of the position of this state relative to the proton threshold is again important for the calculation of stellar reaction rates of hydrogen burning in the $^{20}Ne(p, \gamma)^{21}Na$ reaction. A precision measurement of the Q-value of this reaction was carried out with a gas mixture of natural neon and oxygen. The DC transitions in the $^{20}Ne(p, \gamma)$ ^{21}Na and $^{16}O(p, \gamma)$ ^{17}F reactions to the threshold states at 2425 and 495 keV, respectively were observed concurrently at $\theta_\gamma = 90°$ and $E_p = 1.20$ MeV. From the energy difference of $\Delta E_\gamma = 84.6 \pm 0.5$ keV of both transitions (fig. 4), the Q-Value for $^{20}Ne(p, \gamma)$ ^{21}Na is obtained from the equation

$$Q(^{20}Ne+p) = Q(^{16}O+p) + E_p(\frac{16}{17} - \frac{20}{21}) + \Delta E_x - \Delta E_\gamma$$

where $\Delta E_x = 1929.9 \pm 0.5$ keV is the difference in excitation energies of the two states and $Q(^{16}O+p) = 600.3 \pm 0.28$ keV (ref.[3]). Since both γ-ray transitions have similar energies and identical γ-ray angular distributions[1,4], no further corrections due to Doppler-shift effects have to be applied. The resulting Q-value of 2432.4 ± 0.7 keV is in fair agreement with the value of 2431 ± 2 keV obtained from the mass tables (ref.5), by Endt and van der Leun[6] and a recent measurement, Q = 2430.7 ± 0.5 keV, of Dubois et al.[7]. The 2425 keV state is therefore bound by 7.1 ± 0.7 keV against proton decay.

REFERENCES:

1) C. Folgs, Nucl. Phys. A 217 (1973) 29
2) N. B. Gove and A. H. Wapstra, Nucl. Data Tables 11 (1972) 127
3) C. Rolfs, W. S. Rodney, S. Durrance and H. Winkler,
   Nucl. Phys. A 240 (1975) 221
4) C. Rolfs, W. S. Rodney, M. H. Shapiro and H. Winkler,
   Nucl. Phys. A 241 (1975) 460
5) A. H. Wapstra and N. B. Gove, Nuc. Data A 9 (1971) 267 and 303
6) P. M. Endt and C. van der Leun, Nuc. Phys. A 214 (1973) 1
7) J. Dubois, H. Odelius and S. O. Berglund, Phys. Scripta 5 (1972) 16

SUPERALLOWED β-DECAY: FROM NUCLEAR MASSES TO THE Z VECTOR BOSON

J.C. Hardy and I.S. Towner

Atomic Energy of Canada Limited

Chalk River, Ontario, Canada K0J 1J0

INTRODUCTION

According to the conserved vector current (CVC) hypothesis, the vector coupling constant, G_V, for nuclear β-decay is the same for all nuclei. This has the consequence that all ft values for superallowed β-transitions between $0^+(T=1)$ analogue states should be equal, regardless of the specific nuclei involved, providing that small electromagnetic corrections are properly accounted for. Furthermore, G_V is related through Cabibbo theory to the corresponding constants for the weak decays of hyperons and mesons. Thus, the determination of ft values – for which accurate nuclear mass differences are a prerequisite – can lead directly to a test of both CVC theory and the Cabibbo universality of weak interactions.

For the nuclear transitions, ft values are given by

$$ft(1 + \delta_R) = K/(G_V'^2 |M_V|^2) \tag{1}$$

with $G_V'^2 = G_V^2 (1 + \Delta_R)$

$|M_V|^2 = 2(1 - \delta_c)$

and $K = 1.23062 \times 10^{-94}$ $erg^2 cm^6 sec$. Here f is the statistical rate function, which depends upon Z and the total decay energy (mass difference) W_o; t is the partial half life for the transition; and M_V is the vector (Fermi) matrix element, which for exact $T = 1$ analogue states should equal 2.

The small correction terms δ_R, Δ_R and δ_c in eqn.(1) arise because the decaying nucleon and the emitted positron interact with the electromagnetic field. The first two are conventionally referred to as radiative corrections, one of which (δ_R) varies from nucleus to nucleus and the other (Δ_R) is nucleus-independent; the third (δ_c) reflects the change in the Fermi matrix element caused by charge-dependent forces - forces that alter the simple analogue relationship between the initial and final nuclear states. All three terms have been calculated (for a summary, see ref.1) and are $\sim$ 1%, although Δ_R depends strongly on the model chosen to describe the weak-interaction process.

By defining a corrected ft value, $\mathscr{F}t$, such that

$$\mathscr{F}t = ft\ (1 + \delta_R)(1 - \delta_c) \tag{2}$$

the CVC prediction is verified if all $\mathscr{F}t$ values for superallowed T=1 nuclear transitions are the same. The average $\mathscr{F}t$ value then yields an experimental value for G_V' and a point of comparison with other weak decays.

MASS MEASUREMENTS

In principle, for each superallowed transition of interest here, three quantities must be measured: the decay energy of the $0^+ \to 0^+$ transition, its branching ratio and the total half-life of the β-emitter. To date, as the result of measurements by a number of different groups, the ft values for 14 transitions are known to better than ± 3%, and ten of those to better than ± 1%. In contributing to these studies, we have measured decay energies for ten superallowed β-transitions, half-lives for thirteen and branching ratios for five.

For the purpose of decay energy measurements the transitions fall into two categories depending upon whether or not the daughter nucleus is stable. If so, the energy can be directly determined from the measured Q-values for (p,n) or (^{3}He,t) reactions. If not, the parent and daughter masses must be determined independently from reaction Q-values. We have made both types of measurement, determining the Q-values for a number of (p,t) and (^{3}He,t) reactions.

The masses of the superallowed β-emitters ^{22}Mg, ^{26}Si, ^{30}S and ^{34}Ar were obtained by measuring the Q-values for (p,t) reactions on ^{24}Mg, ^{28}Si, ^{32}S and ^{36}Ar targets[2]. A 40.2 MeV proton beam from the Michigan State University cyclotron was used to initiate the reactions, with reaction products being analyzed in the split pole magnetic spectrograph. The method used was to observe the triton peak corresponding to one reaction of interest, then to vary the magnetic field of the spectrograph until tritons from the calibration reaction, ^{16}O(p,t)^{14}O, were detected at the same position along the focal plane; from both field measurements the difference in triton energies could be accurately determined. No measured Q-value differed by more than 1.6 MeV from that for the calibration reaction so only relatively small changes in the magnetic field were required.

To relate the difference in triton energies to the Q-value difference it was necessary to determine the reaction angle precisely. This was done by scattering protons from a formvar ($C_5H_8O_2$) target and comparing the inelastic protons leading to the 4.439 MeV state in ^{12}C with the elastic protons scattered from H; at the chosen scattering angle (θ_{lab}=22.5°) both proton groups had nearly the same energy, the difference between them varying at the rate of $\sim$ 450 keV/deg, and the angle could easily be determined to ± 0.02°. Ultimately, the (p,t) Q-values were quoted[2] with an uncertainty of ± 3 keV, which includes the effects of all known sources of error: angle, beam energy, peak centroids, spectrometer magnetic field, energy loss in the target, and calibration Q-value. This was a significant improvement over all previous measurements.

In contrast with these spectrometer experiments, the Q values for (^3He,t) reactions leading to ^{26}Al, ^{34}Cl, ^{42}Sc, ^{46}V, ^{50}Mn and ^{54}Co were determined using two counter telescopes to observe the reaction tritons[3]. A variety of two-component composite targets were bombarded by a magnetically analyzed 33.0 MeV ^3He beam from the Chalk River MP tandem accelerator. The targets used were enriched ^{26}Mg, Cd^{34}S, ^{42}Ca, ^{46}Ti, ^{50}Cr and ^{54}Fe, each $\sim$ 100 µg/cm^2 thick and evaporated onto the same thickness of ^{27}Al. In addition, targets of ^{42}Ca + ^{26}Mg and ^{42}Ca + ^{50}Cr were employed. Several sample spectra appear in Fig. 1.

For all target combinations, four separate measurements were made in which the telescopes(denoted A and B) and target normal (T) were oriented with respect to the nominal beam direction as follows: (1) A = +30°, B = -30°, T = 0°; (2) A = +30°, B = -30°, T = 180°; (3) A = -30°, B = +30°, T = 180°; (4) A = -30°, B = +30°, T = 0°. The eight spectra thus accumulated were each analyzed to obtain the energy difference between the ground state groups corresponding to the two targets involved. The average of all eight results yields the Q-value difference rigorously independent of uncertainties in target thickness or beam direction. From the known (± 0.9 keV) Q-value for the reaction $^{27}Al(^3He,t)^{27}Si$, the Q-values (decay energies) for the 6 superallowed emitters were determined[3] with uncertainties ranging from ± 2.0 to ± 3.0 keV.

More recently we have been repeating these (^3He,t) measurements with the new Chalk River Q3D spectrometer to analyze the tritons. Again utilizing composite targets, we are making use of the fact that two triton groups, one from each target, can be brought within a few tens of keV of one another by a judicious choice of observation angle. The energy-separation of the doublet can be measured accurately, and then directly related to the Q-value difference. Ultimately, it is anticipated that this method can yield relative Q-values accurate to ± ¼ keV, an order of magnitude improvement over previous techniques. Preliminary results[4] already confirm our earlier counter measurements for all nuclei except ^{46}V; in that case the ground state cen-

troid was distorted by a ^{48}V contaminant peak that could be resolved
only by the spectrometer, with the result that the Q-value must be
increased by the equivalent of several standard deviations.

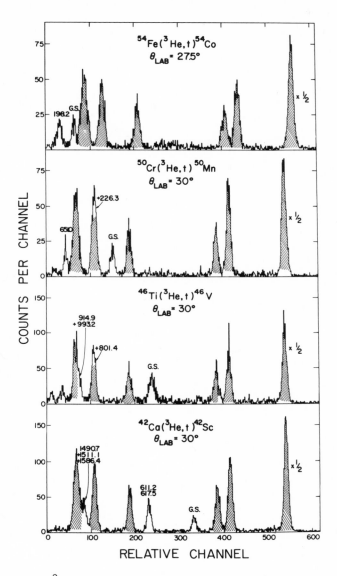

Fig. 1: Sample (^{3}He,t) energy spectra obtained using a counter
telescope. The shaded peaks are from the reaction ^{27}Al(^{3}He,t)^{27}Si;
all others are from the identified reactions. (From ref.[3]).

HALF LIVES AND BRANCHING RATIOS

Half lives have been measured by two different techniques. Where possible, the superallowed β-emitters were selectively produced - by (t,n)[5] or (p,n)[6] reactions - and decay positrons detected in a plastic scintillator. The decay rate was measured as a function of positron energy to eliminate the effects of weak background activities.

In those cases where the emitters could only be produced in the presence of strong competing activities, a gas transport system was used. Nuclei produced by the accelerator beam recoiled out of the target and were thermalized in gas (usually helium). The gas and contained recoils were periodically swept through 4 m of tubing to a shielded counting chamber. En route, the gas passed through a variety of traps and filters that removed most of the contaminant activities. Very clean spectra of β-delayed γ-rays could be obtained from a Ge(Li) detector viewing the counting chamber[7]. Half lives were derived from the decays of individual peaks, and branching ratios obtained by measuring the intensity of γ-rays relative to the 511 keV annihilation radiation.

VECTOR COUPLING CONSTANT AND CABIBBO UNIVERSALITY

Combining our results with those of others (notably the Harwell group[8]) we derive the corrected $\mathscr{F}t$ values shown in Fig.2. The statistical rate function f and correction terms δ_R and δ_C were all calculated in the manner described in reference 1. The weighted average is $\mathscr{F}t = 3082.4 \pm 2.1$ sec with a corresponding χ^2 per degree of freedom of 0.5. Clearly the data are in complete concordance with the CVC prediction of constant $\mathscr{F}t$ values.

Having verified the internal consistency of the experimental data and theoretical corrections, we now use the average $\mathscr{F}t$ value to derive the effective vector coupling constant $G_V{}'$ and to examine its implications for weak interaction theories. From eqs.(1) and (2) we obtain:

$$G_V{}' = G_V(1 + \Delta_R)^{\frac{1}{2}} = (1.4129 \pm 0.0005) \times 10^{-49} \text{ erg.cm}^3. \qquad (3)$$

In Cabibbo theory the coupling constant for nuclear Fermi beta decay relates to that for muon decay, G_μ, through the universality relation:

$$G_V = G_\mu \cos_\theta \qquad (4)$$

where θ_V is the Cabibbo angle.

Recently, the parameters of Cabibbo theory have been determined[11] from a fit to experimental data on hyperon β-decays. The result for θ_V (= 0.234 ± 0.003 rad) when combined with [12] G_μ = (1.43560 ± 0.00016) $\times$ 10^{-49} erg.cm^3 in eqs.(3) and (4) yields a value

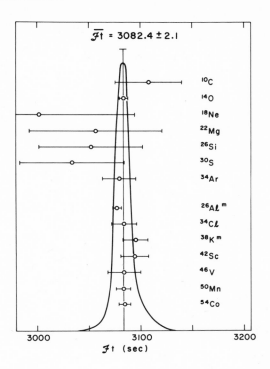

Fig. 2: "Ideogram" for all accurately known $\mathcal{F}$t-values. The significance of "ideograms" is explained in detail on page 93 of ref.[9]).

for the radiative correction Δ_R, viz:

$$\Delta_R = (2.37 \pm 0.17)\% \tag{5}$$

The significance of this result can be appreciated by examining
Fig.3 in which the calculated[10] Δ_R is plotted against M_Z, the mass
of the (as yet unobserved) Z vector boson, for two different quark
models. The experimentally determined Δ_R sets restrictive limits of
$25 \leq M_Z \leq 40$ GeV for a model with four quarks of integral charge (ρ
= 1) and $M_Z \geq 100$ GeV for three coloured quartets ($\rho = 1/3$).

The experiments at Chalk River were done with the collaboration,
at various times, of H.R. Andrews, G.C. Ball, W.G. Davies, J.S.Geiger,
R.L. Graham, K.P. Jackson, J.A. Macdonald and H. Schmeing. Those involved at Michigan State were W. Benenson, G.M. Crawley, E. Kashy and
H. Nann; and at Brookhaven, D. Alburger.

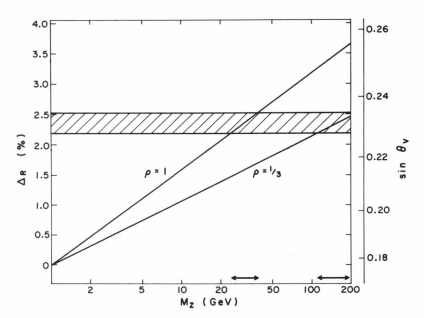

Fig. 3: Radiative correction Δ_R plotted as a function of M_z, the mass of the Z vector boson for two different quark models[10], $\rho = 1$ and $\rho = 1/3$. The cross hatched area corresponds to the value of Δ_R in eqn.5.

REFERENCES

1. I.S. Towner and J.C. Hardy, Nucl. Phys. <u>A205</u> (1973) 33.
2. J.C. Hardy, H. Schmeing, W. Benenson, G.M. Crawley, E. Kashy and H. Nann, Phys. Rev. <u>C9</u> (1974) 252.
3. J.C. Hardy, G.C. Ball, J.S. Geiger, R.L. Graham, J.A. Macdonald and H. Schmeing, Phys. Rev. Lett. <u>33</u> (1974) 320.
4. W.G. Davies, J.C. Hardy, G.C. Ball, K.P. Jackson, J.A. Macdonald and H. Schmeing, to be published.
5. J.C. Hardy and D.E. Alburger, Phys. Lett. <u>42B</u> (1972) 341.
6. J.C. Hardy, H.R. Andrews, J.S. Geiger, R.L. Graham, J.A. Macdonald and H. Schmeing, Phys. Rev. Lett. <u>33</u> (1974) 1647.
7. for example, J.C. Hardy, H. Schmeing, J.S. Geiger and R.L. Graham Nucl. Phys. <u>A223</u> (1974) 157.
8. for example, J.M. Freeman, R.J. Petty, S.D. Hoath, J.S. Ryder, G.T.A. Squier and W.E. Burcham, submission to session 11 of this conference.
9. Particle Data Group, Rev. Mod. Phys. <u>42</u> (1970) 87.
10. A. Sirlin, Nucl. Phys. <u>B71</u> (1974) 29.
11. M. Roos, Nucl. Phys. <u>B77</u> (1974) 420.
12. J. Duclos, A. Magnon and J. Picard, Phys. Lett. <u>47B</u> (1973) 491; Particle Data Group, Phys. Lett. <u>50B</u> (1974) 1.

MASSES OF $T_z = + 5/2$ NUCLEI IN THE s-d SHELL FROM β-DECAY MEASUREMENTS[†]

D. E. Alburger, D. R. Goosman,[*] C. N. Davids,[+] and
J. C. Hardy[‡]

Brookhaven National Laboratory
Upton, New York 11973

1. INTRODUCTION

One of the research programs planned for the Tandem Van de Graaff facility at Brookhaven National Laboratory was the study of radioactivities that could be produced either by higher energy light projectiles or, more particularly, by the heavy-ion beams available from this type of accelerator. Many short-lived radioactivities had been investigated previously using the 3.5-MeV Van de Graaff at Brookhaven including a number of nuclides formed by triton-induced reactions. One of these was ^{20}O produced in the $^{18}O(t,p)^{20}O$ reaction. The search for weak β-ray branches in ^{20}O decay was hindered by background radiations from ^{20}F which was made directly in the competing $^{18}O(t,n)^{20}F$ reaction.

In 1970, the Tandem Van de Graaff facility at Brookhaven first came into operation. While preparations for radioactivity studies with this accelerator were being made a paper on ^{20}O decay was published by Mak, Spinka, and Winkler (1). They produced this activity via the $^{18}O(^{18}O,^{16}O)^{20}O$ reaction and found no direct production of ^{20}F, thus allowing a greater sensitivity for detecting weak β-ray branches in the ^{20}O decay by observing γ rays from states in the daughter nucleus. Although ^{20}O was the strongest activity there were also numerous weak peaks in the γ-ray spectrum some of which were not identified by Mak et al. Since it was felt that improvements could be made by using this approach it was decided to begin our program with the study of ^{20}O decay.

Leaving the details aside for the moment, the $^{18}O + ^{18}O$ reaction exhibited not only the strong ^{20}O delayed activity but many weak γ-ray lines due to competing reactions, several of which could not be

associated with known activities. The outcome of the analysis was
the discovery of the new isotope ^{33}Si, formed in the ^{18}O(^{18}O,2pn)^{33}Si
reaction, with measurements of its half-life, γ-ray spectrum, decay
scheme, and mass. ^{33}Si is a T_z = + 5/2 nuclide and one of a series
of such nuclei in the 2s-1d shell extending from ^{21}O through ^{35}P.
At that time very little was known about this sequence occurring in
the region of the table of elements shown in Fig. 1. It seemed
worthwhile to try to establish the existence of some of these nuclides, determine their spectroscopic properties for comparison with
the shell model, and measure their masses to compare with predictions
of mass relations such as that proposed by Garvey et al. (2). Thus
the transverse or T relation of the Garvey-Kelson formulation shown
schematically in Fig. 1 allows one to predict the masses of all of
these T_z = + 5/2 cases from the known masses of nuclei closer to
the line of β stability.

As a preliminary to an investigation of this region of neutron-
excess "exotic" nuclei (defined as those ≥ 3 nucleons beyond the
nearest stable nucleus) a list of reactions, shown in Table 1, was

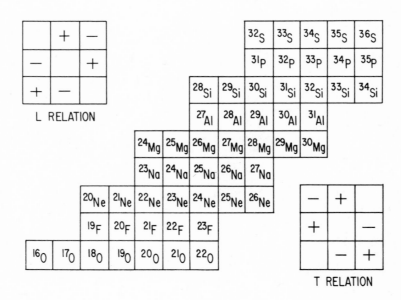

Fig. 1. Region of the table of elements including the eight "exotic"
T_z = + 5/2 nuclei from ^{21}O through ^{35}P investigated in this work.
According to the transverse (T) and longitudinal (L) relations of
the Garvey-Kelson mass formulation the sum of the three (+) masses
should equal the sum of the three (-) masses. Thus the masses of
all 8 of the T_z = + 5/2 nuclei can be predicted using the T relation.

Table 1. Compound reactions making $T_z=+5/2$ nuclides in the s-d shell

Nuclide	Reaction	Nuclide	Reaction
^{21}O	$^{10}Be(^{13}C,2p)^{21}O$	^{29}Mg	$^{18}O(^{13}C,2p)^{29}Mg$
	$^{9}Be(^{18}O,\alpha2p)^{21}O$		$^{14}C(^{18}O,2pn)^{29}Mg$
^{23}F	$^{18}O(^{7}Li,2p)^{23}F$	^{31}Al	$^{18}O(^{18}O,\alpha p)^{31}Al$
	$^{10}Be(^{18}O,\alpha p)^{23}F$		$^{15}N(^{18}O,2p)^{31}Al$
^{25}Ne	$^{9}Be(^{18}O,2p)^{25}Ne$	^{33}Si	$^{18}O(^{18}O,2pn)^{33}Si$
^{27}Na	$^{11}B(^{18}O,2p)^{27}Na$	^{35}P	$^{18}O(^{19}F,2p)^{35}P$

drawn up as possible ways to produce them. The plan for identifying the activities was to depend on the accurate energy measurement of γ rays corresponding to excited states in the daughter nuclei. In all of the cases prior energy level information existed through nuclear reaction studies, although it was sometimes necessary to establish improved level accuracy. Estimates of the expected properties of these nuclei, particularly the half-life, were based in part on theoretical considerations and this allowed the experimental search to be "tuned" for the previously unobserved activity.

During the course of this work the five new $T_z = +5/2$ nuclides ^{23}F, ^{29}Mg, ^{31}Al, ^{33}Si, and ^{35}P were established along with their properties including mass values. Masses, half-lives, and spectro-scopic properties were also measured for ^{25}Ne and ^{27}Na which had meanwhile been discovered at other laboratories. A paper (3) was also eventually published on the decay of ^{20}O which was one of the original intents of the program.

2. EXPERIMENTAL METHODS

For sensitive studies of short-lived radioactivities it is desirable to transport the active product rapidly from the bombard-ment region to a low-background counting station. We have used two commonly known methods, i.e., the "rabbit" transfer of a solid tar-get and the gas transfer system. There are some novel features of these designs which may be described.

The "rabbit" system (4) consists of a stainless steel tubing of either a square (1.0 x 1.0 cm^2) or a rectangular (1.0 x 2.2 cm^2) in-ternal cross section, the former being the more frequently used. Through this tubing passes the "rabbit" which is a light-weight target holder 2.5 or 4.0 cm long made of plastic (delrin), and which is propelled by a short burst of helium gas admitted by a valve. At the bombardment position the rabbit comes to rest against a polyvinyl chloride stopper and its position is checked on a TV monitor by observing a marker line inscribed on the rabbit seen through a window in the wall of the tubing. One of the ways of bringing the beam onto the target is to pass it through a thin foil which isolates the rabbit line from the accelerator vacuum system. But for most of the heavy-ion applications a windowless arrangement must be used and this requires a valve between the accelerator and rabbit line which opens only after the rabbit line is pumped down, remains open during the irradiation, and then closes before helium is admitted for propelling the rabbit. Following the return of the rabbit and the start of pump-down the beam valve can be opened after $\sim$ 2 sec. After the end of the bombardment the beam is cut off by a mechanical chopper upstream, the beam valve closes, and the rabbit moves $\sim$ 5 m through a shielding wall in 0.1-0.4 sec and comes to rest against a stopper positioning it in front of a Ge(Li) γ-ray detector and a plastic scintillator for β-ray measurements. An observation window allows the position of the rabbit to be checked, and on the β detector side a thin Be window minimizes the energy loss for emerging β rays.

All functions of the valves for evacuating and admitting He gas, as well as for operating the counting equipment, are controlled by a timer-programmer (5). This has a crystal controlled clock and ten independent sections whose start and reset times can be fixed at any desired values. Thus in addition to operating the valves of the rabbit transfer line, the timer-programmer can control as many as seven data collecting functions.

The gas transfer system (6) consists of a target cell 10 cm long filled with He gas at $\sim$ 1 atm pressure. The target may be located next to the beam entrance window, or may even be the entrance window material itself. Target thickness, gas pressure, and beam energy are selected so that all of the product nuclei from the target emerge into the gas, slow down, and stop before reaching the end of the cell. After an irradiation a valve is opened to a previously evacuated transfer line connected through a shielding wall to a counting cell. The radioactive products are carried along in the He gas, perhaps passing through liquid nitrogen or charcoal traps, getting to the counting cell in a short time. Up to 70% of the activity can reach the counting cell. Several types of cell can be used (6) depending on whether the particular experiment requires measurements of γ singles or β-γ coincidences.

Techniques for carrying out the experiments and analyzing the data generally follow standard procedures of β- and γ-ray spectroscopy with the exception of deducing masses from the β-ray spectra. The essential types of information needed are the energies and decay rates of the various γ-ray lines and the end-point energies of β rays in coincidence with selected γ-ray peaks. Large volume Ge(Li) γ-ray detectors are used, their output being directed by a timer-programmer into sequential time bins of a Σ-7 computer for obtaining half-life information. Digital selection of peak and background channels in the γ-ray spectrum, and normal coincidence methods, are used to find the net β-ray spectrum in a plastic scintillator (thick enough to fully absorb the expected β-ray energies) in coincidence with a given γ ray.

A convenience in most heavy-ion induced reactions is the production of numerous known radioactivities whose β-γ coincidence spectra can serve as built-in β spectra calibration standards. Stretching or compression of the standard spectrum is carried out by computer techniques so as to fit unknown β spectra thereby finding their end-point energies. This can become very complicated when the selected γ ray is in coincidence with several γ- and β-ray branches other than the β-ray branch feeding the state directly. The complete computer-fitting analysis can contain many terms and requires an accurate knowledge of the response function and absolute efficiency of the β-ray detector for the various coincident γ rays as well as the expected shapes of other coincident β-ray branches. When the example is not too complex, when the statistical accuracy is high, and when there is a nearby calibration β-ray spectrum, the end-point energy of an unknown β ray may be determined to an accuracy no worse than $\pm$ 50 keV, with a corresponding accuracy in the mass value of the unknown nucleus.

3. MEASUREMENTS ON THE T_z = + 5/2 NUCLEI

The delayed γ-ray spectrum following the ^{18}O + ^{18}O reaction was studied (4) using a Ta_2O_5 target made by oxidizing a Ta foil with ^{18}O enriched gas, to an O surface density of 3-4 mg/cm^2. The target was clamped in the rabbit and bombarded with a beam of $\sim$ 300 nA of 42-MeV $^{18}O^{4+}$ ions. In the γ-ray spectrum many peaks due to about 17 radioactivities were observed most of which were assigned to known nuclides by making accurate γ-ray energy and decay measurements. Three unknown peaks decayed with a half-life of 6.3 sec and agreed in energy with excited states in ^{33}P as established accurately by measurements of γ rays in the $^{31}P(t,p)^{33}P$ reaction. It was concluded that the new nuclide ^{33}Si (decaying to the ^{33}P daughter) was being produced via the $^{18}O(^{18}O,2pn)^{33}Si$ reaction. A decay scheme was derived for ^{33}Si having three β-ray branches to states in ^{33}P, and which was consistent with the expected shell-model assignment of J^{π} = 3/2$^+$ to ^{33}Si. Following the procedures described above the β-ray spectrum in coincidence with 1848-keV γ rays was measured (7,8) with the result shown in Fig. 2. From these data a mass excess of -20569 $\pm$ 50 keV was derived for ^{33}Si.

While the discovery of ^{33}Si was more-or-less accidental, the rest of the examples were intentionally pursued. The techniques used were similar to those followed for ^{33}Si, except for ^{25}Ne which employed the gas transfer method. Various targets and heavy-ion beams were used to produce the activities. Aside from ^{33}Si the other activities and the reactions forming them were ^{23}F, ^{25}Ne, ^{27}Na, ^{29}Mg, ^{31}Al, and ^{35}P made in the reactions ^{10}Be(^{18}O,αp)^{23}F, ^{9}Be(^{18}O,2p)^{25}Ne, ^{11}B(^{18}O,2p)^{27}Na, ^{18}O(^{13}C,2p)^{29}Mg, ^{18}O(^{18}O,αp)^{31}Al, and ^{18}O(^{19}F,2p)^{35}P, respectively, (references (9), (6), (10), (11), (12), and (13), respectively). ^{35}P was also studied at the 3.5-MeV Van de Graaff using the ^{36}S(t,α)^{35}P reaction (13). The most unusual target was a 94% ^{10}Be sample of BeO 700 μg/cm^2 thick used in the ^{23}F study (9).

Three attempts were made to find ^{21}O, but without success. In the first of these ^{21}O was sought in the ^{10}Be(^{13}C,2p)^{21}O reaction using a ^{10}Be target in the rabbit system. Another reaction,

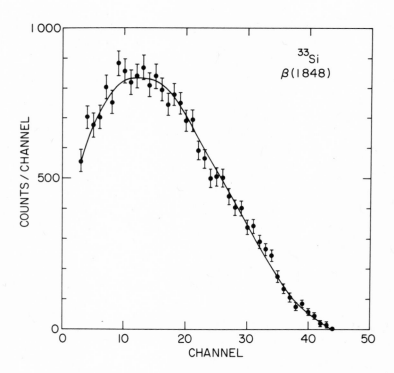

Fig. 2. The net β-ray spectrum of ^{33}Si observed in a plastic scintillator in coincidence with 1848-keV γ rays. The solid curve is a corrected ^{23}Ne calibration spectrum stretched by a factor of 0.9922 to fit the ^{33}Si data. A mass value is derived for ^{33}Si having an accuracy of $\pm$ 50 keV based on a β end-point energy of 3920 $\pm$ 50 keV.

$^{9}Be(^{18}O,\alpha 2p)^{21}O$, was also tried, first with the rabbit method and
finally with the gas transfer system as in the ^{25}Ne study. In none
of the attempts were any γ rays observed that could be ascribed to
the decay of ^{21}O to levels in the ^{21}F daughter nucleus.

4. COMPARISON WITH MASS FORMULATIONS

A comparison of the measured masses of the T_z = + 5/2 nuclides
in the 2s-1d shell with predictions of the Garvey-Kelson mass formulation is shown in the upper part of Fig. 3 which plots the differences between the measured and predicted masses. Results due to the
Lawrence Berkeley Laboratory (LBL), Orsay, and Dubna are included
with those of Brookhaven (BNL). There appears to be a departure between experiment and calculations far larger than the experimental
errors and also of an apparently systematic nature.

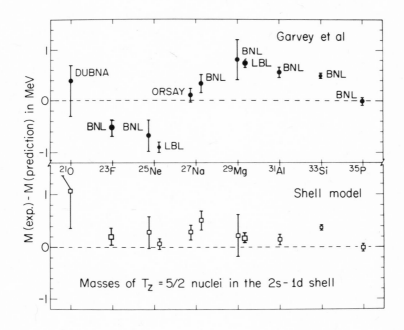

Fig. 3. Upper part - differences between experimental masses and
those predicted by the Garvey-Kelson transverse relation for the
T_z = + 5/2 nuclei in the 2s-1d shell. Lower part - similar comparison of experimental masses with shell-model calculations (Ref. 14).

Simple shell-model relations do well in explaining the observed mass (9) of ^{23}F. Using least squares fitting techniques, recent calculations of masses based on shell-model considerations have been carried out by Wilcox et al. (14). As may be seen in the lower part of Fig. 3 there is considerably better agreement between theory and experiment in this case.

According to either of the formulations it seems that the Dubna mass excess for ^{21}O (15) may be in error and, at least in the shell-model calculation, the deviation of the experimental mass is much larger than for the other members of this series. While it would be of interest to check the mass values of all of these $T_z = + 5/2$ nuclides the mass of ^{21}O is in the greatest doubt and remains as the most important one awaiting a clear-cut mass determination.

REFERENCES

[†] Research carried out under the auspices of the U. S. Energy Research and Development Administration.
[*] Permanent address: Lawrence Livermore Laboratory, Livermore, CA.
[†] Permanent address: Argonne National Laboratory, Argonne, IL.
[‡] Permanent address: Atomic Energy of Canada Ltd.,Chalk River,Ontario.

(1)   MAK, H. B., SPINKA, H., and WINKLER, H., Phys. Rev. C2, 1729 (1970).
(2)   GARVEY, G. T., GERACE, W. J., JAFFE, R. L., TALMI, I., and KELSON, I., Rev. Mod. Phys. Suppl. 41, S1 (1969).
(3)   ALBURGER, D. E. and GOOSMAN, D. R., Phys. Rev. C9, 2236 (1974).
(4)   GOOSMAN, D. R. and ALBURGER, D. E., Phys. Rev. C5, 1252 (1972).
(5)   SCHWENDER, G. E., GOOSMAN, D. R., and JONES, K. W., Rev. Sci. Instr. 43, 832 (1972).
(6)   GOOSMAN, D. R., ALBURGER, D. E., and HARDY, J. C., Phys. Rev. C7, 1133 (1973).
(7)   GOOSMAN, D. R. and ALBURGER, D. E., Phys. Rev. C6, 825 (1972).
(8)   GOOSMAN, D. R., DAVIDS, C. N., and ALBURGER, D. E., Phys. Rev. C8, 1324 (1973).
(9)   GOOSMAN, D. R. and ALBURGER, D. E., Phys. Rev. C10, 756 (1974).
(10)   ALBURGER, D. E., GOOSMAN, D. R., and DAVIDS, C. N., Phys. Rev. C8, 1011 (1973).
(11)   GOOSMAN, D. R., DAVIDS, C. N., and ALBURGER, D. E., Phys. Rev. C8, 1331 (1973).
(12)   GOOSMAN, D. R. and ALBURGER, D. E., Phys. Rev. C7, 2409 (1973).
(13)   GOOSMAN, D. R. and ALBURGER, D. E., Phys. Rev. C6, 820 (1972).
(14)   WILCOX, K. H., JELLEY, N. A., WOZNIAK, G. J., WEISENMILLER, R. B., HARNEY, H. L., and CERNY, J., Phys. Rev. Letts. 30, 866 (1973) and private communication.
(15)   ARTUKH, A. G., AVDEICHIKOV, V. V., GRIDNEV, G. F., VOLKOV, V. V., and WILCZYNSKI, J., Nucl. Phys. A192, 170 (1972).

MASS DIFFERENCES OF PROTON-RICH ATOMS NEAR A = 116 AND A = 190

UNISOR Consortium[*], including
B. D. Kern and J. L. Weil
University of Kentucky, Lexington, KY 40506, U.S.A.
J. H. Hamilton and A. V. Ramayya
Vanderbilt University, Nashville, TN 37203
C. R. Bingham and L. L. Riedinger,
University of Tennessee, Knoxville, TN 37916
E. F. Zganjar
Louisiana State University, Baton Rouge, LA 70803
J. L. Wood, G. M. Gowdy, and R. W. Fink
School of Chemistry,[+] Georgia Institute of Technology,
Atlanta, GA 30332
E. H. Spejewski, H. K. Carter, and R. L. Mlekodaj
UNISOR,[*] Oak Ridge, TN 37830, and
J. Lin
Tennessee Technical University, Cookeville, TN 38501

The members of the UNISOR consortium have been studying the structure of proton-rich nuclei through radioactivity measurements,[1] making use of heavy ion beams of ^{12}C and ^{16}O provided by ORIC, the Oak Ridge Isochronous Cyclotron. As a part of these investigations positron spectra have been measured; in this report we discuss the positron endpoint energies and their interpretation in terms of Q_{EC}.

The UNISOR facilities include a Danfysik mass separator on-line with ORIC, an on-line computer-based data acquisition system, a movable tape source transport system, and ancillary equipment such as Ge(Li), Si(Li), plastic scintillator, and surface barrier detectors and multichannel analyzers. Two additional computer-based data acquisition systems are available.

The UNISOR mass separator provides a beam of radioactive ions which passes through a defining slit before being re-focused onto a movable aluminized mylar tape which is provided to transport the radioactivity to a selected one of three available counting stations. Spectra of gamma rays, beta rays, x-rays, and internal

conversion electrons are typically collected. The x-ray spectra
are especially useful in providing identification of the element
associated with a given half life. For the measurement of life
times and for use in identification, spectra are typically acquired as a function of time (multi-scaled spectra). Coincidences
are acquired in a three-parameter mode, for gamma-gamma, e^--gamma,
β^+-gamma, and x-ray-gamma combinations insofar as conditions permit,
with the time-to-amplitude converter output as the third parameter.

We have used beta-ray scintillation detectors which are cylin-
ders of NE-110 5.08 cm in diameter and 2.54, 3.81, or 5.08 cm in
length, depending on the maximum energy of the positrons which are
to be detected, with the shortest possible length being selected
in order to minimize the summing effect caused by coincident decay
gamma rays and annihilation photons. With the source on the movable tape at 5 mm from the face of the scintillator, this is approximately a 2π detector with "poor geometry". At times, a 9.0
cm diameter X 7.4 cm length scintillator has been used. The electronic amplifiers, coincidence circuitry, and routing circuits are
conventional.

The problem of obtaining an accurate energy calibration of the
beta-ray scintillation detector is a continuing one. Radioactive
sources which give a line spectrum do not exist except for very low
energies, and ones which give gamma rays of energies above 2614 keV
have inconveniently short lifetimes. For the present work, the
electron energy <u>vs</u> pulse height scale was calibrated with Compton
edges and beta-ray endpoints from standard radioactive sources.
Linearity has been verified up to 5 MeV by use of short-lived beta
sources prepared at the University of Kentucky 6-MV accelerator.
At present, the uncertainty in beta-ray endpoint energies is typically ±200 keV.

Radioactive xenon nuclei have been produced through the ^{104}Pd
$(^{16}O,xn)^{120-x}Xe$ reactions using a separated isotope target, and
radioactive iodine nuclei have been produced through the ^{103}Rh
$(^{16}O,xn)^{119-x}I$ reactions. The ^{16}O 5+ beams were varied in energy
from 75 to 142 MeV to produce the desired mass with maximum cross
section. The following results are summarized in Table I.

A = 116. This chain has been studied with the production of
^{116}Xe and ^{116}I with the use of the ^{104}Pd target, and of ^{116}I with
the use of the ^{103}Rh target. We have determined that the lifetime
of the ^{116}I is 2.91 ± 0.15 sec and have confirmed that the lifetime
of the ^{116}Xe is 57 sec.[2] The maximum positron endpoint energy of
6.50 ± 0.20 MeV, determined from a Fermi-Kurie plot, is assigned to
the decay of ^{116}I. Unfortunately the longer lifetime of the ^{116}Xe
permits the higher energy endpoint of the daughter to obscure the
^{116}Xe endpoint; however, a good estimate of the ^{116}Xe endpoint
energy has been made through noting that the positron yield goes

up very sharply below 3.5 MeV. Beta-gamma coincidence spectra have established that the ^{116}I decay proceeds to the ground state of ^{116}Te with Q_{EC} = 7.52 ± 0.20 MeV, consistent with the mass formula predictions[3-6] and the lower limit of Beck et al.[7]

A = 114. A preliminary endpoint energy of 6.5 MeV is tentatively assigned to ^{114}I → ^{114}Te, since the predicted endpoint for the decay of ^{114}Xe is much smaller.

A = 115. A preliminary endpoint energy of 5.0 MeV is tentatively assigned to ^{115}I → ^{115}Te.

A = 117. Our preliminary endpoint energy of 5.3 MeV is tentatively assigned to ^{117}Xe → ^{117}I, since the predicted endpoint for the decay of ^{117}I is much smaller. This is consistent with the lower limit of Ref. 7.

In the A = 190 region, targets of Ta have been bombarded with ^{16}O to produce the Tl isotopes and their daughters, and targets of W have been bombarded with ^{16}O (5+ and 6+) to produce the Pb isotopes and their daughters. The bombarding energies varied from 135 to 204 MeV. In the study of the decay chain Pb → Tl → Hg → Au → Pt, the presence of isomers has complicated the analysis of data leading to Q_{EC} values. In some cases the isomer lifetimes are nearly the same. However, we have had some success in producing an excess of the high-spin isomer directly by the (^{16}O,xn) reaction and an excess of the low-spin isomer by decay from a radioactive parent.

A = 192. Consider first the A = 192 chain which is more simple than the others. With a significant amount of the low-spin isomer being produced, beta spectra were accumulated off-line. One of these spectra is shown as a Fermi-Kurie plot in Fig. 1; the endpoint energy is 4.94 ± 0.2 MeV. The time decays of all pulses above bias energies of 2.25 and 3.11 MeV were determined. For the higher bias there was only the decay with a half life of 8.7 minutes. With the lower bias, there was observed the growing in of another component which we have not identified. The ^{192}Hg has a 5-hour half life and is predicted to decay by electron capture and with very low energy positrons which would not have been observed. The decay scheme of Fig. 1 shows the results. Other even-A low-spin isomers of Tl are known to decay to the 2^+ first-excited level of their Hg daughters, so we make a similar assignment. The resulting Q_{EC} = 6.38 MeV is consistent with predictions.

A = 187. Our preliminary endpoint energy of 4.2 MeV is assigned to the decay of ^{187}Tl; a ground state to ground state transition is assumed.

A = 188. Our endpoint energy of 3.3 MeV is assigned to the 71-sec decay of ^{188}Tl in a transition tentatively assumed to proceed

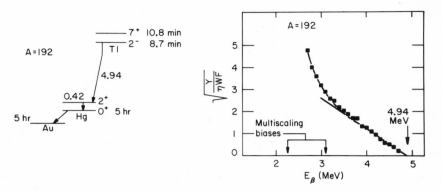

Fig. 1.   A partial decay scheme for the A = 192 chain and a typical
          Fermi-Kurie plot of a positron spectrum in which each point
          is a five channel sum.

from a high-spin isomer at low excitation energy to a high-spin
state at 1.51 MeV in ^{188}Hg, with Q_{EC} = 5.8 MeV.

    A = 189.   Along with the ^{189}Tl, a substantial amount of ^{189}Hg
was formed. A series of 20 spectra were acquired at intervals of
30 seconds each. The decay of selected portions of these spectra
is shown in Fig. 2 in which it can be seen that the decay has been
fitted with lifetimes of 1.4 and 8.0 minutes, except at the high
energy end where the decay follows the 1.4 minute lifetime. The
endpoint energy for the first three minutes of the decay is 4.14
MeV; it is assigned to the decay of a 9/2$^-$ isomer of ^{189}Tl on the
evidence from the decay of x-rays and of gamma rays identified with
transitions in ^{189}Hg. The energy of this isomeric level in ^{189}Tl
is estimated to be ≈0.27 MeV from systematics. As illustrated in
Fig. 2, the lowest energy high-spin level in the daughter ^{189}Hg to
which this decay might go is at 0.215 MeV as shown by the gamma-ray
studies; however, several other nearby levels are also populated by
beta decay with comparable intensities, so we have taken the average
excitation to be ≈0.30 MeV. The resulting Q_{EC} is 5.19 MeV. The
spectra which were accumulated at the end of the 10 minute period
gave a lower energy endpoint of 3.18 MeV which we assign to 7.7-
minute ^{189}Hg,[8] decaying from its ground state to the ground state
of ^{189}Au. There is tentative evidence for the decay of the 4.6-
min[8] isomeric level of ^{189}Au with endpoint energy of 2.14 MeV.

    A = 190.   The decay of two isomers of ^{190}Tl, have been studied.
From multi-scaled gamma-ray spectra taken on-line, the lifetime of
the high-spin (7$^+$) isomer has been deduced to be 3.7 minutes, and
the lifetime of the low-spin isomer (probably 2$^-$), 2.6 minutes.
We deduce from data taken under conditions in which different

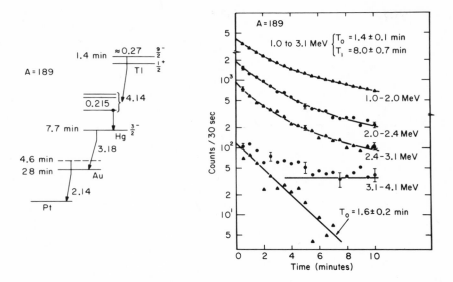

Fig. 2. A partial decay scheme for the A = 189 chain and typical
         time decays of several energy intervals of a positron
         spectrum. Both Tl and Hg were initially produced.

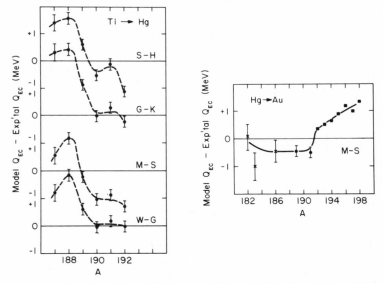

Fig. 3. Comparisons of preliminary Tl and Hg decay results with
         theoretical predictions. On the left, the model Q_{EC} are
         taken from Ref. 3-6. On the right, present results are
         shown by circles. Other data are from Ref. 9.

Table I. A summary of positron decay data for the A = 115 and A = 190 mass regions. Parentheses indicate uncertainty in a value or an assignment.

A	Element	$T_{1/2}$	E_{max} (MeV)	Initial State	Final State	$Q_{EC,expt.}$[a] (MeV)	Q_{EC}, predicted W-G[b]	G-K[c]	M-S[d]	S-H[e]
114	(I)		6.5±0.4	(gs)	(gs)	7.5±0.4	8.70	8.94	9.26	8.7
115	(I)	28 sec	5.1±0.3	(gs)	(gs)	6.1±0.3	5.50	5.81	6.49	5.5
116	(Xe)	57 sec	(3.5±0.3)	(gs)	(gs)	(4.5±0.3)	4.50	4.49	5.36	4.6
116	I	2.91 sec	6.50±0.2	gs	gs	7.52±0.2	7.30	7.75	7.83	7.4
117	(Xe)	61 sec	5.3±0.3	(gs)	(gs)	6.3±0.3	6.07	6.43	6.71	6.3
187	Tl		4.2±0.3	(gs)	(gs)	5.2±0.3	6.40	7.51	5.75	6.6
188	Tl	71 sec	3.3±0.3	(≈0)	(1.51)	5.8±0.4	7.60	8.17	6.91	7.3
189	^{m}Tl	1.4 min	4.14±0.2	≈0.27	≈0.30	5.19±0.2	5.70	6.24	4.86	5.7
189	Hg	7.7 min	3.18±0.2	gs	(gs)	4.20±0.2	4.20	4.39	3.75	3.1
189	^{m}Au	4.6 min	2.14±0.3	(gs)	(gs)	3.16±0.3	3.00	3.15	2.61	
190	^{m}Tl	3.7 min	4.20±0.2	≈0.2	≈1.9	6.9±0.2	7.00	7.03	5.99	6.5
190	Tl	2.6 min	5.4±0.3	gs	(0.42)	6.8±0.3	7.00	7.03	5.99	6.5
190	Au	42 min	3.36±0.2	(gs)	(gs)	4.38±0.2	4.40	4.18	3.78	
191	Tl	5 min	3.82±0.2	≈0.33	≈0.30	4.81±0.2	5.00	5.08	3.92	4.7
191	Hg	35 min	2.35±0.2	gs	(gs)	3.37±0.2	3.30	3.46	2.87	4.2
192	Tl	8.7 min	4.94±0.2	gs	0.42	6.38±0.2	6.30	6.10	5.05	5.2

a) Present results.

b) Ref. 3
c) Ref. 4

d) Ref. 5
e) Ref. 6

amounts of the two isomers have been produced that the 7^+ level at approximately 0.2 MeV decays with endpoint energy of 4.14 ± 0.2 MeV to high-spin levels at ≈1.9 MeV in the daughter ^{190}Hg, giving $Q_{EC} \approx 6.9$ MeV. There is evidence that the 2^- level decays with endpoint energy of 5.40 MeV; assuming that this transition terminates on a known 0.42 MeV level we obtain approximately the same Q_{EC}. After the ^{190}Tl had decayed, we measured the spectrum of the 42-minute ^{190}Au and deduced an endpoint of 3.36 MeV.

A = 191. We have observed the decay of a high-spin isomer (probably $9/2^-$) with lifetime of 5 minutes, from an initial level at approximately 0.33 MeV in the parent ^{191}Tl going to one or more of three high-spin levels grouped close to 0.30 MeV excitation in the daughter ^{191}Hg, as shown by the gamma-ray studies. The endpoint energy is 3.82 MeV, leading to $Q_{EC} \approx 4.81$ MeV. The spectra from a 1.7 hour old source have been analyzed to give an endpoint of 2.35 MeV which has been assigned to the 35-minute decay of the ^{191}Hg ground state to the ground state of ^{191}Au.

A portion of the Q_{EC} results of Table I are displayed in Fig. 3 as Q_{EC}(predicted)-Q_{EC}(experimental), making use of mass formula predictions.[3-6] For Tl → Hg, we observe an increase in the plotted Q_{EC} differences with decreasing mass number, starting at A = 189, for all four theoretical predictions. For Hg → Au, our two data, marked by circles in Fig. 3, fill a gap in a curve taken from a report of Westgaard et al.[9] which makes a comparison with the M-S[5] predictions. Here, there is an increase in the plotted Q_{EC} differences with increasing mass number starting at A ≈ 191.

REFERENCES

*UNISOR, a consortium of 14 institutions, is partially supported by the U.S. Energy Research and Development Administration (ERDA).
+Supported in part by ERDA.

1. HAMILTON, J. H., Science 185, 819 (1974).
2. HANSEN, P. G. et al., Phys. Letters 28B, 415 (1967).
3. WAPSTRA, A. H., and GOVE, N. B., Nucl. Data Tables 9, 265 (1971).
4. GARVEY, G. T., GERACE, W. J., JAFFE, R. L., TALMI, I. and KELSON, I., Rev. Mod. Phys. 41, S1, (1969).
5. MYERS, W. D., and SWIATECKI, W. J., UCRL-11980 (1965).
6. SEEGER, P. A., and HOWARD, W. M., Nucl. Phys. A238, 491 (1975).
7. ISOLDE COLLABORATION. BECK, E., CERN 70-30, p. 353.
8. ISOLDE COLLABORATION. ERDAL, R. B. et al., CERN 70-30, p. 1045.
9. ISOLDE COLLABORATION. WESTGAARD, L., ZYLICZ, J., and NIELSEN, O. B., Fourth International Conference on Atomic Masses and Fundamental Constants, Teddington, England, September 1971, (Plenum Press, New York, 1972), p. 94.

TOTAL β-DECAY ENERGIES OF NUCLEI FAR FROM STABILITY:

EVIDENCE FOR THE WIGNER SYMMETRY ENERGY IN Rb AND Kr MASSES

K. Aleklett*, G. Nyman**, E. Roeckl**, and L. Westgaard

The ISOLDE Collaboration

CERN, Geneva, Switzerland

INTRODUCTION

At the preceding Conference in this series, data from ISOLDE were presented [1] on total decay energies of medium- and high-A nuclei far from stability. We recall here the observation of a "ridge" in the mass surface, unaccounted for in the mass calculations, extending through several Z-values in the A $\sim$ 180 region, 4-5 mass numbers to the proton-rich side of the valley of stability. In continued experiments at the CERN on-line isotope separator [2], similar low-resolution survey-type measurements have been performed on proton-rich Kr and Br nuclei [3], and, more recently, on proton-rich isotopes of Rb and Cs, as well as neutron-rich nuclei in the Fr-Ra-Ac region [4]. In many cases it has been possible to go from mass differences to the atomic masses themselves, by connecting down through a decay chain to a nucleide of known mass.

In the present contribution we describe briefly the experimental technique in the work on isotopes of alkali and alkaline-earth elements; full details will be published in Ref. 4. After summarizing the experimental results, we concentrate the discussion on the masses of Rb and Kr nuclei near the N = Z line, and in particular their relevance for the Wigner symmetry term in nuclear mass calculations.

---

  *)   Chalmers University of Technology, Gothenburg, and the Swedish Research Councils' Laboratory, Studsvik, Sweden.

**)   Gesellschaft für Schwerionenforschung mbH, Darmstadt, Germany.

EXPERIMENTAL METHOD

Short-lived isotopes of Rb, Cs, Fr and Ra were produced in spallation reactions induced by 600 MeV protons in targets of Y (for Rb), La (for Cs) and Th (for Fr and Ra). Continuous mass separation of the products released from the molten targets were performed in the ISOLDE on-line separator. Details of the production method may be found elsewhere [5, 6].

Source-collection and counting techniques in the on-line measurements were the same as described in Refs. 1 and 3. The β-ray spectra were recorded by means of a cylindrical plastic scintillator of dimensions 4 cm diameter × 3.5 cm length, with a source geometry close to 4π. The source was simultaneously viewed by a 35 cm^3 Ge(Li) γ-ray detector. By off-line sorting of the two-dimensional array of β-γ coincidence events through digital gates, β spectra in coincidence with the prominent γ-rays could be evaluated.

In some off-line measurements we used a detector arrangement similar to that described by Rudstam et al. [7], consisting essentially of a 5 mm thick Si(Li) diode for the β particles and two 15.2 cm × 10.2 cm NaI detectors for the γ-rays. Beta spectra in coincidence with two analog windows in the summed γ-ray spectrum could be recorded simultaneously.

Singles and coincident β-ray spectra, after energy calibration and a simple correction for detector response, were subjected to a Fermi-Kurie analysis to extract the β end-point energies. The total β-decay energy is obtained by summing with the corresponding level energy in the daughter nucleus. In several cases, when the knowledge of the decay scheme is insufficient, only a lower limit on the Q-value can be given.

In order to check the experimental procedure a number of cases of known Q-value were treated. Over-all uncertainties were typically 300-500 keV for the on-line and 50-100 keV for the off-line measurements.

RESULTS

The results of the Q-value measurements are summarized in Table 1 together with other recent data [8, 9]. Additional data on the γ-rays in many of the investigated decays are given in Ref. 4.

SYSTEMATICS OF Rb AND Kr MASSES

The experimental masses of the complete series of Rb and Kr isotopes are displayed in Figs. 1a and b. The data are from the present work, the 1971 Mass Table [10] and other recent sources [8, 9, 11-16]. The mass of ^{73}Rb was determined from that of its mirror nucleus ^{73}Kr by adding the Coulomb displacement energy [17] minus the n-H mass difference. In order to emphasize shell and other local effects, the data are plotted as excess over

Table 1

Summary of Q-value measurements

Nucleide	$T_{\frac{1}{2}}$ a)	Gating transition(s) (keV)	Q_{EC} or $Q_{\beta-}$ (MeV)
^{76}Rb	37 sec	424	6.64 ± 0.57
^{77}Rb	3.8 min	179, 394	5.18 ± 0.39 b)
^{78}Rb	17 min	665	5.45 ± 0.37
^{80}Rb	34 sec	617	5.49 ± 0.35 c)
^{121}Cs	2 min	179	≥ 5.64 ± 0.37 d)
^{122}Cs	21 sec	331	7.15 ± 0.70
^{123}Cs	5.6 min	100	4.10 ± 0.31
^{124}Cs	31 sec	354, 495	5.92 ± 0.46
^{126}Cs	1.64 min		4.69 ± 0.14
^{224}Fr	2.7 min	837	≥ 2.62 ± 0.05 d)
^{225}Fr	3.9 min	0	≥ 1.64 ± 0.01 d)
^{226}Fr	48 sec	186, 254	3.77 ± 0.33
^{227}Fr	2.4 min	0	≥ 2.39 ± 0.10 d)
^{229}Ra	4 min	0	≥ 1.76 ± 0.04 d)
^{229}Ac	1.1 h		> 1.2

a)   For reference to the half-lives, see Refs. 4 and 5.

b)   In a recent report Liptak et al. [8] give the value
     Q_{EC} = 4.95 ± 0.15 MeV.

c)   From the measurement of Jaffe et al. [9] of the thresh-
     old of the reaction ^{80}Kr(p,n)^{80}Rb, a value Q_{EC} =
     = 5.701 ± 0.021 MeV is obtained.

d)   The Q-value limit is given as the sum of the β end-
     point energy and the energy of the gating transition.
     A zero in column 3 means that the equality sign is
     valid if the measured β group feeds the ground state.
     The quoted error is the uncertainty in the determin-
     ation of the β end-point.

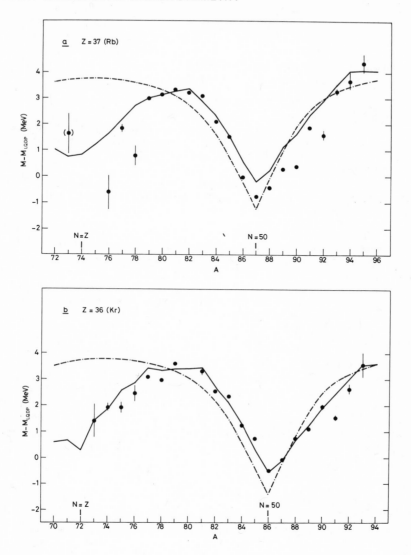

Fig. 1 Experimental (black dots) and theoretical masses of (a) Rb
and (b) Kr isotopes, after subtraction of a smooth liquid-drop term
(see text). The plot is based upon data from Table 1, and work by
Schmeing et al. [11, 12] (^{73,74}Kr), Roeckl et al. [3] (^{75}Kr, ^{73}Br),
Paradellis et al. [13] (^{76}Kr -- see comment in Ref. 4),
Lueders et al. [14] (^{74,76}Br), Clifford et al. [15] (^{91,92,93}Kr,
^{91,92,93}Rb), Macias-Marques et al. [16] (^{94,95}Rb, ^{95}Sr), and the
1971 Mass Table [10]. Calculated masses are after Myers and
Swiatecki [18] (dashed curve), and Seeger and Howard [19] (full-
drawn curve).

a smooth liquid-drop value [18], containing the usual volume, surface, symmetry and Coulomb terms, plus an odd-even term. The influence of the shell closure at N = 50 is evident, as well as an
increasing mass defect relative to the liquid-drop value as the
N = Z line is approached.

Before discussing the general N-dependence of the isotopic
masses, we note the very pronounced odd-even staggering in the
light Rb masses. In fact, the extra binding of the odd nuclei
^{78}Rb and ^{76}Rb appears to be as great as the full pairing energy,
$11/A^{\frac{1}{2}}$ MeV. It is tempting to relate this effect to the presence
of the odd neutron and odd proton in the same state, as the
N = Z line is approached. However, the lack of detailed spectroscopic information on the nuclei in question makes further
speculation along these lines premature.

The experimental data are compared in Fig. 1 with the mass
calculations of Myers and Swiatecki [18] and Seeger and Howard [19].
The "bunching model" shell term of Myers and Swiatecki accounts
well for the N = 50 shell effect in the masses. The same authors
discuss in Ref. 18 the form and magnitude of a phenomenological
correction term to reproduce the increased binding of N ∿ Z nuclei,
but without including such a term in their mass formula.

It has been known from the early work of Wigner [20] that, in
addition to the usual T^2/A term in the symmetry energy, a linear
term proportional to T/A (T = $|T_Z|$ = $\frac{1}{2}|N - Z|$ in the normal case
of totally aligned isospin) is required to account for the isobaric
mass differences in light nuclei. This contribution arises from the
attractive space-exchange interaction of nucleons in identical orbits. Following the arguments of Wigner {see also the discussion in
Blatt and Weisskopf [21]}, Myers [22] arrives at the following expression for the Wigner energy

$$E_{\text{Wigner}} = W(|N - Z| + \Delta)/A \tag{1}$$

where Δ = 1 for N = Z, odd, and Δ = 0 elsewhere. W is a constant
to be determined from experiment.

The mass formula [19] of Seeger and Howard (full-drawn curve
of Fig. 1) contains the slightly simplified Wigner term

$$E_{\text{Wigner}} = W \cdot (|N - Z|/A) , \tag{2}$$

where the value W = 35 MeV is chosen to give a good over-all fit
to the input data. Shell corrections and deformations are determined from a microscopic Nilsson-model calculation, normalized to
the macroscopic part through a Strutinsky smoothing procedure. The
coefficient of the quadratic symmetry energy term in the liquid-drop
expression is taken over in the denominator form suggested by the
more detailed droplet model [22, 23] of Myers and Swiatecki. The
total symmetry energy may be written

$$E_{\text{sym}} = \alpha \cdot T^2/A + \beta \cdot T/A , \tag{3}$$

where α now is slightly A-dependent, and $\beta = 2 \cdot W$. Over the range of A values considered here, the Seeger-Howard mass formula gives a β/α ratio of 0.9 to 1.0. (For A = 20 we would have $\beta/\alpha \sim 1.2$.)

CONCLUSION

It follows from Fig. 1 that the mass calculation represented by the full-drawn curve reproduces well the Rb and Kr mass data (except for ^{76,78}Rb, which are discussed separately), which we believe are the best available to observe this kind of isospin dependence in medium-weight nuclei. Our temporary conclusion is, therefore, that an expression like Eq. (3) above accounts well for the nuclear symmetry energy, with a value of β/α close to 1.

It may be interesting, finally, to compare with expressions in the literature for the symmetry energy in microscopic models of light nuclei. In a model where the full supermultiplet symmetry [20] is conserved, the symmetry energy should go as T(T + 4). If kinetic effects are included the expression is modified to reduce the strength of the linear term, and in this "unified" supermultiplet model Blatt and Weisskopf [21] estimate a value for the β/α coefficient of 1.6 to 2. On the other hand, the extreme independent-particle model demands a symmetry energy of the form T(T + 1). Talmi and Thieberger [24], using a simple jj-coupling shell model involving such a symmetry-energy term, showed that a good fit to experimental binding energies of nuclei through the $1f_{7/2}$ shell could be obtained. A similar shell-model approach has been discussed more recently by Comay et al. [25].

The authors acknowledge with pleasure enlightening discussions with Drs. J. Blomqvist, P.G. Hansen, W.D. Myers and O.B. Nielsen.

REFERENCES

1) L. Westgaard, J. Żylicz and O.B. Nielsen, Atomic masses and fundamental constants 4 (Eds. J.H. Sanders and A.H. Wapstra) (Plenum Press, London-New York, 1972), p. 94.
2) The ISOLDE Isotope Separator On-Line Facility at CERN (Eds. A. Kjelberg and G. Rudstam), CERN 70-3 (1970).
3) E. Roeckl, D. Lode, K. Bächmann, B. Neidhardt, G.K. Wolf, W. Lauppe, N. Kaffrell and P. Patzelt, Z. Physik 266, 65 (1974).

4) L. Westgaard, K. Aleklett, G. Nyman and E. Roeckl, Masses and
Q_β- values of short-lived isotopes of rubidium, caesium,
francium and radium, to be submitted to Z. Physik.

5) H.L. Ravn, S. Sundell, L. Westgaard and E. Roeckl, J. Inorg.
Nuclear Chem. 37, 383 (1975).

6) H.L. Ravn, S. Sundell and L. Westgaard, Nuclear Instrum.
Methods 123, 131 (1975).

7) G. Rudstam, E. Lund, L. Westgaard and B. Grapengiesser, Proc.
Internat. Conf. on the Properties of Nuclei Far from the
Region of Beta-Stability, Leysin, 1970 [CERN 70-30 (1970)],
Vol. 1, p. 341.

8) J. Liptak and W. Habenicht, Dubna preprint P6-8279 (1974).

9) A.A. Jaffe, G.A. Bissinger, S.M. Shafroth and T.A. White,
Atomic masses and fundamental constants 4 (Eds. J.H. Sanders
and A.H. Wapstra) (Plenum Press, London-New York, 1972),
p. 236.

10) A. Wapstra and B.N. Gove, Nuclear Data Tables 9, 267 (1971).

11) H. Schmeing, J.P. Hardy, R.L. Graham, D.S. Geiger and
K.P. Jackson, Phys. Letters 44B, 449 (1973).

12) H. Schmeing, J.P. Hardy, R.L. Graham and D.S. Geiger, Nuclear
Phys. A242, 232 (1975).

13) T. Paradellis, A. Houdayer and S.K. Mark, Nuclear Phys. A201,
113 (1973).

14) D.H. Lueders, J.M. Daley, S.G. Buccino, F.E. Durham,
C.E. Hollandsworth, W.P. Bucher and H.D. Jones, Phys. Rev.
C 11, 1470 (1975).

15) J.R. Clifford, W.L. Talbert, Jr., F.K. Wohn, J.P. Adams and
J.R. McConnell, Phys. Rev. C 7, 2535 (1973).

16) M.I. Macias-Marques, R. Foucher, M. Cailliau and J. Belhassen,
Proc. Internat. Conf. on the Properties of Nuclei Far from
the Region of Beta-Stability, Leysin, 1970 [CERN 70-30
(1970)], Vol. 1, p. 231.

17) J.D. Anderson, C. Wong and J.W. McClure, Phys. Rev. B 138,
615 (1965).

18) W.D. Myers and W.J. Swiatecki, UCRL-11980 (1965) and Nuclear
Phys. 81, 1 (1966).

19) P.A. Seeger and W.M. Howard, LA-5750 (1974) and Nuclear Phys.
A238, 491 (1975).

20) E.P. Wigner, Phys. Rev. 51, 106 and 947 (1937).

21) J.M. Blatt and V.E. Weisskopf, Theoretical nuclear physics
(J. Wiley, New York, 1952).

22) W.D. Myers, preprint LBL-3428 (1974).

23) W.D. Myers and W.J. Swiatecki, Ann. Phys. (NY) 55, 395 (1969),
and 84, 186 (1974).

24) I. Talmi and R. Thieberger, Phys. Rev. 103, 718 (1956).

25) E. Comay, S. Liran, J. Wagman and N. Zeldes, Proc. Internat.
Conf. on the Properties of Nuclei Far from the Region of
Beta-Stability, Leysin, 1970 [CERN 70-30 (1970)], Vol. 1,
p. 165.

TOTAL BETA DECAY ENERGIES OF NEUTRON-RICH FISSION PRODUCTS

Eva Lund and Gösta Rudstam

The Swedish Research Councils' Laboratory

Studsvik, S-611 01 Nyköping, Sweden

INTRODUCTION

Until recently, very little has been known about the shape of the nuclear mass surface far out on the neutron-rich side of beta stability. The reason for this scarcity of data has been, in the first place, experimental difficulties connected to the short half-lives of the nuclides of interest. The introduction of on-line isotope-separator techniques has improved the experimental situation considerably, and it is now possible to study in detail the properties of a wide band of highly neutron-rich fission products.

The isotope-separator facility "OSIRIS" connected to the RII-0 reactor at Studsvik produces isotopes of some 20 fission elements. It is well suited for a systematic study of the decay energies of these nuclides and hence for mapping the mass surface on the neutron-rich side of stability.

The Q_β-values of the most neutron-rich nuclides are of the order of many MeV. The technique has been to measure the total decay energy as the sum of the energy of a beta branch leading to a highly excited state of the daughter and the energy of this state but even so, the beta energy to be measured is usually quite high. Thus reliable energy and efficiency calibrations of the spectrometer have to be performed up to high energies. This is a difficult task and until now it has been done only up to about 4 MeV.

Good knowledge of the level structure of the daughter is al-

ways necessary for unambiguous Q_β-value determinations. In cases
where such information is not available it is only possible to
give a lower limit for the total beta decay energy.

EXPERIMENTAL ARRANGEMENT

The OSIRIS isotope-separator-on-line facility has been
described in detail elsewhere[1]). The ion-source with about 2 g
of ^{235}U as target material is located close to the R II-0 reactor
at Studsvik in a neutron flux of about 10^{11} n/cm^{2}s. The ion
source temperature is 1300 - 1600°C, and all fission products vo-
latile at this temperature are likely to pass through the separator. Activities are found at all mass numbers from 74 to 98 and
from 111 to 147[2]).

Samples strong enough for Q_β-value measurements are available
of fission product elements ranging from Zn to Zr (As, Se, Y and
Zr as decay products only), and from Ag to Ba.

A tape system has been used in order to transport the desired sample from the collector chamber to a shielded measuring
position. A selected mass beam passes through a collimator mounted on a trolley hitting the tape behind. Contamination from
neighbouring masses is minimized to less than 1 % by choosing
a small collimator opening. The transport time is about 3 s
which means that the practical half-life limit for successful determinations is about 5 s. The vacuum chamber is also provided
with an air-lock for the introduction of calibration samples and
off-line sources. Samples with half-lives longer than 10 min
were always measured off-line.

With this experimental arrangement total beta decay energies
have been determined for nuclides with half-lives ranging from
4 s to 20 min in the mass regions 76 - 93 and 115 - 140.

Recently the spectrometer has been slightly modified and
mounted almost on line, i e with a very short tape transport. In
this position determinations of total decay energies are possible
for nuclides with half-lives as short as 0.1 s. With the limitation caused by long transport times removed almost all nuclides
reported in refs.[2-4]) are available for Q_β-determinations. Thus,
in the future, the OSIRIS facility will be capable of producing
a fairly complete map of the mass surface for neutron-rich fission
products.

BASIC PRINCIPLE OF THE Q_β-SPECTROMETER

The total β-decay energy is obtained by measuring beta branches to excited states of the daughter nucleus, the energy of these states being determined from gamma-ray measurements.

The spectrometer used in the present work consists of a system of silicon detectors placed between two sodium iodine detectors which should have a high efficiency for summing pulses of gamma-rays cascading from a given excited level to the ground-state of the daughter nucleus. The Si(Li)-detector system consists of a main 25 mm dia x 5 mm thick transmission detector from which a segment has been cut in such a way that the beta particles from a sample will see a detector with a sensitive depth up to 23 mm, which corresponds to the range of electrons of energy about 10 MeV.

Pulses from the main Si-detector in coincidence with pulses from the NaI-detectors (through a fast-slow coincidence coupling) are registered in a multichannel analyzer with a band selecting system permitting analysis of eight different gamma-gates simultaneously. Large plastic crystals coupled in anti-coincidence with the NaI-detectors cover most of the spectrometer and act as an anti-Compton shield and also as a background shield. Beta pulses from three Si-detectors surrounding the main beta detector block the coincidence circuit when they coincide with pulses from the main detector, thus removing events with beta particles scattered out.

The response function has been measured using conversion electrons from the decay of ^{207}Bi. The energy range thus covered extends to 1.7 MeV. In this range the response function is well represented by a Gaussian full-energy peak and a constant tail down to zero energy, and the peak-to-total ratio is approximately independent of electron energy. The full width at half maximum of the full-energy peak is 10 keV for the beta detector.

For a given response function the efficiency versus energy function has been determined by comparing the electron energy distribution from ^{106}Rh measured with the Si-detector system with the distribution obtained with an electromagnetic beta-spectrometer. The ratio between these distributions gives an efficiency curve. The ^{106}Rh spectrum extends to 3.53 MeV and, consequently, the efficiency can only be accurately extrapolated up to about 4 MeV. Until now, most of the beta end-points used for calculating Q_β-values are below 4 MeV, but for nuclides with shorter half-lives and further out from stability more energetic beta branches will have to be measured. Fortunetely, in the near future it will be possible to make similar comparisons for OSIRIS-produced samples with high decay energies, such as ^{124}In and ^{136}I with beta branches up to 5.5 MeV.

The efficiency function is directly coupled to the response
function as seen from the discussion above. The pulse spectrum
of ^{127}Sn, with 100 % of the beta particles feeding one level at
490 keV has been analyzed using different sets of response and
efficiency functions. Various forms of the response function
with different peak-to-total ratio were investigated, and the
corresponding function was determined as described above. The
results are collected in Table 1, and they show that the beta end-
point energy is not sensitive to changes of the response function.
This can be varied within wide limits as long as it is combined
with the proper efficiency curve. The error of the beta end-point
energy is composed of the statistical error from the Kurie plot
and the calibration uncertainty. The first part, shown in the
third column of Table 1, indicates that a moderately energy-depen-
dent peak-to-total ratio gives the best fit of the experimental
points to a straight line. With the present configuration of the
detectors it has been impossible to check the energy dependence
accurately, however, and an energy independent peak-to-total ratio
of 0.45 has been used in the calculations. The figure 0.45 was
obtained for E_o= 975 keV.

Table 1: Beta end-point energy of ^{127}Sn for various forms
          of the response function

Peak-to-total	Beta end-point energy (MeV)	Statistical error of the Kurie-plot (st dev)
0.45	2.52±0.06	0.035
$0.45 \times (E/E_o)^{-0.5}$	2.47±0.05	0.012
$0.45 \times (E/E_o)^{-1}$	2.47±0.05	0.015
0.2	2.44±0.08	0.067
0.8	2.54±0.07	0.050

The energy calibration was done using conversion electrons
from ^{207}Bi and daughters from ^{228}Th covering the energy interval
from 480 to 2600 keV. In this range the detector was found to be
linear.

EXPERIMENTAL PROCEDURE

The experiments were carried out with optimal sample strengths
by adjusting the reactor power to give satisfactory coincidence
rates and tolerable accidental coincidence rates. The collection
and measuring times were chosen according to the case under study.
As no chemical separation was used in the experiments, the mass-
separated samples would normally contain 1-3 isobaric components.
By adjusting the timing of the experiment properly, one or the

other of the components could be enhanced. Multianalysis was
used to facilitate the assignment of the various beta branches
measured to the different isobars in the sample.

Up to eight beta spectra corresponding to different gamma
gates were measured simultaneously. The gamma gates were carefully chosen according to the knowledge about the decay scheme and
from an inspection of the gamma spectrum. Preference was given to
high energy gamma peaks to reduce the gamma effect in the beta detector (a proper correction for the gamma effect is difficult to
carry out).

The decay of each gate was also examined by multiscaling and
analyzed into its components by means of the HALFLIFE computer
programme[2]. This provides a check of the element assignment of
the gates and also a check of the amount of contamination from
other components. Beside the beta spectrum coincident with the
selected gate, the accidental spectrum was also measured by displacing the time window out of the coincidence range.

EXPERIMENTAL RESULTS

The experimental results are collected in Table 2. The measured Q_β-values are compared to published experimental values and to
three currently used mass formulae, namely those by Myers and Swiatecki,[5] Garvey et al.[6] and Seeger[7]   Furthermore, the number of
gates used for the determination of the Q_β-value is given. In
cases where the gamma gates have been examined by multiscaling,
the resulting half-life is given.

In most cases the determinations are based on well established
decay schemes. No detailed decay scheme has been available for
^{123}In, ^{125}In and ^{129}Sn, however, and then the value given should
be regarded as a lower limit for the total beta decay energy. The
errors are standard deviations except for cases where the beta end-
point has exceeded the limit for accurate calibrations. The errors
have then been somewhat arbitrarily increased (this is done for
^{134}Sn and ^{138}I).

As seen from Table 2 the experimental Q_β-values from this
work is in good agreement with other experiments and also with mass-
formula predictions. An example is given in Fig 1 showing the
Q_β-value as a function of mass number for some Sn-isotopes. The
experimental results seem to lie fairly close to the mass formula
predictions except for ^{130}Sn where especially the value calculated
by Garvey et al.[6] lies 1 MeV below the experimental result. This
is a little surprising as the G-K mass formula agrees very well
with experimental results for ^{128}Sn and ^{132}Sn.

Table 2: Experimental results

Nuclide	$T_{1/2}$	No of gates	Exp Q_β- this work (MeV)	Other determinations (MeV)	Mass formula predictions M/S[5]	G/K[6]	S[7]
[86]Br	55.2±0.5 s	7	8.00±0.10	7.3±0.4[9]	7.9	7.6	7.54
[87]Br	56.3±0.5 s[2]	6	7.06±0.10			6.68	6.43
[88]Rb		2	5.5±0.2	5.30±0.06[10]	5.3	5.1	5.17
[90]Rb	182±5 s	1	6.91±0.12	6.59±0.10[10]	6.8	6.41	6.82
[93]Sr	6.95±0.07 min	3	4.15±0.20	4.15±0.07[11]	4.3	4.1	4.25
[116]Ag	~3 min	4	5.3±0.2		5.3	6.1	5.23
[117]Ag	68±1 s	4	4.18±0.10		4.1	4.3	3.48
[119]Cd	143±1 s	2	3.94±0.13	3.5±0.3[9]	3.8	3.7	4.01
[120]In	3.0±0.5 s	1	5.4±0.3		5.5	5.41	4.83
[121]In	22.6±0.2 s	1	3.40±0.05	3.38 0.04[9]	3.35	3.5	2.83
[123]In	~ 6 s	2	4.5±0.3		4.5	4.4	4.03
[124]In	3.15±0.09 s	4	7.14±0.09		7.0	7.3	6.89
[125]In	3.9±0.3 s, 11.8±0.9 s	4	5.4±0.3		5.5	5.5	5.13
[125]Sn	9.7±0.1 min	2	2.29±0.05	2.36±0.05[12] 2.39±0.01[13]	2.2	2.3	2.50
[127]Sn	3.6±0.2 min	1	3.01±0.06	3.1±0.1[14]	3.3	3.1	3.33
[129]Sn	9.0±0.2 min	2	4.00±0.12		4.4	3.9	4.10
[130]Sn	8.3±0.2 min	3	3.18±0.09		3.0	2.2	2.43
[131]Sn	58±1 s[2]	1	4.8±0.4		5.4	4.6	5.02
[132]Sn	41±2 s	2	3.30±0.12	3.0[15]	4.0	3.3	3.40
[134]Sb	10.8±0.2 s	1	8.6±0.5	8.5±0.4[16]	9.5	8.7	8.79
[135]Te	19.5±0.1 s	2	6.0±0.3		6.5	5.6	5.68
[135]I		2	2.60±0.04	2.73±0.02[9]	3.1	3.0	2.3
[136]I	~75 s	6	6.6±0.2	6.3±0.1[17]	7.5	7.0	6.96
[137]I	24.5±0.2 s[2]	1	5.5±0.2		6.1	5.8	5.84
[138]I	6.62±0.09 s[2]	1	7.3±0.5		8.4	7.8	7.96
[139]Xe	40.8±0.7 s[2]	6	4.57±0.2	4.88±0.07[18]	5.6	4.7	4.90
[139]Cs	~9 min	3	4.44±0.06	4.29±0.07[18]	4.2	4.1	4.13
[140]Cs	65.2±0.6 s	5	5.8±0.2	5.8±0.1[18]	6.6	6.1	6.25

CONCLUSION

This investigation is the first part of a systematic study of the total decay energies carried out for the purpose of mapping the mass-surface for neutron-rich fission products. The continuation of these experiments has already started (together with K.Aleklett. and G.Nyman). Measurements for [125-129]In have already been performed,and the results will be available in the near future, thus completing the series of mass data for the In-isotopes. Recently Q_β-measurements were also carried out for [75-77]Zn and [76-78]Ga as a complement to Q_β-studies on the neutron deficient side of stability[8].

     When the systematic study for all nuclides produced at the
OSIRIS facility is completed we will get a unique possibility to
check and compare the predictions of the various mass formulae in
the fission product region.

Fig. 1: Q_β-value versus mass
number for Sn-isotopes:
◊: experimental value,
this work, ♦ analysis value
by Wapstra and Gove[9]
mass formula predictions
from Δ Myers and Swiatecki,
o Seeger and ▢ Garvey et al.

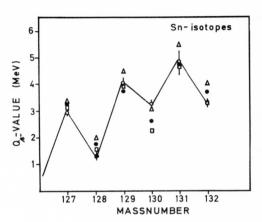

References

1. S. Borg et al., Nucl. Instr. Meth 91 (1971) 109.
2. B. Grapengiesser et al., J. Inorg. Nucl. Chem. 36 (1974) 2409.
3. G. Rudstam and E. Lund, Research Report LF-60 (1974) submitted
        to Phys. Rev.
4. E. Lund and G. Rudstam, Research Report LF-61 (1975) submitted
        to Nucl. Phys.
5. W.D. Myers and W.J. Swiatecki, USA Report UCRL-11980 (1965).
6. G.T. Garvey et al., Rev. Mod. Phys. 41, S1 (1969).
7. P.A. Seeger and W.M. Howard, LA-5750 (1974).
8. L. Westgaard et al., "Masses and Q-values of Nuclei far from
        Beta-Stability, Evidence for the Wigner Symmetry Energy in
        N Z Nuclei" this conference.
9. A.H. Wapstra and N.B. Gove, Nucl.Data Tables 9 (1971) 267.
10. J.R. Clifford et al. Phys. Rev. 7,6,2535 (1973).
11. M.I. Macias-Marques et al., CERN Report 70-30,1,321 (1970).
12. R.L. Auble and W.H. Kelly, Nucl. Phys. 79 (1966) 577.
13. P.A. Baedecker and W.B. Walters, Nucl. Phys. A107 (1968) 449.
14. K.E. Apt and W.B. Walters, Phys. Rev. C.9,1 (1974) 310.
15. A. Kerek et al., Nucl. Phys. A195 (1972) 159.
16. A. Kerek et al., Nucl. Phys. A195 (1972) 177.
17. L.C. Carraz et al., Nucl. Phys. 158A (1970) 403.
18. J.P. Adams et al., Phys. Rev. C8 (1973) 767.

FAR BETA-UNSTABLE ALPHA-PARTICLE EMITTING NUCLEI

Kari Eskola and Pirkko Eskola

Department of Physics

University of Helsinki, Helsinki, Finland

1. INTRODUCTION

The study of α-active nuclei has long been a fertile and effective means of probing the neutron-deficient shoreline of nuclear stability. α-emitting nuclei along with spontaneously fissioning ones also habit the zone, where the present frontier of elements lies. In this report we shall review recent experimental work on new α-active nuclei. The term "new α emitter" refers to nuclides not found to be α emitters at the time of the 1971 Atomic Mass Evaluation by Wapstra and Gove (1), or not included in it.

α-active nuclei are most valuable in the mapping of nuclear mass surface, because 1) α-particle energies can be measured with good precision and 2) α-decay energies exhibit rather smooth dependence on both proton and neutron numbers. The latter feature is often useful in identification of new activities, particularly, when moving in the direction of neutron-deficient isotopes for a given Z. In the realm of the heaviest elements the method to establish genetic links has been used to fix the place of a newly discovered member in the chain of known α emitters.

2. NEW ALPHA-PARTICLE EMITTERS

In fig.1 the upper region of the nuclear chart is displayed. The α-active isotopes are marked by a circle and one can see at a glance their predominance among the known nuclides in the region shown ($Z \geq 70$). The new α-active isotopes are indicated by black circles. A list of these isotopes, with references to original reports, is presented in table 1. A recent compilation of the

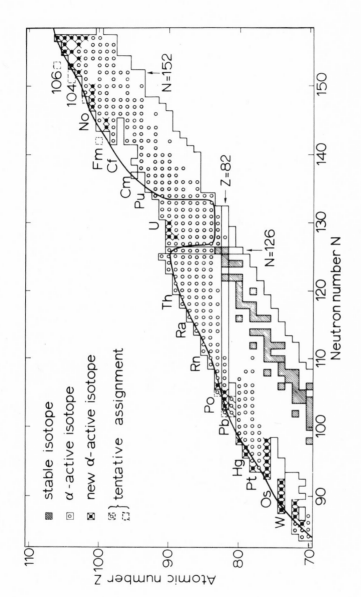

Fig.1. The distribution of α-active isotopes in the upper region of the chart of nuclides. The heavy continuous line is the borderline for the total half-life $T_{1/2}$ less than 1 second.

Table 1

List of new α-active isotopes (reference numbers in parentheses)

^{157}Lu (3)	^{217}Ac (16)
^{159,160,161}Hf (4)	^{218}Th(17,18), ^{219,220}Th(18)
^{162,163,164}W(5), ^{165,166}W(6)	^{243,244}Es (19)
^{169}Os(7),^{170,171}Os(8),^{172,173,174}Os(9)	^{248,249,250,251}Md (20)
^{172}Pt (10)	^{259}No (21)
^{175,176}Au (11)	^{255}Lr(22,23),^{257,259,260}Lr(23)
^{178}Hg (12)	^{257,259}Rf (24), ^{261}Rf (25)
^{183,184,185}Tl (10)	^{260}Ha (26), ^{261,262}Ha (27)
^{184}Pb(10),^{185}Pb(11),^{186,187}Pb(13,14)	^{263}106 (28)
^{188}Bi (10), ^{189}Bi (15,14)	

characteristics of α activities for neutron-deficient nuclei with $65<Z<91$ has been prepared by Gauvin et al. (2). The extensive pioneering work of Macfarlane, Siivola and Valli et al. (see e.g. ref. (2)) in the rare-earth, Ir-Bi, Po-Pa regions, respectively, and the three decades of steady progress in the study of actinide nuclei form the firm background on which the newly-discovered nuclei rest. In fig.1 the contour line closely following the borderline of the known nuclides, except in the region $127<N<135$, corresponds to a total half-life of 1 second. For the elements with $70<Z<90$ and $Z>103$ the line is based on experimental data and for $91<Z<103$ on the predicted values of Keller and Münzel (29). The rate of decrease of half-lives for nuclides to the left of the line mainly depends on how fast the α-decay energies increase as a function of N. In the heavy-element region spontaneous fission appears as a competing and, eventually, as a predominant decay mode. In fig.1 it is seen to happen for neutron-deficient even-even isotopes above californium. The tentative mass assignments refer to the recent work of Oganessian et al. (30), in which lead isotopes were bombarded by heavy ions with $18<Z<24$.

The experimental α-decay energies of the new α emitters have been plotted in fig.2 as a function of N. The α-decay energy has been derived from the measured value of the most energetic α group by adding to it recoil and screening corrections. In addition to the experimental points we have also plotted the Q_α values taken from the 1971 Atomic Mass Evaluation of Wapstra and Gove (1).

The upper part of fig.2 shows that for the even-Z elements, Hg, Pt and Os, the agreement between the experimental and adjusted values is strikingly good whereas for Pb, W and Hf isotopes a sys-

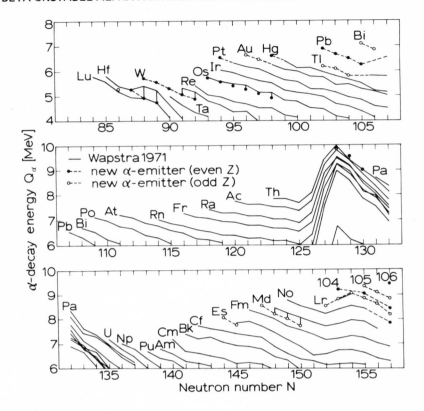

Fig.2.   Experimental Q_α values of new α emitters plotted as a
function of neutron number.

tematic deviation occurs between the two sets of values. The discrepancy is most pronounced, as much as 600-700 keV, for Pb isotopes.

In the middle section of fig.2 there are four new α emitters
^{217}Ac and $^{218-220}$Th. All these α emitters are characterized by half-
lives in the microsecond range. A novel technique, as far as α decay
is concerned, of measuring decay between beam pulses obtained from
a MP tandem accelerator (18) or between natural beam bursts of a
heavy-ion cyclotron (16, 17) has been used in the measurements.

The shortest half-life of 96±7 ns is attributed to ^{218}Th.

It is evident from the lower part in fig.2 that the search for
new α emitters in the heavy-element region has been concentrated on
the far end of the nuclear chart and, in fact, has been motivated
by the quest for new elements. Since, in the identification of α-
active isotopes of new elements (Rf, Ha, 106), genetic links to α-
emitting isotopes of lighter elements have played a crucial role,
it has been necessary to build a broad base of well-known α emitters
adjacent to the front line. A comparison between the adjusted values
and the experimental ones shows a systematic deviation between the
two in the case of the new Md isotopes. A plausible explanation to
this apparent discrepancy is that the observed most energetic α
groups do not populate the ground state (20). None of the known α-
active nuclides beyond nobelium are even-even isotopes. Therefore
it is hazardous to draw conclusions on systematic features based on
the data. However, a reduced spacing between the Q_α lines for successive Z values above nobelium is evident for 155<$\bar{N}$<157. As a consequence the nuclides in question have longer half-lives than predicted earlier (31), a property that has been favourable to their
discovery.

3. PARTIAL HALF-LIVES

Direct experimental measurements of α branches for highly
neutron-deficient nuclei are concentrated in the region 78<Z<86
(32,33). This is in part due to technical difficulties in observing
weak β and α activities in the presence of a host of other reaction
products. At ISOLDE facility, on-line electromagnetic mass separation has provided clean enough conditions for observation of
K x-ray/α ratios from which α-branching ratios are derived. In a
recent report (33) α-branching ratios for $^{188-192}$Pb isotopes are
given ranging from $3.3 \cdot 10^{-2}$ for ^{188}Pb to $7 \cdot 10^{-5}$ for ^{192}Pb. On the
basis of these results Hornshøj et al. (33) calculate the reduced
widths for these Pb isotopes. A comparison of the values with the
reduced widths of other known even-even ground-state to ground-state
transitions reveals a anomalous behaviour. The widths show a strongly decreasing trend as a function of N in contrast to the regular,
rather flat behaviour of Pt and Hg isotopes in this region.

In cases where two or more α emitters are geneticly linked to
each other by α decay, branching ratios can be determined from the
relative intensities of α particles emitted by two successive members in the chain. The method has been applied to the measurement
of the α-branching ratio of the daughter nuclides for several parent-
daughter pairs, e.g. ^{263}106 $\xrightarrow{\alpha}$ ^{259}Rf $\xrightarrow{\alpha}$ ^{255}No $\xrightarrow{\alpha}$ (28), ^{260}Ha $\xrightarrow{\alpha}$ ^{256}Lr $\xrightarrow{\alpha}$
(23), ^{257}Rf $\xrightarrow{\alpha}$ ^{253}No $\xrightarrow{\alpha}$ (34), ^{249}Md $\xrightarrow{\alpha}$ ^{245}Es $\xrightarrow{\alpha}$ (20) in the heaviest-
element region. The genetic link between the parent and the daugh-
ter nucleus has been established either by recoil separation or by
time correlation methods.

A third method to determine α-branching ratios is to compare
the measured yield for production of an α-active nuclide with the predicted total cross section for the particular isotope studied. Such
a method in a somewhat modified form was used by Le Beyec et al. (14)
in their determination of branching ratios for neutron-deficient Bi
and Pb nuclei, and by Eskola (20) for branching ratio determinations
of Md isotopes.

As evidenced by the fig.1 the most neutron-deficient known α
emitters for elements with Z>70 have half-lives of less than a
second. As a consequence, α decay is the predominant decay mode,
except for the heaviest elements where spontaneous fission takes
over, and branching by α decay for the nuclei near the borderline
is nearly 100%.

4.   ALPHA FINE STRUCTURE

Although search for far β-unstable α emitters has resulted in
mapping of vast new areas of the nuclear chart the experiments have
yielded mostly information about the gross properties, and little is
known about the finer details of the decay. In the region Z<82 fine
structure has been observed in the even-even isotopes ^{176}Pt, ^{178}Pt,
^{182}Hg, ^{184}Hg (33,35). It is interesting to note that for these few
known cases the d-wave ($0^+ \rightarrow 2^+$) hindrance factors are roughly an order
of magnitude larger than for the actinides (33), a fact that makes
the observation of the d-wave α groups very difficult. In some odd
α emitters, such as ^{176}Au, ^{177}Au, ^{185}Pb and ^{187}Pb (11), several α
groups have been observed. In the region of α emitters with Z>99
the measured α spectra of new α-emitting isotopes are complex with
several resolved α groups (23,24,26). Experimental difficulties in
producing the nuclides have discouraged detailed spectroscopic studies. However, in the experiments related to identification of the
elements No and Rf by x-rays from the α-decay daughter isotopes
detailed α spectra have been measured (36).

5.   PRODUCTION OF NEUTRON-DEFICIENT ALPHA EMITTERS

A representative sample of the heavy-ion reactions used in
search of far β-unstable α-emitting nuclei is presented in table 2.
Considering first the reactions in the column on the left hand side,
it is seen that the N/Z ratio equals one for the projectiles in each
case, but that the N/Z ratio of the target can be reduced by choice
of more neutron-deficient target isotopes (^{136}Ce, ^{144}Sm). The
lightest compound nucleus that might be reached for each element by
fusion of an optimum combination of stable target and projectile
nuclei lies 2-4 mass units to the left of the most neutron-deficient
nuclei in the region with 70≤Z≤82 (see fig.1). With the advent of
new universal heavy-ion accelerators practically any target-projec-

Table 2

Reactions used in the production of the most neutron-deficient
isotopes of some elements

Reaction	σ (10^{-30} cm^2)	Reaction	σ (10^{-30} cm^2)
^{144}Sm(^{24}Mg,6n)^{162}W	50	^{233}U (^{15}N,5n)^{243}Es	1.7
^{156}Dy(^{20}Ne,7n)^{169}Os	--	^{241}Am(^{12}C,5n)^{248}Md	0.1
^{140}Ce(^{40}Ca,8n)^{172}Pt	--	^{249}Cf(^{12}C,4n)^{257}Rf	0.04
^{150}Sm(^{40}Ca,5n)^{185}Pb	100-1000	^{249}Cf(^{18}O,4n)^{263}106	0.0003

tile combination is available and the question is whether the fusion
reactions will take place with high enough probability. Winn et al.
(37) have predicted limits for production of new proton-rich nuclei
by a heavy-ion compound-nucleus mechanism. On the basis of their
results we have estimated that the contour line representing a yield
of the order of 10^{-30} cm^2 runs roughly parallel to the $T_{1/2}=$ 1 s line
in fig.1, but is 2-3 mass units to the neutron-deficient side.

The data in table 2 pertinent to the heaviest-element region
indicates that the cross sections for the production of nuclides on
the borderline are very low due to severe losses for fission. It
is therefore very difficult to produce new neutron-deficient nuclei
by heavy target - light heavy-ion combinations. The method of
Oganessian et al. (30) to produce cold compound nuclei by fusion of
stable Pb nuclei with medium heavy projectiles (18<Z<24) may lead
the way to a discovery of a number of new α emitters in the very
heavy-element region.

In addition to the fundamental questions relating to the possibility of producing new far β-unstable nuclei by fusion reactions,
a practical experimental difficulty is caused by the rapidly decreasing half-lives when moving further away from the stability line.
The predicted rate of decrease is such that the 10-ms contour line
for $T_{1/2}$ is reached by 2-3 mass units out of the present borderline
(38,29). The short half-lives will make use of the He-jet technique
impracticable, but will be no problem for such effective and fast
tools as recoil spectrometers, which in our opinion will turn out to
be most effective in the future studies of α-active nuclei far from
the line of β stability.

Financial support from the Chancellor of the University of Helsinki and from the National Research Council for Sciences, Academy
of Finland is gratefully acknowledged.

REFERENCES

1. A.H. Wapstra, N.B. Gove, Nucl. Data Tables 9, 265 (1971).
2. H. Gauvin et al., Inst. de Phys. Nucl., Orsay report IPNO-RC-74-05 (1974).
3. H. Gauvin, Y. Le Beyec, N.T. Porile, Proc. European Conf. on Nucl. Phys., Aix-en-Provence, 1972, vol. II.
4. K.S. Toth et al., Phys. Rev. C7, 2010 (1973).
5. D.A. Eastham, I.S. Grant, Nucl. Phys. A208, 119 (1973).
6. K.S. Toth et al., to be published in Phys. Rev. C.
7. K.S. Toth et al., Phys. Rev. C6, 2297 (1972).
8. K.S. Toth et al., Phys. Rev. C5, 2060 (1972).
9. J. Borggreen, E.K. Hyde, Nucl. Phys. A162, 407 (1971).
10. C. Cabot et al., in Inst. de Phys. Nucl., Orsay annual report 1974.
11. C. Cabot et al., Nucl. Phys. A241, 341 (1975).
12. P.G. Hansen et al., Nucl. Phys. A160, 445 (1971).
13. H. Gauvin et al., Phys. Rev. Lett. 29, 958 (1972).
14. Y. Le Beyec et al., Phys. Rev. C9, 1091 (1974).
15. H. Gauvin et al., Nucl. Phys. A208, 360 (1973).
16. T. Nomura et al., Phys. Lett. 40B, 543 (1972); Nucl. Phys. A217, 253 (1973).
17. K. Hiruta et al., Phys. Lett. 45B, 244 (1973).
18. O. Häusser et al., Phys. Rev. Lett. 31, 323 (1973).
19. P. Eskola et al., Phys. Fennica 8, 357 (1973).
20. P. Eskola, Phys. Rev. C7, 280 (1973).
21. R.J. Silva et al., Nucl. Phys. A216, 97 (1973).
22. V.A. Druin, Yadern. Fiz. 12, 268 (1970).
23. K. Eskola et al., Phys. Rev. C4, 632 (1971).
24. A. Ghiorso et al., Phys. Rev. Lett. 22, 1317 (1969).
25. A. Ghiorso et al., Phys. Lett. 32B, 95 (1970).
26. A. Ghiorso et al., Phys. Rev. Lett. 24, 1498 (1970).
27. A. Ghiorso et al., Phys. Rev. C4, 1850 (1971).
28. A. Ghiorso et al., Phys. Rev. Lett. 33, 1490 (1974).
29. K.A. Keller, H. Münzel, Nucl. Phys. A148, 615 (1970); Kern-forschungszentrum, Karlsruhe report KFK-1059 (1969).
30. Yu. Ts. Oganessian et al., Nucl. Phys. A239, 157 (1975); Yu. Ts. Oganessian et al., Dubna report JINR-D7-8099 (1974).
31. V.E. Viola, G.T. Seaborg, J. Inorg. Nucl. Chem. 28, 741 (1966).
32. P. Hornshøj et al., Nucl. Phys. A230, 380 (1974) and references quoted in (33).
33. P. Hornshøj et al., Nucl. Phys. A230, 365 (1974).
34. A. Ghiorso et al., Nature 229, 603 (1971).
35. P.G. Hansen et al., Nucl. Phys. A148, 249 (1970).
36. P.F. Dittner et al., Phys. Rev. Lett. 26, 1037 (1971); C.E. Bemis, Jr. et al., Phys. Rev. Lett. 31, 647 (1973).
37. W.G. Winn, H.H. Gutbrod, M. Blann, Nucl. Phys. A188, 423 (1972).
38. A. Siivola, Lysekil Symposium 1966, Ark. Fys. 36, 413 (1967).

PRECISION MEASUREMENT OF Q-VALUES BY MEANS OF THE MUNICH TIME-OF-FLIGHT SYSTEM AND THE Q3D-SPECTROGRAPH

P.Glässel, E.Huenges, P.Maier-Komor, H.Rösler and

H.J.Scheerer   Beschleunigerlaboratorium der Universität und der Technischen Universität, München

H.Vonach   Institut für Radiumforschung und Kernphysik, Wien

D.Semrad   Physikalisches Institut der Universität, Linz

1. INTRODUCTION

The measurement of Q-values of nuclear reactions is one of the most important methods for determining the masses of unstable nuclei which has contributed much to the present knowledge of nuclear masses. In most cases reactions with charged particles in the entrance and exit channel were used and the Q-value was determined by measurement of the magnetic rigidity of both the bombarding particles and the reaction products in homogeneous field iron magnets. Although the magnetic induction B itself can be measured to about 10^{-6} by means of nuclear magnetic resonance the magnetic rigidities $B\rho$ and thus the particle energies could be obtained only with accuracies of some parts in 10^4 /1/. This is mainly caused by two effects:
1) Even in high quality iron magnets the average field may deviate from the field of the location of the NMR probe by some parts in 10^4. Furthermore this difference may change by about 10^{-4} due to differential hysteresis.
2) In the determination of the magnetic rigidity of the bombarding particles in the analyzing magnets a further uncertainty arises from the finite object and image slit width. Therefore the radius of the trajectories of the particles is usually not defined better than to about $5 \cdot 10^{-4}$. Thus the actual radius of curvature may deviate from the nominal radius defined by the centres of the slits by a few parts in 10^4 because of asymmetries in the beam profiles at the slit positions.

Due to these effects until some years ago most Q-values have been accurate only to about 5-1o keV. In the meantime three approaches have been developped to improve this situation.
1) Using high-resolution γ-spectroscopy with Ge(Li) detectors Q-values for many (n,γ)-reactions have been determined with an accuracy of 1-2 keV. This method can, however, only be used to determine mass differences between different isotopes of one element.
2) Using the so-called Q-value matching technique the mentioned difficulties of absolute measurements of $B\rho$ values could be overcome to a large extent and absolute Q-value measurements can now be performed for many kinds of charged particle reactions with an accuracy of about 2 keV using conventional analyzing magnets and spectrographs /2/.
3) A completely different approach has been followed at the Munich Tandem laboratory. The main idea was to replace the measurement of magnetic rigidity by a direct measurement of the particle velocities. For this purpose a precision time-of-flight measuring system (PTMS) was developped for measurement of the beam energy of the bombarding particles allowing absolute energy measurements to about 1 part in 10^5. For the emitted particles, direct precision measurement of velocity proved not to be possible. This difficulty could be overcome, however, by analyzing the outgoing particles in a high resolution magnet spectrograph which is calibrated with the PTMS via elastic scattering. In this way a system for measurement of Q-values with an accuracy of about 1 keV could be developed. In the following we will describe this system, discuss the factors limiting its accuracy, report on the measurements performed to date and discuss its further possibilities.

2. GENERAL DESCRIPTION OF THE SYSTEM

The time-of-flight system /3/ and the Q3D-spectrograph /4/ are operated in the following way (see fig.1). After passing the 9o° analysing magnet the tandem beam is split by an electrostatic deflector, the so-called peeler. About 1o% of the beam are deflected and subsequently used for an exact energy measurement. The main beam traverses the peeler without deflection. It is then deflected by 5o° in the switching magnet and by 6o° in the additional bending magnet and finally focused on the target in the Q3D scattering chamber. The reaction products are detected in the focal surface of the Q3D by means of a 39oo wire proportional counter, the distance between adjacent wires being o,5 mm. Each particle detected in the multiwire counter produces two signals, the digital position signal indicating the number of the wire activated by the particle and the analog ΔE signal, which is proportional to the energy loss ΔE of the particle in the counter. Simultaneously the absolute energy of the peeled-off beam is determined by the PTMS. The time-of-flight is measured along an exactly measured flight path of about 147 m in time intervals of about 1 msec indicating the average energy of the particles during this measuring period. The system produces an energy fluctuation signal (TE-signal) which is proportional to the deviation of the instantane-

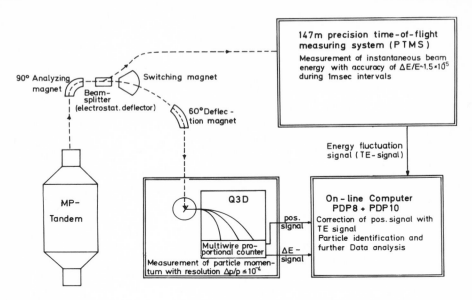

Fig.1: The Munich system for high resolution nuclear spectroscopy
       and precision measurements of Q-values

ous average energy from a preset nominal energy E which is determined
to better than 1.5 : 10^5. The accuracy of the signal is as good as the
absolute accuracy for energy fluctuations not faster than 1oo Hz.
        For any particle detected in the focal surface the following
quantities are measured and fed into the on-line computer.
1) The position signal indicates the position of arrival of the particle in the focal surface of the Q3D.
2) The $\triangle E$ signal which is proportional to the stopping power of the
detected particle. This signal is used to identify the particle.
3) The TE-signal sampled for each event in the focal plane and the
preset nominal energy E. The position signal of each detected particle
is corrected according to these energy fluctuations by software.
        By means of this combined system absolute Q-values can be mea-
sured about half an order of magnitude more accurate than usually performing the experiment in the following way:
1) The reaction, the Q-value of which is to be measured, is studied in
the Q3D; the beam energy is accurately measured by the PTMS and the
position of the reaction products is observed very precisely with the
multiwire counter.
2) Without changing the field the Tandem energy and, if necessary the
kind of the bombarding particles are changed, in order to produce a
beam of the same magnetic rigidity as the reaction products observed
before. Particles of this beam when elastically scattered at forward

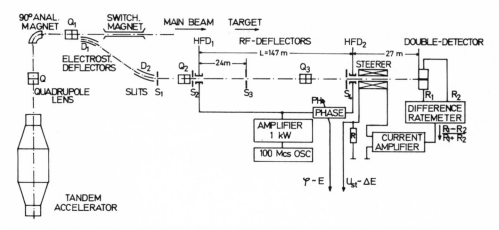

Fig.2: Simplified schematic diagram of the time-of-flight system
       for the precision measurement of the tandem beam energy

angles will arrive approximately at the same position as the reaction
products observed in the first experiment. By measuring the energy of
this beam with the PTMS, the Q3D is calibrated to about 10^{-5}. Therefore the energy of the first reaction products can be measured with
nearly the accuracy of the PTMS. The only demand on the spectrograph
in such experiments is the constancy of the magnetic field, which is
fullfilled to some parts in 10^6.

    As both main components the PTMS /3/ and the Q3D-spectrograph /4/
have already been described in the literature only a brief description
of the properties of the system will be given. The Fig.2 shows a
simplified schematic diagram of the PTMS. A fraction of 100 nA is
peeled off the main beam by two electrostatic deflectors D1 and D2 and
sent through the fields of two identical high-frequency resonators C1
and C2, the centres of which are separated by a carefully measured
distance of L = (14711.89$\pm$0.03) cm. They are fed from a 100 MHz power
supply, the frequency of which is calibrated and stabilized to $1 \cdot 10^{-8}$.
The first rf-field pulses the beam by sweeping it across the slit S.
Particles of rest energy E_0 and kinetic energy E traversing the first
resonator at zero field are not deflected in the second condensor and
cause equal count rates in two scintillators Sc1 and Sc2, if the phase
difference between the two resonators corresponds to their time-of-
flight.

$$T = (L/c)(1 - E_0^2/(E_0 + E)^2)^{-1/2}$$

which is typically some hundred rf-periods. The integer number of rf-
periods is determined from the setting of the 90° magnet while the
exact phase difference is set to the desired value via the phase shif-

ter PH. The phase is stabilized to ± o.1° and is measured to ± o.4°.
From this the nominal kinetic energy, i.e. the energy E of the unde-
flected particles, can be calculated to $1.5 \cdot 10^{-5}$ for 2o MeV protons
and even more accurately for slower particles.

Particles with energies E $\pm \Delta$E are deflected in the second resonator and give rise to an analog signal from the rate meter RM
proportional to the count rate difference between the two detectors.
The signal is normalized to a constant total count rate thus eliminating the influence of beam intensity fluctuations. It is produced
in time interwals of typically 1 ms. A direct calibration of this
signal in terms of ΔE is not possible because for deflections exceeding the size of the beam spot at the detector position the count
rate difference becomes constant. Therefore the RM signal controls
the current in a magnetic steerer which bends the beam back to the
position of zero count rate difference. This steerer current depends
linearly on ΔE. A signal proportional to this current can be registered in coincidence with every event recorded in an experiment,giving
in connection with the preset phase value the momentary beam energy.
In this way ΔE can be measured to about 5%. As typical values of
ΔE/E are $2 \cdot 10^{-4}$, the accuracy of the momentary energy is as good as
that of the nominal energy.

The Q3D spectrograph is ideally suited for the system. It combines a large solid angle ($\approx$12 msr), large dispersion (1o cm pro 1%
ΔE/E) and excellent energy resolution E/ΔE $\approx$ 5ooo (FWHM) for o.5 mm
target width and full solid angle. It therefore allows precise comparison of momenta of the different particle groups.

3. ACCURACY OF THE METHOD

The measurement of any Q-value consists of four single measurements (see fig.3).
1) Measurement of the absolute energy E_1 of the particles used for
the spectrograph calibration with the PTMS.
2) Measurement of the position of the corresponding elastic line in
the focal plane of the Q3D.
3) Measurement of the absolute energy E_3 of the incoming particles of
the reaction, the Q-value of which is to be determined, with the PTMS.
4) Measurement of the position of the corresponding emitted particles.

The total error ΔQ (effective 3δ - error) is obtained by adding
quadratically the maximum errors due to the uncertainties in the velocity resp. position measurements.

$$\Delta Q = (\Delta E_1^2 + \Delta E_3^2 + D\Delta x_{24})^2 + (\Delta D x_{24})^2)^{1/2}$$

D = Dispersion of the Q3D along the focal surface; x_{24} = Position
difference between the particle groups of measurements 2 and 4

The errors in the absolute energy measurements ΔE_1 and ΔE_3 are
about 10^{-5} of the corresponding energy values as discussed in detail
in ref./2/. They are typically at the order of 1oo - 2oo eV.

The error due to the uncertainty in the dispersion D of the Q3D can be neglected, as always the energies E_1 and E_3 are matched so that the position difference x_{24} gets sufficiently small. In determining the position difference x_{24}, however, a number of sources of systematic errors have to be considered, which at present limit the achievable overall accuracy of the Q-value measurements:

1) Difference in the spectrograph field B between measurements 2 and 4. With our NMR system the smallest detectable field change results in a contribution of $\pm$ o.o5 mm to x_{24}.

2) Difference in target position. In general different targets have to be used for measurements 2 and 4. The position of these targets (line targets o.5 x 4 mm) can at present be positioned only to about $\pm$ o.1 mm with respect to the centre of the spectrograph. This results in a contribution of $\pm$ o.2 mm to x_{24} as the horizontal magnification of Q3D is about 2 along the focal surface.

3) Difference in the beam centre on the target. The beam intensity across the o.5 mm width of the target is non uniform as the half-width of the beam is also of about this size. For this reason the effective target centre may deviate from the geometrical target centre. If some care is taken to keep the beam symmetrically on the target the corresponding contribution to x_{24} can be kept below o.1 mm.

4) Energy difference between the particles measured by the PTMS and the average beam energy on the target. Due to the dispersive properties of the Tandem analyzing magnet and the subsequent beam transport system, the average particle energy varies across the beam spot on the target about 1 keV between the right and left edge of the beam. Thus the average energy of the particles may differ slightly from the average energy as measured by the PTMS. This effect has been investigated experimentally by moving the beam across the targets. The maximum error in x_{24} due to this cause is about o.25 mm.

5) Uncertainties in the corrections for energy loss in the targets. Using targets of about 10 $\mu g/cm^2$ on carbon backings of about 5 $\mu g/cm^2$ (both known to about 15%) the uncertainty in the mean energy can be kept less than 1oo eV for reactions involving Z = 1 particles only and less than 3oo eV if Z = 2 particles participate in the reaction.

To these systematic errors the statistical error of the peak position (e.g. of the centroids of the lines) has to be added which, however, is small relative to the systematic errors.

Furthermore kinematic effects introduce errors which can become very large if light and medium target nuclei are studied. In order to determine the Q-values the laboratory energy of the emitted particles has to be converted to the CM-system. In order to do this with a relative accuracy of 10^{-5}, the average reaction angle θ is needed very accurately unless the measurements are performed at an angle of 0° or 180°. Measurements at 0° have proved to be possible if the magnetic rigidity of the reaction products exceeds suffiently that of the bombarding particles e.g. in case of (^{3}He,d) or (^{3}He,t)- reactions.

At present the situation can be summarized as follows. The incident energy in nuclear reaction can be measured with the PTMS with

an absolute accuracy $\Delta E/E$ of about 10^{-5}. The emitted particle energy
can be measured only to about 3-5 10^{-5} due to the effects discussed
above. As both energies are typically of the order of 1o-25 MeV at
present Q-values can be determined with an accuracy of about o.3-1.o
keV. This accuracy has been proved experimentally by measuring excitation energies well known from γ-ray energies /3/.

4. MEASUREMENTS PERFORMED TO DATE

1. Measurement of the α-particle energy of ^{212}Po /5/. The Tandem
was used to produce protons having approximately the same magnetic
rigidity as the α-particles to be measured. The accurate energy of
these protons was measured with the PTMS and its momentum was compared
with the α- particle momentum in the Q3D. In this way the α-energy
was found to be 8784.9 $\pm$ o.36 keV (effective 3δ-error). This value
is about o.5 keV higher than the value of Rytz /6/ the only other value of comparable accuracy.
2. Measurement of β-decay energies for supperallowed Fermi transi-
tions.Recent measurements of the Q-values of the respective (^{3}He,t)-
reactions disagree with ealier values from measurements of (p,n)-
tresholds by up to 1o keV which is several times the combined error
of both measurements /7/. In order to investigate this discrepancy
and to improve the accuracy of these log(ft)-values measurements of
(^{3}He,t)-Q-values were performed for all accessible pairs of nuclei
showing superallowed Fermi transitions.
    Two experiments were performed. In a first run triton energies
of the (^{3}He,t)-reactions on ^{26}Mg, ^{34}S, ^{42}Ca and ^{50}Cr were measured
using 18 MeV deuterons elastically scattered on ^{209}Bi for the Q3D-
calibration (see fig.3). In a second experiment (^{3}He,t) Q-values on
^{26}Mg, ^{42}Ca, ^{46}Ti, ^{50}Cr and ^{54}Fe were measured relative to the ^{14}N
(^{3}He,t) Q-value. Table 1 shows our preliminary results for these
Q-values. A rather large error of 2 keV has been attributed to the
Mg-value of the first run as a rather thick target was used and to the
results of the second experiment as its analysis is very preliminary.

		$^{26}Mg - ^{26m}Al$	$^{34}S - ^{34}Cl$	$^{42}Ca - ^{42}Sc$	$^{46}Ti - ^{46}V$	$^{50}Cr - ^{50}Mn$	$^{54}Fe - ^{54}Co$
Q	a.	-425o.3+2	-551o.4+1	-6441.4+1		-765o.5+1	
	b.	-4251.9+2		-6442.2+2	-7o7o.o+1	-7652.o+2	-826o.8+2

Table 1: (^{3}He,t) Q-values (keV) measured a) with an absolute calibrated spectrograph. b) relativ to the $^{14}N(^{3}He,t)$Q-va-
lue of -5163.4 keV /8/. All errors are effective 3δ-errors

5. FUTURE POSSIBILITIES

After completion of the present upgrading programm of the Munich
Tandem accelerator the energy will be sufficient for Q-value measure-

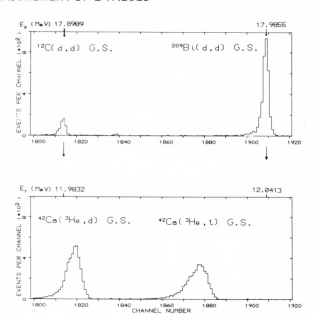

Fig.3: Basic method of absolute Q-value measurement a) calibration of
the Q3D-spectrograph via elastic deuteron scattering $\theta_{Lab}=1o^{\circ}$;
$E_D=17.9913$ MeV; b) Measurement of reaction products $\theta_{Lab}=0^{\circ}$;
$E_{3_{He}}=18.5186$ MeV

ments on all reactions with p, d, ^{3}He, ^{4}He in the entrance and p, d,t,
^{3}He and ^{4}He in the exit channel. An accuracy of better than o.5 keV
should be obtainable. Using (^{3}He,d) and (p,t) reactions which both can
be performed under ideal kinematic conditions at 0°, it should in principle be possible to establish a mass scale from ^{16}O to ^{208}Pb with
an overall accuracy of a few keV.

REFERENCES

/1/ M.J. Le Vine and P.D. Parker, Phys.Rev. 186 (1969) 1021
/2/ J.A. Nolen,jr.,G. Hamilton, E. Kashy and I.D. Parker, Nucl.Instr.
     Meth. 115 (1974) 189
/3/ E. Huenges and J. Labetzki, Nucl.Inst.Meth. 121 (1974) 307
/4/ H.A. Enge and S.B. Kowalski, Proc.3rd Int.Conf.on Magn.Techn.(1970)
     p. 366
/5/ H.Vonach, P.Glässel, P.Maier-Komor, H.Rösler and H.J.Scheerer,
     Sitz.Ber. Öster.Akad.d.Wiss.Abt II 183 (1974) 243
/6/ A. Rytz, B. Grennberg and D.J. Gorman, Proc.4th Int.Conf.on Atomic
     Masses, Plenum Press, London 1972 p 1
/7/ J.C. Hardy, G.C. Ball, J.S. Geiger, R.L.Graham, J.A. Macdonald and
     H. Schmeing, Phys.Rev.Let. 33 (1974) 320
/8/ A.H. Wapstra and N.B. Gove, Nucl.Data A9 (1971) 267

MEASUREMENTS OF NUCLEAR MASSES FAR FROM STABILITY*

Edwin Kashy, Walter Benenson, and Dennis Mueller

Department of Physics and Cyclotron Laboratory
Michigan State University
East Lansing, Michigan 48824   USA

INTRODUCTION

Measurements of properties of nuclei far from nuclear stability provides a challenge to both the experimenter and theorist. To the first, it represents a new frontier in experimentation, and to the second a fertile source of data. There has been in recent years considerable effort on the neutron rich nuclei, and a smaller yet important attack in reaching the limits of particle stability on the proton rich side. Methods for measuring masses of nuclei far from nuclear stability are reviewed with emphasis placed on reaching proton-rich nuclei using particle transfer reactions such as $(p,^6He)$, $(^3He,^8Li)$ and $(^3He,^6He)$.

Neutron-Rich Nuclei: While it is the precise measurement of masses of proton rich nuclei with $A \leq 60$ which is the subject of this paper, we note briefly experiments where an accuracy of better than 100 keV has been obtained for masses of neutron rich nuclei. For example, Scott and collaborators[1] obtained $-10.75 \pm .05$ MeV for the mass excess of ^{29}Mg using the three neutron transfer $(^{11}B,^8B)$ and methods similar to those described here. Goosman and Alburger[2] have obtained $-24.936 \pm .075$ MeV for ^{35}P using $^{18}O(^{19}F,2p)^{35}P$ and measuring the β-end point. A number of $T_z=5/2$ nuclei are reported in this conference by Alburger.[3] Flynn and Garrett[4] used the $^{26}Mg(t,^3He)^{26}Na$ reaction and measured -6.903 ± 0.020 keV for the ^{26}Na mass excess using a Silicon surface barrier counter telescope. At a somewhat lower accuracy, there is the 100-200 keV work of Klapisch and collaborators,[5] who used on-line mass spectrometer for masses of neutron

---

* Research supported by the National Science Foundation.

rich Na isotopes produced by 24 GeV protons on Uranium. In the work of Wilcox, et al.[6] the (^{7}Li,^{8}B) reaction yielded -2.18±.10 MeV for the ^{25}Ne mass excess. Artukh[7] with his collaborators used the ^{232}Th(^{22}Ne,^{21} ^{22}O)^{233} ^{232}U reactions to get the masses of ^{21}O and of ^{22}O.

Proton-Rich Nuclei: Work on nuclei far from β stability includes the highly precise work of Mosher, Kavanagh and Tombrello,[8] who obtained the mass of ^{9}C by measuring the (^{3}He,n) threshold on the radioactive target ^{7}Be. Mendelson and collaborators[9] have used the (^{3}He,^{6}He) reactions to reach Z=N+3 nuclei in the A=13-25 region; Silicon counter telescope techniques were used and 15-40 keV accuracy obtained. Going further to Z=N+4, we have (^{4}He,^{8}He) experiments of Robertson and collaborators[10] who measured ^{8}C and ^{24}Mg with about 200 keV uncertainty. Experiments by Trentelman, Preedom, and Kashy,[11] Proctor, et al.,[12] Benenson, et al.,[13] Nann and collaborators[14] and Mueller and collaborators[15] at our laboratory have yielded what are at present the most precise results for masses of proton rich nuclei from A=13 to A=55, using the reactions (^{3}He,^{6}He), (p,^{6}He,) and (^{3}He,^{8}Li).

Our measurements of masses of new proton rich nuclei far from stability consist of comparing the magnetic rigidities of the reaction products in 2-body reactions where in one reaction a well known mass is reached and in the other, the unknown is produced. The pulsed beam of the Michigan State University Cyclotron induces the reaction. The reaction products emitted at an angle θ enter a magnetic spectrograph and are momentum analyzed and focussed on the focal plane. Detection, location and identification of the particles is done in resistive-wire proportional-counter followed by a plastic scintillator. Since the reactions leading to nuclei far from stability have a very small cross section, a key problem is that of particle identification. Also, since momenta are compared, the incident beam energy must be accurately determined and the magnetic field of the spectrograph calibrated. These three problems will now be discussed.

Particle Identification: Table I shows a set of multinucleon transfer reaction and gives an idea of the Q-values and peak cross section. The values are typical and show strong dependence on the number of particles transferred and on the symmetry of the transferred group. These are very small yields and require a high degree of particle discrimination. The combination of time-of-flight from target to detector in the spectrograph, energy loss in the gas counter and pulse height produced in the backing detector results in that identification. For example, the time-of-flight spectrum measured during a recent investigation using the (^{3}He,^{8}Li), reaction[16] is shown in Fig. 1. It is seen that the (^{3}He,^{3}He), (^{3}He,d), (^{3}He,p), and (^{3}He,^{4}He) are the most prolific reactions. Of the particles groups shown in Fig. 1, the ^{7}Li group is the weakest; yet it is still 20 times stronger than the ^{8}Li yield of interest. Only one ^{8}Li event, located at ∿97ns, would be expected

ΔA	REACTION	ΔN	ΔP	ΔT$_z$	Q-values[†] MeV	~dσ/dΩ[*] μb/sr
3	(p, ^{4}He)	2	1	1/2	−7	50
	(^{3}He, ^{6}Li)	1	2	−1/2	−11	50
	(^{3}He, ^{6}He)	3	0	3/2	−27	1
4	(^{3}He, ^{7}Be)	2	2	0	−8	10
	(^{3}He, ^{7}Li)	3	1	1	−21	2
	(^{4}He, ^{8}He)	4	0	2	−60	0.01
5	(^{3}He, ^{8}B)	2	3	−1/2	−20	0.2
	(^{3}He, ^{8}Li)	4	1	3/2	−33	0.1
	(p, ^{6}He)	4	1	3/2	−37	0.1
	(^{3}He, ^{8}He)	5	0	5/2	−66	0.001 ≤ (not seen)
6	(^{3}He, ^{9}Li)	5	1	2	−51	(0.01?)

[*] 75 MeV ^{3}He
45 MeV p

[†] for ^{24}Mg target

Table I.--Typical peak cross sections for multi-particle transfer reactions

on the same scale. Figure 1 also shows that the time-of-flight for deuterons, ^{4}He, and ^{6}Li is the same for the same magnetic rigidity. Since time does not differentiate between them, and since their energies are in ratios 1:2:3, an energy measurement would easily differentiate them. The plastic scintillator which we use most of the time is however very non-linear; interestingly, for a given Bρ value, its light output is found to be the same for particles with the same Z/A, i.e., d, ^{4}He, and ^{6}Li give about the same pulse height, and likewise for t and ^{6}He. This results in an increased simplicity in setting time gates, since ^{6}He gates can be set using the far more abundant tritons without having to worry about a time walk introduced by pulse height differences between t and ^{6}He. The ambiguity in particle identification is resolved by the difference in specific ionization of the particles in the gas of the proportional counter. Figure 2 shows two-dimentional data where the "ΔE" in the gas is on the ordinate while position in the focal plane is the abscissa. The deuterons, alphas and ^{6}Li, preselected by time of flight, are clearly separated.

Incident Beam Energy: The beam energy can be precisely obtained by the momentum matching method described by Trentelman and Kashy.[17] Table II shows a series of matched reactions for protons, deuterons and ^{3}He.

We note the highly precise Bρ value at the matched point, and the relative lack of sensitivity of that Bρ value on detection angle. At the matched condition, the same B field is seen by both

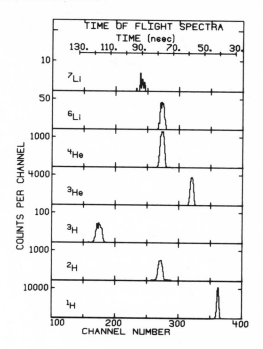

Figure 1.--Time of flight from target to spectrograph focal plane.
Note the increased width of peaks as time increases reflecting the
angular acceptance of the spectrograph.

$$E_{3_{He}} = 76.4 \text{ MeV}$$
$$\theta = 10.°$$

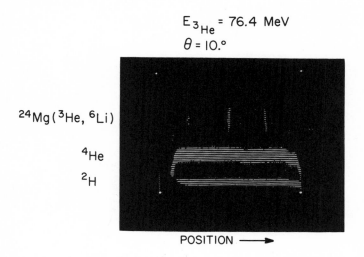

Figure 2.--Specific ionization vs. position in a resistive wire
proportional detector. The beam energy is 76.4 MeV and the angle
10°.

E (MeV)	dE/dθ keV/deg	Bρ Kgauss-cm	dBρ/dθ gauss-cm/deg	Reaction I	E_x I MeV	Reaction II	E_x II MeV
Protons							
14.1471(11)	8.0	449.534(27)	31	$^{11}B(p,p)^{11}B$	4.4451(5)	$^{11}B(p,d)^{10}B$	0.
16.4871(11)	10.4	548.736(21)	34	$^{11}B(p,p)^{11}B$	2.1247(4)	$^{11}B(p,d)^{10}B$	0.
18.6727(11)	12.6	626.614(18)	36	$^{11}B(p,p)^{11}B$	0.	$^{11}B(p,d)^{10}B$	0.
20.9199(20)	7.8	556.523(39)	17	$^{16}O(p,p)^{16}O$	6.1307(2)	$^{16}O(p,d)^{15}O$	0.
27.2454(21)	12.0	758.946(28)	20	$^{16}O(p,p)^{16}O$	0.	$^{16}O(p,d)^{15}O$	0.
28.9045(23)	16.4	718.093(33)	38	$^{12}C(p,p)^{12}C$	4.4391(3)	$^{12}C(p,d)^{11}C$	4.4391(3)
33.0022(25)	18.9	776.813(33)	41	$^{12}C(p,p)^{12}C$	4.4391(3)	$^{12}C(p,d)^{11}C$	2.0000(5)
33.5576(23)	20.6	843.397(28)	41	$^{12}C(p,p)^{12}C$	0.	$^{12}C(p,d)^{11}C$	0.
37.6878(25)	23.2	894.755(28)	44	$^{12}C(p,p)^{12}C$	0.	$^{12}C(p,d)^{11}C$	2.0000(5)
38.7713(33)	22.5	853.017(40)	45	$^{12}C(p,p)^{12}C$	4.4391(3)	$^{12}C(p,d)^{11}C$	4.8042(12)
43.5019(34)	26.9	962.748(36)	48	$^{12}C(p,p)^{12}C$	0.	$^{12}C(p,d)^{11}C$	4.8042(12)
46.9857(37)	29.1	1001.458(38)	50	$^{12}C(p,p)^{12}C$	0.	$^{12}C(p,d)^{11}C$	6.4782(14)
50.3795(42)	31.2	1037.905(41)	52	$^{12}C(p,p)^{12}C$	0.	$^{12}C(p,d)^{11}C$	8.1045(17)
80.5579(143)	50.3	1253.520(49)	92	$^{27}Al(p,\alpha)^{24}Mg$	4.1228(1)	$^{27}Al(p,^6He)^{22}Mg$	0.
86.8029(143)	55.1	1323.760(47)	96	$^{27}Al(p,\alpha)^{24}Mg$	1.3659(1)	$^{27}Al(p,^6He)^{22}Mg$	0.
93.9275(144)	60.2	1387.235(45)	100	$^{27}Al(p,\alpha)^{24}Mg$	0.	$^{27}Al(p,^6He)^{22}Mg$	1.2470(4)
Deuterons							
16.1938(30)	12.0	644.222(50)	65	$^{16}O(d,d)^{16}O$	6.1307(2)	$^{16}O(d,t)^{15}O$	0.
28.6829(32)	26.6	1096.203(30)	86	$^{16}O(d,d)^{16}O$	0.	$^{16}O(d,t)^{15}O$	0.
35.0181(95)	29.8	1099.145(90)	97	$^{16}O(d,d)^{16}O$	6.1307(2)	$^{16}O(d,t)^{15}O$	6.1770(30)
47.7869(99)	44.5	1418.469(70)	113	$^{16}O(d,d)^{16}O$	0.	$^{16}O(d,t)^{15}O$	6.1770(30)
63.4877(64)	84.1	1637.242(39)	241	$^{12}C(d,d)^{12}C$	0.	$^{12}C(d,t)^{11}C$	8.1045(17)
^{3}He							
69.2721(156)	71.8	1236.983(62)	124	$^{27}Al(^3He,^4He)^{26}Al$	3.1594(10)	$^{27}Al(^3He,^6He)^{24}Al$	0.
75.9213(155)	79.9	1317.227(57)	130	$^{27}Al(^3He,^4He)^{26}Al$	0.	$Al(^3He,^6He)^{24}Al$	0.

Table II.—Incident Beam Energy and Reaction Product Momentum Bρ for Momentum-Matched Reaction Pairs at $\theta_L=10°$

particles and thus the results are not dependent upon detailed
knowledge of the spectrograph. We not only get beam energy but also
can calibrate the spectrograph at the same time. Of course, a
knowledge of the dispersion of the spectrograph allows excellent
measurements of the beam energy at points far from the matched
condition. Furthermore, having for example calibrated the analyzing
magnet at 63.488 MeV with deuterons (see Table II) this same magnet
would then pass 127.76 MeV ^{4}He or 167.48 MeV ^{3}He. While the
ultimate accuracy of the method is limited by the knowledge of masses
involved, a number of experimental problems have to be considered
before this limit can be achieved. The angle determination can
usually be accomplished via scattering from the hydrogen almost
always present in the target. Of more importance is the determination of the true position of the particle peaks in the focal plane
as opposed to their electronically derived position. In high yield
reactions, the measurements can be with plates, as by Nolen, et al.[18]
and excellent results are obtained. For very low yield reactions,
care must be taken to insure that the electronically derived positions
are correct, independent of particles involved. In our gas detector,
we simply change the gas gain so pulses from all particles groups
being compared are of the same size when their position is being
determined, thus assuring the accuracy of their relative positions
in the spectrograph focal plane.

Spectrograph Magnetic Field: The momentum analyzing characteristics
of the Enge 90 cm split pole spectrograph have been calibrated for
the region of the focal plane where both calibration and unknown
particle groups are placed. The beam energy was adjusted for momentum
matched reactions, and then a series of other reactions, placed in
turn at the same position on the focal plane. As the field was
increased a proton NMR probe measured the magnetic field in a flat
field region of the spectrograph. In all the measurements, a cycling
procedure was used and the field set on the increasing current side
of the cycle.

It was found that the magnetic field along the path responsible
for the bending, $\int B dl$, was, in spite of the recycling procedure,
sensitive to the rate at which the spectrograph field energizing
current was increased. This could be ascertained by recycling and
then increasing the field directly to the value where the momentum
matched reactions were observed. The procedure was then repeated,
allowing the spectrograph to remain at several intermediate field
values as the field was increased. In all this, the matched reactions insured that the beam energy and the $B\rho$ values remained
unchanged. The problem was clearly the edge fields, and we have
found that by limiting the rate of increase of the current (15 min.
from zero to 16 kG field), reproducibilities at the 1/20,000 level
can be then obtained on successive recyclings. This effect is still
not neglible and we plan to investigate other recycling procedures.

Error Analysis: Table III shows an error analysis, in which the uncertainties in a determination of the [16] ^{19}Na mass relative to that of ^{20}Na are shown. A number of these errors are reduced by increasing the number of measurements.

	Parameter	Uncertainty	Effect on ^{19}Na Mass Excess (keV)
1.	Beam energy(absolute)	30 keV	6.6
2.	Beam energy(fluctuation)	16 keV	8.0
3.	Detection angle	0.15°	3.0
4.	Angular yield dependence	0.25°	5.0
5.	Peak centroids	0.3 channels	6.3
6.	^{20}Na mass	6 keV	4.7
7.	Target thickness	10%	< 1.0
8.	B field scaling		< 1.0
9.	B field reproducibility	1/20,000	4.0
10.	Counter linearity		< 1.0
	Overall uncertainty		15 keV

Table III.--Uncertainties in a determination of ^{19}Na mass excess. The ^{24}Mg(^{3}He,^{8}Li)^{19}Na reaction is compared with ^{24}Mg(^{3}He,^{7}Li)^{20}Na reaction using 76 MeV ^{3}He at a laboratory angle of 10°

Results: Precise measurements of nuclei in the 1p and 2s-1d shell have provided the data for the detailed test of the isobaric multiplet mass equation as reported at this conference by Benenson.[19] Recently measured masses of all nuclei with Z=N+1 in the 1f$_{7/2}$ shell have provided data for obtaining Coulomb displacement energies for these high Z mirror nuclei. At the same time, by using the Kelson-Garvey[20] symmetric mass relation which has proven especially successful when used in a single shell, we can predict the mass of a number of yet unobserved nuclei. (See Table IV). It will be a challenge to produce, observe, and measure properties of exotic nuclei such as ^{48}Ni.

^{54}Ni - 39.28	^{50}Ni - 4.15
^{53}Ni - 29.69	^{49}Ni + 7.58
^{52}Ni - 22.65	^{48}Ni +16.41
^{51}Ni - 12.04	

Table IV.--Predicted mass excess (in MeV) of proton rich Ni isotopes

REFERENCES

1. D.K.Scott, B.G.Harvey, D.L. Hendrie, L.Kraus, C.F.Maguire, J. Mahoney, Y.Terrien, and K.Yagi, Phys. Rev. Lett., 33 1343(1974).

2. D.R.Goosman and D.E.Alburger, Phys. Rev. C6, 820(1972).

3. D.E.Alburger, D.R.Goosman, C.N.Davids and J.C.Harvey,V Conference on Atomic Masses and Fundamental Constants, Paris 1975.

4. E.R.Flynn and J.D.Garrett, Phys. Rev. C9, 210(1974).

5. R.Klapisch, R.Prieels, C.Thibault, A.M.Poskanzer, G.Rigaud and E.Roeckl, Phys. Rev. Letters 31, 118(1973).

6. K.H.Wilcox, N.A.Jelley, G.J.Wozniack, R.B.Weisenmiller, H.L. Harney and J.Cerny, Phys. Rev. Letters 30, 866(1973).

7. A.G.Artukh, G.F.Gridnev, V.L.Mikheev, V.V.Volkov and J.Wilczynski, Nucl. Phys. A192, 170(1972).

8. J.M. Mosher, R.W.Kavanagh and T.A.Tombrello, Phys. Rev. C3, 438(1971).

9. R. Mendelson, G.J. Wozniak, A.D.Bacher, J.M.Loiseaux and J. Cerny, Phys. Rev. Letters 25, 533(1970).

10. R.G.H.Robertson, S.Martin, W.R.Falk, D.Ingahm, and A. Djaloeis, Phys. Letters 32, 1207(1974).

11. G.F.Trentelman, B.M.Preedom and E.Kashy, Phys. Rev. Letters 25, 530(1970); Phys. Rev. C3, 2205(1971).

12. I.D.Proctor, W.Benenson, J.Driesbach, E.Kashy, G.F.Trentelman and B.M.Preedom, Phys. Rev. Letters 29, 434(1972).

13. W.Benenson, E.Kashy, and I.D.Proctor, Phys. Rev. C8, 210(1973).

14. H.Nann, W.Benenson, E.Kashy and P.Turek, Phys. Rev. C9, 1848(1974)

15. D.Mueller, E.Kashy, W.Benenson and H.Nann, Phys. Rev (July 1975).

16. W.Benenson, A.Guichard, H.Nann, E.Kashy, D.Mueller and L. Robinson, to be published.

17. G.F.Trentelman and E.Kashy, Nucl. Instr. and Meth. 82, 304(1970).

18. J.A.Nolen, Jr., G.Hamilton, E.Kashy and I.D.Proctor, Nucl. Instr. and Meth. 115, 189(1974).

19. W.Benenson and E.Kashy, V Conference on Atomic Masses and Fundamental Constants, Paris 1975.

20. I. Kelson and G.T.Garvey, Phys. Letters 23, 689(1966).

SOME (p,n) AND (α,n) REACTION ENERGIES RELEVANT TO

SUPERALLOWED BETA DECAY

J.M. Freeman and R.J. Petty
Nuclear Physics Division, AERE Harwell, Didcot, U.K.
S.D. Hoath and J.S. Ryder
Nuclear Physics Laboratory, Oxford, U.K.
W.E. Burcham and G.T.A. Squier
Physics Department, University of Birmingham, U.K.

1. INTRODUCTION

Continuing interest in superallowed beta decay in relation to weak interaction theory demands ever-increasing accuracy in measurements of the beta decay end-point energies. These in many cases are based on threshold measurements for (p,n), (α,n), or (^{3}He,n) reactions. Since these measurements are also used as a source of precise mass values for the parent nuclei, it seems appropriate to present here the results of recent Harwell work, and to discuss the reliability of these and other similar measurements which have been used to extract mass differences and beta decay data.

The stability and good energy resolution of electrostatic accelerators permit precise relative yield curves to be obtained near threshold for (p,n) and similar reactions, usually in terms of the frequency of the NMR probe of the accelerator's analysing magnet. The yield curve is extrapolated to the axis to give a threshold point, at which the beam energy is then calibrated.

The problems involved in making and calibrating threshold measurements have been reviewed a number of times (e.g. ref. (1)), so we summarise them here only briefly, as they affect the present discussion.

2. DETERMINATION OF THE THRESHOLD POINT

The usual procedure is to assume s-wave neutron emission just above threshold with a reaction cross-section proportional to $(\Delta E)^{\frac{1}{2}}$, where ΔE equals the bombarding energy minus the threshold energy. For a thick target the integrated reaction yield Y is then

proportional to $(\Delta E)^{3/2}$. A plot of $Y^{2/3}$ versus energy is thus
assumed to be linear, and an extrapolation to the energy axis is
taken to give the threshold point. This procedure, if readily
reproducible, is satisfactory for accelerator calibration purposes
but, as has frequently been emphasised (1), its use for deriving
accurate Q-values can lead to significant errors. Sources of error
include: the Lewis effect (fluctuations in energy loss in the
target), finite beam energy spread, and, above all, resonance effects.

The first two effects have been considered in some detail for
thresholds up to a few MeV (2), and we have made some calculations
for higher energy cases (3,4). A small positive correction to the
$Y^{2/3}$ extrapolation point is indicated, of magnitude $\sim$ 0.1 to 0.05%,
depending on beam energy resolution and extrapolation range used.
This correction, frequently neglected, could become a significant
systematic error when a number of separate threshold measurements
for a particular reaction are grouped together to obtain a more
accurate mean value. For example, for $^7Li(p,n)$ the correction is
already twice the standard deviation of the mean value quoted in
the 1971 Mass Tables (5).

The third effect mentioned above, resonances in the compound
nucleus, may seriously affect the shape of the $Y^{2/3}$ curve, or even,
depending on the particular location of resonances in the vicinity
and the detection sensitivity, lead to the true threshold being
missed altogether. To take some examples from thresholds studied at
Harwell: the $^{14}N(p,n)^{14}O$ reaction yield gave a $Y^{2/3}$ plot which
appeared to be linear over a range of at least 30 keV (6), implying
no serious resonance effects. On the other hand, in the case of
$^{26}Mg(p,n)^{26}Al^m$, the shape of the yield curve over a 10 keV range
could be fitted only by assuming a narrow resonance close to
threshold and another about 10 keV higher (3). The $^{54}Fe(p,n)^{54}Co$
yield curve in an early Harwell measurement was apparently linear in
$Y^{2/3}$, but our improved detection method (multiscaler accumulation of
the decays between beam bursts to allow extraction of the ^{54}Co
component from background) has shown (4) that the earlier observation
was of the onset of a sharp resonance and that the yield curve
extends below this, extrapolating to a threshold point about 8 keV
lower in energy. A similar situation was found in the case of
$^{35}Cl(\alpha,n)^{38}K^m$.

3. ENERGY CALIBRATION

Three methods have been employed for determining the absolute
bombarding energy corresponding to the point interpreted as the
threshold point.

a) Absolute calibration, by electrostatic or magnetic deflection,
or by a velocity measurement. Absolute methods are the most
difficult but in principle the most satisfactory, though not immune

from systematic error.

b) Measurement relative to the α-particles from a radioactive
source. This is the Harwell method. It takes advantage of the very
precise absolute measurements by Rytz (8) of the energies of
α-particles from ^{212}Bi(6090.087 + 0.037 keV) and ^{212}Po(8784.37 +
0.07 keV). The comparison is made using a magnetic spectrograph
into which the beam at the threshold energy is scattered from a
thin gold foil, in a well-defined geometry. Since the accepted
values for the α-particle standards changed slightly in 1969,
earlier Harwell threshold values have had to be adjusted accordingly.

c) Measurement relative to other (p,n) thresholds. This method
seems the least satisfactory since the "standards" are frequently
adjusted, and the measurements of the reference thresholds are
subject to the problems mentioned in section 2.

Inconsistencies can arise in threshold calibration, for example
in the case of ^{27}Al(p,n)^{27}Si, which also illustrates the hazards
that beset compilers of mass tables. The eight ^{27}Al(p,n) Q values
used by Wapstra and Gove (5) for the 1971 mass evaluation are
reproduced in column 2 Table 1 with the sources of the threshold
measurements from which they were calculated. The weighted mean was
adopted for the Mass Table. First, two of these data should be
eliminated: number IV is a preliminary Harwell result, quoted only
in the APS Bulletin and subsequently superseded by a Harwell value
given as datum VIII in Table 1; number V does not represent an
additional measurement but is a value recommended by Marion in 1963,
being the average of five results, four of which are already
represented in Table 1. Of the remaining six data the first three
are not expected to change significantly with revised calibration
energies; our attempt to revise VI is shown in column 4. Marion
(1) has adjusted number VII, changing the error to an R.M.S. value
rather than the author's arithmetic sum of experimental errors.
Number VIII, the Harwell measurement, requires adjustment to the
most recent value (8) for the ^{212}Bi α-particle energy. The revised
Q-values given in column 4 Table 1 lead to the weighted mean shown,
but the data are not very satisfactory since the two most precise
and acceptable (1) results differ by 8.8 keV.

4.   THRESHOLD MEASUREMENTS AND Q-VALUES

The results of recent Harwell threshold measurements are given
in Table 2, together with five earlier results adjusted from
published values to take account of the newer values for the ^{212}Po
and ^{212}Bi α-energies (8). On the assumption that the results
represent true threshold points, the equivalent Q-values are also
presented. A result for the Q-value of ^{35}Cl(α,p)^{38}Ar is given
since this was measured at the same time as the threshold for
^{35}Cl(α,n)^{38}K^m. The Q-value for ^{38}Ar(p,n)^{38}K^m, which is relevant

TABLE 1

^{27}Al$(p,n)^{27}$Si Q-values from Wapstra and Gove (5)

Source	$|Q|$ -value used (keV)	Calibration method	Revised $|Q|$ (keV)
I	5583 $\pm$ 10	other thresholds	5583 $\pm$ 10
II	5591 $\pm$ 8	other thresholds	5591 $\pm$ 8
III	5597.5 $\pm$ 6.0	other thresholds	5597.5 $\pm$ 6.0
IV	5596.4 $\pm$ 3.9	^{212}Bi (tentative)	–
V	5588.7 $\pm$ 4.8	recommended mean	–
VI	5593.6 $\pm$ 4.3	other thresholds	5592 $\pm$ 5
VII	5585.1 $\pm$ 4.5	absolute magnetic	5585.3 $\pm$ 2.3
VIII	5593.1 $\pm$ 3.3	^{212}Bi (replaces IV)	5594.1 $\pm$ 3.2

Weighted mean	5592.1		5589.3
int. error	1.7		1.6
ext. error	1.5		2.0

I Kington, Phys. Rev. A99, 1393 (1955); II Bromley, Can. J.Phys.
A37 1514 (1959); III Okano, J. Phys. Soc. Jap. A18 1563 (1963);
IV Montague, Bull. APS, A8 115 (1963); V. Marion, Conf. Nucl.
Masses, Vienna 279 (1964); VI Kuan, Nucl. Phys. 64 524 (1965);
VII Bonner, Nucl. Phys. A86 187 (1966); VIII Murray, Conf. Nucl.
Masses, Winnipeg (1967).

to superallowed Fermi decay, was deduced from the mean of the differences between each pair of $(\alpha,p),(\alpha,n)$ Q-value measurements (7).

In view of the foregoing discussions, how confident can we be
in these Q-value results, from the point of view of (a) resonance
effects, and (b) calibration accuracy? Some comparisons with
other data help to some extent to answer these questions.

In the case of ^{14}N$(p,n)^{14}$O, comparison is possible only with
a similar threshold reaction, ^{12}C$(^{3}$He,n$)^{14}$O. Bardin et al. (9)
coupled this with measuring the Q-value for ^{12}C$(^{3}$He,p$)^{14}$N*, using
^{7}Li$(p,n)^{7}$Be as the standard, and the result gave an equivalent
Q-value for ^{14}N$(p,n)^{14}$O just consistent with our result. Alternatively, averaging three ^{12}C$(^{3}$He,n$)$ measurements and accepting mass
doublet data for ^{12}C-^{14}N gives, according to Wapstra and Gove (5),
the Q-value -5927.24 ± 0.35 keV, in good agreement with our value,
although the error may not adequately reflect systematic uncertainties in the $(^{3}$He,n$)$ threshold determinations and the mass
doublet data.

A good comparison, independent of threshold reactions, and
therefore testing both our method of deriving a true threshold

point and the calibration procedure, is provided in the case of $^{26}Mg(p,n)^{26}Al^m$ by a precision measurement (10) of the Q-value difference between $^{25}Mg(n,\gamma)^{26}Mg$ and $^{25}Mg(p,\gamma)^{26}Al^m$. This leads to the result for the (p,n) reaction shown in Table 2 column 4, and gives satisfactory confirmation of our result.

For $^{34}S(p,n)^{34}Cl$ no such direct comparison is possible, an accurate Q-value being available for $^{33}S(p,\gamma)^{34}Cl$ but not for $^{33}S(n,\gamma)^{34}S$. However reaction data from ref. (11) together with mass doublet data for $^{32}S-^{35}Cl$ from ref. (5) can be used for the chain $^{34}S(n,\gamma)^{35}S(\beta-)^{35}Cl(p,\alpha)^{32}S(n,\gamma)^{33}S(p,\gamma)^{34}Cl$ to give for $^{34}S(p,n)^{34}Cl$ the Q-value -6276.8 + 2.0 keV. Considering the number of data involved, this value is reasonably close to our experimental result (-6271.9 ± 1.9 keV). An important point is that since it corresponds to a threshold value slightly higher than our result it confirms that we did not miss the true threshold region through lack of resonance yield. With this confirmation the $^{34}S(p,n)^{34}Cl$ Q-value result from the threshold measurement can be adopted with reasonable confidence.

The threshold measurement for $^{35}Cl(\alpha,n)^{38}K^m$ can also be compared with other data although the overall accuracy is poorer. The Q-value for $^{35}Cl(\alpha,p)^{38}Ar$ is well determined from mass doublet data for $^{37}Cl-^{35}Cl$ and a $^{37}Cl(p,\gamma)$ measurement (5), in excellent agreement with our result (see Table 2). Reactions connecting ^{38}Ar with ^{38}K via ^{39}K and ^{40}Ca then predict (5,11) a Q-value for $^{35}Cl(\alpha,n)^{38}K^m$. This again corresponds to a threshold higher than our measurement.

There are no other cases (except $^{30}Si(p,n)^{30}P$, see Table 2) where our results can be checked against data independent of threshold measurements. However for the (p,n) thresholds in ^{50}Cr and ^{54}Fe a comparison can be made with work by Hardy et al. (12). They compared the Q-values for $^{50}Cr(^3He,t)^{50}Mn$ and $^{54}Fe(^3He,t)^{54}Co$ with that for $^{27}Al(^3He,t)^{27}Si$, which is based on the $^{27}Al(p,n)$ threshold. Their results (shown in Table 2) indicate that resonance effects have not prevented the correct location of the threshold regions in our recent experiments. (At the same time their work suggests that our early measurements (13) on $^{42}Ca(p,n)$ and $^{46}Ti(p,n)$ are possibly in error on that account.) However, the systematic accuracy of their results is in doubt by a few keV because of uncertainty in the $^{27}Al(p,n)^{27}Si$ Q-value. They combined the mean value from ref. (5) (discussed in section 3) with a value derived from the threshold energy quoted by a Yale group (14). However this, being one of a self-consistent set intended for accelerator calibration purposes, is not reliable as a Q-value standard: it is partly dependent on an earlier $^{27}Al(p,n)$ measurement, partly on the old $^{54}Fe(p,n)$ threshold now shown to be in error, and partly on ^{58}Ni and $^{60}Ni(p,n)$ thresholds which may be subject to resonance effects. However, the Chalk River work can in fact be used to derive an independent result for the $^{27}Al(p,n)^{27}Si$

TABLE 2

Harwell threshold and Q-value results

Reaction	Threshold (keV)	$|Q|$ value (keV)	Comparison data $|Q|$ (keV)	ref.
$^{14}N(p,n)^{14}O$	6355.6 ± 1.6	5927.6 ± 1.5	5929.8 ± 1.2	(9)
			5927.24 ± 0.35	(5)
$^{26}Mg(p,n)^{26}Al^m$	5209.4 ± 1.6	5014.4 ± 1.5	5015.58 ± 0.57	(10)[†]
$^{34}S(p,n)^{34}Cl$	$6458.6 \pm 2.0*$	6271.9 ± 1.9	6276.8 ± 2.0	(11,5)[†]
$^{35}Cl(\alpha,n)^{38}K^m$	6674.8 ± 3.2	5989.3 ± 2.9	5999 ± 8	(5,11)[†]
$^{35}Cl(\alpha,p)^{38}Ar$		837.2 ± 2.4	837.1 ± 0.9	(5)[†]
$^{38}Ar(p,n)^{38}K^m$		6825.7 ± 3.4		
$^{50}Cr(p,n)^{50}Mn$	8586.7 ± 1.9[‡]	8416.1 ± 1.9[‡]	8412.3 ± 2.1	(12)
$^{54}Fe(p,n)^{54}Co$	9192.4 ± 1.8	9023.0 ± 1.8	9025.0 ± 3.1	(12)
$^{10}B(p,n)^{10}C$	4880.8 ± 1.6	4433.5 ± 1.5	4432.8 ± 0.6	(15)
$^{27}Al(p,n)^{27}Si$	5803.7 ± 3.3	5594.1 ± 3.2	see table 1	
$^{30}Si(p,n)^{30}P$	5181 ± 5	5012 ± 5	5012.7 ± 3	(11)[†]
$^{35}Cl(p,n)^{35}Ar$	6942.3 ± 1.6	6747.2 ± 1.6		
$^{58}Ni(p,n)^{58}Cu$	9511.1 ± 4.1	9347.7 ± 4.0	9352.8 ± 3.4	(1)

[†]comparison data independent of a threshold measurement
*mean of $^{34}S(p,n)$ and combined $^{31}P(\alpha,n)$ and $^{31}P(\alpha,p)$ measurements
[‡]preliminary result

Q-value in terms of the well-established (10) measurement for $^{26}Mg(p,n)^{26}Al$. From their published data it follows that $|Q|$ [$^{27}Al(p,n)^{27}Si$] = 5595.3 $\pm$ 2.1 keV. This is 2.8 keV greater than the value 5592.5 $\pm$ 0.9 keV used by Hardy et al.

5. SUMMARY

Threshold measurements, although potentially a very accurate method for deriving Q-values and mass differences, are susceptible to systematic experimental errors, particularly due to resonance effects. However, since improving our method of detecting very low reaction yields close to the threshold points, and making careful studies of the shapes of the yield curves, we have obtained results which, to the extent that they can be checked by other methods, are reliable within their estimated errors. Further checks on measurements (e.g. $^{27}Al(p,n)$) previously made with a less sensitive detection system would be desirable.

References

1.   J.B. MARION, Rev. Mod. Phys. 38, 660 (1966).

2.   R.O. BONDELID and E.E. WHITING, Phys. Rev. 134, B591 (1964); D.W. PALMER et al. Nucl. Phys. 75, 515 (1966).

3.   J.M. FREEMAN, J.G. JENKIN, G. MURRAY, D.C. ROBINSON and W.E. BURCHAM, Nucl. Phys. A132, 593 (1969).

4.   S.D. HOATH, R.J. PETTY, J.M. FREEMAN, G.T.A. SQUIER and W.E. BURCHAM, Phys. Lett. 51B, 345 (1974).

5.   A.H. WAPSTRA and N.B. GOVE, Nucl. Data Tables 9, 357 (1971).

6.   J.M. FREEMAN, J.G. JENKIN, D.C. ROBINSON, G. MURRAY and W.E. BURCHAM, Phys. Lett. 27B, 156 (1968).

7.   G.T.A. SQUIER, W.E. BURCHAM, J.M. FREEMAN, R.J. PETTY, S.D. HOATH and J.S. RYDER, Nucl. Phys. A242, 62 (1975).

8.   A. RYTZ, Nucl. Data Tables 12, 487 (1973).

9.   R.K. BARDIN, C.A. BARNES, W.A. FOWLER and P.A. SEEGER, Phys. Rev. 127, 583 (1962).

10. P. DE WIT and C. VAN DER LEUN, Phys. Lett. 30B, 639 (1969).

11. P.M. ENDT and C. VAN DER LEUN, Nucl. Phys. A214, 1 (1973).

12. J.C. HARDY, G.C. BALL, J.S. GEIGER, R.L. GRAHAM, J.A. MACDONALD and H. SCHMEING, Phys. Rev. Lett. 33, 320 (1974).

13. J.M. FREEMAN, G. MURRAY and W.E. BURCHAM, Phys. Lett. 17, 317 (1965).

14. J.C. OVERLEY, P.D. PARKER and D.A. BROMLEY, Nucl. Instr. Meth. 68, 61 (1969).

15. D.C. ROBINSON and P.H. BARKER, Nucl. Phys. A225, 109 (1974).

Accurate Q-value Measurements and Masses in the Iron region *

P. L. Jolivette, J. D. Goss, G. L. Marolt, A. A. Rollef-
son, and C. P. Browne
Department of Physics, Univ. of Notre Dame
Notre Dame, Indiana

At AMCO 3 I reported on a series of 15 nuclear reaction Q-
values[1] giving values for 9 masses below mass 16. These charged
particle measurements were made with the Notre Dame 50 cm broad-
range spectrograph and 4 MeV accelerator. The problem of differential hysteresis in the spectrograph magnetic field was recognized
and a random error assigned to each measurement to cover the consequent uncertainty. We now have a much larger spectrograph designed especially for energy measurements of high absolute accuracy
and an FN tandem accelerator. Today I shall review the features of
this new spectrograph, describe an advance which allows us to make
much of the differential hysteresis effect systematic and hence
correct for it, give comparisons of results showing improved accuracy of energy measurements, and report on a series of 36 Q-values
and an adjustment of masses from Cr to Zn.

The description of the new instrument, which we call a 100 cm
modified broad range spectrograph, was published[2] and I shall just
summarize the features that are important for this work. Absolute
accuracy was the dominant design consideration so a single dipole
field is used in the energy measurement mode. The trajectory radius
for 90° deflection (ρ_0) is 100 cm, twice that of our older instrument. The exit face of the poles was modified to be partly an
arc of 100 cm and partly an arc of 350 cm. The effect is to maintain the low aberration of the standard broad range at the ρ_0
radius up to much larger radii and to slow the rapid increase in
dispersion above ρ_0 so that we can work out to 1.3 ρ_0 or more,
with high resolution. A useful measure of dispersion is that for a
10 MeV particle, 1 mm along the upper end of the focal surface is
3 keV. We can measure a group position to 0.1 mm and reproduce the
calibration to an accuracy of about this amount ($\Delta\rho = 0.05$ mm).

A photograph of the instrument is given in Ref.2. Total weight is about 31.7t and the length of focal surface is 260 cm, giving $E_{max}/E_{min} = 3.8$. Reference 2 also has a view of the plate holder in the camera box. Two sets of plates and one set of position-sensitive proportional counters may be loaded at one time. Plates are loaded in place and indexed with a fiber-optics indexing system after loading. Positioning of the plate holder for indexing and exposing of the various zones is remotely and semi-automatically controlled.

I now turn to our new calibration procedure. We find that the calibration of the new spectrograph is remarkably stable and reproducible with most calibration group positions reproducing to within 0.1 mm in D. The standard procedure was to set the magnet current at a fixed maximum value for five minutes, then reduce it quickly to the value for the highest-field calibration point, expose the ^{210}Po source, reduce field to the next setting, expose, and so on for the 130 points used to generate the "calibration curve." Q-values measured using this curve are very reproducible. However, when Q-values connecting nuclides in the iron region were combined to form loops, the closure errors although small, were found to be just within the maxima allowed by our estimates of individual errors. When we compared our (d,p) Q-values with accurately measured (n, γ) energies a systematic discrepancy of about our maximum estimated error appeared. We thus sought a small systematic error in our calibration and a way to reduce it. I shall first describe how this was done and then show the resulting improvement in accuracy.

With the 50 cm spectrograph we found[1] that when the magnetic field was recycled, the curve of plate position, D, vs. trajectory radius, ρ, often shifted by nearly the same $\Delta \rho$ at all D. We had been able to show[3] that the shape of the D(ρ) curve was essentially the same for a given set field as it was when the field was stepped during calibration using a ^{210}Po source. With the new instrument the accuracy is improved enough to measure the difference in effective field when a given NMR reading is set immediately from the high current field, as opposed to stepwise setting during calibration. We have shown[4] that the field B, can be related to the NMR frequency f, by

$$B = a f - (bf_i / f_i^2) f^2 \qquad (1)$$

where f_i is the first frequency set for a sufficient time (> 30 min) after recycling, and a and b are constants. The correction term describes two effects, first the fact that the field at the NMR probe is not the same as that averaged along the trajectory and second the fact that this difference depends on the mode of cycling. By cycling to various $f = f_i$, taking the difference between the measured ρ and that given by the stepwise calibration curve, and converting to ΔB, the ΔB vs. f plot shown in Fig. 1. of Ref. 4 was

Table 1.   Reduction of Uncertainties by 2 Source Calibration

| Quantity | Uncertainty | | | ΔE for E_R = 14.2 MeV | |
	One Source Calibration	Type	Two Source Calibration	^{210}Po only (keV)	$^{210}Po+$ ^{212}Po (keV)
Trajectory radius	0.040 mm	R	0.022 mm	0.9	0.48
Trajectory radius (whole-curve shift)	0.1 mm	S	0.050 mm	2.2	1.09
Magnetic field	1.5 G	R	0.25 G	10.5	1.75
Magnetic field	0.0 G	S	0.20 G	0.0	1.40
				10.8	2.54

made ($\triangle B$ is the second term in eq. 1 and the straight line jus-
tifies the form of this term and confirms the " $\triangle\rho$ shift" seen
earlier). To find the $\triangle B$ <u>scale</u> however, the original calibration
curve had to be corrected to eliminate the unknown $\triangle B$ when it was
measured. This was done by simultaneously recording the positions
of two particle groups of well known momenta. Let R = $(B\rho)_1/(B\rho)_2$
= ρ_1/ρ_2. This is independent of B so by displacing the calibra-
tion curve $\rho \rightarrow \rho + \triangle\rho$, we can adjust it to fit the known ρ_1/ρ_2. We used
the ^{212}Po and ^{210}Po α-particle rigidities measured by Rytz et al.[5]
Figure 2 of Ref. 4 shows that the measured R is independent of position. Thus the focal surface is defined by this procedure and the
$\triangle\rho$ in the original calibration curve is determined. The uncertainty in ρ correction is 0.05 mm and in B is 0.2 G. A check of the
validity of the calibration was made by simultaneously measuring
positions of Li^{++} and Li^{+++} ions scattered elastically from Au.
The ratio of the ρ values is the same as the expected ratio with an
uncertainty of 5 parts in 10^5.

The effect of the field correction on the uncertainties in
energy measurements is shown in table 1. Table 2 illustrates the
reduction in closure errors resulting from this two-source calibration. The consistency of our final results is seen in table 3, which
shows closure for reaction chains measured in this work. Clearly

Table 2.   Improvement in Closure Errors

Reaction Chain	Closure Error (keV)
Ni 58 → 59 ← 60	BEFORE Δ = -15.0 ± 7.2
Co 55 57	
Fe 54	AFTER Δ = 2.3 ± 4.5

Table 3.   Summation of Reaction Chain Closures

Reaction Chain	Δ(keV)	σ_T (keV)	σ_R(keV)
Ni ⟵ 62 Co 59⟶60	-1.6	2.8	2.3
Ni 58⟶59⟶60⟶61⟵62 Co 55 ⟶ 59 Fe 54 ⟶55⟵56	-1.1	4.9	3.7
Ni 58⟶59⟵60 Co 55 ⟶ 57 Fe 54	2.3	4.5	3.6
Ni 60⟶61⟵62 Co 58⟵59⟶60	-1.1	4.2	3.2
Ni 60⟶61⟵62 Co 5 ⟶ 59 Fe 54⟶55⟵56	-3.4	3.8	2.8
Co 59 Fe 56⟵57	0.9	3.7	3.0
Ni 60 Co 57 58⟵59 Fe 54⟶55⟵56	-3.9	4.2	3.2
Ni 60⟶61⟵62 Co 58⟵59	0.5	3.3	2.4

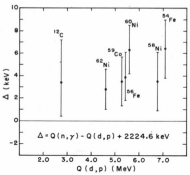

Fig. 1.   Comparison of (d,p) and (n,γ) Q-Values before Field Correlation.

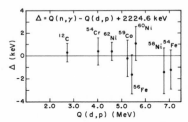

Fig. 2.   Comparison of (d,p) and (n,γ) Q-Values after Field Comparison.

our measurements are consistent and reproducible. The uncertainties in 27 Q-values range from 0.5 keV to 2.6 keV with the average 1.5 keV.

Next, we ask how our results compare with other accurate energy measurements? Figure 1 shows the original discrepancy between our (d,p) and other (n, γ) measurements which suggested the systematic differential hysteresis effect. These are results using the single ^{210}Po calibration and assuming the effective field as set for the

run is the same as that set in steps during calibration. There is
about a 4 keV systematic difference. Figure 2 shows the comparison
when the (d,p) energies are calculated using the field correction
I have just described. The small systematic error has disappeared
(< 0.3 keV).

A second comparison is offered from a set of excitation energies in ^{60}Co, measured by us using the (d,p) reaction and by others[6]
using gamma ray energy measurements. Figure 3 taken from a recently
submitted paper[7], shows typical examples of some 100 or more such
comparisons with accurate gamma ray energies. This also reinforces
the conclusion I stated[8] at AMCO 4 that the Po - α energy scale and
the gamma ray energy scale agree.

We made a further comparison of our energy scale with the recent absolute time of flight measurements done at Munich[9]. The
resonance energy for proton scattering from ^{12}C was measured against
the meter and second by the Munich group. Figure 1 of Ref. 10 shows
our resonance measurements. We kept the accelerator energy and beam
transport fields fixed and as steady as possible and swept over the
resonance by varying the potential of the target with a power supply.
Scattered protons were measured at 160° with a solid state detector.
The constancy of the beam energy was monitored by detecting protons
scattered from a thin backing of gold on the carbon target which
passed through a 1 mm slit on the focal surface of the spectrograph.
A change in momentum of 1/40,000 could be detected. Once the resonance was found the beam energy was measured by using the spectrograph to measure the energy of elastically scattered protons.
Details of fitting the data to find the resonance energy are in our
paper[10]. Our results is 14.2325 ± 0.0022 MeV which agrees with the
Munich result of 14.23075 ± 0.0002 MeV. Thus our energy measurements
with the new spectrograph using two absolute Po - α energy values of
Rytz, agree with an absolute time of flight measurement and also
with many gamma ray measurements based on annihilation energy and
crystal spacing scales. Absolute accuracy of the order of 1 keV is
possible

Finally I present the mass values deduced from our Q-value
measurements in the iron region. For details of the various adjustments see Ref. 4. Figure 4 compares the results with the 1971
mass adjustment[11]. Note the 15 keV shift in the ^{55}Co mass and the
18 keV shift in the ^{61}Co mass.

We also find the 1971 value for ^{55}Cr to be incorrect. The
uncertainties in the ^{53}Fe and ^{61}Co masses are reduced by about a
factor of 5. Very recently we have obtained the Q-values shown in
table 4. The mass of ^{57}Mn is found to be 140 keV different from
the 1971 adjustment and the ^{50}Cr(d,t)^{49}Cr Q-value is 61 keV lower.
It was surmised [11] that the older Ti(d,p) Q-values were systema-
tically high. We find this to be true and note that our values agree
very well with the (n,γ) results, especially those appearing since
the 1971 compilation. Perhaps other earlier (d,p) measurements

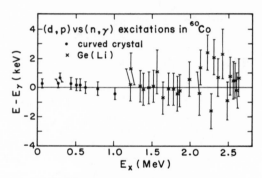

Fig. 3. Differences in ^{60}Co excitations found by gamma ray measurements and in the present (d,p) work.

suffered from the differential hysteresis effect. We note that (d,p) measurements are more sensitive to this than say, (p, α) measurements because usually the magnetic rigidity of the two particles is quite different.

Our new ^{49}Cr mass removes a discrepancy noted by Meuller et al.[12] in Coulomb energies of mirror nuclei in the f7/2 shell.

In conclusion I can say that our new spectrograph is giving unprecedented accuracy for charged particle energy measurements with the absolute accuracy comparable to that of good gamma ray

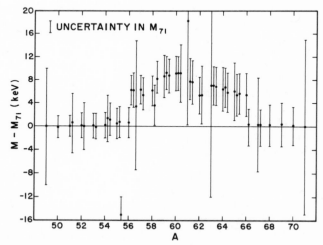

Fig. 4. Differences between 1971 mass adjustment and present mass values. Error bars are those of the 1971 table.

Table 4. Additional Q-values

Reaction	Q (keV)	Q_{71}(keV)	ΔQ (keV)
$^{63}Cu(p,\alpha)$ ^{60}Ni	3754.5± 1.5	3756.1±2.3	-1.6
$^{65}Cu(p,\alpha)$ ^{62}Ni	4344.6±1.8	4351.4±2.7	-6.8
$^{64}Zn(p,\alpha)$ ^{61}Cu	844.0 ±0.9	839.7±3.3	4.3
$^{66}Zn(p,\alpha)$ ^{63}Cu	1544.6 ±1.0	1557 ±5	-12.4
$^{46}Ti(d,p)$ ^{47}Ti	6654.3±1.7	6650.5±1.5	3.8
$^{48}Ti(d,p)$ ^{49}Ti	5918.4±1.7	5918.8±0.8	-0.4
$^{50}Ti(d,p)$ ^{51}Ti	4147.8±1.2	4153 ±5	-5.2
$^{50}Cr(d,t)$ ^{49}Cr	-6743.3±2.2	-6682 ±10	-61.3
$^{54}Cr(\alpha,p)$ ^{57}Mn	-4308 ±8	-4170 ±50	-138

energy measurements. Thirty-six Q-values in the iron region now are more accurate and with smaller uncertainties (mostly 1 to 3 keV). Further evidence for consistency of energy scales used in nuclear reaction work has been given.

References

* Supported by NSF grant GP 27456

1. BROWNE, C. P. and O'DONNELL, F.H. in Proceedings of the Third International Conference on Atomic Masses, edited by R. C. Barber (Univ. of Manitoba Press, Manitoba, 1967) pp. 508-524.
2. GOSS, J.D., ROLLEFSON, A.A., and BROWNE, C.P. Nucl. Instr. and Methods 109, 13 (1973).
3. STOCKER, H., ROLLEFSON, A.A., HREJSA, A.F. and BROWNE, C. P., Phys. Rev. 4, 930 (1971).
4. JOLIVETTE, P.L., GOSS, J. D., MAROLT, G.L., ROLLEFSON, A.A., and BROWNE, C.P. Phys. Rev. C 10, 2449 (1974).
5. RYTZ, A., GRENNBERG, B., and GORMAN in Atomic Masses and Fundamental Constants, 4 edited by J. H. Sanders and A. H.Wapstra (Plenum, London, 1972) p.1; GORMAN, D.T., RYTZ, A., C. R. Acad. Sci. B (Paris) 277, 29 (1973).
6. SMIRNOV, AI., SHAVUROV, V.A., ALEKSEEV, V.L. KAMINKER,D.M., and RYL'NIKOV, A.S., IZV. Akad. Nauk, CCCP 33, 1270 (1969); WASSON, O.A., CHRIEN, R.E., BHAT, M.R., LONE, M.A. and BEER, M., Phys. Rev. 176, 1314 (1968).
7. JOLIVETTE, P.L., GOSS, J.D., and BROWNE, C. P., Phys. Rev. C (submitted).
8. BROWNE, C. P., in Atomic Masses and Fundamental Constants 4, edited by J. H. Sanders and A. H. Wapstra (Plenum, London, 1972) p. 15-25.
9. HUENGES, E., ROSLER, H., and VONACH, H., Phys. Lett. 46B, 361 (1973).
10. GOSS,J.D., BROWNE, C.P., ROLLEFSON, A.A., and JOLIVETTE, P.L., Phys. Rev. C 11, 710 (1975).
11. WAPSTRA, A.H. and GOVE, N.B., Nucl. Data A9, 265 (1971); A11, 128 (1972).
12. MUELLER, D., KASHY, E., BENENSON, W. and NANN, H. Annual Report 1973-74 Cyclotron Laboratory, Michigan State Univ. p.1.

PRECISION ENERGY MEASUREMENTS WITH THE MSU CYCLOTRON[*]

J.A. Nolen, Jr.

Cyclotron Laboratory and Physics Department
Michigan State University
East Lansing, Michigan 48824,   USA

This paper describes the technique currently being used at the
Michigan State University Cyclotron Laboratory for the precise
determination of nuclear excitation energies and nuclear reaction
Q-values. In this work we are using a recently described[1] calibration procedure for nuclear emulsions in an Enge split-pole magnetic
spectrograph to obtain typical uncertainties of 0.1-1. keV in
measurements of particle energies in the 20-40 MeV range. The
choice of energy standards, a description of our calibration procedure, and some recent results are presented in the following
sections.

ENERGY STANDARDS

The measurements to be described here are relative or difference measurements and cannot be referred easily back to a single
absolute energy standard. This is because the magnetic rigidities
of the particles require operation of the spectrograph in a region
where calibration scaling and reproducibility is very difficult at
the level of precision desired. Also beam energy and scattering
angle reproducibility limit the ultimate accuracy in methods require-
ing successive runs at different magnetic fields, etc. Hence, the
ultimate accuracy is obtained when it is possible to observe several
calibration lines simultaneously (in the same spectrum) with the
unknown lines of interest.

It is our desire ultimately to base our measurements on the
precise energy standards now available through the recent work of
Lincoln Smith.[2,3,4] This work provides the opportunity, as discussed in Ref. 4, for the improvement of gamma ray energy standards
by an order of magnitude from ∿20 ppm to ∿2 ppm.

An interim improvement in gamma ray (and excitation energy) standards has been provided recently by the improved ^{14}N-^{15}N mass difference measurements of Kerr and Bainbridge[5] and the reanalysis of the ^{14}N(n,γ)^{15}N gamma-ray energies by Greenwood and Helmer.[6] These excitation energies as reported in Ref. 6 are currently the most well known in the 5-10 MeV excitation range, and hence are very useful standards in the present work.

In an experiment to be described below the objective was to measure the excitation energies of certain states in ^{12}C and ^{16}O. Hence a Uracil target ($C_4H_4N_2O_2$) enriched in ^{15}N was bombarded with 35 MeV protons and the spectrum of inelastically scattered particles was recorded on a nuclear emulsion. A portion of the spectrum showing some of the lines from the various nuclei in the target is displayed in Fig. 1. Using these standards it is possible to measure to $\sim$200 eV the excitation energies of states below 8 MeV in stable nuclei.

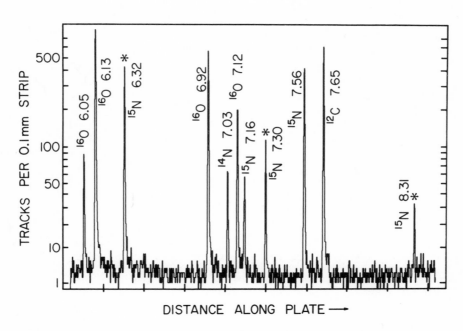

Figure 1.--The spectrum of inelastically scattered protons in the region of 8 MeV excitation from a target containing isotopes of C,N, and O. The lines indicated by the asterisks were used in the calibration.

In this way, the mass doublet measurements provide, relatively directly, the standards for excitation energy measurements. In principle they can similarly provide standards for reaction Q-value measurements, but convenient mass doublets do not always exist. Most of the present interest is in (p,d) and (p,t) reactions to neutron deficient unstable isotopes. Hence the "unknown" Q-values are more negative than standards provided by mass doublet measurements on stable nuclei. A very promising reaction, however, which has not yet been tried is the ^{15}N(p,d) reaction to an excited state of ^{14}N such as the one at 7.9 MeV. The reaction Q-value to this state would be -16.58 MeV which is negative enough to be a very useful reference for neutron deficient isotopes such as ^{11}C, ^{15}O, etc. At the present time, the masses of very few such isotopes are known to 1 keV or better.

CALIBRATION PROCEDURE

The calibration procedure for the MSU spectrograph has been described previously,[1] only the salient features will be repeated here. The procedure combines momentum matching to determine the beam energy, kinematics to determine the scattering angle, and previously known energy levels to determine the focal plane calibration. These concepts are also reviewed by Kashy, Benenson, and Mueller at this conference.

In this method each spectrum is "self-calibrated" via calibration lines in the same spectrum as the unknowns. For a cubic momentum vs. focal plane distance calibration curve there must be a minimum of six calibration lines in the spectrum to determine all of the necessary parameters. The nominal magnetic rigidity $(B\rho)^i_o$ of calibration line i is calculated by reaction kinematics at a nominal beam energy E_o and a nominal scattering angle θ_o. The actual rigidity of the line is then given by:

$$B\rho^i = (B\rho)^i_o + \frac{\partial B\rho^i}{\partial E} \Delta E + \frac{\partial B^i}{\partial \theta} \Delta\theta,$$

where the actual beam energy is $E=E_o + \Delta E$ and the actual scattering angle is $\theta=\theta_o +\Delta\theta$.

The focal plane calibration for the effective radius of curvature ρ vs. focal plane distance D is assumed to be given by the expansion:

$$\rho=\rho_o + \alpha(D-D_o) + \beta(D-D_o)^2 + \gamma(D-D_o)^3$$

where α, β, and γ are treated as unknown coefficients, D_o is an arbitrary point about which the expansion is done, and ρ_o is the (unknown) radius of curvature at the distance D_o. If the line i occurs at the distance D^i in the focal plane then it's rigidity is:

$$B\rho^i = B[\rho_o+\alpha(D^i-D_o) + \beta(D^i-D_o)^2 + \gamma(D^i-D_o)^3].$$

The two equations for $B\rho^i$ can be combined to give:

$$-\frac{\partial B\rho^i}{\partial E}\Delta E-\frac{\partial B\rho^i}{\partial \theta}\Delta\theta+\alpha B(D^i-D_o)+\beta B(D^i-D_o)^2+\gamma B(D^i-D_o)^3+B\rho_o=(B\rho)_o^i$$

which can be treated as a set of simultaneous equations in the six
unknowns $\Delta E, \Delta\theta, \alpha, \beta, \gamma$, and ρ_o. If the calibration lines are chosen
such that the the coefficients are linearly independent then a
solution exists and it can be found by a least squares search (if
there are more than six calibration lines).

Qualitatively, ΔE is determined by relative motion between
different particle types such as protons and deuterons. This is
the essence of momentum matching: $\partial B\rho/\partial E$ for a deuteron is about
twice that for a proton. The angle error $\Delta\theta$ is determined by
relative motion between particles arising from the same reaction
on two different mass targets, such as $^1H(p,p)$ and $^{12}C(p,p')$. The
other coefficients are determined by a set of appropriately spaced
energy levels from the same reaction, e.g. from the $^{15}N(p,p')$
reaction.

For the MSU spectrograph a quadratic focal plane expansion is
normally sufficient in the first 50 cm of the focal plane unless
accuracy 1 keV or better is desired. Also it is sometimes not
necessary to use the cubic term even if sub-keV accuracy is desired.
For example, if the unknowns are very near calibration lines it is
not necessary to accurately fit the portions of the focal plane far
from calibration lines, i.e. it is essentially a "null" measurement.

The coefficients, especially $\Delta E, \Delta\theta, \alpha$, and ρ_o can change slightly
from run to run for several reasons. Firstly, the plate holder must
be shifted and rotated slightly as a function of scattering angle in
order to compensate for kinematic broadening effects. This may
affect α and ρ_o in a way which is not predictable with the accuracy
required for high precision results. They are also affected by non-
linear field scaling (relative to the NMR probe reading). Also ρ_o
and $\Delta\theta$ also change with slight beam spot position variations on
target. The beam energy may also change slightly from run to run
unless the energy analysis system is run with very narrow slits,
which is normally not necessary in the dispersion matching mode.

With the simultaneous self-calibration procedure each line in
the spectrum is subject to exactly the same variables during a run.
Thus the attainable precision with this method is better than the
normal unavoidable magnetic or mechanical uncertainties would allow.

EXCITATION ENERGY MEASUREMENTS

The method described above is particularly useful in determining accurate excitation energies of states in light nuclei. The
sample results presented here were obtained via the (p,p') reaction
on a Uracil target with a 35 MeV proton beam. A portion of the
spectrum obtained at one of the scattering angles is shown in Fig. 1.
The experimental resoltuion in this experiment was about 5 keV (full
width of half maximum) and peak centroids were determined to about
0.1 keV on peaks with adequate counting statistics. The data from

several runs at different scattering angles were averaged to obtain
the final excitation energies.

In these experiments the centroid of the ^{1}H(p,p) peak was used
to determine the average scattering angle during each run with an
uncertainty of ∿0.01°. A knowledge of the beam energy was not
critical in these experiments because only the relative positions
of like particles were needed. Hence the energy calibration (±0.1%)
of the beam transport system was used and beam energy was not a fit
parameter in the data analysis. The resulting excitation energies
determined for several levels in ^{12}C, ^{15}N, and ^{16}O are listed below,
together with previously reported values for these states. In all
cases the accuracy has been improved by the present work but only
the values for the 6.9 and 7.1 MeV states of ^{16}O disagree significantly with the previous numbers.

	Excitation Energies		
Nucleus	Present (keV)	Previous (keV)	References
^{15}N	*	5270.35±0.10	6
^{16}O	6049.6±0.3	6050.2 ±1.0	7
^{16}O	6130.6±0.2	6130.66±0.18	7
^{15}N	*	6323.85±0.12	6
^{16}O	6916.4±0.2	6918.8 ±0.6	7
^{16}O	7117.1±0.2	7118.67±0.35	7
^{15}N	*	7301.09±0.17	6
^{15}N	7564.1±0.2	7566 ±3	8
^{12}C	7654.4±0.2	7655.2 ±0.8	9
^{15}N	*	8312.79±0.14	6

* Calibration lines.
Table I.--The excitation energies of some states in ^{12}C, ^{15}N, and
^{16}O determined in the present experiment listed together with the
calibration lines from ^{15}N.

NUCLEAR REACTION Q-VALUES

The masses of neutron deficient unstable isotopes can be
determined relative to the masses of stable nuclei via measurements
of (p,d) and (p,t) reaction Q-values. We have measured the mass of
^{12}N with an uncertainty of 1 keV using this method. In this case
the ^{16}O(p,t)^{14}O reaction which is known to 0.5 keV was used as a
reference in the determination of the ^{14}N(p,t)^{12}N Q-value.

The experiment consisted of simultaneous recording of the $^{16}O(p,t)$, $^{14}N(p,t)$, and $^{12}C(p,p')$ reactions on nuclear emulsions at a beam energy of 31.2 MeV. At this energy the $^{16}O(p,t)$ ground state group has a magnetic rigidity approximately the same as the elastically scattered protons, while the $^{14}N(p,t)$ ground state group is very near the $^{12}C(p,p')$ 4.439 MeV excited state. Thus the relative momenta of the ^{14}O tritons and the elastic protons fixes the beam energy and the relative positions of the $^{12}C(p,p')$ and ^{12}N tritons fixes the triton energy for that reaction. The mass of ^{12}N determined by analysis of five separate exposures of this type is indicated in Table 2 where previous measurements are also listed.

Reaction	Mass Excess (keV)	Year	Reference
$^{12}N(\beta^+)^{12}C$	17452±60	1957	10
$^{12}N(\beta^+)^{12}C$	17406±15	1963	11
$^{10}B(^3He,n)^{12}N$	17352± 9	1964	12
$^{10}B(^3He,n)^{12}N$	17343±25	1964	13
$^{10}B(^3He,n)^{12}N$	17345±20	1966	14
$^{12}C(p,n)^{12}N$	17340± 9	1966	15
$^{10}B(^3He,n)^{12}N$	17339± 7	1968	16
$^{14}N(p,t)^{12}N$	17338± 1	1975	Present

Table 2.--Comparison of various measurements of the mass of ^{12}N.

REFERENCES

1. J.A. Nolen, Jr., G. Hamilton, E. Kashy, and I.D. Proctor, Nucl. Instr. and Meth. 115, 189(1974).

2. L.G. Smith, Phys. Rev. C4, 22(1971).

3. L.G. Smith, Proceedings 4th International Conference on Atomic Masses and Fund. Constants, J.H. Sanders and A.H. Wapstra (eds.), Plenum Press, p. 164(1972).

4. Lincoln G. Smith and A.H. Wapstra, Phys. Rev. C11, 1392(1975).

5. D.P. Kerr and K.T. Bainbridge, Can. J. Phys. 49, 1950(1971).

6. R.C. Greenwood and R.G. Helmer, Nucl. Instr. and Meth.121, 385(1974).

7. F. Ajzenberg-Selove, Nucl. Phys. A166, 1(1971).

8. F. Ajzenberg-Selove, Nucl. Phys. A152, 1(1970).

9. P.L. Jolivette, J.D. Goss, A.A. Rollefson, and C.P. Browne, Phys. Rev. C10, 2629(1974).

10. F. Ajzenberg-Selove, M.L. Bullock, and E. Almquist, Phys. Rev. 108, 1284(1957).

11. N.W. Glass and R.W. Peterson, Phys. Rev. 130, 299(1963).

12. R.W. Kavanagh, Phys. Rev. 133, 1504(1964).

13. T.R. Fisher and W. Whaling, Phys. Rev. 133, 1502(1964).

14. C.D. Zafiratos, F. Ajzenberg-Selove, and F.S. Dietrich, Nucl. Phys. 77, 81(1966).

15. D.A. Bromley, J.C. Overley, and P.D. Parker, Phys. Rev. Letters 17, 705(1966).

16. J.M. Adams, A. Adams, and J.M. Calvert, J. Phys. A 1, 549(1968).

*Supported by the National Science Foundation.

NEW TESTS OF THE ISOBARIC MULTIPLET MASS EQUATION

R.G.H. Robertson

Cyclotron Laboratory and Physics Department
Michigan State University
East Lansing, Michigan 48824 USA

INTRODUCTION

The isobaric multiplet mass equation (IMME), first propounded by Wigner[1] in 1957, states that the masses M of members of an analog multiplet should be related by an equation quadratic in T_z

$$M = a + bT_z + cT_z^2.$$

A non-trivial test of the equation requires a multiplet of at least 4 states (i.e. $T_z=3/2$), and in 1964 Cerny and his collaborators completed the first isobaric quartet.[2] Since that time, some 18 quartets have been measured, a few with extreme precision, and in only one case, the lowest mass 9 quartet, is there a significant disagreement with the IMME. The present status of isobaric quartets is summarized in this conference by Benenson and Kashy. There is considerable incentive to make new tests of the equation by considering multiplets other than quartets, but only very recently has it become experimentally feasible to test the IMME even in quintets. A number of features make such tests interesting. If one represents deviations from the IMME by additional terms dT_z^3, eT_z^4, etc., only one such term can be determined in a quartet, but two in a quintet. Thus one can test the IMME more rigorously, and, in the event of a violation, gain better insight into the possible mechanisms causing it. Furthermore, many quintets include both particle-stable and unbound nuclei, and if changes in the spatial wave functions across a multiplet can cause deviations from the IMME, then quintets may be rather sensitive to that influence. Finally, if there exists a many-body charge-dependent force, the presence of two determined parameters (d and e) makes quintets an attractive testing ground.

147

ISOBARIC QUINTETS

The first test in an isobaric quintet was made in 1973 with the observation of ^{20}Mg by the ^{24}Mg(α,^{8}He) reaction.[3] However, in that case only four of the five members were observed, and the first complete quintet, mass 8, was obtained in 1974.[5] We shall survey the present status of T=2 multiplets in 4n nuclei and report on some new experiments. Multiplets in 4n+2 nuclei have odd-odd configurations and tend to be very unbound to allowed particle decays.

Mass 8

The mass-8 quintet, consisting of the ground states of ^{8}He and ^{8}C, and the lowest T=2 states in ^{8}Li, ^{8}Be and ^{8}B, is the only one in which all five members are known. The pertinent data have been summarized in Ref. 4, and it was concluded that a small but non-zero e term, 4.4±2.0 keV, was required to describe the data. This rather surprising result is most likely attributable to isospin mixing in the T=2 state of ^{8}Be. In particular, mixing with a T=0 state leads automatically to a quartic contribution but no cubic term. Because of the good spatial overlap between T=0 and T=2 configurations, such mixing may not be unusual, and we shall remark on the consequences of this below.

Two recent measurements improve the experimental situation in mass 8. Geesaman, et al.[5] have obtained a new value for the excitation of the T=2 state in ^{8}Be, 27.500±0.005 MeV, by re-examination of the ^{6}Li+d forbidden resonance reaction.

The mass of ^{8}C obtained via ^{12}C(α,^{8}He) has an uncertainty of 170 keV,[4] far larger than the other members of the quintet. We have therefore begun to remeasure the ^{8}C mass at Michigan State, by means of the ^{14}N(τ,^{9}Li) reaction.[6] Targets of melamine were bombarded with ^{3}He ions from the MSU Cyclotron, and reaction products analyzed in a magnetic spectrograph. In view of the extremely small cross-sections anticipated, a double proportional counter backed by a 5 cm x 1 cm x 1.5 mm Si detector was used. The second proportional counter served as a charge-division position-sensitive detector, and the Si detector provided high-quality energy and time-of-flight information. The spectrum recovered from magnetic tape is shown in Fig. 1. A group of seven events with a Q-value centroid of -42.30±0.09 MeV is interpreted as the ^{8}C ground state. The width of the distribution, 170 keV, is consistent with target thickness contributions alone, suggesting that the ^{8}C ground state may be quite narrow (it is unbound to 2-, 3- and 4- proton decays). The counts at higher excitation probably arise from the breakup continuum, but the event at Q=-41.9 MeV is evidently background. The cross section to the ground state of ^{8}C at 8° (lab) is only 2 nb/sr. The mass excess for ^{8}C from this

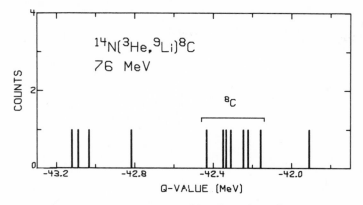

Figure 1.--Spectrum from $^{14}N(^3He,^9Li)^8C$ Reaction.

measurement, 35.14±.09 MeV, lies much closer to the IMME prediction than did the earlier $(\alpha,^3He)$ result, 35.36±.17 MeV.

The present status of the mass-8 quintet is summarized in Table I. When these masses are fitted to a five-parameter IMME, one finds d = 4±7 keV and e = 6±4 keV. A fit to the quadratic form gives χ^2 = 5.6 and residuals shown in Fig. 2. The even-order alternation indicative of a non-zero quartic term is apparent. Fitting to an e term alone gives e = 4.5±2.0 keV and χ^2 = 0.3, an excellent fit. Detailed understanding of the role of isospin mixing in the origin of this term may come from investigations of the decays of the T=2 state in ^8Be, and a search for the perturbing T=0 state.[7]

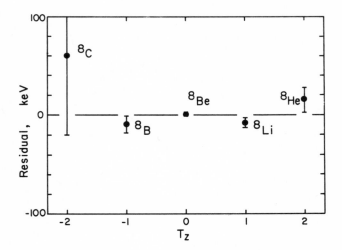

Figure 2.--Amounts by which individual masses exceed those given by a weighted fit to a quadratic IMME for the A=8 quintet.

	T_z	Mass Excess, MeV[a]	Reactions	Reference[b]
^{8}C	-2	35.19(8)	^{12}C(α,^{8}He), ^{14}N(τ,^{9}Li)	4,present
^{8}B	-1	33.542(9)	^{11}B(τ,^{6}He)	4
^{8}Be	0	32.4356(23)	^{6}Li+d, ^{10}Be(p,t)	4,5
^{8}Li	+1	31.7697(54)	^{10}Be(p,τ)	4
^{8}He	+2	31.597(13)	(α,^{8}He)	(4)
^{12}O	-2	32.07(23)	IMME	
^{12}N	-1	29.60(9)	IMME	
^{12}C	0	27.603(14)	^{14}C(p,t), ^{10}Be(τ,n)	8,9
^{12}B	+1	26.080(20)	^{14}C(p,τ)	8
^{12}Be	+2	25.030(45)	^{7}Li(^{7}Li,2p),^{14}C(^{18}O,^{12}Be)	10,11
^{16}Ne	-2	24.09(8)	IMME	
^{16}F	-1	20.81(3)	IMME	
^{16}O	0	17.983(14)	Several	(13)
^{16}N	+1	15.612(7)	^{14}C(^{3}He,p)	(13)
^{16}C	+2	13.695(16)	^{14}C(t,p)	(13)
^{20}Mg	-2	17.79(17)	^{24}Mg(α,^{8}He)	3,(4)
^{20}Na	-1	13.35(3)	IMME	
^{20}Ne	0	9.696(4)	Several	(16)
^{20}F	+1	6.508(8)	^{22}Ne(p,τ),^{18}O(τ,p)	14,(16)
^{20}O	+2	3.800(8)	^{18}O(t,p)	(16)
^{24}Si	-2	10.746(38)	IMME	
^{24}Al	-1	5.901(14)	IMME	
^{24}Mg	0	1.503(3)	^{23}Na(p,γ)	(16)
^{24}Na	+1	-2.4465(17)	^{22}Ne(τ,pγ)	(16)
^{24}Ne	+2	-5.948(10)	^{22}Ne(t,p)	(16)

[a]Uncertainty in last digit given in parentheses.
[b]Bibliographic references in parentheses.

Table I. T = 2 States in 4n Nuclei (n = 2 to 6)

Mass 12

The precision of the masses of the three known T=2 states in A=12 is considerably poorer than in many other quintets, reflecting the fact that resonance reactions apparently do not populate the T=2 states in ^{12}C and ^{12}B (see Table I).

It might reasonably be expected that the T=2 state in ^{12}N could be located, as was that in ^{8}B, by means of the (τ,^{6}He) reaction. The T=1 states in ^{12}N at ~12 MeV excitation should be broad due to allowed particle decay modes and should therefore

contribute only a continuum. A brief preliminary search[12] for
this state has been carried out at MSU using ^{15}N-enriched melamine
targets. A spectrum at 76 MeV and 8° (lab) showed no sign of a
narrow state in the region expected, and an upper limit of 50 nb/sr
can be set on its cross section. This is somewhat puzzling, inasmuch as the cross section to the T=2 state in ^{8}B was 200 nb/sr.
The state may lie underneath the ^{9}C ground state peak at this angle,
if the IMME prediction is wrong by 3 standard deviations.

Mass 16

The experimental situation in mass 16 has not, to our knowledge,
changed since a 1971 compilation[13] (see Table I). Nero and Tribble[14]
have searched for the T=2 state in ^{16}F by ^{19}F(^{3}He,^{6}He), but thus
far without success.

Mass 20

Four members of this quintet have been known since the ^{20}Mg
mass was reported.[3] At that time, it appeared they fitted the IMME
well, but three new measurements of the mass of ^{8}He (see citations
in Ref. 4) and a new measurement[15] of the mass of the T=2 state in
^{20}F have reduced the uncertainties, and there is now a discrepancy.
The results are summarized in Table I. These mass excesses yield
a d (or e) coefficient of -14±8 keV, and the residuals in a quadratic
fit are plotted in Fig. 3. χ^2 is 3.5 for the three-parameter fit.
While this discrepancy seems large, it may be noted that the ^{20}Mg
and ^{8}C masses from the (α,^{8}He) measurements[3] both show a similar
deviation from the IMME, suggesting a systematic experimental effect.

Mass 24

The three most neutron-rich members have been measured quite
precisely (the data are summarized in a compilation[17] see Table I),
and efforts have been made both at MSU and at Princeton[18] to find
the T=2 state in ^{24}Al by ^{27}Al(τ,^{6}He). In ^{24}Al the T=2 state lies
sufficiently low in excitation that nearby T=1 states are narrow.
It is thus a much more difficult task to determine which, if any,
of the multitude of states seen in (τ,^{6}He) is the T=2 state. One
approach has been to examine the correspondence with (τ,t) spectra
to the same levels since that reaction cannot populate T=2 states.
There has been no success in the search to date. There have in the
past been many attempts to produce T_z=-2 nuclides such as ^{24}Si by
(α,4n) or (τ,3n) reactions, but they have been frustrated by very
low cross sections. Nevertheless, the β-decay of T_z=-2 nuclei may
eventually permit mass measurements of T=2 states in T_z=-1 nuclei
through the selectivity of Fermi β-decay.

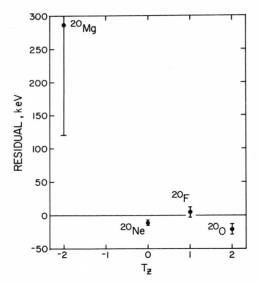

Figure 3.--Amounts by which individual masses exceed those given
by a weighted fit to a quadratic IMME for the A=20 quintet.

CONCLUSION

Tests of the isobaric multiplet mass equation have now been
made in two quintets, and in both appreciable discrepancies are
found. In mass 8, there seems clear evidence that this deviation
is caused by isospin mixing in the inner (T_z=0) member. The widths
and particle decay modes of known T_z=0, T=2 states suggest that
such mixing is by no means the exception. Although we have become
accustomed to using the IMME with confidence in quartets for predicting the masses of T_z=-3/2 nuclei, much greater caution is called
for in quintets. For example, given the masses of the T_z=0, 1 and
2 members, but with a mixing-induced shift of 5 keV in the T_z=0
member, one would calculate a mass for the T_z=-2 member incorrect
by 30 keV. Such shifts may well be typical, and it is only a matter
of time before experimental precision will make them readily
apparent.

REFERENCES

1. E.P. Wigner, in Proceedings of the Robert A. Welch Foundation
    Conference on Chemical Research, Houston, Texas, 1957. Ed. by
    A. Milligan (Robert A. Welch Foundation, Houston, Texas, 1957),
    The Structure of the Nucleus, p. 67.

2. J. Cerny, R.H. Pehl, F.S. Goulding and D.A. Landis, Phys. Rev. Lett. 13, 726(1964).

3. R.G.H. Robertson, S. Martin, W.R. Falk, D. Ingham, and A. Djaloeis, Phys. Rev. Lett. 32, 1207(1974).

4. R.G.H. Robertson, W.S. Chien and D.R. Goosman, Phys. Rev. Lett. 34, 33(1975).

5. D.F. Geesaman, J.W. Noé, P. Paul and M. Suffert, BAPS 20, 597(1975).

6. R.G.H. Robertson, E. Kashy, D. Mueller and W. Benenson, to be published.

7. E.G. Adelberger, S. J. Freedman, A. V. Nero, A.B. McDonald, R.G.H. Robertson and D.R. Goosman, BAPS 20, 597(1975) and to be published.

8. P.H. Nettles, Thesis, California Institute of Technology (1971), unpublished.

9. D.R. Goosman, D.F. Geesaman, F.E. Cecil, R.L. McGrath and P. Paul, Phys. Rev. C10, 1525(1974).

10. H.H. Howard, R.H. Stokes and B.H. Erkkila, Phys. Rev. Lett. 27, 1086(1971).

11. G.C. Ball, J.G. Costa, W.G. Davies, J.S. Forster, J.C. Hardy and A.B. McDonald, Phys. Lett. 49B, 33(1974).

12. W.S. Chien and R.G.H. Robertson, (1975), unpublished.

13. F. Ajzenberg-Selove, Nucl. Phys. A166, 1(1971).

14. A.V. Nero and R.E. Tribble, private communication.

15. H.T. Fortune, private communication.

16. F. Ajzenberg-Selove, Nucl. Phys. A190, 1(1972).

17. F. Ajzenberg-Selove, Nucl. Phys. A214, 1(1973).

18. A.V. Nero and R.E. Tribble, BAPS 20, 87(1975).

ISOBARIC MASS QUARTETS

Walter Benenson and Edwin Kashy

Cyclotron Laboratory and Physics Department
Michigan State University
East Lansing, Michigan 48824   USA

I. INTRODUCTION

In this paper we describe the current experimental status of
isobaric mass quartets. A mass quartet consists of the states
which represent the four projections of isospin T=3/2 on the T_z
axis. Two of the states ($T_z=\pm3/2$) are mirrors of each other, and
the other two are what is called analog states, in this case they
are analogs of the two $T_z=\pm3/2$ levels. In Fig. 1, we show two
illustrative quartets which occur in the A=9 nuclei. From this
figure we can see that it is not necessary for the $T_z=\pm3/2$ levels
to be ground states, and in fact it is possible to have the interesting case of many complete quartets within a given A.

The mass of the T_z member, $M(T_z)$, was shown by Wigner[1] in 1957
to be given by a quadratic equation, the isobaric multiplet mass
equation (IMME)

$$M(T_z) = a + bT_z + cT_z^2$$

Theoretical discussions of the IMME can be found in the original paper as well as in several review articles.[2] In this paper
we will discuss mainly the experimental status of the equation,
but first we will give a short discussion about its foundation.

In the absence of charge dependent effects, the four levels
would, of course, be identical and degenerate in energy. However,
when treating the breaking of the degeneracy, Wigner found, using
first order perturbation theory, that the effect of two-body charge
dependent forces is to produce a quadratic relation. To test this
result one needs therefore multiplets of $T \geq 3/2$ levels. At present
only one multiplet with T>3/2 is known, and it is discussed in
another paper[3] at this conference. Deviations from the IMME could

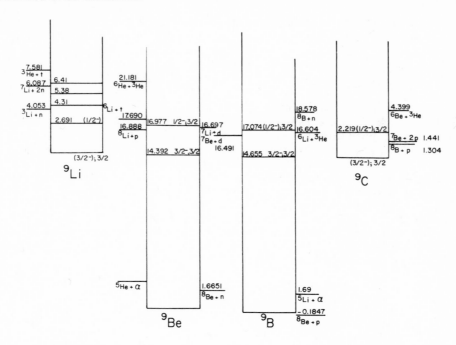

Figure 1.--   T=3/2 levels in the A=9 system

occur if the four wavefunctions are not identical thus violating
the first order treatment. Three body charge dependent forces
would also cause a violation, and one of the hopes of this project
is therefore to be able to obtain an estimate of three-body effects.
After a review of the experimental information, we will discuss the
deviations in terms of the expected wavefunction differences
between the members.

II. EXPERIMENTAL

It will become clear that the IMME is so accurate that very
precise mass measurements are required to reach the level at which
actual deviations are observable. For the most part only the masses
of T_z=3/2 nuclei were already well enough known to be used in this
study. Almost every other level has been experimentally determined
with the IMME in mind.

lla. T_z=-3/2 nuclei

The T_z=-3/2 nuclei are classified as "exotic" or possibly
"far-from-stability" because they lie three nuclei from the stable
targets in the chart of the nuclei. The absence of knowledge of
their masses is what inhibited completion of quartets up until 1964,
when Cerny, et al.[4] were able to measure the mass of ^{9}C by measuring

the Q-value of the $^{12}C(^3He,^6He)$ reaction. Since then nearly all the
T_z=-3/2 nuclei have been measured accurately up to ^{37}Ca by means
of either the $(^3He,^6He)$ or $(^3He,^8Li)$ reactions. As can be seen in
Table I the missing nuclei are ^{27}P and ^{35}K. They are accessible by
either the $(^3He,^8Li)$ or $(^4He,^8He)$ reactions and undoubtedly will be
measured in the next few years. Above ^{37}Ca the absence of T_z=0
targets make the experiments much more difficult, but there is
hope that the T_z=-3/2 masses can be measured via heavy ion transfer
reactions or possibly $(^3He,^{10}Li)$ or $(p,^8He)$. In Table I we also
include in parenthesis some T_z=-3/2 levels which are not members
of a quartet because one of the other members is missing. As will
be pointed out in the next section, the T_z=-1/2 nucleus is often
the one for which experimentally locating more than one T=3/2 level
is most difficult. Multiple levels are an automatic dividend of
transfer reaction mass determinations as can be seen from Fig. 2
in which the excited states of ^{21}Mg, ^{37}Ca and ^{55}Ni are visible in
the $^3He,^6He$ spectra. In Fig. 3 we show the ground and first excited
states of ^{19}Na from the $^{24}Mg(^3He,^8Li)$ reaction.

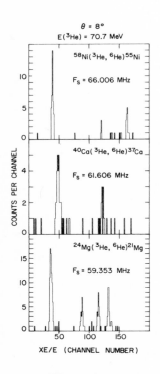

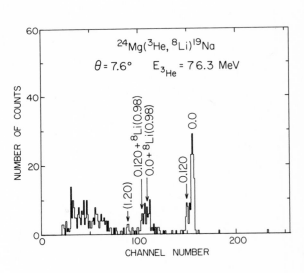

Figure 2   Figure 3

Nucleus	Ex	M.E.	Ref.	Reaction*
^7B	0.0	27.94(100)	a	$^{10}B(^3He,^6He)$
^9C	0.0	28.912(4)	b	$^7Be(^3He,n)$
	2.219(10)	31.131(11)	c	$^{12}C(^3He,^6He)$
^{13}O	0.0	23.105(10)	d	$^{16}O(^3He,^6He)$
^{17}Ne	0.0	16.479(30)	e	$^{20}Ne(^3He,^6He)$
	1.33(15)	17.805(35)		$^{20}Ne(^3He,^6He)$
^{19}Na	[0.0]	12.928(12)	f	$^{24}Mg(^3He,^8Li)$
	0.120(10)	13.048(15)	f	$^{24}Mg(^3He,^8Li)$
^{21}Mg	–	10.910(16)	d	$^{24}Mg(^3He,^6He)$
	0.210(10)	11.120(18)	g	$^{24}Mg(^3He,^6He)$
	[1.08(10)]			
	[1.64(20)]			
	[2.01(20)]			
^{23}Al	–	6.767(20)	f	$^{28}Si(^3He,^8Li)$
^{25}Si	–	3.824(10)	h	$^{28}Si(^3He,^6He)$
	0.040(5)	3.864(11)	h	$^{28}Si(^3He,^6He)$
	[0.815(15)]			
	[1.963(15)]			
	[2.373(10)]			
	[2.606(10)]			
^{29}S	0.0	-3.16(50)	i	$^{32}S(^3He,^6He)$
^{33}Ar	0.0	-9.384(30)	j	$^{36}Ar(^3He,^6He)$
	1.340(20)	-8.040(40)		
	1.790(20)	-7.598(40)		
^{37}Ca	0.0	-13.144(25)	g	$^{40}Ca(^3He,^6He)$
	1.61(20)	-11.534(32)		

Table I.--Mass excesses of states in $T_z=3/2$ nuclei in MeV with errors in parentheses in keV.

* Reaction used for most accurate measurement.

a) R.L. McGrath, J. Cerny, and E. Norbeck, Phys. Rev. Letters 19, 1442(1967).
b) J.M. Mosher, R.W. Kavanagh and T. Tombrello, Phys. Rev. C3, 438(1971).
c) W. Benenson and E. Kashy, Phys. Rev. C10, 2633(1974).
d) G.F.Trentelman, B.M.Preedom and E.Kashy, Phys. Rev. C3,2205(1971)
e) H. Nann, W. Benenson and E. Kashy, to be published.
f) W.Benenson, E.Kashy, H.Nann, D.Mueller, L.W.Robinson, (to be published).
g) W.Benenson,E.Kashy and I.D.Proctor, Phys. Rev. C8, 210(1973).
h) W.Benenson,J.Driesbach,I.D.Proctor,G.F.Trentelman and B.M. Preedom, Phys. Rev. C5, 1426(1972).
i) W.Benenson,E.Kashy, I.D. Proctor and B.M.Preedom, Phys. Letters 43B,117(1973).
j) H.Nann,W.Benenson,E.Kashy and P.Turek,Phys. Rev. C9,1848(1974).

The accuracy of the masses of the T_z=-3/2 nuclei is quite good except in special cases in which there is a target problem e.g., $^{32}S(^3He,^6He)$, or the nucleus is unbound and broad, e.g., ^7B. The A=11 quartet has been completely omitted because ^{11}N is so broad that it can not be positively identified as either the ground or first excited state although agreement with the IMME is good. The experimental technique by which most of the masses were obtained is discussed in another paper[5] at this conference. It involves use of a spectrograph in a mode in which the rigidity of the product of the unknown reaction is compared to the rigidity of a particle from a well known reaction. Time-of-flight mass identification is used to separate the very weak yield of the multinucleon transfer reactions of interest from the products of more prolific reactions. As can be seen in Figs. 2 & 3 the level of background is low, and, in fact, cross sections of 5 nb/sr produce peaks of sufficient quality to measure a mass accurately. This is because the intrinsic resolution is quite good, approximately 40-70 keV.

IIb. T_z=±1/2 Nuclei

For the most part the T=3/2 levels in the T_z=±1/2 nuclei have been observed using the (p,t) and (p,^{3}He) reactions although there are several cases of (p,d) and compound nuclear resonances. Many of the measurements were carried out with the same spectrograph technique described above. The T_z=±1/2 members are of 3 times more importance for the deviation of the IMME. If for example the deviation were of the form dT_z^3, then

$$d=1/6\{M(3/2)-M(-3/2) - 3[M(1/2) - M(-1/2)]\}$$

The most accurately measured quartet is the A=9 ground state. In this case, the T_z=±1/2 levels were studied with 6-10 keV resolution and a final accuracy of 1-2 keV. The analog of the first excited state of ^9C is an example of use of the IMME to find an analog level. The $^{11}B(p,t)^9B$ at 46 MeV was employed, and the spectrograph was set to place the unknown state at channel 110 as is shown in Fig. 4. The resulting level is very high in ^9B yet it is quite narrow, 22 keV, and it is therefore very likely to be the T=3/2 state. The width of the peak in the spectra is almost entirely due to the state itself since the experimental resolution was about 10 keV.

There are many experiments left to be done at the T=3/2 levels in the T_z=±1/2 nuclei either to improve the accuracy of existing quartets or to complete a quartet. Some are reasonably straight-forward such as locating the analog of the ^{19}Na ground state using either $^{21}Ne(p,t)$ or $^{17}O(^3He,n)$ reactions. A particularly interesting case would be in A=25 where many T_z=±3/2 levels are known, but only the first two have their analogs known. In general although reactions like (p,t) and (p,^{3}He) are highly selective, the state of interest stands out from the continuum only for special cases such as L=0 transfer or if it is the only narrow state in the region. This often makes identification very difficult.

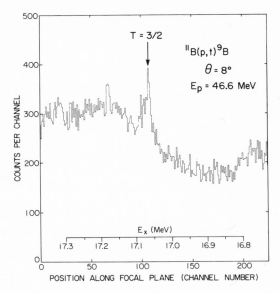

Figure 4.--A spectrum showing the analog of the second excited state of ^{9}C in ^{9}B

IIc. T_z=3/2 Nuclei

For the most part, accurate values for the T_z=3/2 levels can be found in the compilations. However, many such measurements are old and perhaps based on absolete calibrations. When the accuracy of the other members seem to require it, we have checked these masses. In the case of ^{9}Li we have remeasured it[6] using ^{10}Be(d,^{3}He) and ^{7}Li(t,p) from plates loaned to us by F. Ajzenberg-Selove and collaborators. These results changed the ^{9}Li mass by 11 keV.

III.   COMPARISON TO THE IMME

The most convenient form of expressing the deviations from the IMME is via the d-coefficient. The average value of all the measured d-coefficients shown in Table II is slightly positive (~1 keV), but there are only two which have a deviation greater than its error; only the A=9 case is really a significant deviation. No Z-dependence is evident at all, nor is there a dependence on excitation energy within a given A. This is particularly surprising because, as in the case of A=9, the situation can occur in which the first excited state quartet actually includes broad levels whereas the ground state consists entirely of very narrow levels.

IV.   DISCUSSION

IVa.   The d-coefficient

As discussed above the d-coefficient is essentially zero for

18 of the 19 quartets, which indicates a very good agreement with the IMME. There are several mechanisms which can produce a d-coefficient, and they all tend to give a positive d-coefficient. However, they have only been calculated in any detail at all for the A=9 system. The spreading of the wave function due to the repulsion of the last proton in the T_z=-3/2 nucleus was calculated by Bertsch and Kahana.[7] Their d-coefficient, 4 keV, agrees well with the experimentally determined 5.7±1.6 keV. However, the mechanism would be expected to be more important in the excited A=9 quartet because the last proton is actually unbound in this case. The d-coefficient of the A=9 excited quartet is consistent with zero and with the A=9 ground state d-coefficient. The second mechanism, isospin mixing in the T_z=±1/2 levels is negligible for the A=9 ground state quartet.

Shell model calculations of mass multiplets by McGrory, Wildenthal and Benenson[8] have shown the mixing to be larger in the sd-shell. However, the main effect is in the b- and c-coefficients. These calculations therefore are most interesting in discussing the b- and c-coefficients themselves.

IVb. The b- and c-coefficients

The b- and c-coefficients of the IMME carry the same kind of information as do Coulomb energies, which have proved very useful nuclear structure probes. The shell model calculations were carried out in a formalism in which neutrons and protons are treated separately and the single particle energies are taken from experiment. For example, in the lower part of the sd-shell ^{17}O and ^{17}F single particle energies are used for neutrons and protons respectively. The two-body matrix elements are calculated with harmonic oscillator wave functions. The calculated b- and c-coefficients are compared to experiment in Table III. One can see that the overall agreement

Isobaric Multiplet Mass Equation

$$M = a + bT_z + cT_z^2 + [dT_z^3]$$

T_z=-3/2 Nucleus	J^π	E_x(MeV)	d(keV±keV)
^{7}B	3/2$^-$	0.0	-11 ±30
^{9}C	3/2$^-$	0.0	5.8± 1.5
	1/2$^-$	2.691	2.3± 2.9
^{13}O	3/2$^-$	0.0	- 0.5± 2.9
^{17}Ne	1/2$^-$	0.0	5.8± 6.3
	3/2$^-$	1.33	- 5.2± 7.4
^{19}Na	3/2$^+$	0.120	- 7.0± 8.5
^{21}Mg	5/2$^+$	0.0	6.3± 6.9
	1/2$^+$	0.21	- 2.4± 7.0
^{23}Al	5/2$^+$	0.0	- 8.6±13.0
^{25}Si	5/2$^+$	0.0	1.9± 2.9
	3/2$^+$	0.040	4.8± 3.6
^{29}S	5/2$^+$	0.0	.3 ±11
^{33}Ar	1/2$^+$	0.0	2.0± 5.5
	3/2$^+$	1.34	1.0±26.0
	5/2$^+$	1.79	3.8±11.0
^{37}Ca	3/2$^+$	0.0	- 2.4± 4.9
	1/2$^+$	1.61	2.4±15.0

Table II

Shell Model Calculations

Isobaric Multiplet Mass Equation

McGrory, Wildenthal, and Benenson

$$M = a + bT_z + cT_z^2$$

		Experiment		Theory	
A	J^π	b	c	b	c
21	5/2$^+$	-3667	237	-3570	208
	1/2$^+$	-3628	238	-3528	209
25	5/2$^+$	-4382	221	-4382	221
	3/2$^+$	-4371	216	-4417	214
33	1/2$^+$	-5655	211	-5799	195
	3/2$^+$	-5645	200	-5769	195
37	3/2$^+$	-6200	200	-6134	194
	1/2$^+$	-6172	203	-6098	204

Table III

is excellent, but there is a failure to predict the small differences
between quartets of the same A. For example the $5/2^+$-$3/2^+$ energy
spacing in A=25 is predicted to increase slightly with decreasing
T_z, but the experiment shows just the opposite. The b- and c-
coefficients are sensitive to the $s_{1/2}$ component in the wavefunction.
Since it has a different Coulomb energy than the d-particles, small
admixture differences between the two states is expected to be a
mechanism for this effect. The experiment indicates more $s_{1/2}$ in
the $3/2^+$ state than in the $5/2^+$, but the theory seems to predict the
opposite. The shell model calculations of De Meijer, et al.[9] show
the same type of deficiencies.

V. CONCLUSIONS

The remarkable accuracy of the IMME makes it useful in predicting masses or analog state energies with assurance. By the same
token, since it works so well, it is not necessary to complete a
mass quartet in order to study its Coulomb energies. Three members
of a quartet will determine the b- and c-coefficients accurately
enough to compare to Coulomb energy calculations using the shell
model or other methods. A more systematic theoretical treatment is
now called for in which all the known effects are calculated with
maximum accuracy.

REFERENCES

1. E. P. Wigner, Proceedings of the Robert A. Welch Foundation
    Conference on Chemical Research 1957, Edited by W.D. Millikan.

2. D.H. Wilkinson in Isobaric Spin in Nuclear Physics, J.D. Fox
    and D. Robson Eds., Academic Press, New York 1966; G.T. Garvey
    in Nuclear Isospin, J.D. Anderson, S.D. Bloom, J. Cerny and
    W.W. True Eds., Academic Press, New York 1969; J. Janecke in
    Isospin in Nuclear Physics, D.H. Wilkinson Ed., North Holland
    Publishing Co., Amsterdam 1969.

3. R.G.H. Robertson, Proceedings V International Conference on
    Atomic Masses and Fundamental Constants, Paris 1975.

4. J. Cerny, R.H. Pehl, F.S. Goulding and D.A. Landis, Phys. Rev.
    Letters 13, 726(1964).

5. E. Kashy, W. Benenson and D. Mueller, Proceedings V Conference
    on Atomic Masses and Fundamental Constants, Paris 1975.

6. E. Kashy, W. Benenson, R.G.H. Robertson and D. Goosman, Phys.
    Rev. (in press).

7. G. Bertsch and S. Kahana, Phys. Lett. 33B, 193(1970).

8. J.B. McGrory, B.H. Wildenthal and W. Benenson (to be published).

9. R. J. DeMeijer, H.F.J. van Royen, and P.J. Brussaard, Nucl.
    Phys. A164, 11(1971).

THE MASS SPECTROSCOPIC WORK OF L.G. SMITH

A.H. Wapstra

Instituut voor Kernphysisch Onderzoek, Amsterdam and

University of Technology, Delft

DEDICATION

We will dedicate the sessions on mass spectrometry in this Conference to the memory of Lincoln Gilmore Smith, who was born on Febr. 12, 1912 in Irvington, New York, and who died Dec. 9, 1972 in Princeton.

Considering that the first in this series of conferences occured almost twenty years ago, we have lost remarkably few of our most prominent contributers. This is the first time that we feel compelled to devote some words to a friend and colleague who left us.

In doing so, I will not talk about all the contributions that this remarkable man made to science. You can find this in an obituary in the International Journal of Mass Spectroscopy and Ion Physics, 11 (1973) 417, written by Lewis Friedman. I will only mention his truly outstanding work in precision mass spectropy.

Smith started this work in 1948, when the project of accurately measuring atomic masses was adopted as one of the important goals of the newly organized Brookhaven National Laboratory. Characteristically, he decided not to build upon experiences that men as Mattauch, Bainbridge and Nier had gathered in the use of suitably combined electric and magnetic fields for this purpose. Instead, he started to develop a completely different method based first upon measurement of ion flight times, following an idea of S. Goudsmit, but converting a little later to its inverse: cyclotron frequencies of ions in orbits = in a suitable magnetic field. The importance of this decision is that it might have been expected that the systematic errors in measurements with such a

"mass synchrometer" should scarcely be smaller than in that of
more classical mass spectrometers, certainly in the first
measurements with the new technique, but they should very probably
behave in a different way.

Lincoln Smith finished his mass synchrometer in 1952. It had
a 30 cm diameter magnet, and its resolution was 1/60.000. Several
measurements with this instrument were published in 1953-1958.
They belonged at once to the best mass spectroscopical data
available. In addition to the new instrument, this was also due to
his new technique of "peak matching" -subsequently adopted by
several other mass spectroscopists- and to his recognition of the
systematic errors introduced in measurements on fragments of
molecular ions (due to their different velocity distribution).

As things often go, Smith felt that only after finishing and
using his mass synchrometer, he really knew how a good mass
spectrometer should be build. The magnet should be larger and
stronger, and the electrodes should be provided with a high harmonic
of the cyclotron frequency, in the region of radiofrequencies.
Unfortunately, Brookhaven's interest had shifted in other directions.
In 1957 Smith moved to Princeton where he started building his RF
spectrometer supported by the National Science Foundation, and when
this support was terminated, he completed the work without support.
His first measurements were reported to AMCO III in 1969, and his
first mass measurements in 1971. Eventually, a resolution was
reached of slightly better than 1/400.000, and since Smith
demonstrated that he could determine line positions to 1/2500, the
precision of a single measurement in his spectrometer is $1/10^9$.

Shortly after the AMCO IV conference in Teddington in 1971,
where Smith reported a further beautiful set of measurements,
Smith's health started to fail in a serious way. He kept working
until July 1972, and obtained many new results of great importance
that have just been published posthumously.

Already before his death, a place was looked for where work
with Smith's magnificent instrument could be continued. As will
be told in the next talk, this will be done in Delft. We trust that
in this way Lincoln Smith's work will even after his passing keep
contributing to the progress of science.

L.G. SMITH'S PRECISION RF-MASS-SPECTROMETER TRANSFERRED FROM

PRINCETON TO DELFT

E. Koets

Applied Physics Department, Delft University of Technology

Lorentzweg 1, Delft, The Netherlands

1. INTRODUCTION

The operating principles and the essential elements of this very high precision mass spectrometer, as well as the modifications made until August 1971 have been described by Smith (1, 2, 3). Details about improved measurement procedures and instrumental modifications (shielding of insulators, clean-up of the RF-signal, beam blanking during sweep fly back, modification of the optics supplies, etc.) introduced by Smith from August 1971 until July 1972, were given in a recent paper by Smith and Wapstra (4). After completion of these modifications Smith measured the masses of ^{13}C, ^{14}C, ^{14}N, ^{15}N and ^{16}O. Reference (4) also reports on these measurements and on some earlier measurements (masses of ^{1}H, ^{2}H, ^{3}H, ^{3}He, ^{4}He and ^{19}F), partly not published before.

It is the purpose of this paper to give additional information on the error estimates in reference (4) and to cover the events since July 1972.

2. ERROR ESTIMATES

Further study of Smith's huge amount of notes recently revealed additional information for the error estimates of reference (4).

The shielding of the insulators of the Phase Defining slit (the slit located between the two modulator transitions) lowered the correction voltage V_{PD}, which has to be applied to one of the jaws of this slit, by approximately an order of magnitude from typically 250 mV to typically 20 mV. The uncertainty ΔV_{PD} in V_{PD}

has up to now been one of the main contributions (via σ_e in reference (4)) to the mass error estimates. This shielding lowered ΔV_{PD} by at least a factor of five. (The former uncertainty was 5 scale divisions. Probably the uncertainty is now approximately $\frac{1}{2}$ scale division, but because Smith's notes on V_{PD} are in full scale divisions only, the new uncertainty must be taken to be 1 scale division). The improved measurement procedure, mentioned in reference (4), which resulted in better orbit congruence, lowered somewhat the effect of the remaining electrostatic stray fields, and thus ΔV_{PD} was reduced still further. This means that (except for the top four and the very last one) the stated values for σ_e in table I of reference (4) have to be divided by at least a factor of five, while the total errors σ_r have to be divided by factors up to 4.4. As another consequence of the reduced ΔV_{PD} it will be significant to measure somewhat wider doublets now: up to $\Delta M/M=0.03$ for a total error in the doublet spacing of 10^{-9} of the mass of a doublet component.

The standard setting errors $\sigma_{s\,m}$ of the means in table I of reference (4) were calculated from the number of runs, from the precision of the matching procedure and from the peak resolution ρ. For the precision of the matching procedure the value of 1 part in 2500 of the peak half-width from references (2) and (3) was taken. The value for ρ, used throughout, was the value of $214 \cdot 10^3$, that Smith used in reference (3) to calculate the $\sigma_{s\,m}$ for the doublets at masses 3 and 4. However Smith's notes show that for the other doublets ρ was 1.5 to 1.8 times higher.

A review of the ratio of the measured variation σ_m of the means and the corrected $\sigma_{s\,m}$ shows that $\sigma_m/\sigma_{s\,m} > 1$ before the modifications were introduced and that $\sigma_m/\sigma_{s\,m} < 1$ after their completion. The reason why, in the past, σ_m was higher than $\sigma_{s\,m}$ is probably that the uncertainty in the frequently adjusted correction voltages, applied to various electrodes to compensate for charge ups, never has been included in $\sigma_{s\,m}$. Since the shielding of insulators most correction voltages were disconnected and thus their uncertainty became zero.

The fact that in the end σ_m was lower than $\sigma_{s\,m}$ means that by that time Smith was able to match the peaks to better than the above mentioned 1 part in 2500 of the peak half-width. A close review of $\sigma_m/\sigma_{s\,m}$ shows that the improvement is at least 30%, after completion of those modifications which increased the peak stability. Probably it is even possible now to match the peaks to 1 part in 3500 of the peak half-width.

3. TRANSFER TO DELFT AND SPECTROMETER REASSEMBLY

The work of Smith stopped in July 1972. Mainly during his last

months the possibilities for continuation of his work in the United
States were investigated, but without result. In 1973 it was then
decided to move the spectrometer from Princeton to Delft.

Before disassembly of the spectrometer a single reference
measurement was made by the author on N_2-CO, which was consistent
with Smith's results. It was then found that the magnetic stray-field
from the magnet power supply had a magnitude, at the center of the
ion-optics, of approximately 15 μT, which was about an order of
magnitude more than what came from the other sources. In Delft this
power supply will therefore be placed much further away from the
optics.

All parts of the spectrometer arrived in Delft late in August
1974. Remodeling the room where the instrument will be placed has
now been completed. The checking, repairing and tuning of the
electronics is in progress.

Because of the very narrow and long tubes in the pre-vacuum
system Smith had to pump down to 10^{-6} Torr to be able to start the
ion getter pumps. To reach this pre-vacuum took about twelve hours.
These narrow tubes have recently been replaced by new ones, four
times as wide. It is expected that the ion getter pumps can now be
started within a quarter hour. This is of importance because it is
expected that the system will be opened frequently. Furthermore it
is hoped that by means of the faster and better trapped pre-vacuum
system the oil vapor back-streaming and thus part of the
contamination rate, can be reduced.

Measurement of building vibrations revealed that the vertical
component can modulate the magnetic field to a significant extent.
The solid curves in the upper half of fig. 1 give, as a function
of frequency, the maximum during the day time and the minimum during
the night time of the peak value of the floor acceleration caused
by single frequency vibration components on a typical day.
Frequencies higher than 10 Hz are unimportant. The dotted curves
and the left-hand scale represent, in the same way, the resulting
magnet acceleration, when the spectrometer is mounted on a convenient
set of springs. The quality factor Q of this mass spring system is
about four. The magnet acceleration causes a peak value for the
relative variation of the pole piece separation given by the dotted
curves and the right-hand scale. The induced variation of the magnetic
field is reduced by the eddy currents. Measurements confirmed that
in this case the reduction factor G is given by the approximation
$G \approx \sqrt{(1+9f^2)}$, where f is the vibration frequency. The peak value of
the resulting relative variation of the magnetic field is represented
by the curves in the lower half of fig. 1. For the present visual
matching technique the total effect of the psychedelic frequencies
around the resonance peaks is dominating. The resulting total
relative mass error contributions are $2 \cdot 10^{-9}$ and $3 \cdot 10^{-10}$ respectively.

In the case of electronic data reduction these contributions can be made much smaller. Over long periods the building vibrations are expected to stay within a factor of two from the stated values. (In Princeton the vibration effects were probably comparable). Relatively simple means are available to lower the magnet acceleration and the resulting effects by at least a factor of ten.

The spectrometer units are now being (re-)assembled. It will probably not be until autumn before the whole machine can be put together again.

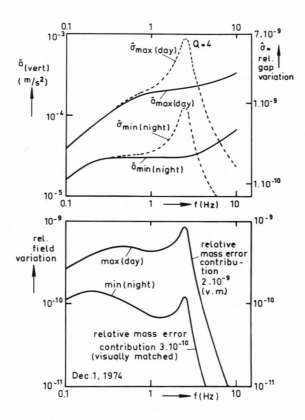

Fig. 1. Top: Peak floor acceleration $\hat{a}$ and the resulting peak relative magnet gap variation $\hat{\sigma}$.
Bottom: Resulting peak relative variation of the magnetic field, with the relative mass error contributions for the visual matching technique.

4. FUTURE RESEARCH

In the future the instrumental research will be focussed on increase in resolution and on reduction of systematic errors.

Fig. 2 shows the FWHM peak resolution as a function of the modulator RF-voltage as measured by Smith. Since Smith usually operated at about 130 V, one likely way to increase the resolution is to increase the modulator RF-voltage. The systematic shift in doublet spacing that could result from an increased RF-voltage can be calculated and thus corrected for.

Smith never reached the ion optical resolving power that he aimed for in 1960. To find out which aberrations will have to be reduced or corrected for to increase this resolving power the caustics will be investigated. For this purpose an ion-electron-transformer with associated optics is being developed by K.J.P. Stigter. To make better use of the active area of this transformer (the channel plate in fig. 3) a projector (f_3=20 mm) magnifies the ion focus ten times. Since the second lens (f_2=(50-∞)mm) is placed in the back focal plane of the first lens (f_1=100 mm) and since the image distance of the second lens has been chosen equal to f_1, the first two lenses form a telecentric object plane selector: the magnification is constant for all possible object planes. Any object plane 0 to 200 mm in front of the first lens can be imaged on the channel plate by adjusting the second lens only. This system will be inserted at different places to investigate the spectrometer foci.

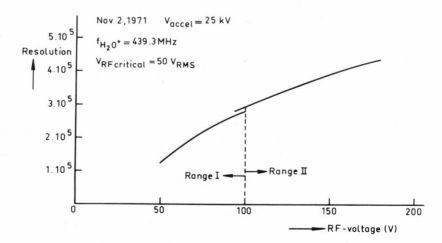

Fig. 2. FWHM peak resolution as a function of the modulator RF-voltage (RMS). Range I and II indicate the voltmeter ranges.

Finally it is hoped to increase the resolution through improved peak matching by means of electronic data reduction. This could also produce information on peak deformations (not caused by mechanical vibrations) and the systematic errors that may result from them.

It will probably be necessary to reduce some systematic errors in the RF frequency ratio by further reduction of spurious signals.

From Smith's notebooks it is clear that the inside walls of the machine must be contaminated with polymers, which together with the unnecessarily wide beam will give stray electric fields. The contamination and the beam width will be reduced as far as possible.

The wide beam will probably also cause stray currents in the high impedance voltage dividers which define the voltages for the electrodes. If so, both these currents and the impedances will be reduced to lower the corresponding systematic errors.

5. REFERENCES

1. L.G. Smith, in "Proceedings of the Third International Conference on Nuclidic Masses", edited by R.C. Barber (University of Manitoba Press, Winnipeg, Manitoba, 1967), p. 811.
2. L.G. Smith, Phys. Rev. C 4, 22 (1971).
3. L.G. Smith, in "Proceedings of the Fourth International Conference on Atomic Masses and Fundamental Constants", edited by J.H. Sanders and A.H. Wapstra (Plenum, New York, 1972), p. 164.
4. L.G. Smith and A.H. Wapstra, Phys. Rev. C 11, 1392 (1975).

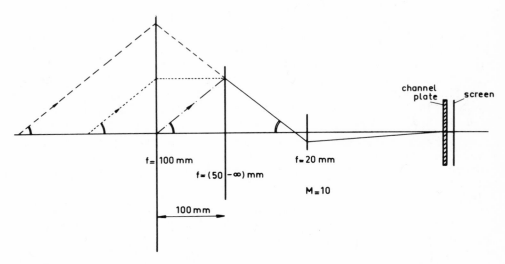

Fig. 3. Ion-electron-transformer with telecentric object plane selector and magnifying projector.

PRESENT STATUS OF THE PROGRAM OF ATOMIC MASS DETERMINATIONS AT THE UNIVERSITY OF MANITOBA

R.C. Barber, J.W. Barnard, S.S. Haque, K.S. Kozier,
J.O. Meredith, K.S. Sharma, F.C.G. Southon, P. Williams
University of Manitoba, Winnipeg, Canada
H. E. Duckworth
University of Winnipeg, Winnipeg, Canada

At previous International Conferences we have described the 9' radius ("Manitoba I")[1] and 1 m radius ("Manitoba II")[2] high resolution mass spectrometers at the University of Manitoba and have reported the progress of our program of atomic mass determinations. In this paper we shall comment on the continuing development of experimental technique and summarize recent results and the status of some work in progress.

EXPERIMENTAL TECHNIQUES

To determine the mass difference between two ionic species, the so-called "peak-matching" technique is used,[1,2] by which, on every second sweep of the oscilloscope display, the potentials in the instrument are changed so as to displace the pattern. The potential change needed to bring a peak on the displaced pattern into coincidence with an undisplaced peak can be used to calculate the mass difference in question.

The introduction of signal averaging techniques to peak matching[3] led to a major improvement in the precision attainable by a given mass spectrometer, inasmuch as the signal to noise ratio is improved by the addition of many sweeps of the oscilloscope trace. In our version, each oscilloscope trace is divided into 1024 sections and the voltage in each interval converted to a number which is added to the magnetic memory. Matching is carried out by allowing the two peaks to cancel one another. The matched condition corresponds to a null on the memory display oscilloscope.

More recently[4] we have made extensive use of computer-assisted methods of peak matching. The signal averager memory is divided into quarters which correspond to 4 consecutive sweeps. A reference peak is added to the first quarter and the "unknown" peak is added to each of the three remaining quarters at positions which correspond to three voltages which bracket the matched condition.

In the analysis most frequently used, the centroid positions of the peaks are calculated and fitted by a least squares linear fit to the related voltages. From this expression the voltage corresponding to the matched condition is deduced.

In an alternate analysis, used for some of the results most recently published,[5] the peak in each of the second, third, and fourth quarters is displaced artificially and matched by a numerical procedure to the reference peak in a way which is exactly equivalent to the "null" method described earlier. The values derived for the three displacements are then related to the corresponding voltages as in the centroid method. We propose to develop this latter technique by minimizing the sum of the squares of the differences between the peaks to establish the null condition.

Until recently the spectra accumulated in the signal averager have been obtained by using the "signal averaging" mode. That is, the input signal was converted, point by point, to digital information and added to the memory at the appropriate address. In this mode, however, low frequency variation in the level of the baseline of the signal entering the averager is added to the desired information. This has the undesirable effect of modulating the position of the centroid of the peak by a small artificial amount. In addition, noise is intentionally added to the signal in the digitizer section of the signal averager. It is found that this noise tends to obscure very weak signals unnecessarily. These deficiencies have been corrected by modifying the detection electronics so that each ion pulse from the electron multiplier is recorded as one event directly in the memory.

The improvement in sensitivity of detection has made possible determination of mass differences of doublets having intensity ratios more unfavourable than was previously possible and has also made possible the detection of hitherto unobservable contaminants. In former work, doublets involving known minor contaminants were avoided, but the present analysis permits the correction of the raw spectra for the presence of unresolved, or barely resolved contaminants of known relative intensity.

Finally, it has been shown[4] that a systematic correction must be made to the value of a doublet. When a known wide doublet of one or two mass units spacing is determined, the uncorrected value is found to differ from the known value by systematic amounts. We ascribe this behaviour to the presence of surface charges; in Manitoba II these accumulate on the plates of the

electrostatic analyser, while in Manitoba I they accumulate both
there and on the walls of the vacuum chamber in the magnetic
analyser. The size of the effect appears to depend on the composition of the ion beam and, less strongly, on the residual pressure in the instrument. In practice, the magnitude of the correction is determined by a wide calibration doublet ($\Delta M/M \sim 1/100$)
and then applied as a fractional correction to the narrower doublet under study. All new doublet values have been corrected in
this way by amounts from 0 to 250 ppm.

RECENT RESULTS

In our systematic study of mass differences between nuclides
in the rare earth region, we have made extensive use of doublets
of the first three types shown in Table I. These doublets, which

TABLE I - Types of Mass Doublets Studied

				Spacing ($\Delta M/M$)
1.	$^{A+2}_X{}^{35}Cl$	$- {}^{A}_Y{}^{37}Cl$	$= \Delta M_1$	$\sim 1/40,000$
2.	$^{A+4}_X{}^{35}Cl_2$	$- {}^{A}_Y{}^{37}Cl_2$	$= \Delta M_2$	$\sim 1/20,000$
3.	$^{A}_X$	$- {}^{A}_Y$	$= \Delta M_3$	$1/40,000-1/115,000$
4.	$^{A+3}_X{}^{16}O_2$	$- {}^{A}_Y{}^{35}Cl$	$= \Delta M_4$	$\sim 1/10,000$
5.	$^{A+1}_X{}^{16}O_2{}^{37}Cl-$	${}^{A}_Y{}^{35}O_2$	$= \Delta M_5$	$\sim 1/10,000$
6.	$^{186}_W{}^{16}O$	$- {}^{12}C{}^{13}C{}^{35}Cl_4{}^{37}Cl$	$= \Delta M_6$	$\sim 1/1,200$

exist in the spectra of rare earth chlorides, are all relatively
narrow and, when combined with the well known $^{37}Cl-^{35}Cl$ mass difference, provide mass links between nuclides X and Y which differ
in mass number by 0, 2, or 4 mass numbers.

The first measurements[6] made with Manitoba II in the region
Z = 59 to 69 were chosen to complement earlier measurements made
with Manitoba I.[7,8] We subsequently combined these data with previous precise mass spectroscopic data, and with the significant
nuclear reaction and decay Q-values in a least squares evaluation[9]
in order to derive a set of best values for the atomic mass differences. We note that these data were not corrected for the effect of the systematic bias which we mentioned above. Since they
largely determine the calculated mass differences between stable
nuclides in this region, these differences are biased by a small
amount ($\sim .5$ μu for a change of 1 u).

The more recent doublet measurements[5] have been corrected for systematic bias by the procedure described earlier. We have combined these data with existing Q-values in a small least squares evaluation for Z = 68 to 72, the results of which are given in Tables II and III.

It would appear from the later set of results in the rare earths,[5] and from studies of wide doublets done during an overlapping time interval, that a reliable estimate of the systematic correction may be made in retrospect and that the earlier determinations (reference 6) should be increased by 250±100 ppm. Such a change would increase the doublet values by, in most cases, an amount only slightly larger than the standard deviation associated with the doublet measurement.

A further constraint against the possibility of accumulated bias in the narrow doublets may be introduced by the determination of a single doublet which gives a mass link equivalent to the sum of several close doublets. Two such doublets are shown in Table IV.

TABLE IV - Wide Doublets

Doublet	This Work (μu)	1971 Mass Table[10] (μu)	Previous Manitoba[5,9] (μu)
$^{175}Lu^{37}Cl-^{142}Nd^{35}Cl_2$	61 249.5±2.5	61 226±18	61 218.5±5.4
$^{176}Lu^{37}Cl-^{143}Nd^{35}Cl_2$	61 067.2±1.4	61 045±16	61 036.3±5.3

Using: $^{143}Nd-^{142}Nd$ = 1.002 089 9 ± 15 u (Manitoba)[9]

$^{176}Lu-^{175}Lu$ = 1.001 909 2 ± 13 u (1971 Mass Table)[10]

Loop closure is 1.6±3.5 μu

Evidently both the 1971 Mass Evaluation[10] and the Manitoba differences[5,9] as published give somewhat lower values for the total mass differences than does the single doublet. In the latter comparison, the correction for the systematic bias amongst the first group of rare earth determinations accounts for ~20 μu of the 31 μu discrepancy. As indicated in the table, the new wide doublets appear to be self-consistent.

Using doublets of the types shown in Table I we have extended our systematic study to W and Re and have obtained new values for the mass differences as indicated in Fig. 1. In each case the numbers shown give the errors in keV. We propose to extend these measurements via similar links to Hf and Ta.

TABLE 2 - Nucleon Separation and Pairing Energies

A	EL	Z	S2N	SN	PN	S2P	SP	PP
161	ER	68	*	*	*	10619 +- 12	6110 +- 50	*
	TM	69	*	*	*	*	*	*
162	ER	68	*	9215 +- 10	*	11244.3+- 1.7	6430 +- 50	1050 +- 39
	TM	69	*	7940 +-140	*	9750 +-110	3640 +-100	*
163	ER	68	16116 +- 12	6901 +- 17	1065 +- 6	11688 +- 5	6437 +- 30	*
	TM	69	17220 +-100	9280 +-100	830 +- 80	10130 +- 60	3701 +- 21	*
164	ER	68	15748.9+- 1.2	8848 +- 7	1036 +- 5	12342.8+- 2.1	6857.7+- 3.5	980 +- 8
	TM	69	16580 +-100	7303 +- 29	930 +- 33	10540 +- 36	4103 +- 21	*
165	ER	68	15498 +- 7	6650.3+- 0.7	1005.9+- 1.9	12726.1+- 4.7	6884 +- 21	*
	TM	69	16350 +- 37	9047 +- 36	949 +- 28	11170 +- 30	4303 +- 30	1055 +- 40
	YB	70	*	*	*	9446 +- 32	5343 +- 37	*
166	ER	68	15126.4+- 1.1	8476.1+- 1.3	966.5+- 0.8	13588.4+- 2.4	7313.5+- 1.4	873 +- 21
	TM	69	16043 +- 23	6995 +- 32	948 +- 25	11550 +- 50	4648 +- 11	*
	YB	70	*	9625 +- 32	*	10223 +- 7	5921 +- 31	1080 +- 29
167	ER	68	14912.2+- 1.3	6436.1+- 0.5	843.8+- 0.4	12461.5+- 2.8	7506.3+- 1.6	*
	TM	69	15733 +- 35	8738 +- 22	911 +- 18	12222 +- 20	4909 +- 19	953 +- 25
	YB	70	16695 +- 34	7070 +- 14	1135 +- 13	10643 +- 12	5995 +- 17	*
	LU	71	*	*	*	9140 +- 80	3220 +- 70	*
168	ER	68	14207.2+- 0.7	7771.1+- 0.6	775.8+- 0.3	14978 +- 5	7961 +- 20	833 +- 15
	TM	69	15574 +- 18	6836 +- 23	775 +- 18	12620 +- 60	5309 +- 14	*
	YB	70	16125 +- 7	9055 +- 12	1043 +- 9	11221.9+- 3.2	6313 +- 18	976 +- 16
	LU	71	*	7770 +-110	*	9910 +- 80	3910 +- 80	*
169	ER	68	13774.2+- 0.6	6003.0+- 0.3	755.6+- 0.4	*	8520 +-100	*
	TM	69	14872 +- 19	8036 +- 14	661 +- 11	*	5574.2+- 0.8	897 +- 23
	YB	70	15922 +- 12	6867.3+- 1.1	947.7+- 3.2	11653.1+- 3.2	6344 +- 14	*
	LU	71	16720 +- 80	8950 +- 80	710 +- 70	10123 +- 34	3811 +- 28	*
	HF	72	*	*	*	8720 +-200	4810 +-220	*
170	ER	68	13260.6+- 1.3	7257.5+- 1.3	707.6+- 0.7	*	8580 +-100	*
	TM	69	14629 +- 14	6592.8+- 0.9	583.7+- 3.6	13532 +- 20	6164.0+- 1.2	*
	YB	70	15337.7+- 3.1	8470.4+- 3.1	864.8+- 1.7	12352.4+- 1.1	6778.2+- 0.9	*
	LU	71	16240 +- 80	7291 +- 31	*	10579 +- 19	4235 +- 13	*
	HF	72	*	*	*	*	*	*
171	ER	68	12939.2+- 1.4	5681.6+- 0.5	686.3+- 3.4	*	9100 +-100	*
	TM	69	14077.5+- 1.2	7484.6+- 1.3	531.6+- 2.4	14960 +-100	6391.1+- 1.5	780 +-100
	YB	70	15084.7+- 3.1	6614.3+- 0.7	815.8+- 0.9	12963.7+- 1.0	6799.7+- 0.9	*
	LU	71	*	*	*	*	*	*
	HF	72	*	*	*	*	*	*
172	ER	68	12532 +- 14	6851 +- 14	750 +- 80	*	*	*
	TM	69	13735 +- 9	6250 +- 9	475 +- 10	19060 +-100	6959 +- 9	*
	YB	70	14635.7+- 0.9	8021.4+- 0.9	765.4+- 0.6	13727.6+- 1.5	7336.4+- 1.3	847 +- 8
	LU	71	*	*	*	*	*	*
	HF	72	*	*	*	*	*	*
173	ER	68	11860 +-300	5010 +-300	*	*	*	*
	TM	69	13166 +- 30	6916 +- 31	464 +- 26	*	7024 +- 33	*
	YB	70	14388.2+- 0.8	6366.8+- 0.5	688.6+- 0.5	14412.7+- 1.5	7453 +- 9	*
	LU	71	*	*	*	12231 +- 30	4894 +- 30	913 +- 23
	HF	72	*	*	*	*	*	*
174	TM	69	12642 +- 41	5730 +- 50	486 +- 40	*	7740 +-300	*
	YB	70	13833.4+- 0.8	7466.7+- 0.9	686.0+- 0.5	15029 +- 13	8004 +- 30	868 +- 23
	LU	71	*	6785 +- 32	*	12766 +- 13	5313 +- 10	*
	HF	72	*	*	*	11000 +- 14	6106 +- 33	*
175	TM	69	12210 +- 60	6480 +- 60	530 +- 50	*	*	*
	YB	70	13289.3+- 1.0	5822.6+- 0.5	672.4+- 0.5	15840 +-300	8100 +- 40	*
	LU	71	14450 +- 30	7664 +- 10	563 +- 11	13515 +- 30	5510.6+- 1.3	920 +- 8
	HF	72	*	6708.8+- 0.5	*	11342 +- 14	6029 +- 17	*
176	TM	69	11590 +- 70	5110 +- 80	*	*	*	*
	YB	70	12690.6+- 1.1	6868.0+- 1.2	586.7+- 0.7	5322 +- 21	8490 +- 50	*
	LU	71	13958 +- 40	6293.2+- 1.1	537.9+- 2.6	14081 +- 40	5981.3+- 1.7	*
	HF	72	15041 +- 13	8332 +- 13	893 +- 7	12207.5+- 2.5	6696.8+- 2.2	*
177	YB	70	12434.7+- 1.7	5566.7+- 1.2	*	3586 +- 20	8940 +- 60	*
	LU	71	13366.8+- 1.8	7073.6+- 1.4	495 +- 13	14670 +- 50	6186.9+- 2.4	863 +- 13
	HF	72	14717 +- 13	6385.1+- 1.0	796.9+- 3.4	12769.9+- 2.4	6788.7+- 1.7	*
178	LU	71	12950 +- 50	5870 +- 50	582 +- 39	15440 +- 80	6490 +- 50	*
	HF	72	14010.7+- 0.8	7625.6+- 0.6	691.3+- 0.6	13527.6+- 2.6	7340.7+- 1.2	*
179	LU	71	12874 +- 40	7000 +- 60	640 +- 50	3670 +- 50	*	*
	HF	72	13726.7+- 0.6	6101.0+- 0.6	703.0+- 0.5	14061.9+- 2.9	7570 +- 50	*
180	LU	71	12570 +- 70	5570 +- 60	*	-389 +- 50	*	*
	HF	72	13489.3+- 0.5	7388.3+- 0.5	745.3+- 0.5	*	7956 +- 40	*
181	HF	72	13082.5+- 1.4	5694.2+- 1.3	650 +-200	*	8080 +- 50	*
182	HF	72	12280 +-200	6590 +-200	530 +-200	*	.*	*
183	HF	72	11974 +- 27	5390 +-200	*	*	*	*

TABLE 3 - Decay Energies and Masses

A	EL	Z	Q(B-) KEV	Q(ALPHA) KEV	MASS EXCESS KEV	MASS EXCESS MICRO U
161	ER	68	-3520 +-100	1795 +- 13	-65204 +- 12	-69998 +- 13
	TM	69	*	*	-61680 +-100	-66220 +-110
162	ER	68	-4790 +-100	1636.8+- 2.7	-66347 +- 6	-71226 +- 6
	TM	69	*	2480 +-100	-61550 +-100	-66080 +-100
163	ER	68	-2417 +- 20	1562 +- 8	-65176 +- 9	-69969 +- 10
	TM	69	*	*	-62759 +- 22	-67374 +- 24
164	ER	68	-3962 +- 20	1301.0+- 2.0	-65952 +- 6	-70802 +- 7
	TM	69	*	1960 +- 50	-61990 +- 21	-66549 +- 22
165	ER	68	-1565 +- 30	1098.7+- 4.8	-64531 +- 6	-69276 +- 7
	TM	69	-2922 +- 10	1800 +- 60	-62966 +- 31	-67596 +- 33
	YB	70	*	2735 +- 33	-60044 +- 32	-64459 +- 35
166	ER	68	-3046 +- 11	825.8+- 2.3	-64936 +- 6	-69710 +- 7
	TM	69	-292 +- 12	1640 +- 60	-61890 +- 13	-66441 +- 14
	YB	70	*	2325 +- 8	-61597 +- 10	-66127 +- 10
167	ER	68	-744 +- 19	665.9+- 2.4	-63300 +- 6	-67955 +- 7
	TM	69	-1960 +- 14	*	-62556 +- 20	-67156 +- 21
	YB	70	-3070 +- 70	2156 +- 14	-60596 +- 14	-65051 +- 15
	LU	71	*	2810 +- 70	-57530 +- 70	-61760 +- 80
168	ER	68	-1679 +- 14	548.5+- 2.5	-62999 +- 6	-67632 +- 7
	TM	69	259 +- 14	1220 +- 60	-61320 +- 15	-65829 +- 16
	YB	70	-4360 +- 80	1948.5+- 3.4	-61579 +- 7	-66107 +- 7
	LU	71	*	2350 +- 80	-57220 +- 80	-61430 +- 90
169	ER	68	353.6+- 0.9	258.6+- 2.9	-60931 +- 6	-65411 +- 7
	TM	69	-909.7+- 3.2	1212.1+- 1.8	-61284 +- 6	-65791 +- 7
	YB	70	-2274 +- 28	1731.5+- 3.4	-60375 +- 7	-64814 +- 8
	LU	71	-3370 +-200	2441 +- 41	-58101 +- 29	-62373 +- 31
	HF	72	*	2880 +-200	-54740 +-200	-58760 +-220
170	ER	68	-311.1+- 1.6	56 +- 6	-60117 +- 6	-64537 +- 7
	TM	69	967.8+- 0.6	850.7+- 3.2	-59806 +- 6	-64203 +- 7
	YB	70	-3453 +- 13	1737.3+- 1.1	-60773 +- 6	-65242 +- 7
	LU	71	*	2145 +- 17	-57320 +- 14	-61535 +- 16
	HF	72	*	*	*	*
171	ER	68	1492.0+- 1.5	*	-57727 +- 6	-61971 +- 7
	TM	69	97.5+- 0.9	687 +- 21	-59218 +- 6	-63573 +- 7
	YB	70	*	1559.0+- 0.9	-59316 +- 6	-63678 +- 7
	LU	71	*	*	*	*
	HF	72	*	*	*	*
172	ER	68	891 +- 10	*	-56506 +- 15	-60661 +- 16
	TM	69	1869 +- 9	*	-57397 +- 11	-61617 +- 12
	YB	70	*	1308.7+- 1.0	-59266 +- 6	-63624 +- 7
	LU	71	*	*	*	*
	HF	72	*	*	*	*
173	ER	68	2800 +-300	*	-53440 +-300	-57370 +-320
	TM	69	1320 +- 30	*	-56241 +- 31	-60376 +- 33
	YB	70	-690 +- 30	945.0+- 1.0	-57561 +- 6	-61793 +- 7
	LU	71	*	1989 +- 30	-56871 +- 31	-61053 +- 33
	HF	72	*	*	*	*
174	TM	69	3060 +- 40	*	-53896 +- 41	-57859 +- 43
	YB	70	-1371 +- 10	735.9+- 1.7	-56956 +- 6	-61144 +- 7
	LU	71	103 +- 17	1796 +- 10	-55584 +- 12	-59672 +- 13
	HF	72	*	2661 +- 14	-55687 +- 15	-59782 +- 16
175	TM	69	2400 +- 50	*	-52310 +- 50	-56150 +- 50
	YB	70	470.5+- 1.3	594.9+- 1.8	-54707 +- 6	-58729 +- 7
	LU	71	-853 +- 14	1616.4+- 1.7	-55177 +- 6	-59234 +- 7
	HF	72	*	2567 +- 14	-54324 +- 15	-58319 +- 16
176	TM	69	4160 +- 60	*	-49340 +- 60	-52970 +- 60
	YB	70	-104.3+- 1.9	578 +- 14	-53503 +- 6	-57437 +- 7
	LU	71	1186.1+- 1.9	1573 +- 9	-53399 +- 6	-57325 +- 7
	HF	72	*	2256.0+- 2.4	-54585 +- 7	-58599 +- 7
177	YB	70	1402.6+- 2.7	20 +-300	-50998 +- 7	-54748 +- 7
	LU	71	497.5+- 1.0	1415 +- 30	-52401 +- 7	-56254 +- 7
	HF	72	*	2237.7+- 2.3	-52898 +- 7	-56788 +- 7
178	LU	71	2250 +- 50	1270 +- 60	-50200 +- 50	-53890 +- 50
	HF	72	*	2078.7+- 2.4	-52452 +- 7	-56309 +- 7
179	LU	71	1350 +- 40	750 +- 60	-49131 +- 41	-52744 +- 44
	HF	72	*	1800.3+- 2.4	-50481 +- 7	-54193 +- 7
180	LU	71	3170 +- 50	290 +- 80	-46630 +- 50	-50060 +- 50
	HF	72	*	1280.0+- 2.6	-49798 +- 6	-53460 +- 7
181	HF	72	*	1152.5+- 3.2	-47421 +- 7	-50907 +- 7
182	HF	72	*	*	-45950 +-200	-49320 +-210
183	HF	72	*	*	-43258 +- 27	-46439 +- 29

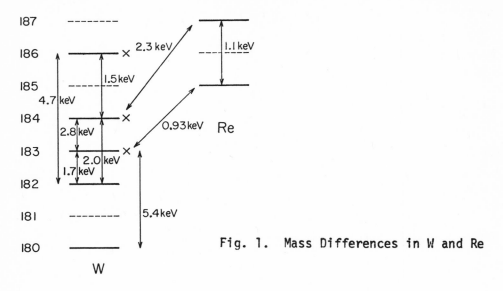

Fig. 1.   Mass Differences in W and Re

In Fig. 2, we show a plot of S_{2n} vs. N for even-N nuclides for N = 94 to 114.   The values shown for Os, Re, and Ta are, with

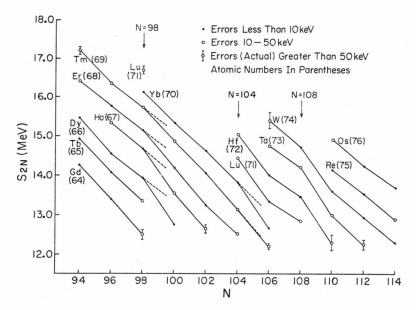

Fig. 2.   Double Neutron Separation Energy Systematics

one exception, from the 1971 Mass Table. The remainder were derived from all presently available data including our new measurements. In particular, the mass of ^{180}W has been revised by keV
in the light of a new measurement under considerably improved
experimental conditions.

As we have noted previously, these curves exhibit a remarkably
systematic behaviour throughout the region, viz, that when viewed
over an interval of two neutron numbers, the segments of adjacent
curves are almost parallel. Thus irregularities in the curve of
one element are reproduced for other elements at the same neutron
number.

On the basis of this regular behaviour we note that well-
defined downward breaks in the curves occur at N = 98, 104, and
108. These changes are reminiscent of the large changes at the
major shell closures and suggest that relatively large energy
gaps exist above the corresponding Nilsson levels (9/2+ (624)
level, N = 108 especially and smaller gaps for 5/2- (523), N = 98
and 1/2- (512), N = 104).

In addition to these doublets which provide mass differences
we are currently studying three wide doublets formed by WO and
C_2Cl_5, one of which is given as 5 in Table I. From these one may
derive the total mass of ^{183}W, ^{184}W, and ^{185}W, rather than the
mass differences. We anticipate that a least squares adjustment
of these masses and mass differences should yield a group of very
precisely known atomic masses for this immediate region.

Finally, using Manitoba I, we have determined doublets in
ZnCN and in ferrocene from which single neutron separation energies are derived directly. The precision associated with these
measurements is high (0.25 to 0.5 keV), although we are not yet
satisfied that the possible sources of systematic error at this
level of precision have been eradicated.

1. BARBER, R.C. et al., Proceedings of the Second International
   Conference on Nuclidic Masses (Springer Verlag, Vienna, 1964)
   p. 393.
2. BARBER, R.C. et al., Proceedings of the Third International
   Conference on Atomic Masses (University of Manitoba Press,
   Winnipeg, 1968) p. 717.
3. BENSON, J.L. and JOHNSON, W.H., Phys. Rev. 141, 1112 (1966).
4. BARBER, R.C. et al., Atomic Masses and Fundamental Constants 4
   (Plenum Press, London, 1972) p. 141.
5. BARBER, R.C. et al., Can. J. Phys. 23, 2386 (1974).
6. BARBER, R.C. et al., Can. J. Phys. 50, 1 (1972).
7. MACDOUGALL, J.D. et. al., Nucl. Phys. A145, 223 (1970).
8. WHINERAY, S. et al., Nucl. Phys. A151, 377 (1970).
9. MEREDITH, J.O. and BARBER, R.C., Can. J. Phys. 50, 1195 (1972).
10. WAPSTRA, A.H. and GOVE, N.B., Nuclear Data Tables A9, 267 (1971).

RECENT DOUBLET RESULTS AND MEASUREMENT TECHNIQUE DEVELOPMENT AT THE UNIVERSITY OF MINNESOTA*

David G. Kayser, Justin Halverson and Walter H.Johnson,Jr.

School of Physics and Astronomy, University of Minnesota

Minneapolis, Minnesota, U.S.A.

This paper will describe some recent doublet results and describe further peak matching technique development which has occurred as part of the continuing program of atomic mass measurements at the University of Minnesota. At AMCO 4, David Kayser presented a brief review of a technique we have named generalized peak matching (1). This technique is particularly useful in circumstances in which the two members of the doublet to be measured are not completely resolved. Dr. Kayser has developed the general theory of this process which deals with the determination of the mass differences for a multi-component spectrum. In order to be brief in this part of the presentation, the technique will be illustrated only with a mass doublet. Suppose that we assume a two component spectrum with individual peak shapes $f(t)$. The two component spectrum may be written $g(t) = f(t) + A f(t-b)$ in which b is the spacing and A is the amplitude ratio between the peaks. We may use Laplace transform theory to transform this equation to $g(s) = (1+Ae^{-bs})f(s)$. This process isolates the quantities A and b from the function $f(s)$.

At this point it is convenient to define a quantity $h(s,B,\beta)$ which will equal $f(s)$ when $B=A$ and $\beta=b$. In this way we may "resolve" $f(s)$. The function $h(s,B,\beta)$ is defined as

$$h(s,B,\beta) = \frac{(1+Ae^{-bs})}{1+Be^{-\beta s}} f(s).$$

*Supported by Contract N00014-67-A-0113-0018 with the Office of Naval Research, a grant from the Graduate School of the University of Minnesota and a grant MPS74-19408 from the National Science Foundation.

Figure 1.   Partial Sums of $h(t,B,\beta)$.

By expanding the denominator in an infinite series, this may be written as

$$h(s,B,\beta) = \left(f(s)+Ae^{-sb}f(s)\right)\sum_{n=0}^{\infty} (-1)^n(Be^{-s\beta})^n.$$

If the inverse transform of $h(s,B,\beta)$ is now taken, the result is

$$h(t,B,\beta) = \sum_{n=0}^{\infty} (-1)^n\left(f(t-\beta n)+Af(t-b-\beta n)\right) B^n.$$

In order to understand the usefulness of this function, let us plot a few partial sums with $B=A$ and $\beta=b$, shown in Figure 1B.

We call $h(t,B,\beta)$ the generalized error signal because the two term series is nothing more than the error signal which we have employed for some time for peak matching. By adjusting the value of β and B one can minimize the value of $h(t,B,\beta)$ in the interval between the two peaks in the sum. This matching may be done as a real-time process or as a computer calculation using a stored spectrum. We have studied both of these possibilities and I wish to spend the remainder of this paper discussing the results we have obtained.

For his Ph.D. thesis, David Kayser has adapted our 16" magnetic radius mass spectrometer to store the proper spectrum which provides the input for a computer mass determination program. The adaptation involved changing to an electrostatic sweep by adding highly linear sawtooth voltages to the usual square wave voltages supplied to the electrostatic analyzer and the ion source. A simple mass scale is produced with this sweep system. Figure 2A shows a doublet spectrum for which the doublet pair is swept and then translated on alternate sweeps by a known mass using the precision voltage divider which we have employed earlier for conventional peak matching. Calibration of the mass scale is provided from the knowledge of the $\Delta R/R$ value after measuring the spacing between the similar pairs of peaks in this spectrum. Ion

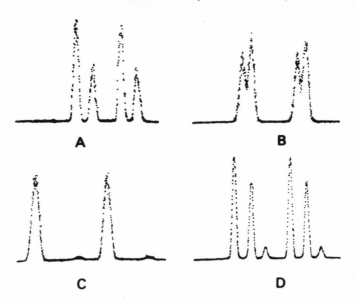

Figure 2.   Spectra of narrow doublets and triplets.

signals were recorded as individual ion pulses and stored in the
1024 channel memory of a FabriTek Model 1062 signal analyzer.
After a suitable number of sweeps are stored, the information is
transferred to a computer compatible magnetic tape.   These data
provide input to a computer matching program whose output is the
mass difference of the stored doublet.

   The experimental work for Dr. Kayser's thesis involved the
determination of masses and mass differences for stable isotopes of
samarium and gadolinium.   Wide doublets reported in this work were
measured using conventional techniques previously described(2).
Narrow doublets reported were measured with the new computer matching technique just described.   Many of these narrow doublets employed metallic chloride ions similar to those measured by the
Manitoba group(3).   Several cases of narrow doublets are shown in
Figure 2.   Figure 2B is the very narrow doublet ^{154}Gd - ^{154}Sm
measured at a resolution of 200,000.   The mass difference measured
for this doublet is 1.343 mu.   Figure 2C shows a wider doublet
$^{155}Gd^{35}Cl_3$ - $^{149}Sm^{37}Cl_3$ for which the intensity ratio is extreme.
Each small peak contains approximately 350 counts.   A more typical
spectrum is shown in Figure 2D which is the triplet $^{160}Gd^{35}Cl_2$ -
$^{158}Gd^{35}Cl^{37}Cl$ - $^{156}Gd^{37}Cl_2$.

The wide doublet results which lead to mass determinations for ^{152}Sm and ^{154}Sm are listed in Table I. The hydrogen mass difference calculated from the two ^{154}Sm doublets is in good agreement with the accepted value. By comparing the value of the ^{154}Sm − ^{152}Sm mass difference calculated from these doublets with the more precise narrow doublet value, a difference of 5μu is obtained. This difference is within the experimental error of the wide doublet measurements. Narrow doublet results of mass differences for most stable isotopes of samarium and gadolinium are reported in Table II. Note that the smallest quoted error, ± 0.4 μu for the ^{154}Sm^{35}Cl − ^{152}Sm^{37}Cl doublet, is comparable to the precision obtained for (n,γ) reactions.

The present results were compared with mass spectroscopic measurements made in other laboratories as well as with nuclear reaction data. Good agreement was generally found in these comparisons. In particular, the agreement with mass spectroscopic measurements made by the Manitoba group(3) is excellent. The only discrepancy of a serious magnitude is for the doublet ^{158}Gd^{35}Cl − ^{156}Gd^{37}Cl for which the difference is 5.8 ± 1.5μu. Comparison with earlier McMaster values which were quoted by the Manitoba group in reference(3) are also in disagreement at about the 2σ level with the present results. As a further test of the present results, a least squares adjustment was constructed for the region praseodymium to gadolinium. The results of the adjustment agree well with the Manitoba evaluation(4). A comparison with the earlier evaluation of Wapstra and Gove(5) shows that the Wapstra and Gove values are generally higher than the present values by 15 or 20 μu with a typical

TABLE I − SAMARIUM MASS DETERMINATIONS

Doublet	Result u	Adopted Error μu
$C_{12}H_8$ − ^{152}Sm	0.142 867 0	5
^{154}Sm − $C_{12}H_9$	0.851 789 0	8
$C_{12}H_{10}$ − Sm^{154}	0.156 035 7	4
^{1}H	1.007 824 6	
^{154}Sm − ^{152}Sm	2.002 482	Δ = 5 μu

TABLE II – NARROW DOUBLET RESULTS

Doublet	Mass Difference μu	Error μu
$^{149}Sm^{35}Cl - ^{147}Sm^{37}Cl$	5239.8	0.8
$^{150}Sm^{35}Cl - ^{148}Sm^{37}Cl$	5404.8	0.6
$^{152}Sm^{35}Cl - ^{150}Sm^{37}Cl$	5402.7	0.8
$^{152}Sm^{35}Cl_2 - ^{148}Sm^{37}Cl_2$	10807.9	1.4
$^{154}Sm^{35}Cl - ^{152}Sm^{37}Cl$	5427.2	0.4
$^{154}Sm^{35}Cl_2 - ^{150}Sm^{37}Cl_2$	10832.9	5.2
$^{154}Sm - ^{154}Gd$	1342.8	0.8
$^{156}Gd^{35}Cl - ^{154}Gd^{37}Cl$	4203.0	1.0
$^{157}Gd^{35}Cl - ^{155}Gd^{37}Cl$	4289.0	0.7
$^{158}Gd^{35}Cl - ^{156}Gd^{37}Cl$	4930.8	0.7
$^{160}Gd^{35}Cl - ^{158}Gd^{37}Cl$	5900.0	0.5
$^{155}Gd^{35}Cl_3 - ^{149}Sm^{37}Cl_3$	14282.4	6.3

quoted error of ± 12 μu. The details of this adjustment as well as a more detailed account of the complete experiment should be published in the near future in Physical Review C.

We will now turn to the third part of this paper, a discussion of a system which employs the generalized error signal technique in real time. Justin Halverson has constructed the circuitry to per-

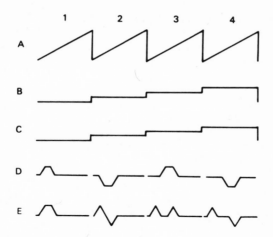

Figure 3.   Wave forms for a 4-step peak matching.

form a 4-step peak matching as his MS thesis project.   Figure 3
shows the necessary wave forms needed for this process.   Wave form
(A) is a sawtooth magnetic sweep which remains the same throughout
the steps.   Wave form (B) is the electrostatic analyzer voltage
generated as V, V + ΔV, V + 2 ΔV and V + 3 ΔV.   Wave form (C)
is a similar system of square waves for the accelerating voltage.
Wave form (D) shows a set of single sweeps for a pair of unresolved

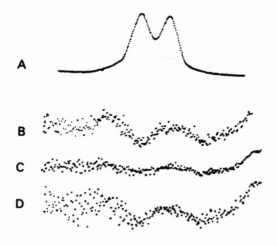

Figure 4. C_3H_8 - CO_2 doublet with error signals.

triangular peaks. On alternate sweeps the polarity of the signal
is reversed. Finally, wave form (E) is the stored signal which results from summing all previous wave forms shown in (D). Wave form
(E) is thus the generalized error signal. Circuitry was constructed
to provide this set of wave forms and was tested on a small double
focusing mass spectrometer with a resolution of about 800. Figure 4
shows the results of the matching of an unresolved C_3H_8-CO_2 doublet.
Figure 4A shows the unresolved doublet and Figures 4B, C and D indicate the central stored data for a series of ΔV values. Sweep B
corresponds to a $\Delta V/V$ ratio of 0.001658, sweep C to 0.001649 and
sweep D to 0.001637. It is clear that the sweep C has the smallest
error signal. With this system, a matching precision of at least
5 parts in 1000 is obtained. Using a measurement technique of
this sort with a high resolution instrument should therefore provide useful information for narrow doublets.

1.   D. C. Kayser, R. A. Britten, and W. H. Johnson, Jr.,
     Atomic Masses and Fundamental Constants 4, edited by
     J. H. Sanders and A. H. Wapstra (Plenum Publishing
     Corporation, New York, 1972), p.172.

2.   J. L. Benson and W. H. Johnson, Jr., Phys. Rev. 141,
     1112 (1966).

3.   R. C. Barber and R. L. Bishop, Can. J. Phys. 50,
     34 (1972).

4.   J. O. Meredith and R. C. Barber, Can. J. Phys. 50,
     1195 (1972).

5.   A. H. Wapstra and N. B. Gove, Nuclear Data Tables
     9, Nos. 4 and 5, (1971).

DOUBLE FOCUSING MASS SPECTROMETERS

OF SECOND ORDER

H. Matsuda

Institute of Physics, College of General Education

Osaka University, Toyonaka, Japan

1. Introduction

The resolution of the on-line mass spectrograph is normally several hundred and this is sufficient to separate nuclei having different mass number A. If the nuclei to be analyzed have the same A but different charge number Z, they can not be separated by this amount of resolution. However, the complete separation of nuclei (both A and Z) can be possible if the resolution is raised by about hundred times, that is, up to several ten thousands. Fig.1 shows an example of the estimated mass excess curve of nuclei with the same A plotted as a function of Z. The length of the bars in the figure corresponds to the separable mass differences for the resolution indicated. It is seen that the resolving power of about 30000 would be sufficient to resolve all nuclei far from the valley of beta-stability. Besides, the direct mass measurement of short lived nuclei would be possible with such a high resolution mass spectrograph. The determination of masses of nuclei far from beta-stability is also a very interesting and important problem.

For this purpose, a mass spectrograph which can collect many ions and still has a high resolution is necessary. In order to satisfy such conditions, it is essentially necessary to realize good focusing. Therefore, the possibility of correcting for second order image aberrations of a double focusing instrument is investigated and several suitable designs are found by computer calculation (2). It is possible to realize a complete second order double focusing mass spectrometer consisting of a toroidal electric field and a homogeneous magnetic field or of a cylindrical electric field plus an electrostatic quadrupole lens and a homogeneous magnetic field. In order to examine the focusing nature experimentally, both types of mass spectrometers are constructed and will be described here.

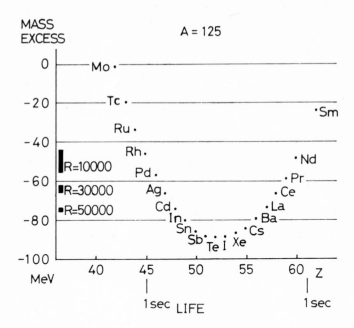

Fig.1. Mass excess as a function of Z for A=125.
(estimated by G.T. Garvey et. al. (1))

2. Image Aberrations

The final position of an ion at the detector slit is given by
the equation,

$$x_f = A_x x + A_\alpha \alpha + A_\delta \delta + A_\gamma \gamma + A_{\alpha\alpha} \alpha^2 + A_{\alpha\delta} \alpha\delta + A_{\delta\delta} \delta^2 + A_{yy} y^2 + A_{y\beta} y\beta$$

$$+ A_{\beta\beta} \beta^2 + A_{\alpha\alpha\alpha} \alpha^3 + A_{\alpha\alpha\delta} \alpha^2\delta + A_{\alpha\delta\delta} \alpha\delta^2 + A_{\alpha yy} \alpha y^2 + A_{\alpha y\beta} \alpha y\beta$$

$$+ A_{\alpha\beta\beta} \alpha\beta^2 + A_{\delta\delta\delta} \delta^3 + A_{\delta yy} \delta y^2 + A_{\delta y\beta} \delta y\beta + A_{\delta\beta\beta} \delta\beta^2$$

in the third order approximation, where α and β denote the radial
and axial inclination of the incident beam, x and y the coordinates
within the object slit, δ and γ the relative energy and mass deviation respectively. The coefficient A_x gives the magnification of
the image and the coefficient A_γ the mass dispersion. The first
order double focusing condition requires that $A_\alpha = A_\delta = 0$. The coefficients A_{ij} are the second order and A_{ijk} the third order aberration
coefficients. Here those aberration terms that include x are neglected because x is expected to be very small in order to get high

TABLE I.

Examples of parameters which satisfy the conditions $A_\alpha = A_\delta = A_{\alpha\alpha} = A_{\alpha\delta} = A_{\delta\delta} = 0$ for the apparatus with toroidal electric field and values of A_x, A_y, A_β, A_{yy}, $A_{y\beta}$ and $A_{\beta\beta}$.

ϕ_e	ϕ_m	r_e/r_m	l_e'/r_m	c	c'	$Q1$	ε'	ε''	A_x	A_y	A_β	A_{yy}	$A_{y\beta}$	$A_{\beta\beta}$
80	92.9	1.223	1.588	0.55	-0.55	-2.0	30	-15	0.39	-1.4	-1.2	-0.17	0.49	-1.23
85	90	1.058	1.179	0.625	-0.625	-2.5	30	-13.2	0.49	-1.9	-1.6	-0.44	0.42	-0.51
90	90.9	0.958	0.985	0.525	-0.525	-2.5	30	-10	0.66	-2.2	-1.5	-0.12	0.46	-0.99
95	92.1	0.862	0.793	0.55	-0.55	-3.0	30	-10	0.76	-2.7	-1.7	-0.18	0.29	-0.72
100	93.6	0.800	0.632	0.55	-0.44	-3.5	30	-10	0.87	-3.0	-1.6	-0.13	0.25	-0.54
105	91.0	0.709	0.458	0.6	-0.36	-4.5	30	-10	1.05	-3.7	-1.7	-0.13	0.19	-0.25
112	90	0.665	0.299	0.59	0	-5.0	32	-10.5	1.25	-4.4	-2.0	-0.03	0.10	-0.07

TABLE II.

Examples of parameters which satisfy the conditions $A_\alpha = A_\delta = A_{\alpha\alpha} = A_{\alpha\delta} = A_{\delta\delta} = 0$ for the apparatus with cylindrical electric field and electrostatic Q-lens and values of A_x, A_y, A_β, A_{yy}, $A_{y\beta}$ and $A_{\beta\beta}$.

ϕ_e	ϕ_m	r_e/r_m	l_e'/r_m	QK	$Q1$	$Q2$	ε'	A_x	A_y	A_β	A_{yy}	$A_{y\beta}$	$A_{\beta\beta}$
70	65.8	1.343	1.934	1.84	-3.5	0	0	0.47	-2.1	-3.8	-0.09	0.03	-0.01
75	68.8	1.322	1.757	1.85	-3.5	0	-3.5	0.46	-1.9	-2.8	-0.09	0.10	-0.10
80	70.8	1.314	1.564	1.86	-3.5	0	-10	0.43	-1.7	-1.8	-0.14	0.14	-0.10
80	67.9	1.317	1.550	1.89	-3.5	-2.0	-4	0.50	-2.0	-2.9	-0.08	0.03	-0.01
85	72.5	1.272	1.290	1.91	-4.0	0	-15	0.44	-1.7	-1.4	-0.20	0.13	-0.04
85	70.7	1.290	1.303	1.91	-4.0	-2.5	-9.5	0.49	-1.8	-2.0	-0.09	0.10	-0.03
90	73.7	1.276	1.173	1.92	-4.0	0	-18	0.43	-1.5	-1.0	-0.19	0.17	-0.17
95	74.1	1.270	1.048	1.92	-4.0	0	-21	0.42	-1.4	-0.6	-0.20	0.25	-0.20

resolution.

If we assume that the magnitudes of α, β, δ and y are all equal to 0.01 and that the mass dispersion A_y is equal to 1, then all the second order aberration coefficients A_{ij} should be at least less than 0.3 and all the third order aberration coefficients A_{ijk} less than 30 in order to get a resolving power of 30000.

The aberration coefficients are calculated by the matrix method. All coefficients up to third order are obtained by making the product of the 49×49 transfer matrices for the sector fields (3-5) and the free spaces. The influence of the fringing field is accounted for by a suitable transformation at the respective ideal field boundaries(6-8).

The conditions for complete second order focusing are sought by changing the following design parameters: ϕ_e, ϕ_m, r_e/r_m, l'_e, c, c', Q1, Q2, ε' and ε'', where Q is the ratio r_e to the radius of curvature of the condenser electrodes and the numbers 1 and 2 denote entrance and exit respectively. The straight boundaries of the magnetic field are adopted for practical reasons. The distance $d = l''_e + l'_m$ and l''_m are adjusted to satisfy the first order double focusing conditions $A_\alpha = A_\delta = 0$.

It was not possible to find suitable parameters for the apparatus consisting of a cylindrical condenser and a homogeneous magnetic field. But many suitable combinations of parameters are found if the cylindrical condenser is replaced by a toroidal condenser or if an electrostatic quadrupole lens is placed between the cylinder and the magnet. The Q-lens is used as a defocusing mode on the median plane. Some of examples are given in TABLE I and in TABLE II. The values of coefficients in the tables are normalized to those for $r_m = 1.0$. The third order aberration coefficients calculated for the mass spectrometers constructed are shown in TABLE III together with those of some famous instruments.

3. Apparatus with Toroidal Electric Field (TH-Type)

A mass spectrometer consisting of a toroidal electric field and a homogeneous magnetic field is constructed at Hitachi Central Research Laboratory. The ion path is shown in Fig.2 and the dimensions are as follows:

$r_e = 212$ mm, $\phi_e = 85.2°$, c= 0.5, c'= -0.5, Q1=-2.0

$r_m = 200$ mm, $\phi_m = 90°$, $\varepsilon'= 30°$, $\varepsilon''= -9.6°$

$l'_e = 248$ mm, $l''_e = 67$ mm, $l'_m = 169$ mm, $l''_m = 273$ mm

The aberration coefficients calculated are given in TABLE III. In the preliminary experiment a resolution of 50,000 and a total transmission of 43% were obtained.

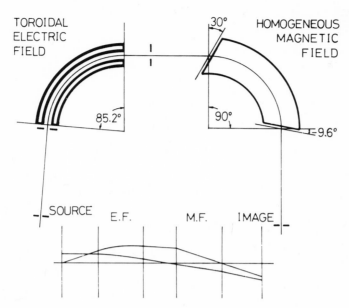

Fig.2.   Ion path of the apparatus with toroidal electric field.
         Vertical trajectory is shown in the lower part.

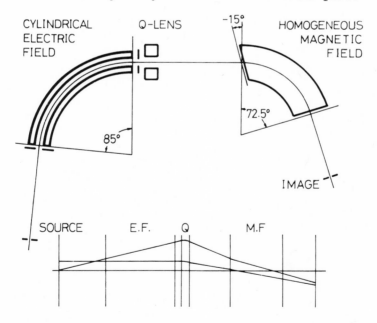

Fig.3.   Apparatus with cylindrical electric field and Q-lens.
         Vertical trajectory is shown in the lower part.

4. Apparatus with Cylindrical Electric Field and Q-Lens (CQH-Type)

    This type of mass spectrograph is suitable for a big instrument because the cylindrical condenser is easier to machine than the toroidal condenser. The dimensions of the apparatus constructed for the test of focusing are as follows:

r_e= 636 mm, ϕ_e= 85°, Q1= -4.0, Q2= 0

r_m= 500 mm, ϕ_m= 72.5°, ε'= -15°, ε''= 0

l_e'= 645 mm, d_{eq}= 75 mm, d_{qm}= 565 mm, l_m''= 447 mm

QK= 0.0382/cm, QL= 110 mm

The ion path is shown in Fig.3 and the aberration coefficients calculated are given in TABLE III.

    The condenser electrodes are made of pure iron. The length of cylinder is 220 mm and the gap distance 40.65 mm. Above and below the cylinder two plates called Matsuda plates are placed and applied 1/63 of total potential, so that very accurate cylindrical electric field is produced in a sufficient region to accept wilde angle beam. The front face of the condenser electrodes is made concavely curved with a radius of 159 mm. The angle defining slit (α-slit) of 0~22 mm is attached at the entrance and the energy defining slit (δ-slit) of

TABLE III. Aberration coefficients up to third order calculated for several instruments.

A	CQH	TH	MINNESOTA (9)	MANITOBA (10)	OSAKA (11)
(x)	0.44	0.55	0.68	0.49	1.34
(y)	-1.65	-1.86	1.10	-0.10	1.22
(β)	-1.39	-1.52	8.04	2.49	9.12
(γ)	0.77	1.06	0.84	0.85	1.86
($\alpha\alpha$)	0	0	0.13	0.12	1.51
($\alpha\delta$)	0	0.04	-0.06	0.19	-6.47
($\delta\delta$)	0	-0.10	0.32	0.31	1.44
(yy)	-0.20	-0.08	-0.87	-0.52	-1.89
($y\beta$)	0.13	0.43	-9.95	-5.04	-14.3
($\beta\beta$)	-0.04	-1.50	-28.7	-13.4	-27.4
($\alpha\alpha\alpha$)	9.1	5.3	-1.5	2.0	66.5
($\alpha\alpha\delta$)	-11.2	-5.6	-2.8	-15.8	-28.8
($\alpha\delta\delta$)	16.3	17.1	10.6	21.0	13.5
(αyy)	-0.8	-1.1	-0.4	0.9	-2.0
($\alpha y\beta$)	5.3	14.9	8.6	-2.8	-30.7
($\alpha\beta\beta$)	9.2	19.0	-33.0	-30.1	-76.5
($\delta\delta\delta$)	-5.9	-4.5	-0.1	-5.8	-2.1
(δyy)	0.7	-0.5	0.7	0.0	2.3
($\delta y\beta$)	-3.5	-6.8	3.8	2.7	14.6
($\delta\beta\beta$)	-8.2	-9.2	-2.2	7.1	18.2

0~28 mm at the exit of the electric field. The maximum acceptable
angle and energy spread are both about 0.04.

The electrostatic Q-lens consists of four cylinder electrodes
of diameter 80 mm and length 80.5 mm. The diameter of the inscribed
circle is 70 mm. The fringing field distribution is assumed to be
similar to that measured for a magnetic Q-lens. The effective length
of the Q-lens is estimated to be 110 mm from this distribution.

The magnet gap is 30 mm and the cross-sectional area of the
vacuum chamber in the gap is 23 mm×110 mm, which is sufficient to
pass ion beams having $|\alpha|\leq0.02$ and $|\beta|\leq0.015$.

A pair of sweep coils of Helmholtz type is placed between the
magnet and the detector and the mass spectrum is observed on a oscilloscope screen or recorded by a multi-channel scaler. The multiscaler is connected to a small electronic computer so that the
position of the peak center is easily calculated and indicated. A
resolution of 30,000 is observed under the condition of the widest
width of α and δ slits.

References

1. G.T. Garvery, W.J. Gerace, R.J. Jaffe, I. Talmi and I. Kelson,
    Rev. Mod. Phys. 41, S1 (1969).

2. Matsuda, Int. J. Mass Spectrom. Ion Phys., 14, 219 (1974).

3. T. Matsuo and H. Matsuda, Int. J. Mass Spectrom. Ion Phys. 6,
    361 (1971).

4. T. Matsuo, H. Matsuda and H. Wollnik, Nucl. Instr. Methods,
    103, 515 (1972).

5. Y. Fujita and H. Matsuda, Nucl. Instr. Methods, 123, 495 (1975).

6. H. Matsuda and H. Wollnik, Nucl. Instr. Methods, 77, 40, 283
    (1970).

7. H. Matsuda, Nucl. Instr. Methods, 91, 637 (1971).

8. H. Matsuda and H. Wollnik, Nucl. Instr. Methods, 103, 117
    (1972).

9. E.G. Johnson and A.O. Nier, Phys. Rev., 91, 10 (1953).

10. R.C. Barber, R.L. Bishop, H.E. Duckworth, J.O. Meredith, F.C.
    G. Southon, P. Van Rookhuyzen and P. Williams, Rev. Sci. Instr.,
    42, 1 (1971).

11. K. Ogata and H. Matsuda, Z. Naturforsch, 10a, 843 (1955).

ON A TWO-STAGE DOUBLE-FOCUSING MASS SPECTROSCOPE UNDER CONSTRUCTION AT OSAKA UNIVERSITY

K. Ogata, H. Nakabushi and I. Katakuse

Department of Physics, Faculty of Science, Osaka Univ.

Toyonaka-shi, Osaka, 560 Japan

Introduction

Planning to construct a huge tandem type mass spectroscope at Osaka University was begun in 1967 (1), in order to determine atomic masses to an accuracy of nearly one order of magnitude higher than we had obtained so far. The new mass spectroscope consists of two sets of double focusing analyzers, each of which has a energy selector with cylindrical electric field and a momentum selector with uniform magnetic field. These sets are aligned in tandem with turned over S-shape.

The machining of the main parts was finished in 1969, and their assembly and the precise aligning had been finished at the end of 1971 (2). In the middle of 1972, we detected the first ion beam at the final focusing point. The mass dispersion was nearly the expected one. However, the convergency of the ion beam was seriously distorted (3).

Since then we concentrated on the focusing properties at the intermediate focusing point with only the first stage. The focusing behaviour was greatly improved and the half-height mass resolution at the intermediate focusing point is now 800,000 or somewhat better (4).

Brief Description on the Mass Spectroscope

The basic calculation on the ion-optical properties of the mass spectroscope has been reported elsewhere (1,2). The machine is of the two-stage double-focusing type. In the first stage, the electric field is followed by the magnetic field, and in the second stage the alignment order of these fields is inverted. The double focusing

point of the first stage is located between the magnets of the two
stages (see fig. 1). The main characteristics are shown in Table 1.

The calculated mass dispersion for 1% mass differences is about
35 cm, and the resolution for an 1μm main slit width is calculated
to be about 9×10^6. The total length of ion-path is about 37m. The
electric fields of two stages are similar. The radii of the central
ion-path in these electric fields are both 3,100mm and the distances
between the electrodes and their heights are 50mm and 300mm
respectively. Thus the same voltage divider can be used
simultaneously for supplying the potentials to the deflecting
electrodes of both fields and measuring doublet mass differences
(5). Fine corrections to the shape of electric field are made with
auxiliary electrodes installed up and down between the main
electrodes.

As for the main magnetic fields, the central radii of the ion-
path of the first and the second stages are 2,600mm and 1,100mm
respectively, and the deflecting angles 106.2° and 100.0°. The gap
widths of these magnets are 12mm both. The surfaces of the pole-
pieces plated with copper, form part of the vacuum chamber.
Some auxiliary coils can be used for changing the magnetic field
strength slighty.

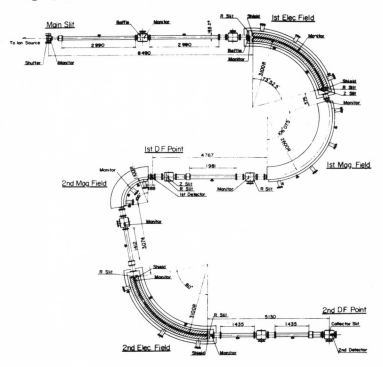

Fig. 1. Schematic view of the two-stage double-focusing
mass spectroscope.

For alignment, the magnets can be moved $\pm$ 20mm in the radial direction along the symmetry line. And all four fields are put on heavy iron tables which can be adjusted by up to about 5cm in both horizontal and vertical positions with three pairs of oil-jacks and screws. These fields including tables weigh over 10 tons. Fig. 2 shows the whole machine.

The ion source part is shown in Fig. 3. Ions are produced by a low-voltage arc type with axial magnetic field, and the ions are extracted by an immersion lens and converged by an einzel lens on the main slit. The ionizing chamber and the lens system can be adjusted against the main slit in the lateral position and direction so as to make as many ions as possible pass through the slit. And the whole system including the main slit can be adjusted against the electric field of the first stage in both position and direction. To ease such adjustments, the source system as a whole is suspended by a pair of beam balances. The ion-accelerating voltage is 50 ～ 60 kV, the stabilized high voltage power supply is described elsewhere (6).

The evacuating system for the main part of the machine consists of fourteen sets of 6"-2" oil diffusion pump system, and the ion source part is evacuated with two sets of 4"-2" oil diffusion pump system. All pump systems have liquid nitrogen traps. The fore-vacuum system, installed in an other room, has two lines including a set of 4" oil diffusion-150 liter oil rotary pump and a mechanical booster pump backed by a 300 liter oil rotary pump. The vacuum presently attained is estimated to be in the range of 10^{-7} Torr.

TABLE I

	E_1	M_1	M_2	E_2
a	310cm	260cm	110cm	310cm
Φ	73.8°	106.2°	100°	80°
l'	850cm	57cm	18cm	0cm
l"	0cm	477cm	315cm	514cm
G	0.25	2.04	2.99	2.51

Dispersion D=34.9cm (for 1% mass difference)
G (total) = 3.80
Resolution R=9.2x10^6 (for 1μ main slit)
Total ion-path L=37.3m

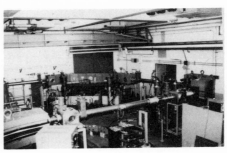

Fig. 2. (A) The first stage of
the mass spectroscope.

(B) The second stage of the
mass spectroscope.

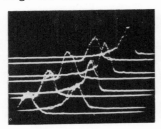

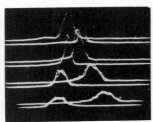

Fig. 4. (N_2-CO) doublet lines taken
at the intermediate focus-
point on photographic plate.

Fig. 3. The ion source system.

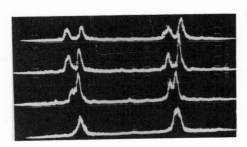

Fig. 5. Image profiles taken by electric method with ion beam having
different incident angles to the magmetic field. Two profiles
in each figure are those taken by changing their energy by
the fractional amount of six parts in 10^4.

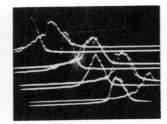

Fig. 6. Image profiles of (C_2H_4-
C_2D_2) doublet taken with
the divergent-angle defining
slit which was deviated into
two parts by 0.3mm tungsten
wire.

Focus Adjustment and Preliminary Experiment

As mentioned above, the focus of the first stage has now been adjusted. The image profiles were checked photographically and electrically. The photographic method demonstrated that the spectral lines taken at various position on photographic plate had different inclinations and wedge-shapes as seen in Fig. 4. The reason for this was investigated by electronic techniques. In this case, a pair of parallel plates deflecting ion beam vertically was installed behind the exit boundary of the magnet. A staircase voltage was supplied to this deflector, and the voltage had a repeating period consisting of four steps, each step synchronized with the saw-tooth voltage for scanning the ion beam horizontally. With such a device, four points of a spectral line along its length can be seen on the long-persistence oscilloscope screen as shown in Fig. 5 and 6.

Fig. 5 shows the image profiles thus taken, changing the incident angle to the magnetic field slightly. The shifted image was taken after changing the ion accelerating voltage by about six parts in 10^4. For shifting the ion accelerating voltage, the square signal is transmitted by a wireless device set from the ground to the high voltage level. Fig. 6 shows the image profiles of a $(C_2H_4-C_2D_2)$ doublet. In this case, a tungsten wire of 0.3mm is inserted around the center of the direction defining slit of about 0.8mm width. The upper traces are split into two parts, but the lowest one seems to be in the focusing point.

Such effect may be caused by an inhomogeneity of the magnetic field strength along the radial and azimuthal directions and/or from a slight lack of orthogonal alignment between the electric and the magnetic field strenghts. And the shape of stray fields of both fields may affect these effects slightly. The homogeneity of the magnetic field was carefully checked, and was improved to within 0.1% (earlier 0.5%). During the focus adjustment, the inner part of the magnet gap was found to be more homogeneous than the outer.

Following the above investigations, the fine focus adjustment was started by changing the incident angles to both fields, shifting the ion collector, making the main and the collector slits better parallel to the direction of magnetic field strength, and other possible adjustments so as to get the finest line.

Fig. 7 shows (N_2-CO) and $(C_2H_4-C_2D_2)$ doublets. Fig. 8(a) shows the same ion peaks taken after changing the electric field strength by 2.4×10^{-6} part. Judging from this separation, the half height resolution is estimated to be about 800,000 or somewhat higher. However, as seen in the figure, the image broadening seems to be comparable with the image width, and such broadening may be ascribed to ion beam oscillation by some oscillating stray field. Fig. 8(b) shows the peaks taken by the same procedure as above mentioned. In this case the shifting amount was 2.2×10^{-6}, so the resolution seems to be better than the above case and estimated to be higher than 1,000,000.

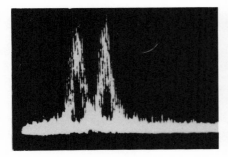
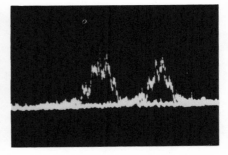

Fig. 7. Doublet peaks.   (A) (N_2-CO) (B) $(C_2H_4-C_2D_2)$

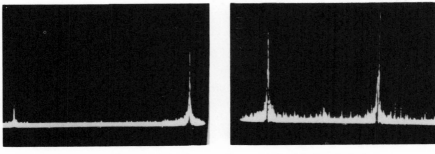

Fig. 8. Peaks of the same ion taken by changing the energy selector
potential. The fraction amount of potential change:
(A) 2.4×10^{-6} (B) 2.2×10^{-6}.

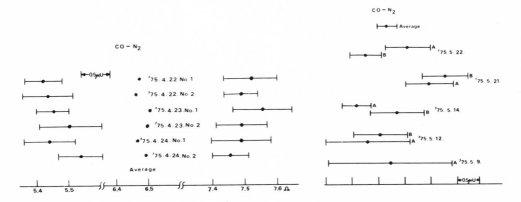

Fig. 9. Mass differences of   Fig. 10. Mass differences of
(N_2-CO) doublet measured (N_2-CO) doublet measured
by visual method.   by digital method.

The reliability of doublet mass difference measurements with this new mass spectrometer was checked by measuring the (N_2-CO) doublet. The measuring techniques adopted was digital peak-matching, quite similar as succesfully used in our previous work reported at the Teddington Conference (7). Prior to this digital measurements, the visual method was used tentatively. Then, the mass differences measured were separated widely into two groups (Fig. 9). This discrepancy was attributed to whether the high or low mass component of the doublet was scanned with the mercury relay for switching the energy selector potential excited. In order to minimize this effect, the relay coil was replaced by a light-optical device. For switching the mercury relay on the high voltage level of the matching circuit, the square signal was transmitted by switching light quanta from the ground level. As shown in Fig. 10, the discrepancy was then greatly reduced.

From these measured values, the mass difference of (N_2-CO) doublet was tentatively estimated as follows: $11,232.64\pm0.21\mu u$. The value seems to be smaller than that adopted at present by about 0.01%, and moreover the statistical deviation is still rather large. Systematic error investigations have not yet been carried out and the stability of the magnetic field strength seems to be not sufficient. Also, a disturbance of the a.c. magnetic field from outside has not yet been completely eliminated. So, these points will have to be improved further in the next step of our work.

REFERENCES

1. K. Ogata, S. Matsumoto, H. Nakabushi, H. Yasuda and I. Katakuse: OULNS 67-1, p.1.

2. H. Nakabushi, I. Katakuse and K. Ogata: OULNS 70-1, p.9; OULNS 72-1, p.48.

3. I. Katakuse, H. Nakabushi and K. Ogata: OULNS 73-2, p.74.

4. H. Nakabushi, I. Katakuse and K. Ogata: OULNS to be published.

5. H. Nakabushi and K. Ogata: Mass Spectroscopy, 22, 121 (1974).

6. H. Nakabushi, T. Hattori and K. Ogata: Mass Spectroscopy, 22, 1 (1974).

7. H. Nakabushi, I. Katakuse and K. Ogata: Proc. Int. Conf. on Mass Spect. Kyoto, p.482 (1970);
   I. Katakuse and K. Ogata: Atomic Masses and Fundamental Constants 4, p.153.

ATOMIC MASS MEASUREMENTS USING TIME-OF-FLIGHT MASS SPECTROSCOPY*

D. F. Torgerson, R. D. Macfarlane and L. S. Spiegel

Cyclotron Institute, Texas A&M University, College Station,
Texas 77843

I. INTRODUCTION

We describe herein a new approach to precise mass measurements
using time-of-flight mass spectroscopy. The spectrometer has been designed to operate in the "on-line" mode coupled with the He-jet system,[1] or to be used off-line with stable species.[2] Although time-of-
flight mass spectroscopy is not normally associated with precision
work, accurate mass measurements can in principle be accomplished
by the use of long flight paths and modern timing instrumentation.
It is the long-range goal of this work to directly measure the masses
of nuclei far from stability.

II. ON-LINE BETA RECOIL MASS SPECTROSCOPY

A schematic diagram of the time-of-flight mass spectrometer is
shown in Fig. 1. In the "on-line" mode, cyclotron-produced radioactivity is deposited onto a thin collector foil in the mass spectrometer using the He-jet method and a skimmer/nozzle system to
effect separation of the activity from the He gas. The thin collector foil is aligned with an 837 cm-long drift tube and a radiation counter. For detecting beta-decay events an NE102 scintillation counter is used. If a beta-decaying nucleus produces
an ionized recoil, the recoil is accelerated through a gridded
lens system to an energy of 10 keV into the drift tube.

For precision work, the drift tube must be of sufficient
length to reduce the $\Delta T/T$ value, where T is the time-of-flight and
ΔT is the peak width. To overcome solid angle effects, therefore,
the drift tube is an electrostatic particle guide consisting of an

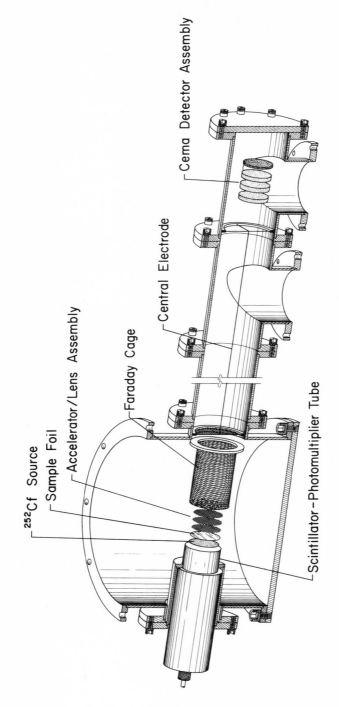

Fig. 1   MAGGIE time-of-flight mass spectrometer using a fission fragment ion source.

electrode wire centered on the axis of the tube. A potential of
-10 to -50 volts is maintained on the electrode wire, producing a
radial field. Ions are captured into a stable orbit about the
electrode and are detected at the end of the tube using a CEMA
detector array.

Time intervals are precisely measured between pulses derived
from detection of the initial beta decay and those derived from
detection of the recoil ions in the CEMA array using the EG&G TDC100
time digitizer. The resolution of the digitizer is better than 125
psec and the integral non-linearity is zero. Time to mass conversion is accomplished using an off-line computer.

An example of a beta recoil mass spectrum is shown in Fig. 2
for reaction products produced by bombardment of ^{40}Ca deposited on
an ^{27}Al foil with 25 MeV protons. The peaks in Fig. 2 are labeled
with the beta-decay parent. Our original intention was to use
beta recoil mass spectroscopy to identify new isotopes, measure
half-lives, and obtain spectroscopic information by replacing the
scintillation counter with a GeLi detector. However, it was decided that more meaningful information on new isotopes could be
obtained if we concentrated our efforts on exact mass measurements.
This involved the need for extensive developmental experimentation
which meant that an off-line method was required, leading to the
discovery of "plasma desorption" (PD) mass spectroscopy.

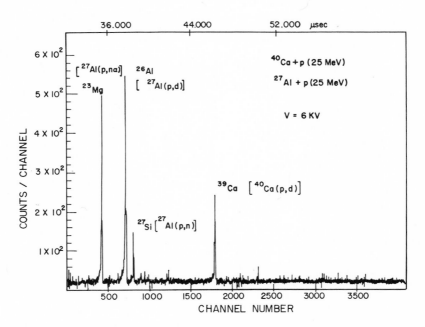

Fig. 2   Beta-recoil mass spectrum using skimmer/nozzle separation.

III. PLASMA DESORPTION MASS SPECTROSCOPY

In the course of the above studies, it was noted that background peaks were present in the beta-recoil spectra which did
not correspond to radioactivity. Closer examination revealed
that beta and alpha radiations were responsible for ionizing impurities adsorbed on the collector foil. However, fission fragments were observed to be the most effective ionizing radiation.
It is believed that ionization and volatilization of the sample
is effected by the high-temperature, short-lived plasma created
by the fission fragment as it passes through the foil. Power
densities of the order of 10^{12} watts/cm^2 are generated in the foil
by a single fission fragment.

For operation in the PD mode, a ^{252}Cf source was positioned
between the scintillation counter and the thin foil on which the
sample was plated as shown in Fig. 1. "Start" pulses were derived
from the detection of one fission fragment in the scintillation
counter while the complementary fragment passed through the sample.

A portion of the PD mass spectrum of ^{148}Sm is shown in Fig. 3.
For this spectrum, ^{148}Sm was vacuum-evaporated to a thickness of
200 μg/cm^2 onto a 1 mg/cm^2 Ni foil. The most interesting observation was the presence of intense mass peaks corresponding to
combinations of Sm, O, and H up to ~2000amu. These peaks undoubtedly originated from the oxide layer on the surface of the Sm
sample.

To study the feasibility of measuring precise atomic masses,
the mass of ^{148}SmO was determined by using the ^{148}Sm, ^{148}SmOH,
and ^{148}Sm(OH)$_2$ groups as standards. Centroids were found by fitting Gaussian distributions to only the most intense 5 channels
in the peak, thereby avoiding any asymmetries due to tailing.
The standards were then fit to a calibration curve to utilize
(and test) the linearity of the time digitizer. Nineteen spectra were analyzed in this manner, and the results are summarized
in Table I.

The object of this experiment was to demonstrate that both
accuracy and precision could be obtained, and that systematic
errors were small. The full width at half maximum of the
peak was ~20 nsec which corresponded to a mass width of ~50 x 10^{-3}
mass units. The resolution could be improved a factor of 4 by
increasing the acceleration potential and decreasing the central
electrode voltage to reduce the acceptance angle.

Our goal in this work is to measure the masses of radioactive nuclei to a precision of ±200 keV. Future developments
include installation of the mass spectrometer on a cyclotron beam

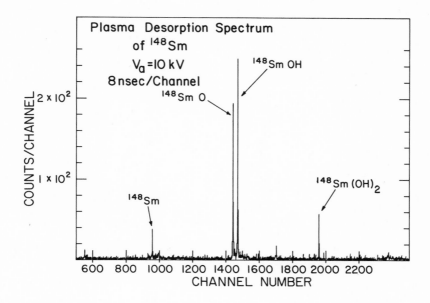

Fig. 3   Mass spectrum of ^{148}Sm vacuum-evaporated onto a 1 mg/cm^2
Ni foil.

TABLE I.   Mass of ^{148}SmO.

Experiment	Mass	Experiment	Mass
1	163.90725	11	163.91365
2	163.90575	12	163.91167
3	163.91136	13	163.90877
4	163.91192	14	163.91168
5	163.91006	15	163.90942
6	163.90730	16	163.91202
7	163.90660	17	163.90907
8	163.91062	18	163.90814
9	163.90943	19	163.90919
10	163.90796		

Average = 163.90958±0.00049
Literature value (Ref. 3) = 163.90976±0.00001
Difference = 0.00018

line for "on-line" plasma desorption studies. In this mode,
the cyclotron beam will be used to produce nuclear recoils, some
of which will be thermalized in a thin foil through which the
beam will pass. Ionized recoils ejected from the thin foil by
the beam plasma will be accelerated into the time-of-flight
tube and mass-analyzed as discussed previously.

References

*Work supported in part by the U. S. ERDA and the Robert A. Welch
Foundation.

1.   R. D. Macfarlane, D. F. Torgerson, Y. Fares, and C. A. Has-
      sell, Nucl. Instr. Meth. 116, 381 (1974).

2.   D. F. Torgerson, R. P. Skowronski, and R. D. Macfarlane,
      Biochem. Biophys. Res. Comm. 60, 616 (1974).

3.   A. H. Wapstra and N. B. Gove, Nucl. Data Tables 9, 265 (1971).

MASS SPECTROMETRY OF UNSTABLE NUCLEI

C. Thibault, R. Klapisch, C. Rigaud, A.M. Poskanzer[*],
R. Prieels[+], L. Lessard[x], and W. Reisdorf[++]
Laboratoire René Bernas du Centre de Spectrométrie
Nucléaire et Spectrométrie de Masse du C.N.R.S.
91406 Orsay, France

INTRODUCTION

The measurement of atomic masses of stable elements by mass spectrometry is a scientific achievement that is well documented in the entire series of AMCO Conference. As will be reported at this Conference, the precision now attained is of the order of kilovolts (or better) throughout the periodic table.

The situation is much less satisfactory when one goes to nuclei at some distance from stability where the data are either not precise or non existent. At the AMCO IV Conference, the possibility of measuring directly the masses of far unstable nuclei by on line mass spectrometry was outlined (1). The purpose of the present paper is to report the first measurement of the masses of short-lived neutron excess nuclei.

These nuclei can be produced (together with many other nuclear species) when GeV energies protons interact with a uranium nucleus (2). We have thus measured the masses of ^{11}Li and of the sodium isotopes 26 to 32 with a special mass spectrometer on line with an external beam of the CERN proton synchrotron.

---

[*] Lawrence Berkeley Laboratory, Berkeley, California.
[+] Institut de Physique Corpusculaire, Université de Louvain, Louvain-la-Neuve, Belgium.
[x] France-Québec post-doctoral fellow 1972-1973, present address : Foster Radiation Laboratory, McGill University, Montreal,Canada.
[++]Gesellschaft für Schwerionenforschung mbH, Darmstadt, Germany.

At the present state of knowledge, an accuracy of 100 keV (i.e. 10^{-5} at mass 10) is already valuable that far from stability (in contrast with the 10^{-8} that are commonly achieved for stable masses). By splitting the uncertainty in the measurement according to the equation :

$$\frac{dM}{M} = \frac{dM}{\Delta M} \times \frac{\Delta M}{M}$$

it was conjectured that it could be accomplished by a single stage mass spectrometer of resolving power $(M/\Delta M) \sim 500$, if the statistics are adequate to determine the centroid of a peak with an accuracy of $dM/\Delta M \sim 5.10^{-3}$.

MASS SPECTROMETER

A schematic view of the mass spectrometer is shown on Fig. 1. The target which serves directly as ion source is made of an array of graphite foils covered with uranium and wrapped in a rhenium foil. The relative thicknesses are such that the nuclei produced in the reaction U + protons are stopped in the graphite. As the source is heated around 1800°C, the alkalis diffuse very quickly out of the graphite and are selectively ionized by thermoionic effect on the rhenium foil. The target is kept at a positive potential of about 10 kV DC. The ion beam is focused on a defining slit S_1 at ground potential by two intermediate electrodes. S_1 acts as an object for a 90° magnetic sector with a weakly inhomogeneous field. Because of the background related to the proton beam burst, the ions passing across slit S_2 have to be transported through a shielding wall and refocused on the detector M_2. This detector is an electron multiplier

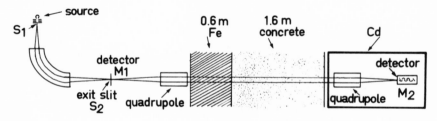

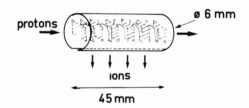

Figure 1. Schematic view of the on-line mass spectrometer.

capable of counting single ions. During the experiment, a rough check
of the efficiency of the transportline could be done by comparing
count rates on detectors M_1 and M_2 at each end of the line. Because
this efficiency was essentially 100 %, one will thus assume that
the exit slit of the spectrometer is S_2 only.

EXPERIMENTAL METHOD

In order to measure masses, we make use of the well known
theorem that if two ions of masses M_A and M_B are accelerated under
potentials V_A and V_B and if they follow the same trajectory between
slits S_1 and S_2, the following relation holds :

(1) $M_A/M_B = V_B/V_A$

In our case, A and B are always two isotopes of the same element.
Because of the small number of available ions, we cannot expect to
match exactly the peaks A and B. Then, in order to determine the potentials that would match the trajectories, Fig. 2 shows that a periodic, calibrated triangular modulation is added to the well stabilized
DC power supply U. The peaks are recorded on a multiscaler synchronized with the modulation. The centers of gravity of the two peaks
corresponding to the ion beam sweeping back and forth across slit S_2
are called A_1 and A_2 (or B_1 and B_2). Then, the voltage V_A (V_B) is the
sum of the DC component U_A (U_B) and of an increment v_A (v_B). This
increment is calculated by comparing the distance between the two
peaks $A_2 - A_1$ ($B_2 - B_1$) to the distance corresponding to phase zero,

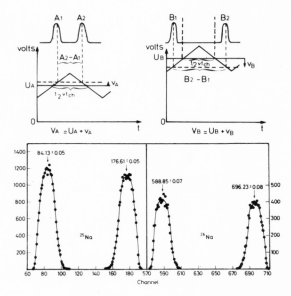

Figure 2. Principle of mass measurements. $M_B = M_A \, V_A/V_B$

which is equal to one half of the period of the triangular modulation. In order to determine U_A (or U_B), the 6 to 10 kV DC potential is divided by 10^4 through a special divider bridge and measured by an accurate six digit digital voltmeter.

Data are recorded on the two masses A and B by switching two physically distinct DC power supplies and accumulating the corresponding counts in different sections of the multiscaler memory. A measurement would consist of a number of such alternate mass sweeps until the statistics is adequate for both masses. This could last from 10 to 30 minutes or longer for low yield isotopes, and during this time, the digital voltmeter readings, interfaced to a PDP 15 computer, are averaged out. The stability of the DC voltages as measured during a run was 10^{-5} and does not seem to be the limiting factor in the precision of the measurements. At the end of a run, the data were transferred from the multiscaler to the computer and could be stored on magnetic tape and processed while data for the next run were accumulated.

As systematic errors are always possible due, for example, to the heating voltage, we modified formula (1) and introduced a parameter δ in order to make the method self calibrating:

(2) $M_A/M_B = (V_B + \delta) / (V_A + \delta)$

δ is measured by using pairs of known masses of other isotopes of the same element, produced under the same experimental conditions. During our first measurements the values of δ were found to be several volts and were fluctuating with time more than predicted by statistics. Different procedures were thus adopted (2) which gave good results but were time consuming. However, using a new divider

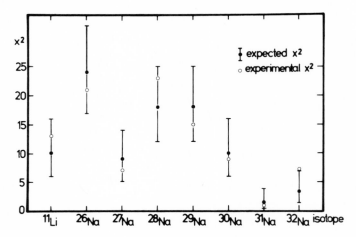

Figure 3. Values of $X^2 = \sum_1 \left[(M_i-\overline{M})/\sigma_i\right]^2$ for each isotope.

bridge for U measurement and recording 3 masses M_A, M_B, M_C in succession, we finally succeeded to be freed of the time fluctuation of δ.

The error calculation of each measurement takes into account the dispersion of the δ values when necessary and the statistical errors on the centroïds of the peaks. Then, the final experimental mass of each isotope is the weighted mean value of all single measurements. In order to check the consistency of the method, Fig. 3 shows the values of X^2 for each isotope. It appears that all the measurements are compatible while the experimental conditions could be very different. It means also that our error estimation is realistic and that our assumption on the similarity of the shape of the peaks of the different isotopes is valid.

RESULTS - DISCUSSION

All our results are reported in Table I and compared with previous measurements when available.

The particular case of ^{11}Li was a very interesting puzzle as it was generally predicted unbound before it was identified in 1966 by Poskanzer, Hyde and Cerny (2). The binding energy of its last pair of neutrons is finally very small as it is only 170 $\pm$ 80 keV.

In the case of ^{26}Na, it is seen that the accuracy could be up to 10^{-6}, due to a high statistics. This value is in agreement with the measurement of Flynn and Garrett (2) and seems to confirm that the value from Ball et al. could have been influenced by an unresolved excited state at 80 keV. For ^{27}Na, our value is also in good agreement

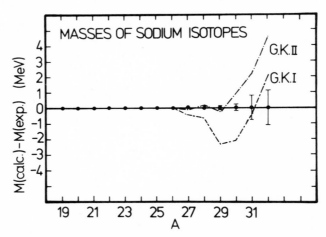

Figure 4. Differences between our experimental results and the mass-excesses calculated by the Garvey-Kelson method.

with the measurement of Alburger et al.(2,3), with a better accuracy.

Taking into account our measurement and previously known masses,
the range of measured Na isotopes covers N/Z ratios from 0.7 (^{19}Na)
to 1.9 (^{32}Na). To our knowledge, this is the biggest excursion to
date away from the bottom of the valley of stability.

We have compared our results with four basically different theoretical approaches to the problem of nuclear masses (2). These are
the liquid drop model, a large scale shell model calculation by Cole,
Watt and Whitehead, the Garvey-Kelson method, and a Hartree-Fock
calculation. We shall only comment the two last calculations as they
finally brought the most striking features. Fig. 4 shows the comparison with the masses calculated according to the Garvey-Kelson method with two different sets of parameters adjusted for light masses.
The first one (GK I) has been fitted using $T_Z \leqslant 2$ experimental masses.
A smooth divergence appears from A = 26 up to A $\sim$ 30 while for
A = 31, 32 there is an abrupt reversal of this trend. With the second
set of parameters (GK II) which includes also masses with $T_Z = 5/2$
and our experimental masses for A $\leqslant$ 30, the agreement is good until
A = 29 but the fast deviation for A = 31, 32 becomes even more apparent. As the Garvey-Kelson approach retains only the smoothly varying
features of the nuclear energy surface, these deviations from the
model could indicate the sudden onset of a change of nuclear shape.

Support for this possibility is in fact obtained by merely plotting the experimental two-neutron binding energies versus neutron

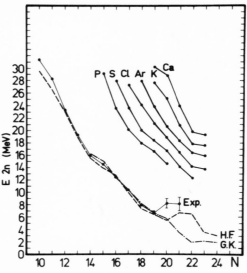

Figure 5. Values of the two-neutron binding energies versus neutron
number, and comparison with a Hartree-Fock calculation.

number (Fig. 5). Our data for ^{31}Na and ^{32}Na indicate a hump at
N = 20, 21 which does not appear in the experimental values for
other elements closer to stability. By comparing with similar E_{2n}
graphs presented by Wapstra and Gove (2), it is concluded that this
hump observed for N = 20, 21 is inconsistent with the classic shell
closure effect and more reminiscent of the behavior one observes
when entering a region of sudden deformation (like the region N = 88
to 92 in the rare earths). As expected, the Garvey-Kelson calculation
which extrapolates the behavior of the higher Z elements do not produce a hump. Another support for this deformation comes from the work
of Campi, Flocard, Kerman and Koonin who have applied to the sodium
case the Hartree-Fock calculations which may provide nuclear radii
and deformations. They find a hump due to an abrupt change in the
shape of the nuclei which become more prolate (4).

In conclusion, the results of these experiments and Hartree-
Fock calculations are quite unusual in that they find a region of
strong deformation at a neutron number of 20 which, in the calcium
region, corresponds to a well known spherical closed shell. The
question of wether or not this is a new region of deformation must
however await for further experimental and theoretical evidence.

TABLE I

Adopted values for the experimental mass excesses of ^{11}Li, $^{26-32}$Na.

Isotope	M-A (keV)	Previous measurements (keV)
^{11}Li	40940 + 80	
^{26}Na	− 6901 + 25	−6903 + 20 Flynn, Garrett (2)
^{27}Na	− 5620 + 60	−6853 + 30 Ball, Davies, Forster, Hardy (2)
		−5650 + 180 Alburger, Goosman, Davids (2,3)
^{28}Na	− 1140 + 80	
^{29}Na	2650 + 100	
^{30}Na	8370 + 200	
^{31}Na	10600 + 800	
^{32}Na	16400 + 1100	

REFERENCES

1 − R. Klapisch and C. Thibault, Proceedings of the 4th Int. Conf.
    on Atomic Masses and Fundamental Constants, Teddington (1971),
    ed.Plenum Press, New York (1972), p. 181.
2 − C. Thibault, R. Klapisch, C. Rigaud, A.M. Poskanzer, R. Frieels,
    L. Lessard and W. Reisdorf, Phys. Rev. C (1975) to be published,
    and references therein.
3 − D.E. Alburger, these proceedings.
4 − X. Campi, H. Flocard, A.K. Kerman and S. Koonin, these procee-
    dings.

THE ACTIVITIES OF THE II. PHYSICAL INSTITUTE OF THE UNIVERSITY OF GIESSEN IN THE INVESTIGATION OF SHORT-LIVED HEAVY NUCLEI

E.Ewald and H.Wollnik

II.Physikalisches Institut der Justus Liebig-Universität

6300 Giessen, Arndtstraße 2

Our first investigations of short-lived heavy nuclei concerned the properties of short-lived fission products. For this purpose we have built around 1960 a double focusing fission product separator which we installed at the reactor in München[1]. A thin foil of ^{235}U placed close to the reactor core yielded about 10 fission products per second for a good mass line behind the separator. This intensity was barely sufficient to measure nuclear charge distributions within isobaric mass chains[2], but for nuclear spectroscopy it was too small. After the decision that a high flux reactor should be built in Grenoble, we designed in cooperation with people from the research center in Jülich a new recoil separator which together with the hundred times higher neutron flux gave us more than 1 000 times higher particle intensities[3]. This instrument uses the principle of the Kaufmann-Thomson parabola spectrograph where, however, the magnetic and electrostatic fields had been separated. Since about two years this instrument -called LOHENGRIN- has worked succesfully and has opened new possibilities of investigations.

One of the most interesting results obtained with this machine is the nuclear charge distribution of fission products. Passing the monoenergetic fission products of one mass behind the exit slit of LOHENGRIN through a thin ΔE Si-surface barrier detector results in an energy loss signal depending on the nuclear charge[4]. The results for the light group of fission products are complete. The resulting average nuclear charge Z_a as well as the half width at full maximum ΔZ is given in fig.1 showing minimal values of ΔZ whenever Z_a is even. We will also report some other work with LOHENGRIN in another session of this conference.

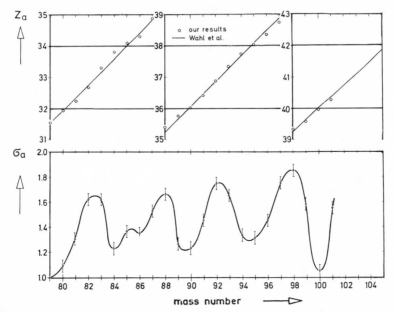

Fig.1
For fission products of the reaction ^{235}U (n_{th},f) the average nuclear charge is plotted as well as the width (fwhm) ΔZ of this distribution as a function of particle mass. It should be noted that
ΔZ is always smallest if the average nuclear charge falls on an
even number since in this case the production of the central even
proton nucleus is favoured over the two neighbouring odd proton
nuclei.

Besides fission work using recoil separators, we have also thermalized fission products in hot graphite, ionized by surface ionization, accelerated and passed them through a usual small 90° magnetic mass separator named OSTIS (This name comes from On-line Separator für thermisch ionisierte Spaltprodukte). Since we used
a surface ion source only, we have investigated so far only Rb
and Cs fission products. Among other results from these measurements, Q_β-values were derived[5] for neutron-rich Cs-isotopes up to
the mass 145. To enable us to investigate even more neutron-rich
and thus less abundantly produced fission products we intend to
move OSTIS to the high flux reactor in Grenoble where we expect
about 100 times higher particle intensities.

At the moment most of our activities are directed towards the preparation of experiments on heavy ion reaction products at the Uni-
lac accelerator in Darmstadt. The largest installation we prepare
there, in close cooperation with people from the GSI in Darmstadt
especially P.Armbruster, is a Wien filter. This system uses sepa-

rated magnetic and electrostatic fields[6]. It should be capable
of separating the beam of primary ions from the beam of compound
nuclei and fusion fission products[7], so that complex experiments can be performed with them. In fig.2 the principle of this
instrument is shown and consists of 4 magnetic and 2 electrostatic dipoles and of 6 magnetic quadrupoles. From fig.2 also the expected beam envelopes can be seen. The system consists of two
Wien-filters put together back-to-back. For the first system the
magnetic quadrupole triplet causes a fucusing behind the first
electrostatic field and the first two magnets. At this point particles of a certain velocity can be selected by a slit independent
of their ionic charges. If the particles of interest cover some
velocity range, this slit may be opened to allow particles to pass
with a velocity deviation of ± 5 % from an average velocity.

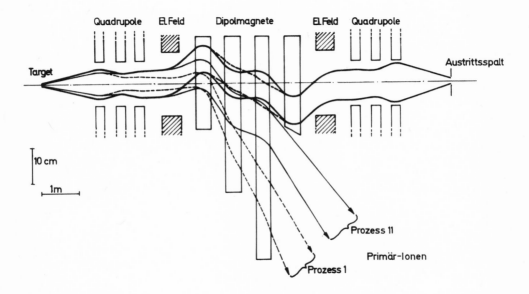

Fig.2
The beam envelopes in the Wienfilter SIS (Schwer Ionen Separator)
under construction at the heavy ion accelerator in Darmstadt. This
system should be capable of separating the beam of primary ions
from the compound nuclei reaction products or from fusion-fission
products. The separated heavy ion reaction products of a velocity
range of ± 5 % and of a broad range of ionic charges are refocused
to an exit slit.

The second Wien-filter then will focus all these particles again
to a point source. The primary beam, however, leaves the system
before the last magnet. The instrument should be put into operation as soon as the first beam of the heavy ion accelerator is

available. A problem remains in the handling of the rather high
voltages (±300kV) required to form the electrostatic field.
After the experience with the fission product separator LOHENGRIN
in Grenoble we feel, however, that this problem should not be
too difficult.

A second project is a He-jet ion source separator for heavy ion
reaction products developed together with a group from the university in Marburg. Here we investigate not only heavy ion reaction products which leave the reaction target at small angles
to the primary beam, but also transfer reaction products which
form large angles with the primary ion beam. In this system the
reaction products are thermalized in some small volume of He
or another gas and then swept through a capillary to some other
place where the decay properties of these nuclei can be studied,
or to some ion source and a conventional mass separator[8]. The
problem here is to transport the reaction products quantitatively through the capillary. Since normally they would diffuse to
the capillary walls we add to the gas specific aerosols to which
the reaction products can attach themselves. The size of such
aerosols is large in respect to the reaction products, so that
they diffuse only slowly towards the capillary walls. The size
of such aerosols, however, should not be too large either since
then possibly a large fraction of them may be lost by sedimen-
tation. If the aerosols are of a specific size, the transport
efficiciences[9] are between 60% and 90%. Using a plasma ion
source the overall efficiencies of the He-jet ion source separator are around 1%. A new magnetic sector mass separator is
being designed. It should have a sufficiently high resolving
power so that it can be used to measure Q_β-values by direct
mass measurements.

We have developed an alternative solution together with a group
from the Institute of Technology in Darmstadt. Here we study mass-
analyzed short-lived heavy nuclei by a He-jet transport system
which deposits α-active particles on some catcher foil. Observing
by delayed coincidence methods the α particle and the recoiling
emitter nucleus after a flight path of perhaps 30 cm, we can determine the mass of this nucleus by knowing the α-energy as well as
the flight time[10]. For a α-decay the recoil energy is in the order of 50 keV so that a flight time of about one microsecond is
to be expected. In the case of a β-decay the recoil energy is only
in the order of a few 10 eV. In this case the charged recoils can
be accelerated to a constant energy, for instance 6 keV as in our
experiment. For fission products from a ^{252}Cf-source this experiment has been performed. The resultant mass spectrum is shown in
fig.3. It can be seen that the obtained mass resolution $M/\Delta M \simeq 200$
is quite sufficient to distinguish between particles of neighbouring masses of fission products.

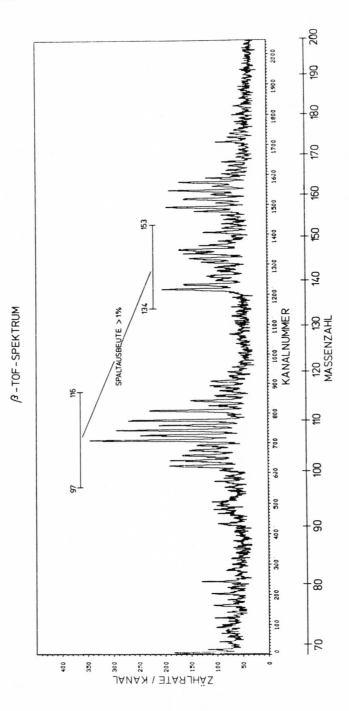

β-TOF-SPEKTRUM

Fig.3
The time of flight spectrum of the β-recoils of fission products of ^{252}Cf accelerated to an energy of 6 keV. It should be noted that no rare gas fission products can be seen in this spectrum since they would not stick to the catcher foil.

With all these efforts we hope to be able to contribute to the investigation of short-lived nuclei especially at the heavy ion accelerator in Darmstadt.

For support of this work we would like to thank the "Gesellschaft für Schwerionenforschung" in Darmstadt and the "Bundesminister for Forschung und Technologie".

References

1) H.Ewald, E.Konecny, H.Opower, H.Rösler; Z.Naturf. $\underline{19a}$ (1974)
   194, 200
2) E.Konecny, H.Gunther, H.Rösler, G.Siegert, H.Ewald; Z.Phys.
   $\underline{231}$ (1970) 59
3) P.Armbruster, H.Ewald, G.Fiebig, E.Konecny, H.Lawin, H.Wollnik;
   Ark.f.Fys. $\underline{36}$ (1967) 305
   E.Moll, H.Schrader, G.Siegert, M.Asghar, J.P.Bocquet, G.Bailleul,
   J.P.Gautheron, J.Greif, G.I,Crawford, C.Chauvin, H.Ewald,
   H.Wollnik, P.Armbruster, G.Fiebig, H.Lawin, K.Sistemich;
   Nucl.Instr.&Meth. $\underline{123}$ (1975) 615
4) G.Siegert, H.Wollnik, J.Greif, G.Fiedler, M.Asghar, G.Bailleul,
   J.P.Bocquet, J.P.Gautheron, H.Schrader, H.Ewald, P.Armbruster;
   Phys.Lett. $\underline{53B}$ (1974) 45
   G.Siegert, J.Greif, H.Wollnik, G.Fiedler, R.Decker, M.Asghar,
   G.Bailleul, J.P.Bocquet, J.P.Gautheron, H.Schrader, P.Armbru-
   ster, H.Ewald; Phys.Re.Lett.,Vol.34, 16, p.1034
5) K.Wünsch, H.Wollnik, G.Siegert; Phys.Rev.C10, (1974) 2523
6) H.Wollnik, G.Münzenberg, H.Ewald; Nucl.Instr.&Meth.III (1973) 355
7) H.Ewald, P.Hinckel, K.Güttner, G.Münzenberg, H.Wollnik;
   GSI-Bericht GSI 73-14 (1973) 1
   H.Ewald, W.Faust, K.Güttner, P.Hinckel, G.Münzenberg;
   GSI-Bericht GSI-J2-74 (1974) 1
8) E.Georg, W.Jung, T.Matsuo, G.Röbig, W.Weitzel, H.G.Wilhelm,
   H.Wollnik, H.Fleischer, W.Kornahl, A.K.Mazumdar, H.Wagner,
   W.Walcher, R.Brandt, Y.Fares, D.Hirdes, H.Jungclas; GSI Bericht
   GSI-J2-74 (1974) 70
9) H.Wollnik, H.G.Wilhelm, G.Röbig, H.Jungclas; Nucl.Instr.&Meth.
   in print
10) Y.Fares, R.Faß, D.Schardt, K.Wien, H.Wollnik; GSI Bericht
   GSI-J2-74 (1974) 91

ATTEMPTS TO OBTAIN HIGHLY RESOLVED MASS SPECTRA OF SHORT-LIVED

FISSION PRODUCTS WITH THE LOHENGRIN SEPARATOR

H. Wollnik*, G. Siegert, J. Greif*, G. Fiedler*, M. Asghar,
J.P. Bailleul, J.P. Bocquet**, M. Chauvin, R. Decker*,
B. Pfeiffer*, and H. Schrader
Institut Laue-Langevin, Grenoble, France
*II.Physikalisches Institut der Universität Giessen, Germany
**DFR, CEN, Grenoble, France

The LOHENGRIN is a large Kaufmann Thomson parabola spectrograph
the two fields of which are not superimposed but separated and
shaped as sector fields [see fig.1] so that the system focuses
stigmatically[1]). Its field boundaries are curved so that the image

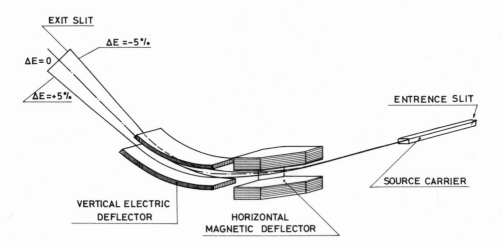

Fig.1
Principle sketch of the LOHENGRIN fission products separator in
Grenoble. The overall path length of the separator is 23 m.

aberrations are minimized[2]. Since it is designed to deflect recoil fission products of about 100 MeV energy which have been
stripped of about 20 electrons the dimensions of the system are
chosen to be large. With 0.2 T the magnet deflects the fission
products on a radius of 4m, while the electrostatic condensor
uses ±300KV on its electrodes separated by 30 cm to deflect the
fission products on a radius of 5.6 m[3]). Using a thin layer of
400 μg/cm^2 of ^{235}U placed close to the reactor core of the high
flux reactor in Grenoble in a neutron flux of about $5 \cdot 10^{14}$ thermal
neutrons per cm^2 and second, sufficient fission products enter
LOHENGRIN, so that at its exit slit over 10^4 particles of one
mass can be recorded per second. The separator has a mass dispersion of 32 mm/% perpendicular to the exit slit and an energy
dispersion of 70 mm/% parallel to the exit slit. It is designed
so that particles whose energies deviate by ±5% from the mean
energy pass through the 700 mm long exit slit.

For some applications the separated particles pass through some
detection device with the high recoil energy[4]. For nuclear spectroscopy measurements the particles are slowed down to thermal
energies in a gas pressure chamber. They are then swept through
17 capillaries to an evacuated chamber together with the sweep
gas, forming a point source on a catcher foil. By recording γ-β
coincidences emitted from this point source, Q_β-values of short-
lived fission products have been determined [5].

A quite different approach to obtain Q_β-values is a precise direct
mass measurement. This procedure is free of systematic or propa-
gated errors that may enter into the derivation of Q_β-values from
β end-point energies. Decreasing the size of the entrance slit
of the LOHENGRIN separator from 5 x 70 mm^2 to 0.1 x 5 mm^2, and
limiting also the aperture angles so that the corresponding image
aberrations are sufficiently small, the resolving power increases
considerably over the usually obtained M/ΔM=800. The fission products are recorded by bombarding with them glass plates which
afterwards are etched[6] thus showing an etch pit for each particle.
As in the old days of recording highly resolved mass spectra by
photographic emulsions, the number of etch pits as a function
of position on the glass plate are then counted.

In fig.2 the best resolved mass peak is shown exhibiting a mass
resolving power of M/ΔM≈15 000 which corresponds to a difference in binding energy of about 6.5 MeV. For neutron-rich fission
products, the difference between neighbouring elements within one isobaric chain should often exceed this value, especially in the light
group. The distances between the different lines should reveal the
corresponding Q_β-values directly and the height of each line the
yield for the corresponding nuclear charge. Since we know the
nuclear charge distribution from other experiments we should be

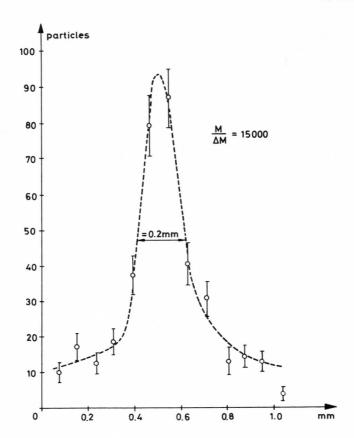

<u>Fig.2</u>
The highest obtained mass resolving power is demonstrated by the quite narrow mass line of ^{90}Kr. The neighbouring ^{90}Br and ^{90}Rb are produced with low yield only so that no mass multiplet is to be expected for the mass 90.

able to calculate the different Q_β-values even if the different lines of a multiplet are not completely resolved.

The production of ions in this experiment is very much different from the usual mass spectrometric work. First, we have particles of different ionic charges, for instance ...18, 19, 20, 21, 22,..., so that we often have multiplets of quite different masses, each line of which is characterized by a different ionic charge. This often allows the determination of one mass of one element relative to quite different masses of other elements. Secondly, we have quite energetic particles with velocities of a few percent that of light so that the relativistic mass change with varying energy is directly abservable and must be taken into account in the interpretation of the measured data.

Actually we had hoped that we would be able to present here a few measured masses of neutron-rich fission products. Unfortunately this is not possible since we encountered several difficulties. We observed line broadenings due to mechanical vibrations of the 23 m long spectrometer as well as due to small angle scattering in the residual gas of the vacuum system. Both these effects, however, could be reduced to a tolerable level. Unsolved is still the problem of the stability of the high voltage supply of the electrostatic sector field. The two ±400 keV generators in principle are capable of delivering constant voltages with deviations smaller than a few times 10^{-5} but both generators are a few years old already and to tune them is rather costly and time consuming. A second difficult problem is that there are very different experiments to be done with this separator in Grenoble with only this requiring the highest possible performance of the instrument. Despite the many difficulties, we have seen some resolved mass doublets. One of them is shown in fig.3. The statistical fluctuations are high, the background is very large due to target difficulties, but still one can see the high Q_β-value of ^{94}Rb which is about 10 MeV. During our next test run, which is scheduled for July this year, we hope to improve the method so that during this or the next run more highly resolved mass spectra can be obtained, and the methods presented here can be used actually to determine the masses of most fission products produced with high yield.

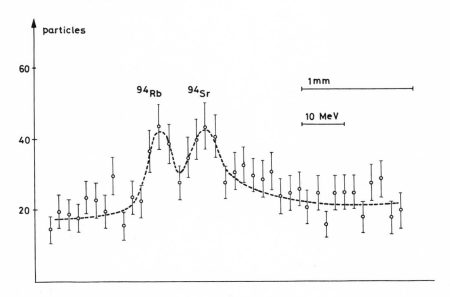

Fig.3
The almost resolved masses of ^{94}Rb and ^{94}Sr are shown which are separated by about 10 MeV.

For support of this work we would like to thank the "Bundesminister for Forschung und Technologie".

References

1) S.Neumann and H.Ewald, Z.f.Phys. 169 (1962) 224

2) G.Fiebig, H.Lawin, H.Wollnik, Int.Rep. KFA Jülich (1971)

3) E.Moll, H.Ewald, H.Wollnik, P.Armbruster, G.Fiebig, H.Lawin,
   Proc.Int.Conf.Elect.magn.Isotope Sep. Marburg/Germany, Rep.
   BMBW-FB K70 (1970) 225

4) G.Siegert, H.Wollnik, J.Greif, G.Fiedler, M.Asghar, G.Bailleul,
   J.P.Bocquet, J.P.Gautheron, H.Schrader, H.Ewald, P.Armbruster,
   Phys.Lett. 53B (1974), 45

5) H.Schrader, R.Stippler, F.Munnich, R.Decker, H.Wollnik,
   E.M.Monnand, M.Asghar, J.P.Bocquet, G.Siegert,
   Spring.Conf.on Nucl.Physics, Den Haag, (1975)

6) J.Aschenbach, G.Fiedler, H.Schreck-Kölner, G.Siegert,
   Nucl.Instr.&Meth., 116, (1974), 389

THE MASS DETERMINATION IN RELATION TO NUCLEAR STRUCTURE

AND TO THE THEORY OF NUCLEAR MATTER

K. Bleuler

Institut für Theoretische Kernphysik

D-53 Bonn, Nussallee 14-16   W-Germany

I. The experimental determination of nuclear masses throughout the periodic table yields - as one of the most important consequences - a well defined value for the binding energy of Nuclear
Matter (Bethe, Swiaticki). In the same time the corresponding density was determined through extended measurements of nuclear radii
and nuclear charge distributions (Hofstaedter). On the other hand,
a comparison of these quantities with the results from a numerical
calculation based on general theoretical principles (Bethe-
Brueckner theory) represents a decisive test of microscopic theories
of nuclear matter and hence of nuclear structure. It is, however,
well-known for many years (compare F. Coester, Phys.Rev. C1, 1970,
769) that all calculations based on any expression for the fundamental nuclear forces, which,in turn,is in reasonable agreement
with the experimental scattering phases of the 2-nucleon problem,
will not yield simultaneously the right experimental values for
binding and density of nuclear matter (i.e. for reasonable binding
values the density is too high and vice versa). Even using the
(otherwise satisfactory) description of the nuclear force based on
a boson exchange model (compare as an example K. Holinde,K.Erkelenz,
R. Alzetta, Nucl.Phys. A 194, 1972, 161), a characteristic discrepancy remains. The main purpose of this short contribution is
to give some general reasons for this disturbing fact:

1) Any explicite expression for the nuclear force, which is
determined through the boson exchange in between two free nucleons,
is changed in a characteristic way (by the order of magnitude of a
few percents) if the nucleons in question are embedded into nuclear
matter. This fact is mainly due to a typical displacement of some
relevant intermediate nucleon states which occur during the ex-

change process. In this way the boson theory of nuclear forces is
- in principle - able to lift the characteristic discrepancy
(apparent   already in the 3-body problem) between the forces to
be used in the two-body (scattering) problem and the nuclear matter
properties. In order to calculate this new effect, bosonic variables must be introduced explicitely in the entire calculation
(compare D. Schütte, Nucl.Phys. A221, 1974, p. 45o-46o, and
K. Kotthoff, Doctor Thesis, Bonn, 1974). While this fundamental
enlargement of nuclear theory yields corrections of the order of
a few percents to all relevant nuclear properties, it might yield
large changes in the case of extremely high nuclear densities
which occur in Neutron Stars where this method will lead in a natural way to the so-called boson condensation.

2) Apart from the introduction of the additional bosonic
variables, it appears of fundamental importance to include also
the inner degrees of freedom of the nucleons themselves. This means
that the virtual excitation (through boson absorption) of the
nucleons into their various higher lying resonance states must also
be considered. For numerical reasons only the first state - the
so-called Δ —resonance (1236 Mev) - must be considered. This kind
of ' polarisation' is in fact inevitable and yields a most important
contribution to the middle-range attractive part of nuclear forces
(compare A.M. Green and P. Haapakoski "The effect of the Δ(1236)
resonance," reprint from Research Inst.f.Theor.Phys., Helsinki) and
is thus - to a large extent - responsible for nuclear binding.
(In addition it replaces the "unphysical" scalar σ-boson which
had so far been used in most boson-theoretical deductions of nuclear
forces). The contribution from the virtual Δ -excitation to the
nuclear force is,again, different in the cases of free and of
"embedded" nucleons. While the introduction of bosonic variables
leads to a needed increase of nuclear binding, the effect of the
Δ -resonance yields an important adjustment (decrease) of nuclear
density through its characteristic strongly repulsive short-range
contribution to the nucleon-nucleon interaction.(There remains,
however, the still unsolved problem of a numerical evaluation of
the effect due to a characteristic change of the mesonic self-
energies of nucleons when embedded into nuclear matter).

II. After this rather theoretical application of nuclear mass
values, I would like to emphasize its importance in the more
practical problem of a systematic representation (as well as the
interpretation) of nuclear spectra. For this purpose it is useful
to determine from these values the so-called Separation Energies
(i.e. mass differences) according to the rules given in earlier
contributions to these conferences (compare "Geometrical Representation of Separation Energies", Proc.Int.Conf.Nuclidic Masses,
Hamilton 196o, p.514, and Winnipeg 1967 "Calculation of Separat-

ion Energies for Spherical Nuclei," p.9). A suitable 3-dimensional
geometric plot of these values (compare also "Geometrical Representation of Nuclear Properties" by M. Beiner and K. Bleuler,
Nucl.Phys. 22, 1961, 589-597) yields at the same time an intuitive
survey of a few typical nuclear properties (characteristic steps
at the position of magic number, typical variations corresponding
to changes of internal structures, i.e. internal α-structure in
light nuclei as well as deformations of the nuclear shapes in
heavier elements).

The main application of this systematics of empirical nuclear
separation energies is, however, the fact that it can be used as
an absolute basis for the comparison of the spectra of successive
nuclides of similar types (i.e. odd neutron with even proton, odd
proton with even neutron number. It was realized that levels with
the similar properties – e.g. single particle levels with the
same assignments – are lying within our 3-dimensional plots on
most characteristic smooth surfaces which, in turn, can be - with
relatively fair accuracy - reproduced with the help of a theoretical calculation within the framework of conventional microscopic
theory (compare for this purpose the two papers by K. Bleuler and
M. Beiner "Pairing Approximation in Spherical Nuclei" I and II,
Nuovo Cimento 52 B, 1967, p.45-62 and 149-186, where the pairing
effects play a decisive role; a few more details are found in
"Nuclear Separation Energies and the Possibility of Extrapolations"
by K. Bleuler et al, Arkiv För Fysik, Band 36, Nr. 46, 1967,p.385).

In view of the tremendous improvement of the accuracy of mass
determinations, as well as some progress in the theoretical methods,
it seems really worthwhile to continue this kind of research on a
larger scale; it seems possible by now to check experimentally
some characteristic nuclear properties which are related to the
extensions of nuclear theory discussed in part I: We think, first
of all, about some effects due to the inner excitation of the
nucleons (i.e. virtual excitation of the Δ-resonance) within
nuclear structure. Such an excitation has typical consequences on
the Pauli-principle which, in turn, plays a decisive role - among
other cases - at the shell closures, i.e. the corresponding characteristic steps of separation energies. (There are slight indications
of an expected "smoothing out" of these steps within our plot). A
second effect might be found in a more accurate numerical determination of the pairing effects which play such a decisive role in
the theoretical interpretation of our 3-dimensional plots. Finally
one might obtain - with the help of a careful and systematic comparison between the theoretical and the empirical behaviour of
separation energies - some information about the (so far unsolved)
problem of the variation of the self-energy of nucleons within
nuclear matter.

III. In order to illustrate these facts we add here a revised 3-dimensional plot (constructed by M. Beiner) of separation energies $B_{\overline{2N}}$ and $B_{\overline{2Z}}$ (e.g. for Neutrons and for Protons) based on experimental nuclear masses throughout the periodic table (see our figures 1 and 2). Our so-called "average" separation energies (compare the earlier publications mentioned in section II) are defined by

$$(1) \quad B_{\overline{2N}} = \frac{1}{2} \Big(B(Z,N+1) - B(Z,N-1) \Big),$$

$$(2) \quad B_{\overline{2Z}} = \frac{1}{2} \Big(B(Z+1,N) - B(Z-1,N) \Big),$$

where B(NZ) represents the (experimental) binding. These values are then plotted as functions of the proton <u>and</u> neutron number N and Z (for technical reasons the values Z and $\overline{N-Z}$ are used as co-ordinates on a "basis plane") along the (negative) vertical axis. We thus obtain a surface within the 3-dimensional space spanned by the 3 axis for $B_{\overline{2N}}$ (or $B_{\overline{2Z}}$ respectively) N-Z and Z. This construction leads (with the help of a suitable projection) to an immediate and intuitive survey of all separation energies which in turn are interpreted theoretically by the position of the Fermi-level with respect to the value of the nuclear potential at infinity; it thus constitutes, so to speak, a natural link between empirical mass values and nuclear theory.

The striking regularity of our "surfaces" was reached through the elimination of the well-known even-odd staggering with the help of the characteristic "double step" (N+1, N-1 and Z+1, Z-1 respectively) introduced in our formulas (1) and (2). The remaining irregularities are, on the other hand, perfectly understandable from well-known properties of nuclear structure:
(i) The characteristic steps in the realm of spherical nuclei are due to the major shell closures; their geometrical shape and their numerical values correspond to a large extent to a theoretical treatment of shell structure including pairing effects (compare the paper by K. Bleuler and M. Beiner mentioned above) and possibly the inner excitation of nucleons.
(ii) The typical periodicity (periods of two units) in the domain of lighter nuclei corresponds to an admixture of α-structure, which is fading out in a natural way in the heavier nuclei.
(iii) A discontinuous increase of separation energies near $A \sim 16o$ is due to the well-known "sudden" deformation in that part of the periodic table whereas the more continuous decrease and subsequent increase of deformation are seen in the domains $A \sim 2oo$ and $A > 2o8$.

In view of these facts it appears relatively easy to represent our surfaces by analytical formulas containing a certain number of parameters which in turn might have natural physical interpretations. With the help of an "integration" one might thus obtain a suitable expression for a phenomenological mass-formula containing a rather restricted number of free parameters. The main advantage of such a procedure is the fact that the parameters are - apart from their physical interpretation - quantities which in principle can be calculated from a detailed microscopical theory: The main para- meters are those for the (non-local) nuclear potential (including deformations) and for the pairing effects (leading to a smoothing out of the subshells and a reduction of the step-hights at the closures of the major shells; compare the various papers by M. Beiner and the author mentioned above). In this context it must, however, be realized that new domains of deformed structures must be taken into account if this procedure is to be used for an extra- polation far from the "stability line".

Concluding this extremely short contribution to the fundament- al domain of nuclear binding, I would like to emphasize the fact that the determination (and a systematical representation) of separa- tion energies from empirical datas represents an important step in an understanding of the typical variations of nuclear structure throughout the periodic table. On the one hand, our "geometrical representation" is shown to be a most suitable tool for surveying typical nuclear properties, and on the other hand the characteristic parameters of an analytic representation of our "surfaces" repre- sent important values for a comparison with theoretical calculat- ions. As far as theory is concerned it might be preferable to consider and to calculate first these general parameters rather than special properties of individual nuclei. In view of the funda- mental theoretical problems related to the understanding of nuclear binding (compare section I) the consideration of the numerical values of these parameters constitutes,so to speak, the next step. Finally, I would like to observe, that the parametrisation proposed here has some similarity with the interesting phenomenological approach presented to this conference by Prof. E.R. H i l f .

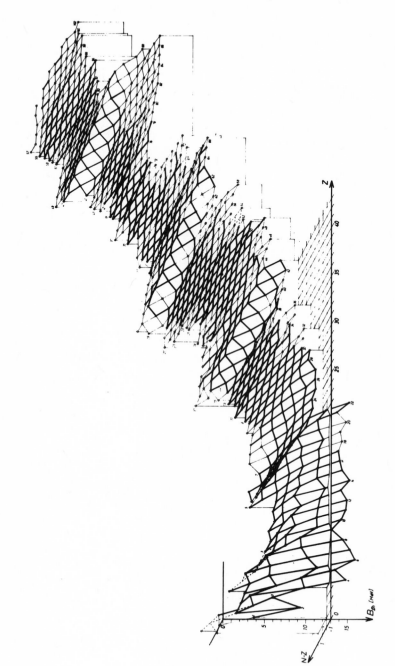

Fig. 1 : Three-dimensional representation of the empirical Separation Energies for Neutrons

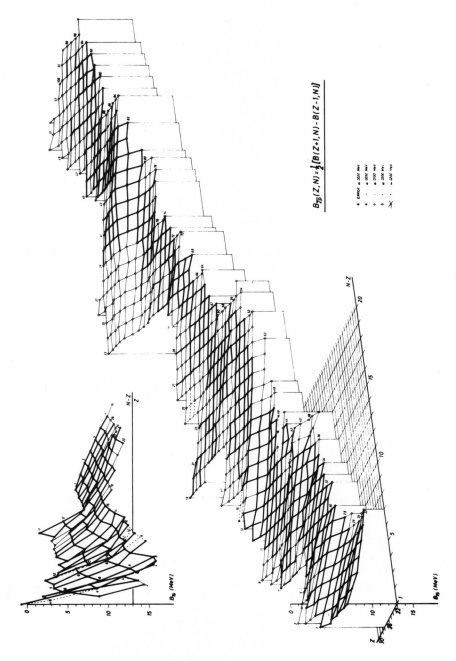

$$B_{7p}(Z,N) = \frac{1}{2}[B(Z+1,N) - B(Z-1,N)]$$

Fig.2 : Three-dimensional representation of the empirical Separation Energies for Protons
(lower part of the periodic table on separate diagram)

SELF-CONSISTENT CALCULATIONS OF NUCLEAR

TOTAL BINDING ENERGIES

M. Beiner, H. Flocard, Nguyen Van Giai,
and Ph. Quentin
Institut de Physique Nucléaire
Division de Physique Théorique*
91406, Orsay, France

I INTRODUCTION

The aim of this paper is to show that it is possible
to obtain nuclear binding energies with a sufficient
accuracy within the framework of the Hartree-Fock (H.F.)
approximation. The calculations reported here have been
done with the effective Skyrme interaction in the simplified form proposed by Vautherin and Brink [1]. This
phenomenological, density-dependent interaction contains
six parameters which have been determined by requiring
to fit as well as one could the binding energies and
radii of magic nuclei. This procedure gives a whole family of acceptable parameter sets which can be classified according to their density dependence. This density
dependence will be labelled by the value of the parameter t_3 of the force, larger values of t_3 corresponding
to forces which have a stronger density dependence.

A detailed discussion of various ground state properties of nuclei calculated with these parameter sets
has been given in ref.[2] where the spherical shape was
assumed for the studied nuclei. Calculations of deformation energy curves of spherical, well deformed or soft
nuclei have been performed [3-5]. For some nuclear systems the symmetric fission barrier or heavy ion fusion
potential curve have also been obtained [6-7]. Here we

---

* Laboratoire associé au C.N.R.S.

shall mostly discuss results concerning total binding
energies of stable or nearly stable nuclei and some
attempts to extrapolate towards superheavy elements. For
non-closed shell nuclei the pairing effects are taken
into account by H.F.+B.C.S. calculations [2].

II GROUND STATE PROPERTIES IN A SPHERICAL DESCRIPTION

Before discussing the calculated nuclear binding
energies let us review briefly the main features of the
nuclear densities and single-particle spectra predicted
by the Skyrme force.

a) Nuclear Densities

All the parameter sets give r.m.s. charge radii
which are in good agreement (within less than 2 %) with
the experimental data. A more detailed test of the H.F.
charge densities is provided by comparing calculated and
observed electron scattering cross sections. Such a comparison for the nuclei ^{40}Ca and ^{208}Pb shows that the
cross sections obtained from the calculated charge densities agree rather well with experiment at least for not
too large values of the momentum transfer q [2]. The
agreement is improved when the density dependence of the
interaction decreases.

b) Single Particle Spectra

The main property concerning the single particle
level density is that it is closely related to the non-
locality of the H.F. field. In the case of the Skyrme
force, this non-locality is governed by an effective
mass which is related in a very simple way to the parameters of the force. In order to obtain a calculated
level density close to the experimental one around the
Fermi level a strong density dependence in the force is
needed (large values of t_3), whereas the requirement
that the deep levels be deep enough favors a weaker
density dependence (small values of t_3). The parameter
set SIII which gives an effective mass $m^*/m=0.76$ in nuclear matter, realizes a reasonable compromise between
these two conditions. Most of the results we discuss here
have been obtained with the interaction SIII.

c) Total Binding Energies

Two parameter sets, SIII ($t_3=14,000$ MeV.fm^6) and
SIV ($t_3=5,000$ MeV.fm^6) have been adjusted to fit the ra-
dii of the magic nuclei ^{16}O, ^{40}Ca, ^{48}Ca, ^{56}Ni, ^{90}Zr,
^{140}Ce and ^{208}Pb and also their total binding energies
within an average error less than 1 MeV (SIII) or 2 MeV
(SIV). Two other sets, SV ($t_3=0$) and SVI ($t_3=17,000$ MeV.

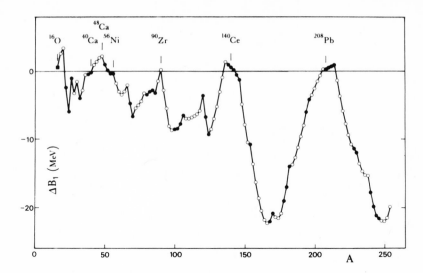

Fig.1 - Differences ΔB_1 between spherical HF+BCS and experimental total binding energies. Series of black and white circles correspond respectively to successive isotones and isotopes. The calculated values correspond to interaction SIII.

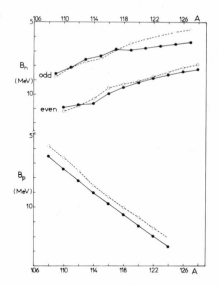

Fig.2 - Experimental (open circles) and HF+BCS (black circles) neutron and proton separation energies B_n and B_p of the Sn isotopes. The calculated values correspond to interaction SIII.

fm^6) have been obtained by a linear extrapolation of the
sets SIII and SIV for two arbitrary values of t_3. Despite
the simplicity of this procedure, it is remarkable that
the good agreement for the binding energies of magic
nuclei is nevertheless preserved.

In order to explore in more detail proton and neutron shell and subshell effects we have calculated about
120 even nuclei lying on a line which connects alternatively isotope and isotone series. Each isotope (isotone)
series contains all the nuclei the masses of which have
been measured [8]. Pairing correlations have been taken
into account by performing spherical H.F.+B.C.S. calculations. The results are shown in Fig.1. Except for
strongly deformed nuclei (they will be discussed in the
next section) the discrepancy never exceeds 8 MeV, which
is less than 1 % of the total binding energy thus demons-
trating that a spherical H.F.+B.C.S. approximation can
explain for nearly all of the nuclear binding energy.
One can also notice that the rather large slope of the
curve ΔB1 around magic nuclei shows that our H.F.+B.C.S.
calculations overestimate the magnitude of shell effects.

For the purpose of checking the neutron-proton
asymmetry properties of the interaction as well as the
pairing matrix elements we used it is interesting to
compare calculated and experimental separation energies
in the Sn isotopes (see Fig.2). The proton and neutron
separation energies are defined by $B_p(Z,N)=B(Z,N)-$
$B(Z-1,N)$, $B_n(Z,N)=B(Z,N)-B(Z,N-1)$. The correctness of
the asymmetry properties of the interaction is illustrated by the lower part of Fig.2 which shows that the calculated and experimental curves for B_p are parallel. The
upper part of Fig.2 shows that, excepted for the neutron-
rich isotopes, we estimate quite correctly the odd-even
differences, i.e. we use reasonable values for the pairing matrix elements.

III   CORRECTIONS DUE TO DEFORMATION EFFECTS

In order to see how the inclusion of deformation
effects could improve our results for nuclear ground
state properties, deformed H.F.+B.C.S. calculations for
some nuclei of the S-D shell, the rare earth region and
the actinide region have been performed. As an example
we show in Table 1 a comparison between experimental
and calculated total binding energies for some rare

	B_{exp}	$B_{HF + BCS}$	ΔB
^{152}Sm	1253.13	1249.81	-3.32
^{158}Gd	1295.93	1293.37	-2.56
^{162}Dy	1324.12	1321.61	-2.51
^{166}Er	1351.60	1349.63	-1.97
^{174}Yb	1406.63	1405.88	- .75
^{178}Hf	1432.85	1432.89	+ .04
^{184}W	1472.96	1472.80	- .16
^{190}Os	1512.83	1509.96	-2.87

Table 1 - Experimental and calculated (deformed HF+BCS with SIII) total binding energies of some rare earth nuclei. The last column gives the differences between the two values. Units are in MeV.

	$^{294}110$	$^{298}114$	$^{342}114$	$^{348}120$	$^{366}138$
Q_α	7.94	9.66	-.71	3.32	10.87
T_α	(50h)	(2s)		$(10^{41}y)$	(0.2h)
Q_β	.28	- .59	6.95	2.82	-5.59
T_β	(270d)		(16s)	(0.5h)	
$Q_{e.c.}$	-5.16	-2.09	-11.81	-7.88	.66
$T_{e.c.}$					(3.5d)

Table 2 - α-decay, β-decay and electron capture energy releases (in MeV) for some possible superheavy elements. Crude estimates of the corresponding lifetimes are also given.

earth nuclei. These results indicate that deformed H.F.+
B.C.S. calculations with the Skyrme interaction SIII can
be expected to predict the total binding energy of any
nucleus of the chart table with a discrepancy of less
than 5 MeV. Moreover the magnitude of these energy defects can probably be explained by the long range correlation effects which are not included in our calculations.

IV   EXTRAPOLATION TO SUPERHEAVY NUCLEI

The quality of the results obtained with the Skyrme
interaction concerning the ground state of stable or
nearly stable nuclei encourages one to extend the calculations to the region of superheavy elements. The possible magic numbers emerging from our spherical H.F.+
B.C.S. calculations are Z=114, 126 and 138 for protons
and N=178, 184 and 228 for neutrons. In agreement with
many other approaches (see e.g. refs. [9-11]) a sizeable
shell effect is found at N=114 and N=184. The shell
effects at N=178 or Z=120 are smaller. We also find a
large shell effect at N=228.

To determine the fission barrier heights for one
nucleus in this region (^{298}114) we have performed H.F.
calculations with a constraint on the quadrupole moment.
The calculated barrier parameters are with the usual
notation : E_A=7 MeV, E_{II}=1.5 MeV and E_B=5.2 MeV, while
previous microscopic-macroscopic approaches have given
higher values of E_A:10 MeV [10], 11 MeV [11] and 13 MeV
[9].

The Q_α-values $Q_\alpha(N,Z)$=B(Z-2,N-2)+B(2,2)-B(Z,N) have
been computed by performing H.F. calculations for the
daughter nuclei and using the experimental value of
28.297 MeV for B(2,2). The corresponding α decay half-
lives T_α are approximately evaluated by the formula of
Viola and Seaborg [12] valid in the actinide region. In
view of the uncertainties in the calculation of Q_α-va-
lues this estimate of T_α seems to us completely sufficient. The calculated results for Q_α and T_α are shown
in Table 2. Our values of Q_α are larger by more than
2 MeV than those obtained by using the Strutinsky's
renormalization method [13]. In any case there are large
uncertainties in any calculation of α-decay lifetimes.

The energy releases $Q_\beta(N,Z)$ and $Q_{e.c}(N,Z)$ corresponding to the emission or capture of an electron by a

superheavy nucleus have also been computed by performing H.F. calculations with the filling approximation. The corresponding half-lives T_β and $T_{e.c}$ have been estimated by using the approximate formulae (18) and (20) of ref. [13] with the numerical value of the average density of states in the daughter nucleus given in the Table 4 of the same reference. These values of T_β and $T_{e.c.}$ which are reported in Table 2 are only meant as rough estimates since this approximation completely neglects variations in the relevant matrix elements and the degree of forbidness of the transitions. Among the nuclei we have calculated, two are found to be stable against beta decay : $^{298}114$ and $^{366}138$. For the other nuclei however, all the possible transitions but one are forbidden, so that the corresponding lifetimes should be somewhat longer than the estimates of Table 2. The only exception is an allowed Gamow-Teller transition from the $1j13/2$ neutron state to the $1j15/2$ proton state which occurs in $^{342}114$. Except for the $^{366}138$ nucleus, all the calculated nuclei are stable by electron capture.

REFERENCES

[1] D. Vautherin and D.M. Brink, Phys.Rev.C5(1972)626
[2] M. Beiner, H. Flocard, Nguyen Van Giai and Ph. Quentin, Nucl. Phys. A238(1975)29
[3] H. Flocard, Ph. Quentin, A.K. Kerman and D. Vautherin Nucl. Phys. A203(1973)433
[4] M. Caillau, J. Letessier, H. Flocard and Ph. Quentin, Phys. Lett. 46B(1973)304
[5] X. Campi et al., Contribution to this Conference
[6] H. Flocard, Ph. Quentin, D. Vautherin, M. Vénéroni and A.K. Kerman, Nucl. Phys. A231(1974) 176
[7] H. Flocard, Phys. Lett. 49B(1974)129
[8] A.H. Wapstra and N.B. Gove, Nuclear Data, Table 9 (1971)265
[9] M. Bolsterli, E.O. Fiset, J.R. Nix and J.L. Norton, Phys. Rev. C5(1972)1050
[10] S.G. Nilsson, C.F. Tsang, A. Sobiczewski, Z. Szymanski, S. Wycech, C. Guftafsson, I.L. Lamm, P. Möller and B. Nilsson, Nucl. Phys. A131(1969)1
[11] M. Brack, J. Daamgard, A.S. Jensen, H.C. Pauli, V.M. Strutinsky and C.Y. Wong, Rev.Mod.Phys.44(1972)320
[12] V.E. Viola and G.T. Seaborg, J. Inorg. Nucl. Chem. 28(1966)741
[13] E.O. Fiset and J.R. Nix, Nucl. Phys. A193(1972)647

SHELL AND PAIRING EFFECTS IN SPHERICAL NUCLEI

CLOSE TO THE NUCLEON DRIP LINES

M. Beiner and R.J. Lombard

Institut de Physique Nucléaire
Division de Physique Théorique*
91406, Orsay, France

I INTRODUCTION

Until recently, the ground state binding energy of
nuclei was most frequently calculated by using mass formulae of different types or by integrating finite difference equations derived from local mass relations.
Typical approaches belonging to the first class are based on liquid drop model expressions with shell and
pairing corrections [1-3]. The second methods [4-6] have
been made possible by the considerable improvement of
the experimental knowledge of the atomic masses during
the two last decades [7-9]. Apart from least squares
fits occuring at some stage, the above treatments are
numerically rather simple accurate and rapid. Unfortunately, the reliability of their predictions for nuclei
situated away from the valley of β-stability is ques-
tionable. This is connected to their high degree of
phenomenology which does not allow to predict structural
changes. On the other hand, as pointed out many times,
several parameters of the liquid drop model cannot be
determined with a sufficient accuracy (or even depend
on the sample chosen for their adjustement) and consequently mass extrapolations are rather delicate.

Self-consistent approaches are supposed to do better
in this respect although specific choices concerning
either the many-body trial wave functions or the effective interactions tend to reduce the domain of reliability of variational calculations as well. Efforts made

---

* Laboratoire associé au C.N.R.S

237

toward a better understanding of effective interactions
and their connections to the free nucleon-nucleon force
[10-12] have allowed to develop, during the last few
years, accurate as well weakly parametrized models closely related to more fundamental treatments. Among such
models the Brueckner formulation of the energy density
formalism [13] has been extended in order to take
into account shell structure and pairing correlations
[14] . In this case, H.F.-B.C.S. like calculations reproduce bulk properties of measured spherical nuclei
with a high accuracy and we thus believe this particular
approach to be of practical interest for computing several properties of exotic nuclei.

The main characteristics of our microscopic calculations will be summarized. The results concerning more
than 300 spherical light and medium nuclei located in
the nucleon drip regions will be presented and
discussed.

II MAIN FEATURES OF THE APPROACH

A complete derivation of the formalism, the numerical determination of the eleven parameter (of which
seven describe nuclear matter, two refer to finite size
effects and the two last one to spin-orbit coupling),
the treatment of the pairing correlations and a summary
of results can be found in ref.[14]. Here, we shall merely emphasize three points which are significant for
the present application.

i) The fine adjustment of the parameter on properties of few selected magic spherical nuclei was done
under the constraint of fitting the neutron matter calculations by Buchler and Ingber [15].

ii) The gap matrix is calculated by means of an
energy independent long range interaction derived from
the Hamada-Johnston potential by using a Moskowski-Scott
type separation method[16]. This interaction fits accurately the nucleon-nucleon data up to $\simeq$ 150 MeV [17].

iii) For a large sample of spherical nuclei distributed throughout the whole nuclidic chart (see Section V
and in particular fig. 3,7 and 15 of ref.[14]), the distribution of the differences between calculated and measured total binding energy has a mean value $\overline{\Delta B}$=-2.3 MeV*

---

* In the present approach, the parameter have been adjusted in such a way that the calculated total binding
energy of doubly magic nuclei never exceeded the measured one and not to constraint $\overline{\Delta B}$ to be zero as it is
usually done in the case of mass formulae.

and a mean quadratic deviation of 1.7 MeV, shell closures, separation and pairing energies, single particle
spectra as well as rms charge radii are also in close
agreement with experiment.

In the present work, we shall also restrict
ourselves to spherical calculations. In regions of
actually deformed exotic nuclei, they will give useful
reference results. Studying the structure of the nucleon
drip lines requires obviously accurate predictions of
the one and two nucleon separation energies up to unusual values of the neutron excess $(N-Z)/A$. This implies
an approach able of reproducing correctly shell-,pairing-,
symmetry- and deformation effects in a large (Z,N)-domain,
i.e., a fortiori, in the narrow band of measured nuclei
for which they are well known from a detailed analysis
of the experimental nuclear mass surface [18,19]. Figure
1 illustrates a general comparison between the important
first and second order finite differences of the measured and calculated binding energies

(1) $B_{2n}(Z,N) \equiv B(Z,N+1) - B(Z,N-1)$

(2) $B_{2p}(Z,N) \equiv B(Z+1,N) - B(Z-1,N)$

(3) $\Delta_{pn}(Z,N) \equiv B(Z+1,N+1) + B(Z-1,N-1) - B(Z+1,N-1) - B(Z-1,N+1)$

in the case of medium and heavy even nuclei. We see that
the agreement is quite good in the case of the spherical
nuclei (see also fig.7 of ref.[14]). The most important
deviations occur around $N=82$ for B_{2n} as well as around
$Z=50$ for B_{2p}, where we overestimate the shell effects,
and for $71 < N < 81$ for Δ_{pn}, where the experimental values
present a rather deep local minimum. More systematic
discrepancies appear as expected in the deformed regions.
In the case of the rare earths, we overestimate B_{2p} while
we underestimate both B_{2n} and Δ_{pn}. In the actinides, we
further underestimate B_{2n} and Δ_{pn} but do not find systematic missfit to B_{2p}.

This sensitive check, together with the good results
of recent extended calculations of bulk properties of
light stable and unstable nuclei $1 < Z < 8$ and of Na-,K- and
Rb isotopes [20], give us confidence in investigating the
impact of shell structure and pairing correlations on the
nucleon drip lines.

III   STRUCTURE OF THE NUCLEON DRIP LINES

At a fixed value of Z, a nucleus $(Z,N+1)$ is neutron
unstable if the one neutron separation energy $B_n(Z,N+1) \equiv$
$B(Z,N+1) - B(Z,N)$ is negative and is unstable against the
emission of a neutron pair if the two neutron separation

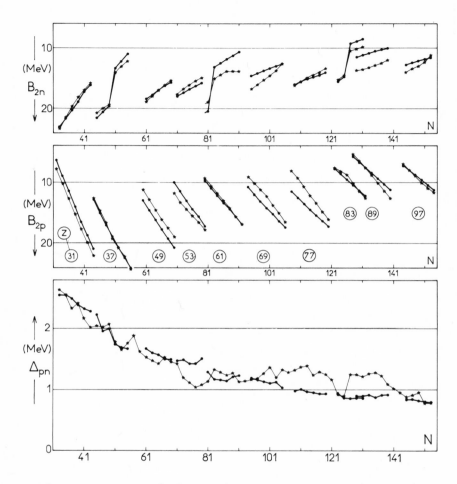

Fig.1- Measured (stars) and calculated (dots) finite differences of the total binding energy of medium and
heavy nuclei (see text) plotted as functions of N for
ten selected values of Z (encircled in the middle part
of the drawing). Upper part : two neutron separation
energy, $B_{2n}(Z+1,N)$, middle part : two proton separation
energy, $B_{2p}(Z,N)$, and lower part : variation with respect to N (or to Z) of the two proton (two neutron) separation energy, $\Delta_{pn}(Z,N)$.

energy (1) gets negative[*]. Approximate relations exist
between the chemical potential λ_n and the quasineutron
energies E_ν on one hand, and the neutron separation energies on the other hand. For instante, we have for N odd
and N±1 non magic

$$B_{2n}(Z,N) \simeq -2\lambda_n(Z,N)$$

(4)
$$B_n(Z,N) \simeq -\lambda_n(Z,N) + \frac{1}{2}\partial_N\lambda_n(Z,N) - E_\nu^{min}(Z,N)$$

where $E_\nu^{min}(Z,N)$ is the smallest quasineutron energy in
the odd-N nuclei (Z,N) and $\partial_N\lambda_n(Z,N)$ the slope of λ_n
along isotope lines. The chemical potentials and the
quasiparticle excitations are closely related to the single particle energies in the neighbourhood of the Fermi
levels. Typical examples of the evolution of the neutron
single particle energies and chemical potentials as functions of N in the case of the Na-K- and Rb isotopes are
given in figs.7,9 and 10 respectively of ref.[20]. The
(positive) slope $\partial_N\lambda_n$ around $\lambda_n=0$ is small, less than
.15 MeV in the case of the K isotopes for instance (fig.9).
Note also the vanishing of the magic character of N=50
and the appearance of a new spherical shell closure at
N=58 in these isotopes.

The estimates (4) are useful to analyse the structures of the nucleon drip lines, i.e., the conditions of
the emission of one or two nucleons. When λ_n tends to
zero and except at neutron shell closures, the slope
$\partial_N\lambda_n$ is small compared to the quasineutron energies E_ν^{min}
which are of the order of one MeV. This explains the
existence of long series of neutron stable even-N isotopes adjacent to unstable odd-N ones. In the proton case, the coulomb interaction increases the slope of λ_p
along isotone lines. This slope is not small compared to
the quasiproton energies even as λ_p tends to zero and
consequently the even-Z isotones become much more rapidly unstable in the proton drip region.

The stability properties against nucleon emission
of light and medium spherical nuclei is summarized in
Fig.2. The most stricking features of this survey are
a) for nuclei close to the neutron drip line : i) the
vanishing of the magic character of the neutron number
20 and 28 as well as its strong weakening at N=50. ii)
the appearance of new neutron spherical shell closures,
especially at N=16, 34 and 58, and iii) the existence
of very long series of even-N isotopes stable against
particle emission and adjacent to odd-N isotopes which
are unstable against the emission of one neutron (see
for instance the Z=30 and 32 series), and b) for nuclei

---

[*] Of course, corresponding expressions and comments hold
    for the protons.

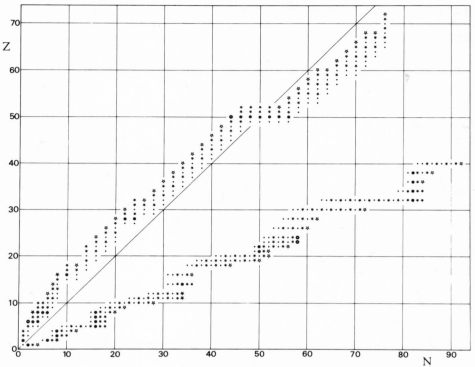

Fig.2 - Structure of the nucleon drip lines for
light and medium spherical nuclei obtained in the present
microscopic self-consistent approach. •Nucleus (● Z or
N magic nucleus) stable against proton (or neutron)
emission. ✦Nucleus (✦ Z or N magic nucleus) unstable
against one proton (or one neutron) emission. ✫ Nucleus
(✪ Z or N magic nucleus) unstable against two proton
(or two neutron) emission but stable against one nucleon
emission.

close to the proton drip line : i) the relatively weak
influence of the proton spherical shell closures at
Z=16 and 28 but the very strong one at Z=50, and ii) the
narrow transition domain separating the nuclei which are
stable from those which are unstable against one proton
emission is limited by rather long series of nuclei
having constant values of N-Z.

The definition of "magic character" or "shell
closure" contains necessarily some arbitrariness. Here
we have adopted a definition (see ref.[20]) taking
advantage of the particular property of the BCS gap

equations to have no non trivial solution for certain
nucleon numbers if the level spacing around the chemical
potentials is large compared to some average value of
the pairing matrix elements. In such case the nucleon
number corresponds to the filling up of the Fermi orbital.

IV CONCLUSIONS

Previous work by Brueckner et al.[21] using the
Thomas-Fermi approximation has shown the influence of the
self-consistency on the drip lines. The present results
underline fine and even stronger effects due to the
pairing correlations and to shell closures at new magic
numbers which cannot be predicted by any mass formula or
mass extrapolations so far developed.

REFERENCES

[1]   W.D. Myers and W.J. Swiatecki, Nucl.Phys.$\underline{81}$(1966)1
[2]   V.M. Strutinsky, Nucl. Phys. $\underline{A95}$(1967)420
[3]   P.A. Seeger and W.H. Hovard,$\overline{LA5750}$ UC-34c (Oct.74)
[4]   G.T. Garvey et al, Rev. Mod. Phys. 41(1969)S1
[5]   A. Sorenson, preprint (1971)
[6]   J. Jänecke and H. Behrens, Phys.Rev. $\underline{C9}$(1974)1276
[7]   L.A. König, J.H.E. Mattauch, and
       A.H. Wapstra, Nucl. Phys. $\underline{31}$(1962)18
[8]   J.H.E. Mattauch, W. Thiele and A.H. Wapstra, Nucl.
       Phys. $\underline{67}$(1965)1,32,73
[9]   A.H. Wapstra and N.B. Gove, Nuclear Data Tables,
       Vol.9 (1971) 265
[10]   J.W. Negele, Phys. Rev. $\underline{C1}$(1970)1260
[11]   X. Campi and D.W. Sprung, Nucl. Phys. $\underline{A194}$(1972)401
[12]   J.W. Negele and D. Vautherin, Phys.Rev.$\underline{C5}$(1972)1472
[13]   K.A. Brueckner, J.R. Buchler, R.C. Clark, and
       R.J. Lombard, Phys.Rev. $\underline{181}$(1969)1543; R.J. Lombard,
       Ann. of Phys. (NY) $\underline{77}$(19$\overline{73}$) 380
[14]   M. Beiner and R.J. Lombard, Ann. of Phys.(NY)
       $\underline{86}$(1974)262
[15]   J.R. Buchler and L. Ingber, Nucl. Phys.$\underline{A170}$(1970)1
[16]   S.A. Moskowski and B.L. Scott, Ann. of Phys. (NY)
       $\underline{11}$(1960)65
[17]   K. Bleuler et al., Nuovo Cimento $\underline{B52}$(1967)45,149
[18]   N. Zeldes, M. Gronau and A. LeV, Nucl.Phys.$\underline{63}$(1965)1
[19]   M. Beiner, IPNO/TH- Aussois 1971
[20]   M. Beiner, R.J. Lombard, and D. Mas, Nucl. Phys.
       (in press);IPNO/TH 75-4
[21]   K.A. Brueckner et al., Phys. Rev. $\underline{C4}$(1971)732

HARTREE-FOCK CALCULATION OF NUCLEAR BINDING ENERGY

OF SODIUM ISOTOPES

X. Campi[+], H. Flocard[+], A.K. Kerman[++],
and S. Koonin[++]

[+]Institut de Physique Nucléaire, Division de
  Physique Théorique[*], 91406, Orsay, France
[++]Massachusetts Institute of Technology
  Cambridge, Massachusetts, U.S.A.

Recents measurements [1] of the nuclear binding
energy (B) of exotic sodium isotopes have shown that the
two neutron separation energy B_{2n}, defined as $B_{2n}(A,Z)=$
$B(A,Z)-B(A-2,Z)$ displays a sudden discontinuity for the
mass number A=31 (see fig.1) which is very similar to
those observed in the rare earth region. In the latter
case this phenomenon has been explained in terms of a
shape transition of the intrinsic ground state.

Previous Hartree-Fock (H.F.) calculations using a
Skyrme type effective interaction [2] (called SIII)
have accurately reproduced the binding energies B and
the quadrupole moments Q of a large number of nuclei
along the stability line. Using the same Skyrme interaction S-III we have performed H.F. calculations for the
sodium isotopes A=19 to A=35. For mass number A between
28 and 34 we find that there are two prolate H.F. solutions. This can be seen in the figure 2 where we show
the deformation energy curves of the same sodium isotopes plotted as a function of the proton quadrupole moment. These curves have been obtained by a constrained
H.F. calculation using the quadrupole moment as a constraint. For A smaller (larger) than 31 the less (more)
deformed solution is more stable. The curve SIII in fig.
1 has been calculated using the lowest H.F. solution.
We have corrected this curve by adding the rotational energy

_____

[*] Laboratoire associé au C.N.R.S

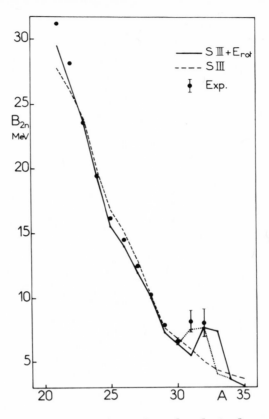

Fig.1 - Experimental and calculated values of
the separation energies of the last pair of neutrons

calculated in an approximate way : $E_{rot}=<J^2>/2I$ where the
rigid body value has been taken for the moment of inertia
I. The corresponding curve (SIII+E_{rot}) as shown in fig.1,
agrees well with experiment. In particular its slope is
very similar to the experimental one. This agreement
shows that the (surface and volume) symmetry energy corresponding to our interaction is correct. Moreover the
discontinuity in the experimental B_{2n} curves is also
reproduced. We also show in fig.1 (dotted line) how the
previous results change when one assumes the shape transition to occur between ^{30}Na and ^{31}Na. In fact this
assumption is reasonable since the calculated rotational

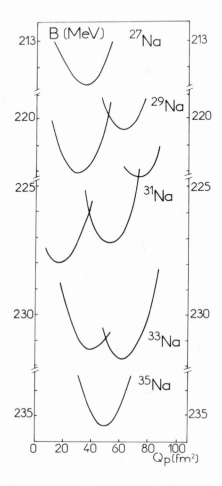

Fig. 2 - Deformation energy curves
of sodium isotopes as a
function of the proton
quadrupole moment

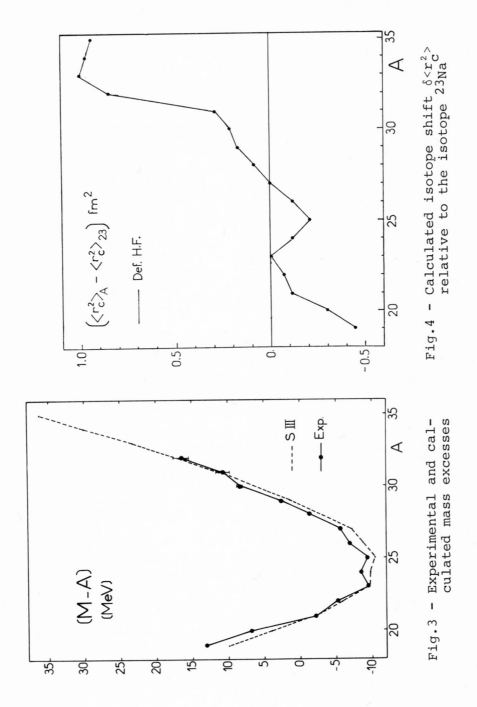

Fig. 4 - Calculated isotope shift $\delta\langle r_c^2\rangle$ relative to the isotope ^{23}Na

Fig. 3 - Experimental and calculated mass excesses

energy (2.1 MeV) of the very prolate state of ^{31}Na compensates the small excitation energy of this state relative to the lowest minimum of the constrained H.F. energy curve (see figure 2).

In Fig.3 we compare the experimental values of mass defects of Na isotopes to the H.F. results. The mean curvature of the experimental curve is correctly reproduced and our calculation accounts for a fraction of the odd-even staggering effect.

In table (1) we give the calculated values of the quadrupole moments of the mass and protons distributions. We have also indicated the values of the deformation parameter β_2 of the liquid drop with the same r.m.s. radius and quadrupole moment. One notices the very large values of β_2 for the isotopes A=32 to A=34, larger than those previously found in H.F. calculations of the ground state properties of rare earth and actinide nuclei [3]. It is likely that this region of large deformation extends to the neutron rich isotopes of neighbouring elements. Calculations in this direction are in progress.

In table (1) one also sees that the values of proton β_{2p} and mass β_{2M} deformation parameters of a same isotope can be very different. This is particularly striking for the quasi-spherical isotopes preceeding the transition region (A=28 to 31) where the ratio β_{2p}/β_{2M} is close to 1.9. This phenomenon is remarkable since Skyrme H.F. calculations in other regions of the chart of elements, including neutron rich nuclei like ^{150}Ce have always led to values of β_{2p}/β_{2M} close to one. It seems then that this strange behaviour is characteristic of light exotic nuclei. It should be noted that in many phenomenological calculations of ground state properties one assumes the equality $\beta_{2M}=\beta_{2p}$. Our calculations show that this assumption must be relaxed for light nuclei.

In Fig.4 we have plotted the variation of the charge radius $<r_c^2>$ (relative to the isotope ^{23}Na) as a function of the mass number A. The structure of this curve is essentially due to the deformation and the comparison of table (1) and figure (4) shows a clear correlation between the changes in β_{2p} and the changes in radii.

From the previous discussion and results the H.F. method appears as a powerful tool to calculate the properties of exotic nuclei. In particular the experimental

A	Q_p	β_{2p}	Ω_M	β_{2M}
19	25.80	.212	33.72	.172
20	36.89	.296	59.68	.276
21	46.55	.365	86.55	.364
22	47.28	.369	93.42	.368
23	48.33	.374	101.14	.373
24	40.67	.319	83.82	.296
25	32.63	.259	64.90	.221
26	32.04	.252	66.86	.213
27	33.51	.260	74.88	.223
28	31.42	.242	69.59	.196
29	29.45	.225	64.30	.170
30	22.91	.175	45.61	.115
31	16.60	.127	28.11	.068
32	55.89	.396	174.12	.367
33	60.37	.421	198.37	.394
34	53.90	.376	177.25	.339
35	47.23	.330	154.77	.286

Table 1 - Calculated values of the quadrupole
moments of the mass (Q_{2M}) and protons
(Q_{2p}) distributions

total binding energies are well reproduced. Contrary to
many models based on extrapolations the H.F. method is
able to predict sudden transitions like the one which
seems to occur in the vicinity of the drip line ($N \simeq 20$).
However further experimental work measuring either the
mass of ^{33}Na (see fig.1) or the radius isotope shift
(see fig.4) is still necessary to definitively prove
the existence of the calculated shape transition.

REFERENCES

[1] C. Thibault, R. Klapisch, C. Rigaud, A.M. Poskanzer,
    R. Prieels, L. Lessard and W. Reisdorf. To be
    published
[2] M. Beiner, H. Flocard, Nguyen Van Giai and
    Ph. Quentin, Nucl. Phys. A238(1975)29 ; H. Flocard,
    Ph. Quentin, A.K. Kerman and D. Vautherin, Nucl.
    Phys. A203(1973)433
[3] H. Flocard, Ph. Quentin and D. Vautherin, Phys. Lett.
    46B(1973) 304

NUCLEAR MASS SYSTEMATICS AND EXOTIC NUCLEI

K. Takahashi[+]

Institut für Theoretische Physik,Universität zu Köln

D-5 Köln 41, Germany

H. v. Groote[++], E. R. Hilf

Institut für Kernphysik, TH Darmstadt

D-61 Darmstadt, Germany

1. INTRODUCTION

The experimental study of nuclei far from stability (exotic nuclei) has been remarkably successful in the past few years as can easily be seen in the rapid data-accumulation of atomic masses, high-energy α- and β-decay half-lives, β-decay strength-functions, delayed-particle emission-rates and spectra, and isotope shifts [1].

On the theoretical side, these new data provide a sensitive check of mass formulae and their extrapolation power, since they lie on the outer limits of the experimentally known nuclear-chart. Here effects which cannot be seen in the β-stability region may become crucial and hopefully give important feedback for our fundamental understanding of nuclear properties. Obviously, for this purpose one should follow guidelines based on physical (not numerical!) arguments. For example, possible shape-transitions in the exotic nuclear region may be explained by the Hartree-Fock calculation [2]. Another example is the droplet-model (DM) mass-formula [3] which besides ground-state masses describes also nuclear radii, ground-state deformations, nuclear compressibility and fission barriers. Having established the

[+]Fellow of the Alexander von Humboldt-Stiftung
[++]Supported by the Gesellschaft für Schwerionenforschung (GSI)

important ingredients for extrapolation, one may expect more reliable predictions of nuclear properties in the unknown region. An in-
ventive application of these nuclear systematics is found in the r-
process nucleosynthesis calculation, which relates predicted properties of nuclei in the unknown region of the nuclear chart to the
abundance of nuclei in the β-valley.

The main aim of the present report is to point out the possible
feedback to the mass predictions from the theoretical studies of nuclei far from stability. In the following, we shall illustrate by
several examples how to check and in which direction to improve the
mass formulae.

2. r-PROCESS

It is generally agreed [4] that the abundances of a bulk of the
heavy elements and all transbismuth elements are due to rapid neutron
capture by very neutron-rich nuclei along the so-called r-process
path and to β-decay into the valley of β-stability. A shell effect
in the r-process path leads to higher abundances of the corresponding β-stable isobars. In these calculations the implication of the
atomic-mass prediction is twofold: (i) the r-process path is mainly
determined by the neutron separation-energies (2∼3 MeV) and (ii)
the time-scale is determined by the β-decay half-lives summed up
along the r-process path. These half-lives can be calculated using
the gross theory of β-decay [5], once the Q-values (10∼15 MeV) are
predicted.

Concerning (i), it has been pointed out [6] that the abundance
curve recently compiled by Allen et al.[7], i.e. the observed abundances corrected for the s-process contributions, suggests a decreasing shell-effect of N =50 magic number when the proton number is
much smaller than that of nuclei in the β-valley. That is, the abundance curve no longer shows a strong peak around A =80 but a plateau
in A ≲80. This can hardly be explained by e. g. the Myers-Swiatecki
shell-correction term [8] having the strong magicity at N =50 pro-
pagated into the region of neutron-rich nuclei. This suggested dis-
appearance or, more generally, the change of the magicity on the
neutron-rich side is supported by recent calculations with either a
Woods-Saxon potential [9] or the energy-density formalism [10].

In the following joint report [11], we present a strategy to embed
these phenomena into the shell-correction term of the mass formula.

As for point (ii), the absolute time-scale of the r-process has
recently been suggested [6] to be of the order of a few seconds
which is considerably longer than implied in the former dynamical
r-process calculations [12], much less than 1 sec. A new dynamical
calculation [13] is in progress which takes into account the energy

release due to β-decays and fissions (as well as its feedback to the hydrodynamics) which might prevent the temperature to decrease too rapidly, since otherwise the synthesis will terminate before the existing heaviest nuclei (e.g. Th, U, Pu) could be produced. However, if the astrophysical environment in which the r-process takes place in fact requires a shorter time-scale, the β-decay half-lives should be shorter than expected. It is thus necessary to reinvestigate the β-decay formula with Q-values predicted by the improved mass-formula.

Another quantity concerning nuclear masses which one has to evaluate in the r-process calculation is the nuclear partition function: Because of the rather high temperature (in excess of 4×10^9K, i.e. 0.34 MeV, at the start of the r-process) the nuclei are partially in excited states. Using the available experimental information of spins and masses of low-lying states and starting from the Gilbert-Cameron level-density formula [14], El Eid [15] has recently obtained a partition-function formula which allows for a reliable extrapolation into the unknown neutron-rich region. We also mentioned that the first step to introduce the spin-parity dependence in the mass systematics has been forwarded by Uno and Yamada [16].

3. DELAYED-PARTICLE EMISSION-RATES

In the region where β-delayed-particle emission occurs, the simplest idea to check the mass formulae is to examine the window-energy i. e. the β-decay Q-value minus the particle separation-energy in the daughter. For delayed protons, $p\,\beta^+$-coincidence measurements are available to determine the window-energy, $\Delta E_p = Q_\varepsilon - S_p$. Table I compares some of the experimental and the mass-formula values.

TABLE I
Delayed-proton window-energy ΔE_p [MeV]. Experimental data are taken from [1]. The mass-formula values are from LDM: [8], DM-vG: [17], DM-M: [18] and LZ: [19].

precursor	$(\Delta Ep)^{exp}$	$(\Delta Ep)^{cal}$			
		LDM	DM-vG	DM-M	LZ
^{73}Kr	4.85 $\pm$ 0.30	3.38	2.58	2.32	4.45
^{109}Te	7.14 $\pm$ 0.10	7.66	6.49	6.60	7.96
^{111}Te	5.07 $\pm$ 0.07	5.50	4.53	4.64	5.64
^{115}Xe	6.20 $\pm$ 0.13	5.57	4.96	5.13	5.88
^{117}Xe	4.10 $\pm$ 0.20	3.43	2.94	3.02	3.67
^{118}Cs	4.7 $\pm$ 0.3	4.83	4.04	3.66	5.66

In the case of delayed neutrons, on the other hand, coincidence experiments are extremely hard to do, and only the sign of the window-energy, $\Delta E_n = \tilde{Q} - S_n$, has up to now been used to check the mass formulae. Here, we present a new technique to derive semi-experimental values of ΔE_n by combining the measured emission-rates with the calculated values. Contrary to the case of delayed protons, one may expect the effects of γ-decay competitions to be small even when the window-energy is not so large. This makes the evaporation calculation simpler.

The experimental total delayed-neutron emission-rate is defined by

$$\lambda_n^{exp} = (\ell n\, 2 / T_{1/2}) \cdot P_n \tag{1}$$

in terms of the β-decay half-life $T_{1/2}$ and the observed branching-ratio P_n of delayed neutrons to the total β-decay. The theoretical values [20] for λ_n can be approximated by

$$\lambda_n^{cal} = (0.5 \sim 2.0) \cdot 10^{-4} \cdot (\Delta E_n)^{4.64} \quad sec^{-1} \tag{2}$$

where ΔE_n is given in MeV. Considering the theoretical uncertainties and the experimental errors (mainly in P_n), we only give the possible range of ΔE_n as

$$\left[\frac{min\ \lambda_n^{exp}}{2.0 \cdot 10^{-4}} \right]^{1/4.64} \leq \Delta E_n \leq \left[\frac{max\ \lambda_n^{exp}}{0.5 \cdot 10^{-4}} \right]^{1/4.64} . \tag{3}$$

A comparison of our results with the values of different mass-formulae and systematics is displayed in Fig. 1 together with available experimental data. In general, the experimental data support the semi-experimental range of ΔE_n given in eq. (3). The mass-formula predicitons reproduce nicely the general trends of the semi-experimental results. Only for Rb, the DM values disagree distinctively. We think that this is due to the shell-correction term which does not take $Z = 40$ to be a magic number.

4. HIGH-ENERGY α- AND β-DECAYS

As shown by Hansen [1], a simple but crucial check of any mass formula is provided by the α-decay kinetic-energy data for light (neutron-deficient) isotopes in the Hg-Pb region. Especially, it is a nice way to examine the propagation of the $Z = 82$ magicity. From Fig. 2 we again see that the shell-correction term of the droplet-model mass-formula has to be improved.

Experimental determinations of β-strength functions have been shown to be rather sensitive to the Q-values assumed. Thus it might be possible to predict reasonable Q-values by comparing the experimental and theoretical β-strength-functions, because the latter is

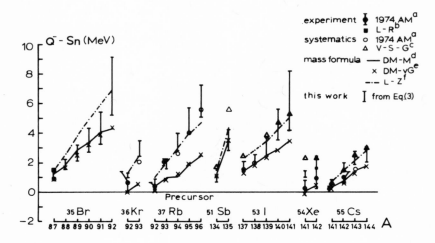

Fig. 1: Comparison of experimental, semi-experimental, and theoretical delayed-neutron window-energies $\Delta E_n = Q^- - S_n$. References are: a:[21], b:[22], c:[23], d:[18], e:[17] and f:[19]. Experimental values of Pn and $T_{1/2}$ needed for eq.(3) are taken from[24].

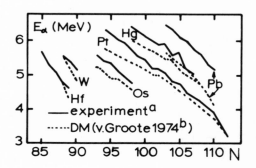

Fig. 2: Alpha-decay kinetic-energy. a: experimental values taken from [1], b: calculated values [17] from the droplet-model.

much less sensitive to the Q-values assumed. As a special example, the ^{87}Br has been discussed [6].

5. CONCLUSIONS

In this report, we have presented several tests for existing mass-formulae or systematics, by using not only experimental data but partly theoretical results. We believe that most of the short-comings we found are due to shell effects not included in these formulae. We hope that the simple tests presented here will provide us with guidelines for further improvements of the mass formulae, especially of their shell-correction term.

Much of this work and of the following joint report is still in progress, and the final results together with their application to the dynamical r-process will be published somewhere else rather soon.

Finally, we wish to take this opportunity to thank W. Hille-brandt, M. El Eid and T. Kodama for their collaborations on special topics, and W. D. Myers and N. Zeldes for making available to us their newest mass-formula results. One of us (K. T.) appreciates the great interest and encouragement by P. Mittelstadt.

REFERENCES

1   P. G. Hansen, Proc.Int.Conf. on Nucl.Structure and Spectroscopy, Amsterdam(1974) Pt. II (Scholars Press. 1974) p. 662
2   M. Cailliau, J. Letessier, H. Flocard and P. Quentin, Phys. Lett. 46 B (1973) 11
3   W. D. Myers and W. J. Swiatecki, Ann.Phys. 55 (1969) 395
4   e.g. D. D. Clayton, Principles of Steller Evolution and Nucleo-synthesis (McGraw-Hill, 1968)
5   e.g. K. Takahashi, M. Yamada and T. Kondoh, Atomic Data and Nucl. Data Tables 12 (1973) 101
6   T. Kodama and K. Takahashi, Nucl. Phys. 239 (1975) 489
7   B. J. Allen, J. H. Gibbons and R. L. Macklin, Advanced in Nucl. Phys. ed. M. Baranger and E. Vogt, Vol. 4 (Plenum, 1971)p.205
8   W. D. Myers and W. J. Swiatecki, Nucl. Phys. 81 (1966) 1
9   E. R. Hilf and H. v. Groote, Proc.Int.Workshop on Gross Proper-ties of Nuclei and Nuclear Excitations III, Hirschegg, 1975 (AED-CONF, 75-009, TH Darmstadt) p. 230
10 M. Beiner and R. J. Lombard, this conference
11 E. R. Hilf, H. v. Groote and K. Takahashi, this conference
12 A. G. W. Cameron, M. D. Delano and J. W. Truran, CERN-report 70-30 (1970), Vol. 2, p. 735
     D. N. Schramm, Astrophys. J. 185 (1973) 293
13 W. Hillebrandt et al., in progress
14 A. Gilbert and A. G. W. Cameron, Can.J.Phys. 43 (1965) 1446

15 M. El Eid, thesis, TH Darmstad (1975)
16 M. Uno and M. Yamada, private comm. (1975)
17 H. v. Groote, to be published
18 W. D. Myers, private comm. (1975)
19 S. Liran and N. Zeldes, private comm. (1975)
20 K. Takahashi, Prog. Theor. Phys. <u>47</u> (1972) 1500
21 The Oak Ridge Nuclear Data Group, 1974 Atomic Masses (tabulated
    by F. Serduke)
22 E. Lund and G. Rudstam, private comm. (1975)
23 V. E. Viola Jr., J. A. Swant and J. G. Graber, Atomic Data
    and Nucl. Data Tables <u>13</u> (1974) 35
24 L. Tomlinson, Atomic Data and Nucl.Data Tables <u>12</u> (1973) 179
    K. L. Kratz and G. Herrmann, Z.Phys. <u>263</u> (1973) 435

THE VALIDITY OF THE STRUTINSKY METHOD FOR THE

DETERMINATION OF NUCLEAR MASSES

M. Brack [+] and P. Quentin [++]

The Niels Bohr Institute

Blegdamsvej 17, 2100 Copenhagen Ø, Denmark

In theoretical estimates of nuclear masses of experimentally unknown isotopes, one has used two different kinds of methods. The first consists in deducing these masses from some known neighbouring ones with the help of a convenient extrapolation formula. This has been extensively discussed for instance in Ref. [1]. One of the problems of such an approach is that one cannot predict any sudden change (e.g. in deformation) in the unknown region if it has not shown up for any known nuclei in the vicinity. The other methods, on which we will concentrate here, are based on a detailed description of total binding energies within a given model. In the last two or three years, one has been able to parametrize phenomenological effective interactions and to give a very satisfactory systematic reproduction of nuclear masses on the whole chart of nuclides, using both the Hartree-Fock approximation [2-3] and the Hartree-Fock-Bogolyubov approximation [4]. But the oldest and still rather successful approach to nuclear masses is the liquid drop model [5]. For the description of fine details connected to the existence of magic nuclei one has been obliged to introduce some

(+) Present address: Department of Physics, S.U.N.Y. at Stony Brook, N.Y., U.S.A.

(++) Permanent address: Division de Physique théorique, I.P.N., Orsay, France.

shell corrections as done in Ref. [6]. Strutinsky has
given a consistent description of the nuclear binding
energy in terms of a sum of a liquid drop energy plus
first and higher order shell corrections [7]. Such an
expansion relying on the validity of the Hartree-Fock
description of the nuclear ground state is referred to
as the Strutinsky energy theorem [8]. The energy averaging method widely used to extract the shell correction
has been shown to be equivalent to many other possible
prescriptions (such as the so-called temperature method
or various semi-classical expansions) [9]. A consistent
fit of the parameters of both the liquid drop model and
the single particle potential needed in the Strutinsky
method has been done by Seeger and co-workers (see e.g.
Ref. [10]) for nuclei with $A \geqslant 40$. Such approaches to
nuclear masses are met with two kinds of difficulties:
i) how reliable is the extrapolation of single particle
potential parameters? ii) What is the accuracy of the
Strutinsky method itself?

The aim of this contribution is to provide an
answer to the second question pertaining to the validity of the Strutinsky expansion of the energy when stopped
after the first order terms. Starting from a microscopic hamiltonian with some effective nucleon-nucleon interaction, one can write [7] the total energy E in the
Hartree-Fock approximation as:

$$E = \bar{E} + \delta E_1 (\hat{\epsilon}_i) + \delta E_2 . \tag{1}$$

In (1), $\bar{E}$ is a "liquid drop" energy which depends only
on the average part of the density matrix. The quanti-
ty $\delta E_1 (\hat{\epsilon}_i)$ is the usual shell correction energy evaluated for the spectrum $\hat{\epsilon}_i$ which will be defined below,
whereas δE_2 is the sum of higher order shell corrections which is neglected in the usual shell correction
approach. The single particle energies $\hat{\epsilon}_i$ are eigenvalues of the average part of the Hartree-Fock potential.

Using the effective interaction of Skyrme (in the
parametrization SIII [2]) we have performed Hartree-Fock
calculations leading to the knowledge of the energy E .
From such self consistent solutions we have obtained the
quantities $\bar{E}$ and $\delta E_1 (\hat{\epsilon}_i)$, thus leading via Eq. (1) to
the corrective term δE_2 . Some results have already
been discussed in Ref. [11]. Let us summarize the main
conclusions.

i) The average energy $\bar{E}$ does behave like a liquid drop energy. This is illustrated in Fig. (1). The
variation of $\bar{E}$ as a function the constrained quadrupole moment Q is smooth with the exception of some
wiggles around the ground state and the first fission

barrier of the considered ^{240}Pu nucleus (Hartree-Fock
solutions are taken from Ref. [12]). These wiggles are
in fact due to a rapid variation (with respect to Q) of
the hexadecapole moment h of the considered solutions.
This is ascertained by the comparison of $\bar{E}$ with the
energy of a liquid drop [6] having the same multipole
moment Q and h , since the latter presents the same
wiggles as $\bar{E}$. One could try to extract from such
curves the liquid drop parameters associated with a
given effective interaction. This is somewhat difficult
[11] with a good accuracy, and a fit of $\bar{E}$(A) for spherical solutions would be far better. Preliminary results
are sufficiently realistic to confirm the relevance of
the used interaction for testing the shell correction
approach.

     ii) The obtained first order corrections $\delta E_1(\hat{\bar{\epsilon}}_i)$
are close to those found in the usual calculations using

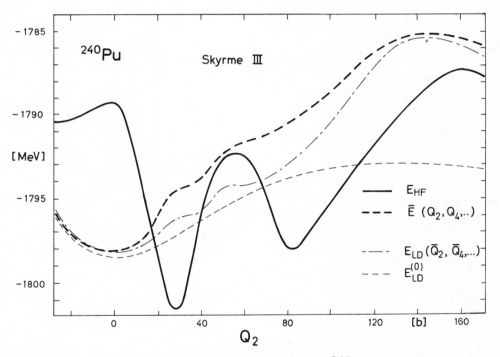

Figure 1: Deformation energy curves for the ^{240}Pu nucleus. Hartree-
Fock (E_{HF}), Strutinsky smoothed ($\bar{E}$) and liquid drop model (E_{LD})
energies are shown. The curve $E_{LD}(\bar{Q}_2, \bar{Q}_4,...)$ corresponds to a
liquid drop having the same moments $\bar{Q}_2$ and $\bar{Q}_4$ as in $\bar{E}$, whereas the
curve E_{LD} is obtained along the liquid drop fission valley. (Note
that on the figure the argument in $\bar{E}$ must be read as $\bar{E}(\bar{Q}_2,\bar{Q}_4)$!).

phenomelogical single particle potentials. Some detailed
differences may be due to small deficiencies in the parameters of one or the other approach (explaining e.g. the
bad first fission barrier of ^{240}Pu as calculated with the
interaction of Skyrme III).

iii) The sum of the higher order shell corrections
δE_2 is found to be relatively small ($\sim$1-2 MeV) for medium
and heavy nuclei. As expected, for light nuclei the convergence of the expansion (1) is less rapid than for
heavier nuclei. Indeed in the ^{40}Ca nucleus, both $\delta E_1(\hat{\epsilon}_i)$
and δE_2 have the same order of magnitude, as can be seen
on Fig. 2. (Pairing correlations with 3 different
strengths have been included, corresponding to a constant average gap [7] of $\tilde{\Delta}$ = 0, 1 or 2 MeV). One
should particularly note here the wrong value of the
first order shell-correction of $\delta E_1 \sim$ -5 MeV for the
ground state of ^{40}Ca. Inclusion of the higher order

corrections brings
the value of the
shell-correction
close to zero.
This is in much
better agreement
with the small
empirical value of
$E=E_{exp}-E_{LD}$ de-
duced from the ex-
perimental mass
E_{exp} of ^{40}Ca and
its liquid-drop
fit E_{LD} of Myers
and Swiatecki [6].
However these last
results have been
found to be rather
dependent on the
way in which one
defines the
smoothed energy
and the single par-
ticle spectrum en-
tering the defini-
tion of δE_1 . We
have recently pro-
posed [13] a
slightly diffe-
rent version of
the energy theo-
rem (1) in which
a selfconsistent

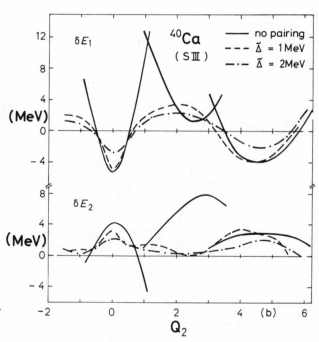

Figure 2: First order (δE_1) and second
order (δE_2) shell corrections in the ^{40}Ca
nucleus as a function of the mass quadrupole moment Q_2 (in barn). Three different
values of the average (constant) pairing
gap (see ref. [7]) have been used: $\tilde{\Delta}$= 0,
$\tilde{\Delta}$ = 1 MeV and $\tilde{\Delta}$ = 2 MeV.

average density matrix $\tilde{\rho}$ is introduced; the average
energy $E(\tilde{\rho})$ and the eigenvalues $\tilde{\varepsilon}_i$ of the average one
body potential are then derived selfconsistently. It
was found [13] that the alternative form

$$E = E(\tilde{\rho}) + \delta E_1(\tilde{\varepsilon}_i) + \delta E_2' \qquad (2)$$

for the expansion of the exact H.F. energy converges
much more rapidly than eq. (1). For ground states of
nuclei as light as ^{16}O or ^{40}Ca, as well as in heavy nuclei,
the remaining corrections $\delta E_2'$ range from 0 to 0.6 MeV.
On Fig. 3, the approximate deformation energy $E(\tilde{\rho})$
+ $\delta E_1(\tilde{\varepsilon}_i)$ for ^{40}Ca is compared to the exact energy E.
The difference between the two energies is for all defor-
mations as small as for the ground state.

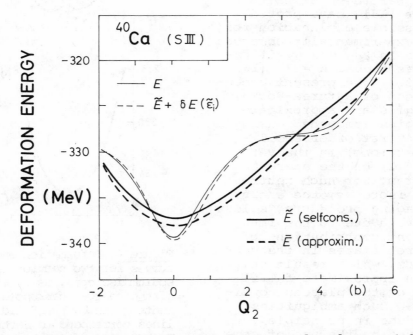

Figure 3: Deformation energies versus mass quadrupole moment Q_2
(in barn) for the ^{40}Ca nucleus. Hartree-Fock (E), normal smoothed
($\bar{E}$) and selfconsistently smoothed ($\tilde{E}$) energies are shown. The
approximation $\tilde{E} + \delta E_1(\tilde{\varepsilon}_i)$ to the energy E is also plotted. Pairing correlations are included ($\tilde{\Delta} = 1$ MeV).

More stringent tests of theoretical calculations
are provided by so-called transitional nuclei where
ground state properties (among others) are crucially dependent on the difference in the binding energies of two
distinct intrinsic states. Close to the isotopes where
a change from one intrinsic state to another is to be
expected, the binding energy differences are rather
small, (i.e. ~ 0-1 MeV). Therefore all possible sources
of errors have to be in-
vestigated. Among others
(related to e.g. pairing
correlations, vibrational
or rotational zero point.
energies etc.) one should
not forget possible fluc-
tuations of higher order
shell corrections. This is
studied in the particular
case of the neutron rich
sodium isotopes where re-
cent Hartree-Fock calculations [15] have provided
a possible explanation for
the experimentally observed
raise in B_{2N} energy differences around A = 31 [16].
On Fig. 4 we present a comparison of Hartree-Fock [15]
energies and approximate
energies (smoothed energy
plus first order shell
corrections) in the ver-
sion (2) of the energy
theorem. For such neutron
rich isotopes, the energy
averaging procedure is less
reliable due to the im-
portant contributions of
neutron states in the continuum, which result only
in an approximate fulfil-
ment of the plateau condition. Such ambiguities
could be in principle
avoided by the use of some
semi-classical expansion
method [17] . Bearing these
limitations in mind, it is
however shown on Fig. 4,
that the approximated

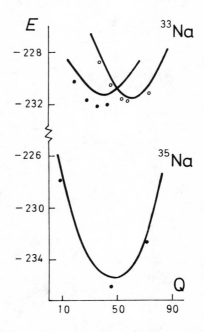

Figure 4: Deformation energy
curves for two neutron-rich
sodium isotopes, as functions
of the charge quadrupole moments Q (in fm^2) . Solid
lines correspond to Hartree-
Fock energies, dots to their
approximation by $\widetilde{E} + \delta E_1 (\widetilde{\epsilon}_i)$.
Different branches of the
curve are obtained for different neutron configurations.

energy differences reproduce the exact ones with a precision of $\sim$.5 MeV. Of course, in many cases a higher
accuracy may be required.

In summarizing, it should be stressed that we have
shown how the shell correction method in principle can
be very accurate. As to its practical application with
phenomenological liquid drop parameters and single particle potentials, the importance of the neglected higher
order shell corrections depends on the region of nuclei.
For light nuclei in the s-d shell we have demonstrated
the importance of using selfconsistently obtained average
potentials and liquid drop energies. On the other hand,
in medium and heavy nuclei the higher order corrections
were found not to contribute more than $\sim \pm$ 1 MeV in any
case, which is an order of magnitude less than the amplitude of the first order shell corrections.

[1] G.T.Garvey, W.J.Gerace, R.L.Jaffe, I.Talmi and I.
     Kelson, Rev.Mod.Phys. 41 (1961) S1
[2] M.Beiner, H.Flocard, Nguyen Van Giai and P.Quentin,
     Nucl.Phys.A238 (1975) 29; see also these proc.
[3] M.Beiner and R.J.Lombard, Ann. of Phys.86 (1974)262;
     see also these proc.
[4] D.Gogny, Proc. Int.Conf. on Hartree-Fock and selfcon-
     sistent theories in nuclei, Trieste, (North-Holl.,
     Amsterdam, 1975) in press
[5] C.F.V.Weiszäcker, Z.f.Phys.96 (1935) 431;
     H.A.Bethe and R.F.Bacher, Rev.Mod.Phys.8 (1936)82
[6] W.D.Myers and W.J.Swiatecki, Nucl.Phys.81 (1966)1
[7] V.M.Strutinsky, Nucl.Phys.A95 (1967)420; A122(1968)1;
     M.Brack, J.Damgaard, A.S.Jensen, H.C.Pauli, V.M.
     Strutinsky and C.Y.Wong, Rev.Mod.Phys.44 (1972)320
[8] H.Bethe, Ann.Rev.Nucl.Sci.21 (1971)93
[9] See e.g. M.Brack, Proc.of Int. Summer School on
     Nucl.Phys., Predeal, Romania, 1974
[10] P.A.Seeger and W.M.Howard, Nucl.Phys.A238 (1975)491
[11] M.Brack and P.Quentin, Proc.Int.Conf. on Hartree-
     Fock and selfconsistent theories in nuclei, Trieste,
     (North-Holl., Amsterdam, 1975) in press
[12] H.Flocard, P.Quentin, D.Vautherin, M.Veneroni and
     A.K.Kerman, Nucl.Phys.A231 (1974)176
[13] M.Brack and P.Quentin, Phys.Lett.B, in press
[14] G.Leander and S.E.Larsson, Nucl.Phys.A239 (1975)93
[15] X.Campi, H.Flocard, A.K.Kerman and S.Koonin, these
     proc.
[16] R.Klapisch, priv.com.
[17] see e.g. B.K.Jennings, R.K.Bhaduri and M.Brack,
     Phys.Rev.Lett.34 (1975)228

We thank the N.Bohr Inst. for warm hospitality, Statens Nat.Vid.
Forskn.Råd and Japan World Exposition Comm.Fund for financial support.

NUCLEON CORRELATION EFFECTS IN MASS SYSTEMATICS

N. Zeldes
Racah Institute of Physics, Hebrew University of Jerusalem
Institut für Kernphysik, Technische Hochschule Darmstadt

S. Liran
Racah Institute of Physics, Hebrew University of Jerusalem

I. EFFECTIVE INTERACTIONS

The effective interactions between nucleons correlate their motions in various ways. Nuclear mass systematics results from these correlations.

We first summarized pertinent features of the effective interactions and then relate them to the systematics.

1. Shells and Subshells

The major outcome of internucleon interactions is the nuclear potential, in which nucleons are bunched in subshells within major shells. Shells are filled at the magic numbers. Shell structure is most clearly manifested by discontinuities in separation energies and sudden slope changes of the nuclear mass surface at the magic numbers (Fig. 1).

2. Two-body effective interactions

Schiffer[1] recently extracted experimental effective-interaction matrix elements from two-nucleon spectra. Plotted as function of the angle θ_{12} between the angular momenta of the two nucleons, the data smoothly divide into two groups, as shown for equivalent nucleons in Fig. 2. There is only one strongly bound T = 1 state, namely the $J = | j_n - j_p | = 0$ pairing state. On the other hand, for T = 0 both the $J = j_n + j_p = 2 j$ aligned state and the $J = | j_n - j_b | + 1 = 1$ state are considerably bound. This reflects the overall short-range attractive nature of the effective two-nucleon interaction, which gives more binding for higher overlap of the nucleons orbital planes. Similarly, the center of mass of the T = 0 space-spin-symmetric states is lower than that of the T = 1 anti-symmetric states, where the average distance between the nucleons is larger.

II. CORRELATIONS AND MASS SYSTEMATICS

3. Pairing Correlations

According to the above, the ground state of singly-closed-shell nuclei is expected to have the maximum possible number of $J_{12} = 0$ identical-nucleon pairs. Such states have the lowest possible value of the seniority quantum number. On the other hand, in nuclei with both neutrons and protons outside closed shells, aligned neutron-proton $J_{12} = j_n + j_p$ pairs should exist as well, and higher seniority components mix with the main lowest seniority wave function.

Configuration mixing in the ground state of singly closed shell nuclei is mainly due to $J_{12} = 0$ identical nucleon pair excitations between subshells, resulting in a pairing wave function[2]

$$|\text{ground state}>_{\text{pairing}} = \alpha|\ j_1^{n_1}\ j_2^{n_2}>_{\text{low sen}} + \\ + \beta\ |j_1^{n_1-2}\ j_2^{n_2+2}>_{\text{low sen}} + ... \tag{1}$$

where each subshell is in a state of lowest seniority. In mixed major shells single-nucleon and also neutron-proton pair excitations occur as well.

In the lowest seniority approximation the nuclear mass surface consists of four parallel sheets according to parity type. Masses of nuclei with odd numbers of nucleons are higher, because they have less pairing energies. In a given shell region with neutrons and protons in different shells, each of the four mass surfaces is a quadratic function of N and Z[3]. Parabolic sections of the mass surfaces through the odd-A and even-A isotopes of Cl, Ar, Pb and Bi are shown in Fig. 1.

4. Isopairing Correlations

According to sect. 2, for nuclei with both neutrons and protons outside closed shells, the ground state should contain the maximum possible number of isopaired neutron-proton pairs with $T_{12} = 0$. Consequently, the total isospin of the ground state should have the lowest possible value, $T_{min} = \frac{1}{2}|N - Z|$.

The dependence of the two-nucleon effective interaction on T is most clearly expressed by an isoscalar term $V(r_{12}, \sigma_1, \sigma_2)(t_1 . t_2)$. This leads [4] to a term proportional to $T(T + 1)$ in the total energy expression.

For $T = \frac{1}{2}|N - Z|$ this results in each "diagonal" shell region in the (N,Z) plane in a different parabolic isobaric section of the

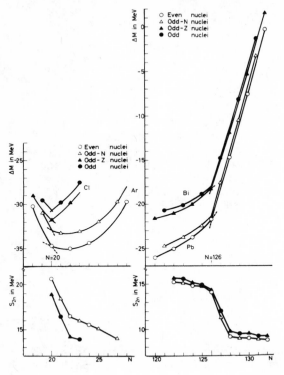

Fig. 1 - Mass Parabolas and
         the corresponding
         separation energies
         below for Cℓ, Ar,
         Pb and Bi.
         Notice the sudden
         changes of slope
         associated with
         N = 20 and 126.

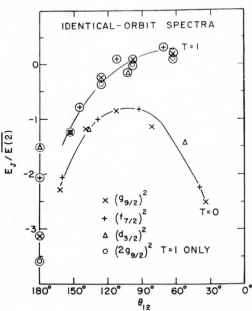

Fig. 2 - Experimental
         particle-particle
         matrix elements
         for both particle
         orbits identical,
         normalized to the
         average two-body
         energy of each
         multiplet. Taken
         from ref. 1.

mass surface on each side of the N = Z line. The two intersecting parabolas form a cusp at N = Z. This is illustrated in Fig. 3.

5. Competition between Pairing and Isopairing in Odd-Odd Nuclei

In even-even and odd-A nuclei, pairing and isopairing are mutually consistent, and the ground state has both lowest T and a major component of lowest seniority. On the other hand, in a mixed j^a configuration of odd-odd nuclei, the state $v = 0$, $T = T_{min}$ is excluded by the Pauli Principle[4]. There is thus a competition between pairing and isopairing. The ground state can either have $T = T_{min}$, $v = 2$ or $T = T_{min} + 1$, $v = 0$. The energy difference between these consists of two parts, resulting from differences of symmetry and of pairing energies, respectively :

$$E_{v=0, T=T_{min} + 1} - E_{v=2, T=T_{min}} = a(|N-Z| + 2) - \lambda \qquad (2)$$

where a and λ are, respectively, symmetry-and pairing-energy[†] coefficients.

For given mass number A, eq. (2) is linear function of $|N- Z|$. Consequently, the excitation energy of isobaric analog $v = 0$, $T = T_{min} + 1$ states in odd-odd nuclei should increase linearly with T. Such a general trend is indicated in Fig. 4. Fig. 3 shows for A = 48 and 50 isobaric ground state mass parabolas together with masses of $v = 0$, $T = T_{min} + 1$ states in the odd-odd nuclei. One observes the increase in relative energy of the $v = 0$ states with increasing distance from the cusp at N = Z.

The lower part of Fig. 3 shows sections of the mass surface through ground states of N = Z and N = Z + 2 nuclei, as well as masses of $v = 0$, $T = T_{min} + 1$ states in the odd-odd nuclei. In light nuclei a is so much larger than λ, that the ground state is always $T = T_{min}$ and $v = 2$. However, a seems to decrease with A faster than λ, and the excitation energy of the $v = 0$ state decreases with A for constant I = N - Z. For I = 0, the ground state first becomes 0^+ at $C\ell^{34}$, and then again in the $1f7/2$ shell, as seen more clearly in the negative part of the T = 1 line in Fig. 4. For I = 2, the coefficient of a in eq. (2) is larger, and the $v = 0$ and $v = 2$ lines do not cross, the ground state always having $T = T_{min}$. This is even more so for higher I's.

The odd-odd ground state T value is relevant when calculating masses by extrapolation. It seems that in N = Z nuclei beyond the $1f7/2$ shell pairing correlations are more important, and $T_{g.s.} = T_{min}^{+1}$. On the other hand, for $N \neq Z$ isopairing wins over, and $T_{g.s.} = T_{min}$. Detailed study[5] of extrapolated shell model masses essentially confirms these predictions.

[†] λ actually differs somewhat from the pairing energy, and is also larger in N = Z than in $N \neq Z$ nuclei.

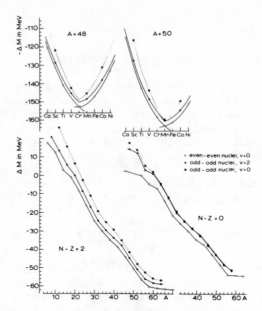

Fig. 3 - Upper part : Isobaric parabolas through the even-even
         $v = 0$ and odd-odd $v = 2$, $T = T_{min}$ masses (thick line)
         and through the odd-odd $v = 0$, $T = T_{min} + 1$ masses (thin
         line) for $A = 48$, 50.
         Lower part : Isodiapheric sections of the mass surface as
         above, for $I = 0,2$.

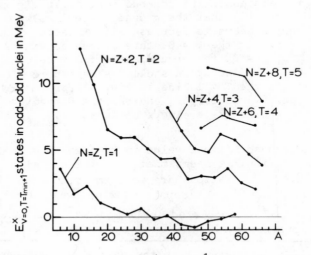

Fig. 4 - Excitation energy of 0^+, $T = \frac{1}{2} |N - Z| + 1$ levels in odd-
         odd light nuclei.

6. Pairing Violation in Heavier Nuclei

In the pairing approximation (1) neutron-proton interactions cannot contribute to configuration mixing. Thus the neutron and proton states are defined independently, and deviations from the quadratic lowest-seniority mass equation are due to the intrinsic neutronic and protonic interactions. They can be represented as an additive function, $f(N) + g(Z)$.

Such an additive function cannot contribute to mixed mass differences like $\Delta_{np}(N,Z) = S_{2n}(N,Z) - S_{2n}(N,Z-2)$. Thus the latter should be given by the coefficient of NZ in the lowest seniority quadratic mass equation.

Experimental values of Δ_{np} are shown in Fig. 5. One observes oscillatory structure superposed on an overall monotonous decrease with A. Oscillating neutron-proton correction terms were indeed found[5] essential for a quantitative description of nuclear masses. Such terms can only result from breakdown of the pairing approximation (1), due to isopairing.

The above oscillating terms are mainly symmetric with respect to particle-hole conjugation between magic numbers. This is also seen in the $\Delta_{nn}(N,Z) = S_{2n}(N,Z) - S_{2n}(N - 2, Z)$ graph between $N = 82$ and 126 in Fig. 5. This symmetry is a general consequence of the bunching of nucleons in shells.

7. Four-Periodicities and Quartets

In addition to pair correlations, correlated four-nucleon structures called quartets have been considered[6] as nuclear sub units. In the simplest version a quartet is defined as the ground state of a two-neutron two-proton group in a shell, like in the $|j^4 T = 0$, $J = 0 >$ ground sate of Ti^{44}. According to sect. 2 it should consist of major $v = 0$ component of paired nucleons, and a smaller isopaired $v = 4$ component. In detailed $1f7/2$ shell model calculations[7] an $f 7/2^4$ quartet is 77 % $v = 0$ and 23 % $v = 4$.

Fig. 6 shows section of the mass surface and binding energies along the $N = Z$ line. Four-periodicities are clearly observed. Additionally, the binding energy of "alpha-particle" nuclei increases almost linearly with the number of alpha particles (thin B line in the figure). Both these regularities have sometimes[6] been considered as evidence for quartet structure.

However, as actually noted[6,8], such regularities are as well direct consequence of the lowest seniority pairing approximation. As a matter of fact, since the major component of the quartet consists of paired nucleons, both quartet and pairing correlations should predict similar results.

We looked for possible quartet structure by looking for four-periodicities in the residuals remaining after subtracting from

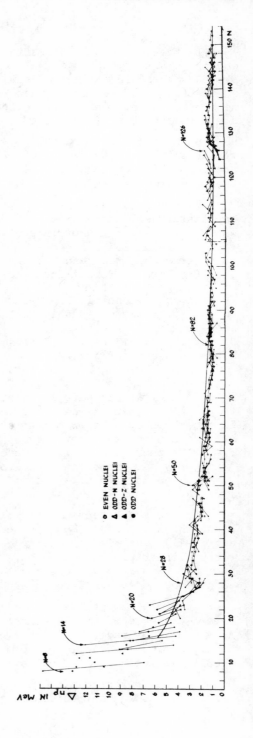

Fig. 5 — — Δ_{nn} (upper part) and Δ_{np} (lower part) as function of N.
Isotopes of the same parity type are connected by line.
Taken from ref. 9.

the experimental masses their lowest-seniority adjusted values[5].
Particle-hole-symmetric oscillations between magic numbers are
evident, but no four-periodicities.

It seems, that no clear evidence for quartet structure is
obtained from mass systematics.

REFERENCES

(1) J.P. Schiffer, Annals of Physics 66 (1971) 798 ;
    N. Anantaraman and J.P. Schiffer, Phys. Lett. 37B (1971) 229
(2) B.R. Mottelson, in the Many Body Problem (Dunod Paris 1959) p. 283
(3) N. Zeldes, Nuclear Physics 7 (1958) 27
(4) A. de Shalit and I. Talmi, Nuclear Shell Theory (Academic Press
    New York - London 1963)
(5) S. Liran, Calculation of Nuclear Masses in the Shell Model
    (Ph. D. Thesis, Jerusalem 1973)
(6) M. Danos and V. Gillet, Z. Physik 249 (1972) 294 and previous
    works quoted therein
(7) J.D. Mc Cullen, B.F. Bayman and L. Zamick, Phys. Rev. 134
    (1964) B 515
(8) A. Arima and V. Gillet, Annals of Physics 66 (1971) 117
(9) N. Zeldes, A. Grill and A. Simievic, Mat. Fys. Skr. Dan. Vid.
    Selsk. 3 n° 5 (1967)

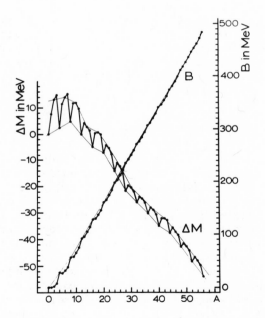

Fig. 6 - ΔM and B of N = Z nuclei. The points for odd-A values are
         averages of $T_z = \pm \frac{1}{2}$ nuclei, as in ref. 6.

INTERPOLATION MASS FORMULAE

E. Comay and I. Kelson

Department of Physics and Astronomy
Tel-Aviv University
Tel-Aviv, Israel

ABSTRACT

Critical remarks are made concerning mass tables in general and the mass table based on mass relationships in particular. Specific criteria for acceptability of a table are formulated. Two methods - which form a logical sequence - of using these relationships to predict masses, are described. The first involves the construction of regional mass tables with a very limited number of free parameters to be adjusted. The second involves the prediction of each unknown mass individually and independently and uses no parameters whatsoever.

I. INTRODUCTION

Attempts to express the masses of nuclides as functions of Z and N are among the oldest endeavours in nuclear physics. They serve, in principle, a dual purpose. First, to try and extract basic physical quantities; second, to extrapolate and predict unknown masses, needed for analysis of nuclear reactions, astrophysical study of the periodic table evolution, identification of fission products, or for sheer intellectual curiosity. The meaningful physical quantities to be extracted are very few indeed. Thus, all mass formulae are nothing but extrapolation devices, and should be judged as such. Most formulae use parameters (few or many) as intermediate agents - to be determined through a fitting procedure to known masses and to be used to predict unknown ones - and the success of the entire operation depends to a large extent on the skill and luck in the choice of parameters.

272

All existing tables suffer from one fundamental flaw: they fail to provide a reliable estimate of the uncertainties in the predictions they make. The quality of the fit over the known masses, is by no means a valid relevant indicator. In fact, the least square fitting procedure (rather arbitrary in the first place) to known masses, effectively discriminates against unknown masses. These, so to speak, appear in the fit with infinitely large experimental uncertainties. A cursory examination of various mass tables reveals shocking discrepancies between predictions - which far exceed the average deviations obtained by each for fitted masses. One might argue that such uncertainties could be determined for the intermediate parameters and through them for the predicted masses. Even if this was done, the result would be unsatisfactory for we expect the quality of the predictions to deteriorate with the distance of the extrapolation made, rather than merely reflect a changing combination of the parametric variations.

The introduction (1) of mass relationships (rather than formulae) was a step towards the potential alleviation of the intermediate parameters. These relationships provided a simple means of expressing one mass directly in terms of others, while retaining globally their validity and statistical rigour. We consider it regrettable that in constructing (2) a mass table based on them, we fell back on the same general technique - the extensive utilization of parameters - whose weaknesses we try to expose, and whose shortcomings are so readily apparent.

In this paper we report on two stages towards constructing a mass table which is based on the mass relationships, but conforms to the criteria which we deem so essential. Namely, the absence of artificial parameters and the inherent calculation of the proper uncertainties in the predicted masses. Detailed tables - the results of the presently sketched approach - will be published elsewhere.

II. REGIONAL TABLES

We deal explicitly, but not exclusively, with the transverse, homogeneous equation, which relates the masses $M(Z,N)$ of six neighbouring nuclei

$$M(Z,N+1) + M(Z-1,N) + M(Z+1,N-1) - M(Z,N-1) - M(Z+1,N) -$$

$$- M(Z-1,N+1) = 0 \tag{1}$$

It has been shown that the most general solution to such a difference equation in two variables is of the form

$$M(Z,N) = f(Z) + g(N) + h(A) \tag{2}$$

where f,g,h are arbitrary functions of Z,N and $A = Z+N$ respectively. In reference (2) the values of these functions at all points were treated as free parameters and a rather cumbersome fitting procedure followed.

The present analysis starts by defining the shift operators S_Z and S_N through

$$S_Z F(Z,N) = F(Z+1,N) \qquad\qquad S_N F(Z,N) = F(Z,N+1) \tag{3}$$

In terms of these operators and their inverses, we can rewrite eq. (1) in operator form

$$\hat{O}M(Z,N) = 0$$

where

$$\hat{O} = S_N - S_N^{-1} - S_Z + S_Z^{-1} - S_N S_Z^{-1} + S_N^{-1} S_Z . \tag{4}$$

We now note that $\hat{O}$ can be decomposed into a product of three operators:

$$\hat{O} = (1-S_Z)(1-S_N^{-1})(1-S_N S_Z^{-1}) . \tag{5}$$

The usefulness of this decomposition is demonstrated by applying the product of any two of them to the representation of the masses in eq. (2).

Thus,

$$(1-S_N^{-1})(1-S_N S_Z^{-1})M(Z,N) = -g(N-1) + 2g(N) - g(N+1) =$$

$$= -g''(N) \tag{6a}$$

and, similarly,

$$(1-S_N S_Z^{-1})(1-S_Z)M(Z,N) = -f''(Z) \tag{6b}$$

$$(1-S_Z)(1-S_N^{-1})M(Z,N) = -h''(Z+N) \tag{6c}$$

We are therefore able, in principle, to evaluate the second differences of the functions f,g,h <u>directly from experimentally measured masses</u>. However, since the relationship on which (2) is based is only approximate, f'', g'' and h'' at each point can be calculated in a variety of ways, each giving a somewhat different result. What we obtain are the <u>average</u> values of the second differences, which are, explicitly

$$f''(Z) \quad = \quad <M(Z+1,N)+M(Z-1,N+1)-M(Z,N+1)-M(Z,N)>_{Z\ \text{constant}}$$

$$\tag{7a}$$

$$g''(N) \quad = \quad <M(Z,N-1)+M(Z-1,N+1)-M(Z-1,N)-M(Z,N)>_{N\ \text{constant}}$$

$$\tag{7b}$$

$$h''(Z+N) \quad = \quad <M(Z,N-1)+M(Z+1,N)-M(Z+1,N-1)-M(Z,N)>_{Z+N\ \text{constant}},$$

$$\tag{7c}$$

as well as the spreads in their distributions. Clearly, if eq. (1) were exact, there would be no spread in eqs. (7) and a double quadrature would yield $f(Z), g(N)$ and $h(Z+N)$. The only point to remember is that the arguments of the three functions are dependent, so that only three (out of six) integration constants need be determined, and these may be chosen as a constant term and the linear coefficients of Z and N.

In reality, of course, eq. (1) is not exact, and we encounter in f'', g'', h'' spreads of the order of 200 keV. If we allow the values of the second derivatives (or of the functions f, g, h themselves) to fluctuate around their averages, and treat these fluctuations as free parameters - we regenerate the procedure of ref. (2). Obviously, we wish to avoid that. Instead of treating each fluctuation individually, we wish to analyze their combined effect. Appending the second derivatives by $\delta f'', \delta g'', \delta h''$ may be easily seen to add a special term δM to M,

$$\delta M(Z,N) \quad = \quad \sum_{Z'} \delta f''(Z')(Z-Z')\theta(Z-Z') + \sum_{N'} \delta g''(N')(N-N')\theta(N-N')$$

$$+ \sum_{A'} \delta h''(A')(A-A')\theta(A-A'). \tag{8}$$

The cooperative effect of many such fluctuations can be expanded as a power series in Z, N or A respectively, where we expect the long-wavelength dependence to be dominated by the third power. Also, it is clear, that if we restrict the summation in (8) to smaller regions of the periodic table the size of such systematic effects will diminish. This brings out another crucial point: All mass formulae treat the entire periodic table at once - an example of theoreticians' megalomania - while for purposes of efficient local extrapolation we should consider only the relevant segment of the periodic table, and this segment may be very limited.

The following procedure thus emerges:
1. Calculate according to eqs. (7) the values of f'', g'', h'' by averaging over experimental masses.

2. Find any particular solutions for $f(Z), g(N)$ and $h(Z+N)$ by double quadrature, and freeze them.

3. For a given set of experimental data (to which we refer as a "region", extending between mass numbers A_{min} and A_{max}) perform a least square fit to the masses with the expression:

$$M(Z,N) = f(Z) + g(N) + h(Z+N) + \zeta_o + \zeta_1 Z + \nu_1 N +$$

$$+ \sum_{i=2}^{n} \{\zeta_i Z^i + \nu_i N^i + \alpha_i (Z+N)^i\} \tag{9}$$

The free parameters are ζ_o, ζ_1, ν_1 (the three independent constants of quadrature) and $\{\zeta_i, \nu_i, \alpha_i; i=2,n\}$ which describe the cooperative statistical effects. We can make the general statement, that for large regions the primary contribution of the sum is exhausted by $n \leqslant 3$. •

Table 1 gives average absolute deviations between experimental and fitted masses, for various regions, among which is the entire table with $A \geqslant 26$, and for various n.

The choice of a regional table (namely A_{min} and A_{max}) to be used for predicting a particular nuclidic mass $M(Z,N)$ is dictated by the extrapolative structure of the difference equation: the range $[A_{min}, A_{max}]$ should include all nuclei with either Z protons or with N neutrons, but - consistent with this - should be as small as possible.

III. INDIVIDUAL MASS TABLE

In the previous section we discussed a method for using the difference equation, which involves only a small number of parameters (essentially - for small regions - the quadrature constants). Although the results are very useful, the procedure is still unsatisfactory from the conceptual point of view. Known masses are known and should not be fitted. Here we sketch briefly what we consider a logical extension of that procedure. It has the following characteristics: no parameters are used and hence known masses are used directly; each mass to be predicted is treated individually; estimates of the uncertainties in the predictions are automatically obtained.

Let us consider a particular unknown nuclidic mass $M(Z,N)$ which is to be predicted. Through a repeated application of the difference equation - regarded here as a recurrence relation - it is always possible to express $M(Z,N)$ as a linear combination, with integral coefficients, of a set of known masses:

$$M(Z,N) = \sum_i n_i M(Z_i, N_i) \tag{10}$$

At each application of the difference equation a random deviation is known to be introduced, and the combined result of the entire process is an intrinsic uncertainty in the prediction. We express-ly ignore the uncertainties in the experimental values of the masses used, which may always be chosen so that they have a negli-gible effect. The explicit evaluation of the uncertainty is never-theless not very easy, and it depends on the particular choice of the masses in eq. (10). We therefore use a different prescription. We consider an ensemble of expressions of this form

$$M^{(j)}(Z,N) = \sum_i n_i^{(j)} M(Z_i^{(i)}, N_i^{(j)}) \qquad j = 1,\ldots \tag{11}$$

and we identify as the canonical prediction the average of the values $M^{(j)}$, and as the uncertainty - the width of their distrib-ution.

The number of combinations that can be used in (11) is stag-gering indeed. We estimate, for example, that for a nucleus removed just one atomic mass unit away from the body of known masses, where this valley is 5 units thick, about 10^9 different combinations are possible. Obviously, an efficient (and intell-igent) sampling and weighing procedure is needed. We shall, in practice, limit ourselves to what we call "simple skeletons" (the term skeleton refers here to the group of known nuclides used in the expansion): two adjacent nuclei for each successive mass number, where the successive pairs are also adjacent, and follow monotonically non-decreasing sequences both in Z and in N. Of particularly simple - and useful - form are those skeletons that include 4 sets of nuclei with successive constant values of neutron-proton number difference.

In Table 2 we give a number of randomly selected examples obtained with such "diagonal skeletons". For each case we quote the average value and the uncertainty, as well as the number of combinations in the ensemble, and the distance from the body of known masses. Perhaps the most important point to pay attention to, is the way the uncertainty tends to grow as the extrapolation is more far-reaching, thus fulfilling what - to our mind - is an essential criterion for a predictive mass table.

REFERENCES

1. G.T. Garvey and I. Kelson, Phys. Rev. Letters 16, 197 (1966).
2. G.T. Garvey, W.J. Gerace, R.L. Jaffe, I. Talmi and I. Kelson, Reviews of Modern Physics 41, S1 (1969).

A_{min} - A_{max}	Number of Nuclides	$\bar{\Delta}(n=1)$ (3 par.)	$\bar{\Delta}(n=2)$ (6 par.)	$\bar{\Delta}(n=3)$ (9 par.)	$\bar{\Delta}(n=4)$ (12 par.)
26 - 254	1077	2524	1870	702	644
41 - 70	128	255	202	164	140
111 - 140	161	126	91	88	79
181 - 210	129	498	174	111	109
61 - 120	303	617	417	222	211
131 - 190	312	231	183	139	130
161 - 220	277	930	396	214	191

Table 1. Absolute average deviations, in keV, between experimental masses in some regions $[A_{min}, A_{max}]$ and various expressions using the expansion of eq. (9). The number of nuclides in each region is also given, where only masses known to better than 200 keV were included. Note the dependence of the fit on the size of the region, and the quality of fitting the cooperative statistical effects with higher n-values.

N	Z	A	Units Removed from Data	Number of Combinations	Predicted Mass Excess (MeV)	Uncertainty (MeV)
75	61	136	1	8	-71.38	±0.37
74	62	136	2	7	-67.11	±0.73
73	63	136	3	7	-56.74	±0.96
112	72	184	1	11	-41.56	±0.19
113	71	184	2	10	-35.40	±0.49
114	70	184	3	9	-32.13	±0.67
115	69	184	4	9	-23.61	±0.90
116	68	184	5	8	-17.82	±1.29
145	89	234	1	7	-45.30	±0.20
146	88	234	2	6	-48.05	±0.51

Table 2. Examples of masses predicted directly from experiment-ally measured masses, along with estimated uncertainties stemming from the spread among different combinations of extrapolation bases.

Work reported by the Israel Commission for Basic Research

NUCLIDIC MASS RELATIONSHIPS AND MASS EQUATIONS (III)

J. Jänecke and B. P. Eynon

Department of Physics, The University of Michigan

Ann Arbor, Michigan 48105, U.S.A.[*]

Mass equations $M(N,Z)$ are generally obtained by establishing the analytical form of the equation on the basis of nuclear structure considerations and by subsequently minimizing the differences $M(N,Z) - M_{exp}(N,Z)$ for all known masses by adjusting parameters. An alternate approach is described in this contribution. It makes use of nuclidic mass relationships which represent partial difference equations. Some aspects of this approach have been reported earlier[1,2].

The objective of any procedure is, of course, to find a mass equation $M(N,Z)$ which describes the known and unknown nuclear masses for all N and Z. Thus, if $M_{exact}(N,Z)$ represents the exact masses of known and unknown nuclei, we want $M(N,Z) \approx M_{exact}(N,Z)$ to hold in the best possible way over the widest possible range of N and Z values. From this equation it also follows that $D\ M(N,Z) \approx [D\ M(N,Z)]_{exact}$. Here, D represents any partial difference operator. If $[D\ M(N,Z)]_{exact}$ were known exactly from nuclear structure theories and if the partial difference equation

$$D\ M(N,Z) = [D\ M(N,Z)]_{exact} \qquad (1)$$

has unique solutions when required to reproduce the known experimental masses $M_{exp}(N,Z)$, then $M(N,Z) = M_{exact}(N,Z)$. Thus, limited information about $M_{exact}(N,Z)$ may be sufficient to derive an exact mass equation. Approximate theories or assumptions about $[D\ M(N,Z)]_{exact}$ based on nuclear structure considerations will lead to approximate solutions $M(N,Z)$. The earlier boundary conditions have to be replaced by a χ^2-minimization of the differences between the experimental and calculated masses. Another important aspect of this treatment is that if <u>two</u> operators D_T and D_L can be found with theoretical predictions $[D_T M(N,Z)]_{theor}$ and $[D_L M(N,Z)]_{theor}$, then the resulting solutions $M_T(N,Z)$ and $M_L(N,Z)$ must satisfy

279

$M_L(N,Z) - M_T(N,Z) \approx 0$ for all values of N and Z for which unique solutions can be derived. This is a necessary condition, and the degree to which it is violated, particularly for neutron-rich and proton-rich nuclei, makes it possible to judge the reliability of the underlying theoretical assumption $[D_T M(N,Z)]_{theor}$ and $[D_L M(N,Z)]_{theor}$ and with it the reliability of mass extrapolations. It should be noted that the standard deviation σ_m for reproducing the experimental masses is not necessarily a good measure for judging the reliability of mass extrapolations.

Partial difference operators D_T and D_L can be obtained by defining recursively

$$mn_\Delta \equiv \underbrace{10_\Delta 10_\Delta \ldots 10_\Delta}_{m} \underbrace{01_\Delta 01_\Delta \ldots 01_\Delta}_{n}. \tag{2}$$

Here, $10_\Delta f(N,Z) \equiv f(N,Z)-f(N-1,Z)$ and $01_\Delta f(N,Z) \equiv f(N,Z)-f(N,Z-1)$. The quantity

$$I_{np}(N,Z) \equiv {}^{11}_\Delta B(N,Z) = -{}^{11}_\Delta M(N,Z) \tag{3}$$

represents essentially the effective neutron-proton interaction[3] which is responsible for the symmetry and the neutron-proton pairing energy. Additional small contributions to I_{np} result from the Coulomb energy and from collective effects. Partial difference equations in accord with the above general considerations can now be obtained by considering the A- and T_z-dependence of I_{np}. We therefore define the transverse and longitudinal partial difference operators

$$D_T \equiv ({}^{10}_\Delta - {}^{01}_\Delta) \, {}^{11}_\Delta, \tag{4a}$$
$$D_L \equiv ({}^{10}_\Delta + {}^{01}_\Delta) \, {}^{11}_\Delta. \tag{4b}$$

Expressing D_T and D_L as partial differential operators gives

$$D_T \approx (\partial/\partial N - \partial/\partial Z)(\partial^2/\partial N \partial Z) = \partial/\partial T_z(\partial^2/\partial A^2 - \tfrac{1}{4}\partial^2/\partial T_z{}^2) \tag{5a}$$

$$D_L \approx (\partial/\partial N + \partial/\partial Z)(\partial^2/\partial N \partial Z) = 2\partial/\partial A(\partial^2/\partial A^2 - \tfrac{1}{4}\partial^2/\partial T_z{}^2). \tag{5b}$$

The partial difference equations $D_T M(N,Z) = 0$ and $D_L M(N,Z) = 0$ are the well known transverse and longitudinal Garvey-Kelson nuclidic mass relationships[4].

General solutions of

$$D_T M(N,Z) = [D_T M(N,Z)]_{theor} \tag{6a}$$
$$D_L M(N,Z) = [D_L M(N,Z)]_{theor} \tag{6b}$$

consist of a special solution of the inhomogeneous equation and the most general solution of the homogeneous equation. Such solutions can easily be obtained if certain simple assumptions are made about the A- and T_z-dependence of I_{np} (such as CON, GK-T, GK-L; see below). Also, if the $I_{np}(N,Z)$ are derived by differentiation from a given mass equation $M_{eq}(N,Z)$, then $M_T(N,Z) = M_L(N,Z) = M_{eq}(N,Z)$ are special solutions of the inhomogeneous equations (6). The underlying theoretical assumptions for I_{np} are, of course, those used in the derivation of the respective mass equation. The most general solutions are

$$M_T(N,Z) = M_{eq}(N,Z) + G_1(N) + G_2(Z) + G_3(N+Z) \qquad (7a)$$
$$M_L(N,Z) = M_{eq}(N,Z) + F_1(N) + F_2(Z) + F_3(N-Z) \qquad (7b)$$

where the $G_i(k)$ and $F_i(k)$ are arbitrary functions. These functions can be determined uniquely[2] from a χ^2-minimization of the differences between experimental and calculated energies for those regions of N, Z, N+Z and N-Z for which experimental masses are known. Solutions which satisfy eqs. (6a) and (6b) simultaneously have been discussed earlier in ref. [2].

The following assumptions and theories about $I_{np}(N,Z)$ have been used (see also ref. [2]) to obtain solutions $M_T(N,Z)$ and $M_L(N,Z)$:

CON $I_{np}(N,Z) = I_0 + I_1(-1)^A$;

GK-T $I_{np}(N,Z) = I_{np}(N+Z)$;

GK-L $I_{np}(N,Z) = I_{np}(N-Z)$;

LZ1   Liran and Zeldes [5], shell model, seniority scheme;

BW1   Bethe Weizsäcker, liquid-drop model;

BW2   Bethe Weizsäcker, liquid-drop model, Coulomb energy parameters optimized independently;

MS   Myers and Swiatecki [6], droplet model;

C   Cameron et al. [7], liquid-drop model;

S   Seeger [8], liquid-drop model; S-T is the transverse solution for S.

The χ^2-minimizations were carried out for the over 1000 experimental masses with $N,Z \geq 20$ from ref. [9] by using certain sparse matrix computer programs. Table 1 lists for the above cases the standard deviations σ_m for reproducing the experimental masses. The σ_m are smaller for the transverse solutions $M_T(N,Z)$ ($\sim$ 110 keV) than those for the longitudinal solutions $M_L(N,Z)$ ($\sim$ 190 keV) due to the increased number of adjusted parameters (functions $G_i(k)$ and $F_i(k)$). They are nearly the same for the various cases even though some of the extrapolations into the regions of very neutron- and proton-rich nuclei are entirely different. The experimental Coulomb displacement energies are not well described by $M_T(N,Z)$ and $M_L(N,Z)$. This fact indicates correlations between the symmetry and Coulomb energy terms as already observed earlier [2]. It will eventually require the introduction of the constraints $G_1(k)=G_2(k)$ and $F_1(k)=F_2(k)$.

Table 1   Standard deviations σ_m in keV for the experimental masses for several mass equations $M_{eq}(N,Z)$ and associated solutions $M_T(N,Z)$ and $M_L(N,Z)$ of partial difference equations ($N,Z \geq 20$)

	CON	GK-T	GK-L	LZ1	BW1	BW2	MS	C	S	S-T
M_{eq}	-	109	188	759	2932	3589	3273	2870	4097	109
M_T	109	109	125	135	110	110	110	112	109	109
M_L	188	105	188	209	211	203	186	179	185	105

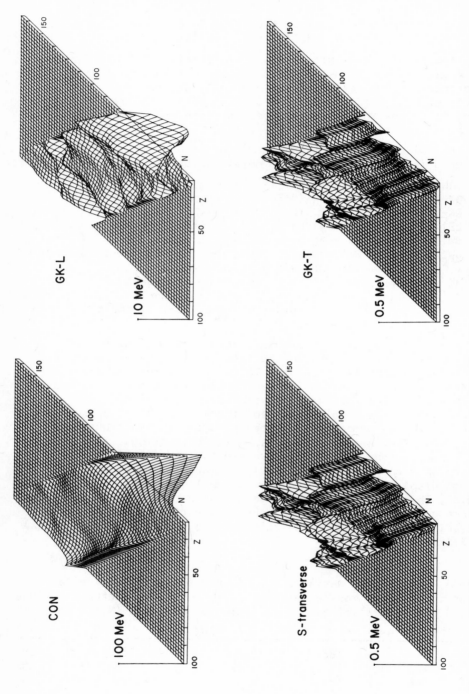

Fig. 1   Differences between the solutions $M_L(N,Z)$ and $M_T(N,Z)$ based on four different assumptions or theories about the effective neutron-proton interaction $I_{np}(N,Z)$ (see text).

Fig. 1 shows plots of the differences $M_L(N,Z)-M_T(N,Z)$ for a few examples for those regions of N and Z for which unique solutions exist for both equations. The differences labeled CON represent the differences between the longitudinal and transverse Garvey-Kelson mass relationships. The consistency tests for GK-T and GK-L (right hand side; note different scales) clearly prove that GK-T is superior to GK-L. The test for the transverse solution based on the liquid-drop model equation of Seeger [8] (lower left) leads to differences practically identical to those for GK-T. This result and the (N,Z)-dependence of the differences point to the need for an explicit shell dependence of I_{np}. The limits of neutron and proton stability differ slightly for S-T and GK-T. Results for S-T are shown in fig. 2.

Predictions of unknown masses of very light nuclei are of particular interest. Therefore, a shell-model mass equation was constructed with emphasis on the symmetry and Coulomb energy terms. This equation was subsequently used as a basis for the partial difference equations (6a) and (6b).

For the regions where neutrons and protons fill the same major shell (1p, 1d2s, etc.) the seniority coupling scheme in the isospin formalism [10] was used. This approach is very similar to one used by Liran and Zeldes [5]. The Coulomb energy parameters for the seniority equations of Hecht [11] were obtained independently from

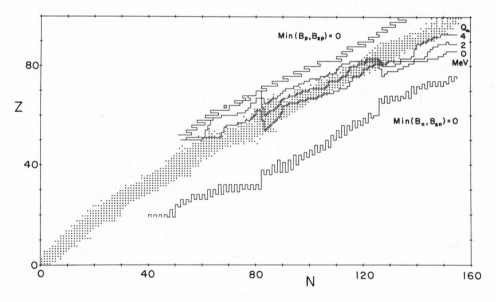

Fig. 2   Predicted lines of neutron, proton and alpha-particle (Q_α = 0, 2, 4 MeV) instability from the transverse solution S-T. Nuclei with known masses [9] are marked.

Table 2   Functions $G_1(N)$, $G_2(Z)$ and $G_3(N+Z)$ in keV for the transverse Garvey-Kelson nuclidic mass relationships obtained from a χ^2-adjustment to all experimental masses [9] with $N \geq 4$, $Z \geq 2$, $N-Z \geq 0$ ($N=Z=$odd is excluded). The equation is written in the form

$$\Delta M(N,Z) = N\Delta M_n + Z\Delta M_H - 13000\,A + 120\,Z^2 + 197.7907\,(N-Z)^2 + G_1(N) + G_2(Z) + G_3(N+Z)$$

G1(N)

```
 152297.3  135333.8  122229.9       0.0       0.0  308848.6  282350.8  253330.6  230038.9  206064.8  188472.1  168338.0    10
  50589.4   45443.3   43595.5   39001.9   37773.6   95949.7   83704.6   75852.1   66209.0   60798.9   53130.3   30373.4    20
  32391.6   32364.6   35238.3   35856.4   39565.1   40586.5   34215.6   33429.1   30319.0   31072.4   30373.4   52701.9    30
  57626.0   59614.2   64550.7   66660.0   71580.6   73606.3   44804.9   46224.0   50877.6   52701.9   86788.7   86788.7    40
  92823.9   97149.0  103006.4  107055.2  112692.9  116480.3   78207.3   80196.6   84766.0   86788.7  131492.5  134492.6    50
 139789.5  142373.5  147297.5  149569.0  154026.8  156008.7  122318.1  128884.3  131492.5  134492.6  165187.4  166039.2    60
 169199.2  165565.2  165678.1  171846.2  173625.4  172723.1  160140.2  161541.1  165187.4  166039.2  172201.0  169829.3    70
 159930.4  166249.3  146527.0  165008.1  164642.6  162323.7  173669.1  172055.9  172201.0  169829.3  156739.3  152371.2    80
 149678.9  144627.6  149035.5  135520.5  131387.3  125184.6  161110.3  158319.1  156739.3  152371.2  108172.7  105558.4    90
  94774.5   87035.6   80292.4   71969.9   64677.7   55708.7  120375.7  113623.0  108172.7  105558.4   29955.5   19961.1   100
  11114.6     437.8   -8460.0  -20416.4  -30356.0  -42490.9   47897.3   38276.0   29955.5   19961.1  -77199.1  -93471.9   110
-102562.6 -116340.6 -123348.2 -143393.5 -156962.1 -171677.9  -53186.5  -65800.2  -77199.1  -93471.9 -211359.6 -226441.9   120
-240379.0 -256109.7 -271116.4 -287745.0 -303375.3 -320651.8 -184067.0 -198177.6 -211359.6 -226441.9 -372283.9 -390922.4   130
-408948.3 -428320.3 -446961.3 -467183.1 -486632.3 -507552.8 -336957.3 -355071.9 -372283.9 -390922.4 -570782.6 -593490.0   140
-615294.8 -638795.7 -661271.4 -685326.0 -708596.9           -527806.3 -549621.7 -570782.6 -593490.0       0.0       0.0   150
                                                                                                                           160
```

G2(Z)

```
  45552.5   25750.4  308948.8  272195.9  237226.3  197226.5  160832.5  134279.5  106038.5   85820.0   62712.4    10
 -57648.1  -62033.4   11473.8   -5367.4  -15425.8  -27235.8  -27235.8  -34497.1  -43247.5  -46261.8  -55234.6    20
 -67120.6  -65598.1  -64372.5  -68499.2  -69433.1  -72400.0  -72400.0  -71909.7  -73743.3  -74422.8  -70622.6    30
 -27764.2  -24201.0  -61020.8  -58994.4  -53991.3  -51637.5  -51637.5  -46263.9  -43447.9  -37579.8  -34118.8    40
  20706.4   25526.2  -19053.5  -14729.4   -6663.2   -5196.6   -5196.6     895.6    4305.4   10332.0   13635.5    50
  69493.3   71997.4   32201.2   36554.4   42892.6   46726.3   46726.3   52661.5   56046.8   61537.3   64462.1    60
  97054.4   98717.0   76509.8   78604.5   82710.0   84574.7   84574.7   80255.3   89944.8   93177.1   94549.6    70
 105318.3  109210.9  101197.5  101804.1  103918.6  104324.7  104324.7  106559.8  106602.5  108158.0  107818.7    80
 116050.0  114410.9  114133.8  112795.9  114947.0  115004.2  115004.2  116477.2  119869.4  116774.5  115702.8    90
                     111104.0  108338.0  108338.0  103411.4  103411.4  101233.4  101233.4   97277.6            100
```

G3(A)

```
 -386976.7 -354634.7       0.0       0.0       0.0 -570437.2 -531110.8 -493887.3 -456656.8 -421790.0    10
 -122219.8 -102689.8 -830139.2 -295297.4 -266915.4 -239411.0 -213152.5 -188661.7 -164975.0 -143610.6    20
   44987.5   57128.9   69031.8   79641.9   90120.2   99450.1  108661.1  116943.4  117740.3   31543.7    30
  139070.0  145777.9  151178.2  156016.1  161107.4  165078.3  169258.2  172600.0  125020.7  132021.4    40
  182801.4  184997.6  188221.2  190170.3  192116.4  195078.1  195075.0  195913.8  196307.0  179449.8    50
  197708.3  197396.0  197361.8  196554.1  196017.0  194739.7  193921.1  192237.3  196827.8  197294.2    60
  187103.5  184753.9  182582.5  179653.8  177064.7  173931.9  171110.7  190822.8  190822.8  188824.4    70
  158118.2  154404.5  151115.6  147184.2  143479.5  139176.1  134996.8  167893.3  164833.2  161335.8    80
  177192.1  112559.0  109203.9  103469.7   98932.2   94189.4   89552.0  130496.8  126003.5  121507.1    90
   22304.2   65856.5   61135.7   56131.4   51438.3   41804.4   36874.3   84983.4   80437.2   75568.8   100
   22364.2   17905.5   13331.0    8412.4    3844.2    -989.3  -10186.6  -54154.2  -14656.1  -27391.7   110
  -23766.6  -28457.4  -32849.4  -37339.3  -41578.1  -45997.0  -50007.5  -57954.2  -14656.1  -19364.8   120
  -65351.9  -68177.7  -72287.7  -75541.6  -78669.5  -81709.7  -84481.7  -87380.6  -89938.9  -61778.6   130
  -95153.3  -97695.1 -100028.0 -102472.5 -104580.9 -106066.9 -108810.3 -110995.8  -57954.2  -92624.7   140
 -127996.3 -117854.1 -119355.2 -120209.1 -123585.9 -124649.8 -125763.4 -110995.8 -112575.5 -114455.3   150
 -129233.7 -120703.3 -129121.1 -125919.0 -122270.5 -129982.4 -130052.1 -125763.4 -124692.8 -127377.4   160
 -120542.8 -129041.9 -128337.6 -127803.4 -126923.0 -126086.3 -125068.3 -130095.0 -129829.7 -126662.1   170
 -103479.3 -101575.4 -117756.4 -116202.1 -114481.9 -112918.3 -111086.6 -124113.7 -122921.7 -121825.3   180
  -80200.3  -77576.7  -99426.5  -97199.7  -95116.0  -92958.2  -90592.8 -109127.8 -107241.7 -105497.2   190
  -43079.3  -44039.1  -74772.9  -73553.8  -68967.9  -66015.8  -62644.0  -88217.4  -85613.8  -63121.0   200
   -3418.9    1353.7  -39768.7  -37955.2  -31133.8  -26040.6  -22212.7  -59465.4  -55811.2  -52089.8   210
   49347.6   55070.7    6462.0   11411.8   16649.0   21748.2   27188.5  -17763.5  -12907.3   -8352.9   220
  114036.4  121390.5  128976.3  136621.8  144512.4  152436.8  160648.3  -32557.0   30072.9   43530.3   230
  194850.3  203757.8  212927.2  222225.8  231763.5       0.0       0.0   86344.0   99825.2  106808.6   240
                                                                       168941.7   92948.3  186011.5   250
                                                                             0.0  177401.7  186011.5   260
```

the experimental Coulomb displacement energies [12] prior to the adjustment of the few nuclear parameters. For the regions where neutrons and protons fill different major shells, the particle-hole monopole interaction $H_{ph} = -a + b\ \underset{\sim}{t}_p \cdot \underset{\sim}{t}_h$ of Zamick [13] was employed. The expression for the ground state energies of nuclei becomes particularly simple

$$M(N,Z) = M(N,Z_0)+M(N_0,Z)-M(N_0,Z_0)+(\tfrac{1}{4}b-a)(N-N_0)(Z-Z_0). \qquad (7)$$

Here, $(N-N_0)(Z-Z_0) \leq 0$. Thus, only a single additional parameter is required. Eq. (7) accounts for the Coulomb energy as well as for charge-symmetric nuclear forces. Furthermore, it ensures continuity of mass predictions at shell crossings.

Solutions $M_T(N,Z)$ and $M_L(N,Z)$ for the neutron- and proton-rich light nuclei with $2 \leq N,Z \leq 50$ were obtained from a χ^2-adjustment to the experimental masses [9]. The standard deviations σ_m are 643 keV for $M_{s.m.}(N,Z)$, 303 keV for $M_T(N,Z)$, 411 keV for $M_L(N,Z)$. These values are preliminary. Charge symmetry of nuclear forces requires the constraints $G_1(k)=G_2(k)$ and $F_1(k)=F_2(k)$ which have not yet been introduced.

*   Work supported in part by USERDA Contract E(11-1)-2167.
[1] J.Janecke and H.Behrens, Z.Physik 256(1972)236; Phys. Rev. C9(1974)1276.
[2] J.Janecke and B.P.Eynon, Nucl.Phys. A243(1975)326.
[3] A.de-Shalit, Phys.Rev.105(1957)1528.
[4] G.T.Garvey, W.J.Gerace, R.L.Jaffe, I.Talmi and I.Kelson, Rev.Mod.Phys. 41(1969)S1.
[5] S.Liran and N.Zeldes, Proc.Int.Conf. on Nuclear Physics, Munich 1973, p.232; S.Liran, Ph.D. thesis 1973, Hebrew University of Jerusalem.
[6] W.D.Myers and W.J.Swiatecki, Ann. of Phys. 55(1969)395; Ann. of Phys., to be published; report LBL-1957.
[7] J.W.Truran, A.G.W.Cameron and E.Hilf, Proc.Int.Conf. on properties of nuclei far from the region of beta-stability, Leysin 1970, Vol.1, p.275.
[8] P.A.Seeger, ibid. p.217.
[9] A.H.Wapstra, K.Bos and N.B.Gove, 1974 Supplement to "1971 Atomic Mass Adjustment, Nucl. Data Tables A9(1971)265."
[10] A.de-Shalit and I.Talmi, Nuclear Shell Theory (Academic Press, New York, 1963).
[11] K.T.Hecht, Nucl. Phys.A102(1967)11; Nucl. Phys.A114(1968)280.
[12] W.J.Courtney and J.D.Fox, Atomic Data and Nucl. Data Tables 15(1975)141.
[13] L.Zamick, Phys.Lett. 10(1965)580; R.K.Bansal and J.B.French, Phys.Lett. 11(1964)145; R.Sherr, R.Kouzes and R.DelVecchio, Phys.Lett. 52B(1974)401.

THE LIQUID DROP MASS FORMULA

AS A SHELL MODEL AVERAGE [*]

M. Bauer

Instituto de Física

Universidad Nacional Autónoma de México

1. Introduction

It has been previously shown[1] that an expression for the
binding energies of nuclei with the structure of the semi-
empirical Bethe-Weizsäcker mass formula can be derived from
a shell model with residual interactions. Pending quantitive
agreement-which is the subject of this paper -this approach
provides a microscopial foundation for the exceptionally good
semi-empirical mass formula. To our knowledge, this constitutes the first verification of the commonly accepted, but otherwise
unproven, statement that the liquid drop mass formula should
emerge from a suitably averaged shell model, this being the basic
assumption in the mass formulae constructed by using the liquid
drop expression as a base line to which shell corrections are added.

The analytical expression in terms of the mass number A and
the atomic number Z is obtained by evaluating the Hartree-Fock
ground-state energy with a conveniently parameterized set of phenomenological single-particle energy levels. The coefficients of the
mass formula thus obtained - volume, surface, symmetry and pair -
ing, as well as new ones which are predicted - are given as definite functions of the shell model parameters. The fit to the experimental data is done through a search of the parameters associated
with the single particle potential and the residual interactions. The
search is limited to the range of phenomenological values consistent
with nuclear spectra calculations and optical model analysis of

---

*Work supported by the Instituto Nacional de Energía Nuclear, México nuclear reactions. It is found that such values also yield an acceptable fit to the nuclear binding energies. We can finally note that the phenomenological shell model has received by now ample justification from first principles, i.e., Hartree-Fock calculations with realistic nuclear forces.

2. The Method

In this section we review briefly the construction of the mass formula and point out the essential conceptual points (the details can be found in Ref. 1, to be denoted hereafter as I). One is considering the ground state energy of a system of N neutrons and Z protons as given in a constrained Hartree-Fock formulation with residual interactions. The constraints correspond to having the correct quadrupole moment and number of particles. The first constraint will lead to deformed self-consistent potentials (taking into account the long range part of the interaction) while the second is introduced because the pairing interaction (short range part) does not conserve the number of particles. The main part of the total energy is then given by an expression of the form

$$E = \sum_\alpha \langle \alpha | t + \tfrac{1}{2} u | \alpha \rangle \, V_\alpha^2 \equiv \sum_\alpha \epsilon_\alpha \, V_\alpha^2 \tag{1}$$

where V_α^2 is the occupation probability of the single-particle state $|\alpha\rangle$ with energy η_α:

$$\eta_\alpha = \langle \alpha | t + u | \alpha \rangle = \epsilon_\alpha + \tfrac{1}{2} \langle \alpha | u | \alpha \rangle \tag{2}$$

where t is the kinetic energy and u the self-consistent potential. We have for simplicity omitted in eq. (1) the contribution of the pairing interaction (given in I), which adds little to the total energy although it is important for the detailed structure of the mass surfaces. Besides yielding the pairing term in the mass formula [1], the pairing interaction diffuses the occupation probability around the Fermi level (i.e., V_α^2 is not a step function) and smooths out the small discontinuities within a major shell. Following Belyaev's formulation [2], the sum within each major shell in eq. (1) is substituted by an integration with a continuous energy level density which is taken to depend linearly on the energy in the following way: for heavy nuclei, it increases with energy if the nucleus is spherical (the levels are bunched towards the top of the shell) but remains uniform for deformed nuclei. This is what one expects from a Nilsson scheme if one applies an averaging procedure of the Strutinsky type. We obtain

$$E = \sum_k \bar{\epsilon}_k^{(n)} N_k + \sum_k \bar{\epsilon}_k^{(p)} Z_k \tag{3}$$

where N_k and Z_k are the numbers of neutrons and protons in the kth major shell. The $\bar{\epsilon}$'s are average energy contributions of the form $\bar{\epsilon} = \frac{1}{2}(\epsilon'' + \epsilon') + \frac{\zeta}{6}(\epsilon'' - \epsilon')$, where ϵ'' and ϵ' are respectively the upper and lower bound in each major (neutron or proton) shell of the quantities ϵ_α (eq. 1). The "bunching" parameter ζ characterizes the continuous level density assumed in order to carry out the integration, with $\zeta = 0$ corresponding to uniform level density, and $\zeta > 0$ to increasing level density.

In constructing the single particle level sheme it is important to consider that the self-consistent potentials are nonlocal. This gives rise to a larger spacing between levels and therefore the nonlocal potential must be deeper than the local one that binds the same number of particles. Only when this is taken into account, does one obtain correct binding energies. The expression for the single particle energies is of the Nilsson type:

$$
\eta_{n\ell jm}(\beta) = -\frac{m_0}{m^*}V + \sqrt{\frac{m_0}{m^*}}\,\hbar\omega\,(n+\tfrac{3}{2}) - D\,\ell(\ell+1) -
$$

$$
-\left[1 - \frac{m^*}{m_0}\right]\frac{(\hbar\omega)^2}{8V}\left\{n(n+3) + \ell(\ell+1) + \tfrac{9}{2}\right\} -
$$

$$
-\frac{C}{2}\left\{j(j+1) - \ell(\ell+1) - \tfrac{3}{4}\right\} + \beta\sqrt{\frac{m_0}{m^*}}\sqrt{\frac{5}{4\pi}}\,\hbar\omega\,(n+\tfrac{3}{2})\frac{3m^2 - j(j+1)}{4j(j+1)}
$$

$$(4)$$

In this expression the parameters V and ω correspond to the equivalent local potential. The effect of the nonlocality is explicity exhibited through an effective mass m* (m$_0$ is the free nucleon mass) and through the 4th term in eq. (4)[3]. All the other terms are the familiar ones. The quantities ϵ_α needed in eq. (1) are obtained from eq. (4) by means of the virial theorem.

Finally the total energy E(A, Z) is evaluated and shown to have the structure of liquid drop formula, once the known A-dependence of the shell model parameters is taken into account. The coefficients are, however, still dependent on deformation and shell-filling. The detailed expressions are given and discussed in I.

3. The Fit to Experimental Data

In I only a few binding energies were calculated. We present here a fit to experimental data consisting of 1319 nuclei in the range $87 \leqslant A \leqslant 257$. We have used an expression limited to six terms:

$$
B(A, Z) = \left\{a_0 - b\left(\frac{A - 2Z}{A}\right)^2\right\}A - a_1 A^{2/3} +
$$

$$
+ a_2 A^{1/3} - a_3 - a_c(Z^2/A^{1/3})\left\{1 - \tfrac{1}{5}\beta^2 - 0.763\,Z^{-2/3} - 1.641\,A^{-2/3}\right\}
$$

$$(5)$$

where the coefficients a_i (i=0,3) and b are functions (given in I) of the following parameters:

1) V_0 and V_1 , characterizing the neutron and proton potential well depths, mamely $V=V_0 + \mathcal{T} V_1 (N-Z)/A$, with $\mathcal{T}=+1$ for protons and $\mathcal{T}=-1$ for neutrons,

2) r_0 , nuclear radius parameter in $R=r_0 A^{1/3}$,

3) m* , average effective mass (eq. I-33),

4) λ* , related to the spin-orbit coupling strength (eq. I-28),

5) γ , related to the Nilsson coefficient $D= \gamma (h^2/2m_0 R^2)$,

6) β , equilibrium deformation,

7) $\mathfrak{z}$, level bunching parameter.

The Coulomb term was not derived. The form proposed here in - cludes a deformation dependence, as well as exchange and diffu - seness corrections [4]. The coefficient was treated as an inde - pendent parameter.

The adjustment of the parameters was carried out as follows. The coefficients were taken as given by their closed shell expres- sions $\bar{a}_i$, corresponding to having set $\alpha_m = \alpha_p = 1$, $\beta = 0$ and $\mathfrak{z}$ = maximum value in eqs. I 46. This obviates the problem of having to parameterize the deformation β and the level bunching parameter $\mathfrak{z}$ as functions of shell filling. The values for λ*, γ and $\mathfrak{z}$ were fixed at phenomenological values [4] and the search was made over the other parameters. Also, because the contribution of the $D\ell^2$ term is not included explicitly in the expression for a_3 given in I, we treated this term as a parameter. Thus, we had six adjustable parameters, namely, V_0 , V_1 , r_0 , $m*/m_0$, a_c and a_3 . The search was restricted to ranges acceptable from the phenomenological and theoretical point of view. The fit shown in Fig. 1 was achieved with the following values:

$$V_0 = 53.45 \text{ MeV}, \qquad V_1 = 53.0 \text{ MeV}, \qquad r_0 = 1.25 \text{ F}$$
$$m*/m_0 = 0.57, \qquad a_3 = 252.0 \text{ MeV}, \qquad a_c = 0.536 \text{ MeV} \quad (6)$$
$$\mathfrak{z} = 0.3, \qquad \lambda* = 0.016, \qquad \gamma = 0.174, \qquad \beta = 0 .$$

The chosen values for λ* and γ give-orbit coupling and $D\ell^2$ terms within 10% of those used in ref. (4). In relation to the val- ues obtained for the other parameters, we refer to some recent work[5] in which the Hartree-Fock self-consistent potential is subs- tituted by a phenomenological Woods-Saxon well to carry out binding energy calculations with Skyrme and Negele interactions. The re - sult of this shell model approach agrees well with the full self- consistent calculation and with experiment. The potential well parameters are $V_0 = 53.6$ MeV, $V_1 = 30$ MeV, $r_0 = 1.16$ F, in agree-

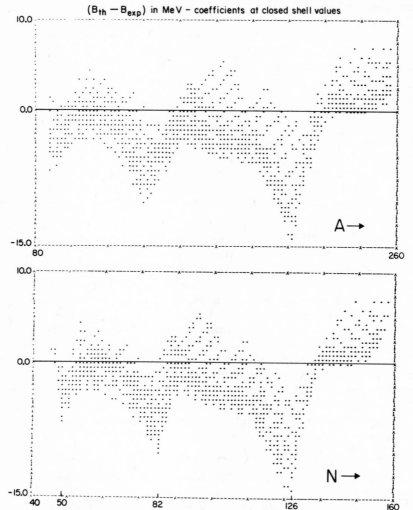

Fig. 1.   Difference between theoretical and experimental
binding energies, as function of A and N.

ment with the phenomenological optical model extensively used in
the study of nuclear reactions [6]. The effective mass ranges from
0.91 to 0.58 depending on the interaction. In another paper[7], the
equivalent local potential to the Hartree-Fock self-consistent nuclear
field obtained with the Skyrme interaction is shown to have a depth
$V_0 = 43$ MeV and a radius $R = (1.1A^{1/3} + 0.75)F$. The effective mass
at the origin is $m*/m_0 = 0.76$. A nuclear matter calculation with

hard core Reid interactions yields V_o = 56 MeV , V_1 = 14 MeV and $m*/m_{o\,k=0}$ = 0.63. Lower values for the effective mass have appear-ed many times in the literature [9].

We note that the largest discrepancy with accepted values in the symmetry potential V_1 , and the Coulomb coefficient: V_1 is too high and a_c is somewhat low compared with the range 0.6 to 0.7 found in most mass formulae. Yet, we have the correct value (~30) for the ratio b/a_c which determines the valley of stability [10] as we get b = 15.43 MeV[*].

The fit shown is comparable to the ones obtained with the liquid drop formula when the coefficients of the different terms are varied independently. The expected systematic deviations associated with the magic numbers emerge clearly. The values we obtain for the coefficients, using the parameters given in eqs. (6), are, in MeV:

$$\bar{a}_o = 12.57 \qquad \bar{a}_1 = 17.01 \qquad \bar{a}_2 = 93.97 \qquad (7)$$

$$\bar{a}_3 = 252.0 \qquad \bar{b} = 15.43 \qquad \bar{a}_c = 0.536$$

The previous results show that one can write the binding energy as

$$B = B(\bar{a}_i) + \left[B(a_i) - B(\bar{a}_i) \right] \qquad (8)$$

where the first term is a liquid drop type expression and the second is then the shell correction. Insofar as $B(a_i)$ is a Hartree-Fock ground-state energy, eq. (1), this shell correction does not corres-pond to Strutinsky's prescription which is based on the summation of single-particle energies. The assumption that the Strutinsky correction has nevertheless the same fluctuations as the actual one, eq. (8), has been critically examined and shown to be inaccurate [11]. Therefore, if one wants to construct a mass formula with the liquid drop expression as a base line, the shell correction has to be obtained by applying the averaging procedure in the Hartree-Fock scheme.

---

[*] A case for a lower Coulomb coefficient can be made on the fol-lowing basis. The proton optical potential depth is actually given by $V_p = V_o + \frac{N-Z}{A} V_1 + \Delta V_c$ where $\Delta V_c \sim Z^2/A^{1/3}$ is a Coulomb correction due to nonlocality which increases the well depth [6],[7],[12] Included in the proton single particle energies, this term gives a contribution to the total energy which has the same structure as the Coulomb energy but of opposite sign.

4. Conclusion

It has been shown, both qualitatively and quantitatively, that the liquid drop mass formula emerges from an averaged or smoothed shell model formulation of the ground state energy of nuclei. Additional terms are predicted [1], which should be taken into account for a more detailed fit including the parameterization of the deformation as a function of shell filling. Furthermore the present analytical approach gives a microscopical insight into the nature of the coefficients of the liquid drop formula as functions of the shell model structure. This will be helpful in the process of making meaningful extrapolations.

The author would like to express his thanks to Dr. R. Jastrow and to Dr. V. Canuto for their hospitality at the Institute for Space Studies, NASA, where this work was carried out.

References

1. Bauer, M. and Canuto, V., Ark f. Fys. 37, 393 (1967)
2. Belyaev, S. T., Mat. Fys. Meeld. Dan Vid. Selsk. 31 , No. 11 (1959)
3. Frahn, W. E. and Lemmer, R. H., Nuovo Cimento 6 , 1221 (1957)
4. Nilsson, S. G., et al., Nuc. Phys. A131, 1, (1969)
5. Ko, C. M., Pauli, H. C., Brack, M. and Brown, G. E., Nuc. Phys. A236, 269 (1974)
6. Perey, C. M. and Perey, F. C., Nuc. Data Tables 10, 539 (1972)
7. Dover, C. B. and Nguyen Van Giai, Nuc. Phys. A190, 373 (1972)
8. Jeukenne, J. P., Lejeune, A. and Mahaux, C., Phys. Rev. C10, 1391 (1974)
9. Vautherin, D. and Brink, D. M., Phys. Rev. C5, 626 (1972)
10. Evans, R. C., "The Atomic Nucleus", Ed. McGraw Hill (1955) p. 393
11. Bassichis, W. H. , Kerman, A. K. and Tuerpe, D. R., Phys. Rev. C8, 2140 (1973)
12. Perey, F. C., Phys. Rev. 131, 745 (1963)

MAGIC NEUTRON-RICH NUCLEI

AND A NEW SEMI-EMPIRICAL SHELL-CORRECTION TERM[+]

E. R. Hilf, H. v. Groote[++], K. Takahashi[+++]

Institut für Kernphysik, TH Darmstadt

D-61 Darmstadt, Germany

1. INTRODUCTION

We aim at a semi-empirical mass formula which has the properties
1) to be analytic, and simple to calculate,
2) to have its analytic structure being based on a theoretical micro-scopic understanding,
3) to be macroscopically self-consistent, that is, to be derived by a variational procedure, and
4) to reproduce the known experimental masses well, e.g. with a small rms-deviation, without using too many parameters.

Within the valley of β-stability, the droplet model (DM) [1] plus the simple shell-correction term of Myers and Swiatecki [2] already meets the requirements giving a rms of 0.96 MeV [3]. However, away from the β-valley a comparison [4] with experimental data sometimes shows that the MS shell-correction term does not have the right (N,Z)-dependence. We try to remove this shortcoming by constructing a new shell-correction term.

2. STRATEGY

The nuclear mass $M(N,Z)$ is separated into a smooth part $\bar{M}$ which describes the average properties and a "shell-correction term" $\widetilde{M}$

[+] Supported in part by the German Bundesministerium für Forschung und Technologie BMFT
[++] Supported by the Gesellschaft für Schwerionenforschung (GSI)
[+++] Fellow of the Alexander von Humboldt Stiftung

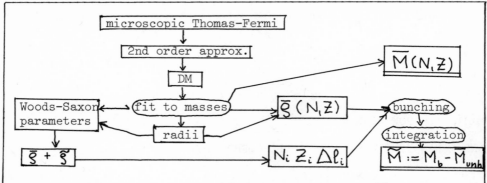

Diagram 1: The procedure to gain a shell-correction $\widetilde{M}$ to the mass formula $M = \overline{M}(N,Z) + \widetilde{M}(N,Z)$ from the microscopic self-consitent Thomas-Fermi model.

which takes care of the individual properties of the nucleus in question. For $\overline{M}$, we adopt the droplet model. It has been shown [1] that it approximates to second order the analytic structure of the microscopic Thomas-Fermi approach. Because of its macroscopic self-consistency, it reproduces the radii $R_n(N,Z)$, $R_p(N,Z)$ of the nucleon-distributions. These we shall use to construct an analytic expression for the shell-correction-part $\widetilde{M}$. First, we create a Woods-Saxon potential of a corresponding size, both for the protons and neutrons, respectively: Its depths $V(N,Z)$, radii $R(N,Z)$, and surface diffuse-nesses $a(N,Z)$ are known from the DM [5] and the Thomas-Fermi calculations [5,6]. These WS-potentials provide us with the magic numbers and the structure of the (N,Z)-dependence of the magic gaps of the spectrum. The shell-correction term $\widetilde{M}$ is then derived from the analytic Fermi-gas level-density by bunching it according to the information obtained from the WS-calculation.

3. LEVEL DENSITY

For the smooth, average level-density $\overline{g}(t)$, we take the Fermi-gas expression [7,8], which has been shown to be an approximation to second order in $1/\sqrt{t}$ of the WS-potential;

$$\overline{g}(t;N,Z) = \left(\frac{2m}{\hbar^2}\right)^{3/2} \cdot \frac{V}{2\pi^2} \cdot \sqrt{t} \cdot \left\{ 1 + \alpha\frac{\lambda(t)}{R_0} - \frac{\pi}{4}\sqrt{\frac{\hbar^2}{2m}} \cdot \frac{S}{V} \cdot \frac{1}{\sqrt{t}} \right\} \quad (1)$$

where t is the kinetic energy of the nucleon, and V and S are the volume and the surface-area of the neutron and proton Woods-Saxon potentials, respectively, given by

$$V = \frac{4\pi}{3}R_0^3(1+\xi)^3 \cdot \left\{ 1 + (2.46 \mp 0.645 \cdot \overline{\delta} + 4.0 \cdot \overline{\delta}^2)/R_0 \right\} \quad (2a)$$

$$\frac{S}{V} = \frac{3 \cdot (1 - \bar{\varepsilon})}{R_o} \cdot \left\{ 1 - \left(0.82 \mp 0.22 \cdot \bar{\delta} + 1.33 \bar{\delta}^2 \right) / R_o \right\} , \qquad (2b)$$

where $R_o = r_o \cdot A^{1/3}$, and $\bar{\varepsilon}$ and $\bar{\delta}$ are the DM-quantities describing the neutron-excess of the nucleus and the density-deviation in the nucleus from the saturation-density, respectively. In eq.(1), $\lambda(t)$ is the tunneling distance of a single particle wave-function; it takes care of the potential-broadening for $t \to t_F$. The third term in eq.(1) including S is due to the finite size of the system (surface term). This is quite an important correction, changing both magnitude as well as the effective t-dependence drastically. By integrating eq.(1), one obtains the number of states below t as $n(t) = \int_o^t dt \cdot \bar{g}(t)$. Inverting this, leads to the single-particle spectrum $t(n; N, Z)$ which rather nicely reproduces the smooth trends of the single-particle spectrum of the Woods-Saxon [9] or even of the microscopic Brueckner-Hartree-Fock K-matrix [10] approaches. In order to gain our shell-correction term, this level-density has to be bunched between the gaps at the magic numbers.

4. MAGIC GAPS

A nucleus is denoted as magic with regard to neutrons (or protons) if the gap $G_i(N, Z)$ between the last occupied single-particle level and the next one is distinctively above average. The Woods-Saxon potential with the size determined by the DM yields [11] the magic numbers

$$\begin{aligned} N_i &= 2, \ 8, \ 20, \ 28, \ 40, \ 50, \ 82, \ 126, \ 184, \\ Z_i &= 2, \ 8, \ 20, \ 28, \ 40, \ 50, \ 82, \ 114. \end{aligned} \qquad (3)$$

Since the (N, Z)-dependence of the gaps should reflect the average level-density, we propose the ansatz

$$G_i(N, Z) = \gamma_i / \bar{g}(t(N_i); N, Z) . \qquad (4)$$

The magic gap also depends on the difference $\Delta \ell_i$ between the angular momenta of the two single-particle levels defining the gap, because the neutron (proton) wave-function with the higher ℓ is pushed out and explores more likely the neutron-(proton)-skin of the nucleus. Due to the lack of protons (neutrons) as interaction partners there, they will be less bound. For the magic numbers given in eq.(3) the WS-calculation [11] yields to be

$$\begin{aligned} \Delta \ell_i(N_i) &= -1, \ -1, \ -1, \ +2, \ -3, \ +2, \ +2, \ -3, \ -2, \\ \Delta \ell_i(Z_i) &= -1, \ -1, \ -1, \ +2, \ -3, \ 0, \ -5, \ +3(0). \end{aligned} \qquad (5)$$

Allowing for the above-mentioned effects we make the ansatz

$$\gamma_i = \gamma_i^o \cdot (1 - \xi \cdot \Delta \ell_i \cdot d\nu) \qquad (6)$$

where $d\nu$ stands for the number of neutrons (protons) in the neutron (proton) skin given by the DM to be

$$d\nu = A \cdot (I - \bar{\delta}) / (1 - \bar{\delta}) \tag{7}$$

with $I := (N-Z)/A$ and parameters γ_i^0 and ξ. If γ_i goes down to 1, the gap reduces to average and no longer is magic.

5. BUNCHING OF THE LEVEL-DENSITY

With the magic numbers and the corresponding $\Delta\ell_i$ assumed to be known we construct a shell-correction term $\tilde{M}$ for the mass-formula by bunching the smooth level-density (3): The $(N_i - N_{i-1})$ levels of a shell which in the smooth spectrum are spaced by $\{\bar{g}(t)\}^{-1}$ are now compressed (see Fig. 1), so that the top level $t(N_i)$ of the shell is lowered by $_i\Delta_2$ and the bottom-level $t(N_{i-1})$ is raised by $_i\Delta_1$. The bunches are separated by the magic gaps

$$G_i := {}_i\Delta_1 + {}_{i-1}\Delta_2 \tag{8}$$

The precise location of the bunch, or say, the location of the gaps is still open and might be taken to be a free parameter,

$$b_i := {}_i\Delta_1 / {}_{i-1}\Delta_2 \tag{9}$$

Thus a bunch i of $(N_i - N_{i-1})$ levels is constructed with a level-density $g(t) > \bar{g}(t)$ the profile of which may be assumed to be proportional to $\bar{g}(t)$. The numbers of levels in the shell and in the bunch have to be the same, so we have the constraint

$$\int_{shell\,i} dt \cdot \bar{g}(t) = \int_{bunch\,i} dt \cdot g(t) = (N_i - N_{i-1}) \quad \text{with } g(t) = p_i\,\bar{g}(t). \tag{10}$$

Instead of taking b_i to be a free parameter one may fix it in such a way that the center of gravity (c.o.g.) of the bunch i and that of the shell coincide:

$$\int_{shell\,i} dt \cdot t \cdot \bar{g}(t) = \int_{bunch\,i} dt \cdot t \cdot g(t) \tag{11}$$

6. SHELL-CORRECTION TERM FOR THE MASS-FORMULA

The mass-formula is

$$M(N,Z) = \overline{M}(N,Z) + \tilde{M}(N,Z) \tag{12}$$

with

$$\tilde{M}(N,Z) = {}_n\tilde{M}(N,Z) + {}_p\tilde{M}(N,Z) \tag{13}$$

where the smooth part of (12) is taken to be the DM-expression. The shell-correction term is now gained by integrating the single-particle energies over the bunched spectrum up to the N-th level and subtracting the same expression for the smoothed spectrum:

$$_n\widetilde{M}(N,Z) = \sum_i {}_n\widetilde{M}_i(N,Z) ,$$ (14)

the contribution of the i-th shell being

$$_n\widetilde{M}_i(N,Z) = \int_{\text{bunch } i} dt \cdot t \cdot g(t) - \int_{\text{shell } i} dt \cdot t \cdot \bar{g}(t) .$$ (15)

By inserting the (N,Z)-dependent expressions $\bar{g}$ and g, defined above, we have constructed a simple, analytic shell-correction term. I contains only up to two free adjustable parameters per magic number (location and magnitude of the magic gaps). A detailed description together with a fit to experimental masses and applications will be given elsewhere [11]. The here presented mass-formula meets all requirements mentioned before.

Our original strategy was to gain an inherent consistency of the (N,Z)-dependence. This, however, is more and more spoiled by successive simplifications of (20) as discussed now.

First, we neglect the surface- and potential-broadening-corrections to $\bar{g}(t)$, and assume the level-density to be constant for each bunch. Then

$$_n\widetilde{M}_i(N,Z) = E_i(N) - \tfrac{3}{5} \cdot \tfrac{a}{A^{4/3}} (N^{5/3} - N_{i-1}^{5/3}) + \sum_{j=1}^{i-1} \left\{ \tfrac{1}{2} (e_2 - e_1)(N_j - N_{j-1}) - \tfrac{3}{5} \tfrac{a}{A^{4/3}} (N_j^{5/3} - N_{j-1}^{5/3}) \right\}$$ (16)

with

$$E_i(N) := \tfrac{1}{2} \tfrac{N - N_{i-1}}{N_i - N_{i-1}} \cdot \left\{ (N - N_{i-1}) {}_i e_2 + (2N_i - N - N_{i-1}) {}_i e_1 \right\} ,$$ (17)

where the expressions

$$_i e_1 := \tfrac{a}{A^{2/3}} N_{i-1}^{2/3} \cdot \left\{ 1 + x_{i-1} f_{i-1} \gamma_{i-1}^0 \right\} , \quad x_i := (1 + b_i^{-1})^{-1}$$

$$_i e_2 := \tfrac{a}{A^{2/3}} \cdot N_i^{2/3} \cdot \left\{ 1 - (1 - x_i) \cdot f_i \cdot \gamma_i^0 \right\}$$ (18)

give the kinetic energies of the levels bordering the bunch i. The parameter a is defined by $\bar{g} = 3 \cdot \sqrt{t} / (2\, a^{3/2} A)$. Finally

$$f_j := \tfrac{2}{3 N_j} \cdot \left\{ 1 - \xi \cdot \Delta l_j \cdot (I - \delta) \cdot A / (1 - \delta) \right\} .$$ (19)

In this ansatz we still keep the assumed Δl-dependence and the changes in the magicities. The adjustable parameters of this shell-term are

$$\gamma_i^0 \quad , \quad b_i \quad , \quad a \quad , \quad \xi \quad .$$ (20)

Reasonable results for light elements were obtained with the estimates $\xi \simeq 10^{-3}$, $b_i \simeq 2 - 4$, $a \simeq 5 - 10$. However, a full least-squares fit has not yet been completed.

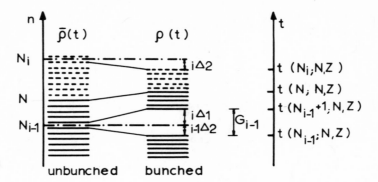

Fig. 1   Bunching of the average single-particle level-density
         according to the magic gaps.

As the next step of simplification we fix b_i by eq.(11), mock
up the (N,Z)-dependence of the smooth level-density by allowing a
to vary from shell to shell, and put $\xi = 0$. We then fitted the remaining parameter a_i and γ_i° together with the DM-parameters to the
experimental masses and obtained a rms of 0.46 MeV.

If one no longer specifies the level-density profile for the
bunches, but mocks up the bunching by a linear interpolation between
the energies of the complete and the no-bunching limit, then one is
left with only one parameter for each shell. For this ansatz we got
by a fit an rms of 0.69 MeV.

Finally, if we assume this parameter to be the same for all
shells, we end up with the MS-shell-correction for which our fit [3]
gives 0.96 MeV.

Even in this most simplified version, the prediction of the
masses of the very neutron-rich sodium-isotopes up to ^{31}Na agree
nicely with the exp. masses recently measured by Thibault [13] (see
Fig. 2). The deviations stay within the exp. error-bars plus the
rms of the mass-formula, and show no blow-up in contrast to the numerical mass-extrapolations [14].

The stability of the DM and MS-shell-correction term against
variations of the parameters was presented in 1971 by Ludwig et al.
[15]. For example, changing the volume-asymmetry coefficient γ by
10 % spoils the rms by only up to 5 %. For the other coefficients
the changes are about the same or smaller.

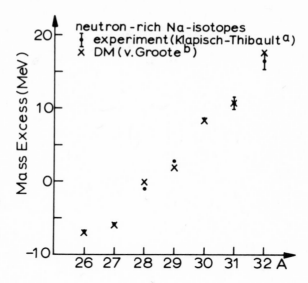

Fig. 2 Exp. neutron rich Na-isotopes compared with droplet
model prediction [3] .

We want to thank N. Zeldes for stimulating discussions, encouragement, and for helping to prepare the manuscript.

REFERENCES

1   W. D. Myers, W. J. Swiatecki, Ann.Phys. 55 (1969) 395
2   W. D. Myers, W. J. Swiatecki, Ark.Phys. 36 (1967) 343
3   H. v. Groote, to be published
4   K. Takahashi, H. v. Groote, E. R. Hilf, this conference
5   W. D. Myers, Nucl.Phys. A 145 (1970) 387
6   H. v. Groote, unpublished (1974)
7   E. R. Hilf, G. Süßmann, Phys.Letters 21 (1966) 654
8   H. v. Groote, E. R. Hilf, K. Takahashi, F. Beck, report
    GSI-J2-74 (1974) 249
9   E. R. Hilf, Proc.Int.Workshop on Gross-Properties of Nuclei
    and Nuclear Excitations III, Hirschegg (1975), AED-Conf. 75
    009 (1975) 247
10 D. Kolb, H. v. Groote, E. R. Hilf, to be published (1975)
11 E. R. Hilf, H. v. Groote, AED-Conf. 75 - 009 (1975) 230
12 H. v. Groote, E. R. Hilf, K. Takahashi, in preparation
13 R. Klapisch, C. Thibault, this conference
14 G. T. Garvey et al., Rev.Mod.Phys. 41 (1969) S1
15 S. Ludwig, H. v. Groote, E. R. Hilf, A. G. W. Cameron,
    J. Truran, Nucl.Phys. A 203 (1973) 627

STRUCTURE OF NUCLEAR ENERGY SURFACE AND THE ATOMIC MASS FORMULA

Kolesnikov, N.N., Vymyatnin, V.M., Larin, S.I.

Moscow State University

We should like to present some results concerning the structure of the formula for binding energies of nuclei with A > 40 [1,2]. Our analysis is based on the experimental data [3] alone and did not use any model assumptions.

In this analysis we proceeded from the condition of the least r. m. s. deviation from experimental masses of nuclei and the best quality of the parameters introduced [1,2]. In doing this we assumed that 1) the binding energy B was a function of two variables x and y, which may be A and Z or their simple combinations, 2) B may be represented as the product (or a sum of the products) of the functions $\Phi(x)$ and $\Psi(x)$, which may be approximated by the following expression:

$$F(x) = \sum_{k} \frac{(a_k + b_k x)^{c_k}}{(1 + d_k x)^{g_k}} \tag{1}$$

The fitting parameters a_k, b_k, c_k, d_k, g_k may be chosen in any combination. Their numerical values were found from the condition of the least r. m. s. deviation σ. The best functions with fixed number of parameters are those providing the least σ. The number of parameters was gradually increased until introducing the new parameter resulted in noticeable σ decreasing. The Coulomb term was not fitted to the experimental masses, but was calculated as a Coulomb energy of nuclei with Hoffstadter's charge distribution.

Thus the formula for binding energy without shell corrections may be written as follows (in MeV) [1,2]:

$$B(A,Z) = 14.7270A - 25.2452A^{0.385} - (0.671 + 0.00029A) \frac{Z^2}{A} - \tag{2}$$

$$- (13.1640 + 0.0445A) \frac{(A - 2Z)^2}{A} + (1.750 + 0.000462A) *$$

$$(A + |A - 2Z|) - \delta(A).$$

Here even-odd coreetion $\delta(A)$ is:

$$\delta(A) = \begin{cases} - 11.71 \cdot A^{-0.49} & \text{for even-odd nuclei} \\ - 11.02 \cdot A^{-0.46} & \text{for odd-even nuclei} \\ - 32.00 \cdot A^{-0.57} & \text{for odd-odd nuclei} \\ 0 & \text{for even-even nuclei} \end{cases} \tag{3}$$

Formula (2) as compared to other smoothed formulae gives the least r. m. s. deviation for the nuclear masses (σ_B), for the β-stability line (σ_{Z_0}) and especially for the curvature of isobaric parabolas (σ_k) (See table 1).

TABLE 1

Formula				
Bethe – Weizsaecker – Green	[4]	3.05	0.459	0.270
Mozer	[5]	5.03	0.430	0.126
Cameron	[6]	2.88	0.455	0.204
Kodama	[7]	2.73	0.443	0.137
Myers – Swiatecki	[8]	2.81	0.432	0.163
Formula (2)		2.34	0.401	0.115

The particularities of formula (2) are: the modification of the surface energy term., the presence of the Wigner's term and the correction to the symmetry term.

By introducing the shell model correction in the form of Myers - Swiatecki [8]:

$$S(Z,N) = 1.265 [S(Z) - S(N)] - 0.208 (N - Z), \tag{4}$$

with

$$S(x) = \sum \frac{(x - M_{i-1}) (M_i - x)}{M_i - M_{i-1}} \theta (x - M_{i-1}) \theta (M_i - x) \tag{5}$$

$[\theta(x) = 0$ for $x < 0$ and $\theta(x) = 1$ for $x > 0]$, formula (2) leads to $\sigma_B = 1.12$ MeV, if the values Z and N = 20, 28, 50, 82, and N = 126, 184, Z = 110 as well, are assumed to be magic number M_i.

The same result may be obtained by introducing the peak-line corr correction.

If the parameters of correction (4) are considered to be different for different shells, it is possible to reduce (with total number of parameters being 34 in binding energy formula) the value of σ_B to 0.35 MeV and the maximum deviation – to about 1 MeV. Nevertheless the further increasing of the number of parameters in the formula does not reduce σ_B noticeably. All this shows the limitation of the shell correction method. The further decrease in σ_B may be obtained by using the magic numbers of protons and neutrons and having a gap at magic N and Z.

References

1. N.N. Kolesnikov, V.M. Vymyatnin, Izv. Vysh. Uchebn. Zaved., Fiz. (12), p. 122, 1974.

2. N.N. Kolesnikov, V.M. Vymyatnin, Izv. Akad. Nauk. SSSR, Ser. Fiz., 39, (3), 637, 1975.

3. A.H. Wapstra, N.B. Gove, Nuclear Data Tables, 9, 265, 1971.

4. A.E.S. Green, Phys. Rev., 95, 1006, 1954.

5. F. Mozer, Phys. Rev., 116, 970, 1959.

6. A.S.W. Cameron, Canad. J. Phys., 35, 1021, 1957.

7. T. Kodama, Progr. Theor. Phys., 45, 1112, 1971.

8. W.D. Myers, W.J. Swiatecki, Nucl. Phys., 81, 1, 1966.

9. V.M. Strutinsky, Yad. Fiz., (Ac. Sci. SSSR) 3, 614, 1966.

10. N.N. Kolesnikov, Zh. Ehksp. Teor. Fiz. (Ac. Sci. SSSR), 30, 889, 1956; Vestn. Mosk. Univ. (6), 76, 1966.

ATOMIC MASS EXTRAPOLATIONS

K. Bos, N.B. Gove[†], A.H. Wapstra

Instituut voor Kernphysisch Onderzoek

Amsterdam

These days ever more data come available on reaction energies connecting pairs of nuclides which are not themselves connected to the system of primaries (nuclides which are connected to each other in many ways). It is the aim of the Atomic Mass Adjustment program[1]) to compare all available data, to check the system for internal consistency and to produce a list of atomic masses and linear combinations of atomic masses. Therefore unconnected data should be made connected in some way to become usefull inputdata for the program. Up till now this was done by adding so called systematics to the inputdata: binding energies of two protons or two neutrons, alpha- and (double) beta-decay energies were used since they will be influenced only slightly by paring energies. Values were calculated by manual extrapolation in the graphs for the four quantities simulteneously. Today some thousand systematic data are necessary to connect all the available data so this procedure has become a very time consuming part of the preparation. In this paper an attempt is presented to computerize this part of the work.

Instead of using the extra data mentioned above, we decided to use the masses themselves for extrapolation since some promising results had already been obtained that way. For evident reasons, rather than using atomic masses directly we made use of differences of experimental masses and calculated values from an atomic mass formula. To this end the smooth part of the mass formula of Meyers and Swiatecki[2]) is used to make a first calculation. The calculated values give a good expression of local effects determining the mass values. As long as the extrapolation also is confirmed to a local

[†] Present address: International Atomic Energy Agency, Vienna

area in terms of N and Z these numbers can now be considered useful
specimen for the further calculations.

We restrict ourselves to extrapolation of cases where a nucleus
can be reached along at least three of the lines of N = constant,
Z = constant, N + Z = constant and N - Z = constant. Along each line
a weighted least squares fit to a parabola is made of the
experimental data. The numbers are weighted with the squares of the
inverse uncertainties and with a factor related to the nearbyness
in terms of N and Z. Along all four lines only values from the same
mass surface are used.

The masses so calculated in one computerrun will now be
considered as actual masses for a next iteration, yielding again a
new set of masses.

Test runs have been made on the stability of the procedure.
Only after four succesive calculations the discrepancies rose over
1 MeV. But at that point already 90% of the necessary masses for
the Mass Adjustment program had been calculated. Besides some 100
massvalues resulted not necessary for this purpose which may however
been interesting for others.

In this stage of the work the unconnected data will be used as
inputdata for the Atomic Mass Adjustment program. An other way to
use these data directly is to extrapolate along linear combinations
of masses as was done by hand formerly. We are now trying to
computerize this too.

References

1) Wapstra, A.H., Gove, N.B., Nucl. Data Tables 9 (1971) 265-468,
   11 (1972) 157-280.

2) Myers, W.D., Swiatecki, W.Y., Nucl. Phys. 81 (1966) 1.

TIME AND FREQUENCY

Helmut Hellwig, David W. Allan, Fred L. Walls

National Bureau of Standards, Frequency & Time Standards
Section
Boulder, Colorado   USA

FREQUENCY STANDARDS AND ACCURACY
In 1967 the General Conference on Weights and Measures adopted
the cesium resonance frequency for the definition of the second.
Universal Coordinated Time (UTC) has used a close approximation to
the atomic second since 1972 (1). Time scales which refer to the
rotation of the earth such as UTC are generated by inserting or
leaving out seconds (leap seconds) at certain specified dates during
the year, as necessary. This process is coordinated worldwide by the
Bureau International de l'Heure (BIH).   UTC is the de-facto basis
for civil or legal time in most countries of the world (2).   In
addition to cesium beam standards, the atomic hydrogen maser has
found use as primary frequency reference and clock.
   Other promising techniques have been developed. The following
table depicts a summary of all those techniques which promise accuracies better than 1×10^{-13} together with their currently reported
accuracies. Accuracy is to be understood here in a very special
meaning: It is a measure of the degree to which--in an experimental evaluation--one can deduce the unperturbed frequency of
the respective transition by subtracting environment and apparatus related effects from the measured frequency (3).

FREQUENCY STABILITY
   More important than accuracy to most frequency & time metrologists is probably the stability of a standard. Stability can be
characterized in the frequency domain or in the time domain. The
instantaneous fractional frequency deviation $y(t)$ from the nominal frequency ν_o is related to the instantaneous phase deviation
$\Phi(t)$ from the nominal phase $2\pi\nu_o t$ by definition

$$y(t) \equiv \frac{\dot{\Phi}(t)}{2\pi\nu_o} \tag{1}$$

Table 1.   Summaries of Devices Promising Accuracies $< 10^{-13}$

CLASS OF DEVICES	DOCUMENTED ACCURACY (ORDER OF MAGNITUDE)		COMMENTS
MICROWAVE BEAMS	(Cs)	10^{-13} (3)	Cs BEST DOCUMENTED; REALIZATION OF THE DEFINITION OF THE SECOND
ATOM STORAGE (H)	(H)	10^{-12} (4)	COATED STORAGE VESSEL ONLY USABLE FOR FEW, SELECTED ATOMS
ION STORAGE (Hg, Ba, He, Tℓ, ETC.)	(He)	10^{-9} (5)	SIGNAL-TO-NOISE PROBLEMS WITH PRESENT TECHNIQUES
SATURATED ABSORPTION (CH_4, I_2, SF_6, ETC.)	(CH_4)	10^{-12} (6)	USEFULNESS AS "TRUE" TIME/ FREQUENCY STANDARDS CRITICALLY DEPENDENT ON PRECISION MICRO- WAVE/INFRARED/VISIBLE FREQUENCY SYNTHESIS
TWO PHOTON TRANSITIONS (Na, H, ETC.)			
INFRARED & OPTICAL BEAMS (Mg, Ca, I_2, ETC.)	(I_2)	10^{-11} (7)	

I.   frequency domain:

In the frequency domain, frequency stability is conveniently defined (8) as the one sided spectral density $S_y(f)$ of $y(t)$.   More directly measurable in an experiment is the phase noise or, more precisely, the spectral density of phase fluctuations $S_\phi(f)$ which is related to $S_y(f)$ by

$$S_y(f) = \frac{f^2}{\nu_0^2} S_\phi(f) \tag{2}$$

For the above, f is defined as the Fourier frequency offset from ν_0.

Of course, $S_\phi(f)$ cannot be perfectly measured; however, useful estimates of $S_\phi(f)$ can easily be obtained.   One useful experimental arrangement to measure $S_\phi(f)$ is given in Fig.1.   If the measured os- cillator and the reference oscillator are equal in their total per- formance, and if the phase fluctuations are small, i.e., much less than a radian, then, for one oscillator (9):

$$S_\phi(f) \approx 2 \frac{V^2(f)}{V_R^2} \tag{3}$$

where V(f) is the rms voltage at the mixer output due to the phase fluctuations within a 1 Hz bandwidth at Fourier frequency f, V_R is a reference voltage which describes the mixer sensitivity and is equivalent to the sinusoidal peak-to-peak voltage of the two oscil- lators, unlocked and beating.   The phase lock loop is only necessary to keep the signals in phase-quadrature at the mixer; it must be a sufficiently loose lock, i.e., its attack time (corresponding to the unity-gain condition) is long enough to not affect all (faster) fluctuations to be measured.   Fig. 2 depicts the measured phase noise performance of a system as in Fig. 1 which uses available state-of-the-art electronic components.   Also shown is the measured phase noise of one of the best available crystal oscillators; its corresponding time domain performance is approximately shown in Fig. 4.   Fig. 2 clearly shows that present measurement capabilities are not yet taxed by available oscillators.

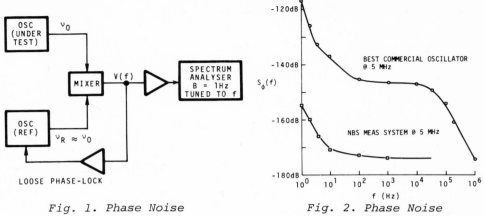

Fig. 1. Phase Noise
Measurement System

Fig. 2. Phase Noise
Plots

II. time domain

The relationship between the frequency and time domain, essentially a Fourier transformation, is extensively covered in Ref. (8).
In the time domain, frequency stability is defined by the sample
variance:

$$\sigma_y^2(N,T,\tau,B) \quad \equiv \left\langle \frac{1}{N-1}\left(\sum_{n=1}^{N} \bar{y}_n - \frac{1}{N}\sum_{k=1}^{N}\bar{y}_k\right)^2\right\rangle \tag{4}$$

where $\langle\ \rangle$ denotes an infinite time average, N is the number of frequency readings in measurements of duration τ and repetition interval
T,B is the bandwidth of the measurement system. Some noise processes
contain increasing fractions of the total noise power at lower Fourier
frequencies; e.g., for flicker of frequency noise the above variance
approaches infinity as $N \rightarrow \infty$. This, together with the practical difficulty to obtain experimentally large values of N led to the convention of using always a particular value of N (10). In recent years, frequency stability has become almost universally understood as meaning
the squareroot of the two-sample or Allan Variance (8) $\sigma_y^2(\tau)$ defined as in Eq. (4) for N = 2, T = τ.

$$\sigma_y^2(\tau) = \left\langle \frac{(\bar{y}_{k+1} - \bar{y}_k)^2}{2}\right\rangle \tag{5}$$

$\sigma_y(\tau)$ is convergent for all noise processes commonly found in oscillators. It should be noted that even for Eq. (5) B remains an important parameter which must be taken into consideration. Fig. 3 depicts
three different measurement systems (11) which may be used to determine $\sigma_y(\tau)$.

In Fig. 4 a typical measurement capabiliy (at 5 MHz) using
Shottky barrier diode mixers is depicted. As in the case of frequency domain measurements, it is obvious that the existing measurement system capability is fully adequate to measure any existing
oscillator. Figure 4 is adapted from Ref. (12). It includes crys-

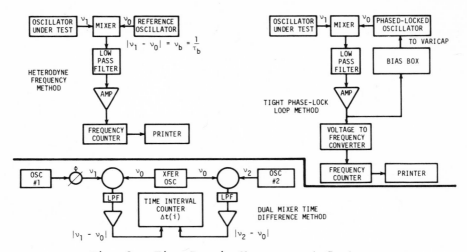

Fig. 3.   Time Domain Measurement Systems

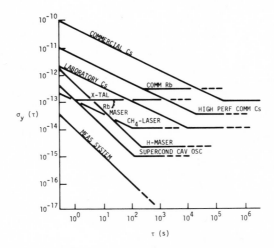

Fig. 4.   Measured Frequency Stability of Various Frequency Standards

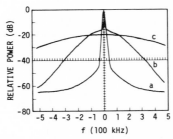

Fig. 5.   Spectral Line Profile Under Multiplication.
Resolution is Limited to 10 KHz Spectrum Analyzer Bandwidth

tal and superconducting cavity oscillators and various types of laboratory and commercial atomic frequency standards. Figure 4 shows
that for short sampling times quartz crystal oscillators, superconducting cavity oscillators, and rubidium masers are the oscillators
of choice. For medium-term stability, the hydrogen maser and superconducting cavity oscillator are superior to any other standard which
is available today. For very long-term stability or clock performance, cesium standards are presently the devices of choice. Rubidium standards are not superior in any region of averaging times, however, they excel in the combination of good performance, cost and
size.

FREQUENCY SYNTHESIS ABOVE THE MICROWAVE REGION

We will later return to the importance of long-term stability
in time generation. We shall now consider the importance of high
short-term stability (or low-phase noise) in the area of frequency
synthesis into the infrared. Precision synthesis is a crucial prerequisite to high resolution absolute frequency measurements, to the
use of frequency standards in the infrared and visible regions
(comp. Table 1), and to the concept of a unified standard for time
and length via the definition of the speed of light (13). Present
successes with frequency synthesis to the 88 THz transition of methane realized only 6×10^{-10} measurement precision (14). One of the
critical limitations is the need for a series of intermediate
oscillators (klystrons, lasers) of inferior stability not only to
compensate for inefficient multipliers but also to serve as spectral
"purifiers". Let us examine the phase noise requirements on an oscillator for single step harmonic generation up to a certain harmonic
number. If a state-of-the-art crystal oscillator at 5 MHz with a
phase noise performance as in Fig. 2 is assumed, then Fig. 5 shows
its spectral line profile at 9.2 GHz (curve a), 150 GHz (curve b),
and 1.5 THz (curve c). Although there still is power available,
the carrier has now totally disappeared, the linewidth has increased
a factor of 10^8 from .006 Hz to 600 KHz. Clearly curve c can no
longer be used as a precision reference signal.

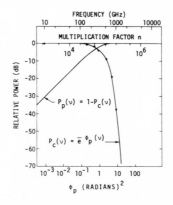

Fig. 6.   Power in Carrier and in the Pedestal Under Multiplication

The relative power in the carrier P_c, and the relative power in the pedestal P_p, (15) are shown in Fig 6.
The exponential loss of power from the carrier when the mean square phase fluctuations from the pedestal $\phi_p(\nu)$ exceed 1 radian is the most serious effect. For the present state-of-the-art 5 MHz crystal oscillators, the power is evenly divided between carrier and pedestal at ~300 GHz while at 1 THz the carrier has only -50 dB of the total power. Therefore, the only way to extend the useful working range of the present 5 MHz oscillators above 1 THz is to reduce the phase modulation due to the pedestal. This can be done either by reducing the white phase modulation level of the pedestal or reducing the pedestal bandwidth B_0, either in the oscillator or somewhere along the multiplier chain. Of course, changing the pedestal height or width affect the spectrum in different ways.

It appears possible (16) to construct crystal oscillators with at least 40 dB reduction in the white phase level of the pedestal. This would provide an oscillator with a possible working range extending to 30 THz without the use of intermediate oscillators or filters. The use of a passive filter with a bandwidth of 6 Hz at 9.2 GHz such as a superconducting cavity filter (15) would make it possible to multiply the present oscillators to 100 THz without the need for intermediate oscillators. The linewidth would be approximately 70 Hz. The realization of oscillators of higher spectral purity than presently available, a prerequisite for precise infrared frequency synthesis, thus is within today's technical possibilities.

TIME AND CLOCKS

One of the principal applications of frequency standards is their use as clocks. In a very real sense, any long-term frequency measurement and astronomical distance measurement is a time measurement. Astronomy has in a de-facto sense relied on a unified time-length standard by measuring distances in units of time using an adopted value for the speed of light.

The time error T at the elapsed time t after synchronization can be written as

$$T(t) = T_0 + R_0 t + \frac{1}{2} D t^2 + \ldots + \epsilon(t) \tag{6}$$

where T is the time of the clock minus the time of the reference (ideal "true" time), T_0 is the synchronization error at $t = 0$ and R_0 the rate (fractional frequency) difference between the two clocks under comparison averaged around $t = 0$. D is the linear (fractional) frequency drift term and $\epsilon(t)$ contains all other fluctuations; e.g., those due to white noise, flicker noise, etc. The time dependence of $\epsilon(t)$ can be calculated or estimated statistically (17), knowing the power laws of the noise processes that model the clocks involved.

In addition to the above considerations it must be noted that time (date) in contrast to time-interval (frequency), cannot be reproduced . This is a very significant difference because accuracy and stability, which are well-defined quantities in discussing frequency, become more complex and elusive with regard to the passing

of time (dates) as can be seen from Eq. (6). For example the question of a time standard and its long-term stability and accuracy is
not trivial. Our time is generated today by an ensemble of coordinated clocks of worldwide distribution (18) which establish the de-
facto time (date) standard. Time (date) accuracy is loosely used as
the degree of conformity of a clock to this de-facto reference which
in reality may not be accurate nor stable but only uniform, the uniformity being assessed by internal comparisons and evaluations of the
ensemble. The ensemble as a whole may have an undetected or unaccounted offset from the unit of time, a frequency drift term, etc.
Thus internal estimates of "accuracy" (really uniformity) which today
are of the order of microseconds per year, may be much too optimistic
with regard to a hypothetical, ideal clock running unchangingly on the
exact unit of time. Undetected frequency offsets of parts in 10^{13} and
drifts of parts in 10^{13} per year are likely. The actual time errors
then may be more like tens of microseconds per year. Such errors can
have critical importance in long-term and frequency measurements
in experiments such as a determination of a possible change in the
fine structure constant with time. Only the availability of primary
standards of sufficient accuracy can reduce such errors. Preferably,
several primary standards which are based on different physical principles, e.g., kind of atom, design of apparatus, etc., should be available to minimize the probability of undetected effects common to one
particular type of device.

REFERENCES

(1) H. M. Smith, Proc. IEEE 60, p. 479 (1972).
(2) CCDS Recommendation S1 (1974).
(3) D. J. Glaze, et al., IEEE Trans. on I&M, IM-23, p. 489 (1974).
(4) H. Hellwig, et al., IEEE Trans. on I&M, IM-19, p. 200 (1970).
(5) F. G. Major and G. Werth, Phys. Rev. Letters, 30, p. 1155 (1973).
(6) This number is an estimate based on information from:
     S. N. Bagaev, E. V. Baklanov, and V. P. Chebotaev, JETP Letters
     16, p. 243 (1972); J. L. Hall, Proc. Fifth Int. Conf. on
     Atomic Masses & Fund. Const., Paris, June (1975) (to be
     published); G. Kramer, PTB, Braunschweig, private comm.
(7) L. A. Hackel, et al., Proc. Fifth Int. Conf. on Atomic Masses
     & Fund. Const., Paris, June (1975) (to be published).
(8) J. A. Barnes, et al., IEEE Trans. on I&M, IM-20, p. 105 (1971).
(9) D. W. Allan, et al., Chapter 8, NBS Monograph 140, p.151 (1974).
(10) D. W. Allan, Proc. IEEE 54, p. 221 (1966).
(11) D. W. Allan, Proc. of Precision Time & Time Interval,
     Washington, DC (1974); NBS Tech. Note 669 (1975) (to be publ).
(12) H. Hellwig, Proc. IEEE, 63, p. 212 (1975); NBS Tech Note 662.
(13) IAU Resolution, Sydney, Australia (1973).
(14) K. M. Evenson, et al., Phys. Rev. Letters, 29, p. 1346 (1972).
(15) F. L. Walls, et al., IEEE Trans. on I&M, IM-25, Sept. (1975).
(16) F. L. Walls et al., IEEE Trans. on I&M IM-24, p. 15 (1975).
(17) D. W. Allan, Chapter 9, NBS Monograph 140, p. 205 (1974).
(18) J. A. Barnes and G. M. R. Winkler, Proc. 26th Ann. Symp. on
     Freq. Contr., p. 269 (1972).

APPLICATIONS OF FREQUENCY STANDARDS

R. F. C. Vessot

Center for Astrophysics
Harvard College Observatory and Smithsonian Astrophysical
  Observatory
Cambridge, Massachusetts 02138

INTRODUCTION

The atomic clock is the most precise instrument ever made by man, and it is natural to seek to use time or frequency as a basis for making other measurements. Consequently, a metrology based on time and frequency seems to be evolving. However, it is obvious that there are more fundamental reasons for this than the fact that clock technology has leaped forward in the last two decades.

Time is a scalar quantity, which can actually be measured by counting some highly stable periodic process. Currently, the stable processes chosen are those of resonances in atomic hyperfine interaction. Our present-day "proper clocks" are quantum-mechanical devices based on processes associated with particles having atomic dimensions, and we can only wonder at the philosophical implications of using such processes in the study of astronomy, where different concepts of the dimensions of space are involved. The numbers that describe the fractional frequency stability of today's clocks range into the 10^{-15} region, and some of us have seen data that enter the 10^{-16} region. It is obvious that many important questions in relativity can be answered by using clocks.

At present, the unit of time is embodied in a definition based on the hyperfine separation of the Cs^{133} atom. Dr. Hellwig, the previous speaker, has shown us the stability performance of cesium, rubidium, and hydrogen clocks (see Figure 4 of his paper). It is clear that applications of these devices depend on their availability, cost, and properties, an excellent survey of which is given by Hellwig [1].

Present applications of atomic standards cover a very wide field of activity and have created a substantial and rapidly growing amount of

commercial activity in atomic frequency standards. Of all the basic standards, frequency or time has become the dominant unit. The practical voltage standard is now tied to frequency through the Josephson effect [2] and
the acceptance of the ratio e/h.

The current standard of length, the meter, is defined by the wavelength of a spectral line of krypton. As a result of the recent remarkable
work connecting microwave radiation to optical laser radiation [3], it is
proposed to redefine the meter in terms of the speed of light, and we see
here the units of "light time" of astronomy entering the terrestrial laboratory through the use of quantum electronics. Very precise measurements
have been made of the magnetic field by using electron and proton resonance. Temperature-measurement techniques utilizing nuclear quadrupole
resonance – for example, in potassium chlorate [4] – offer very high sensitivity and excellent reproducibility.

Table 1 shows a list of the present applications and the types of
standards most often utilized.

It is difficult to estimate the number of atomic standards in the
field; however, several thousand cesium and rubidium standards are
known to be in use, and about 20 hydrogen masers are now in service.

It appears, among those who require highly stable frequencies and
accurate time, that it is less costly to have an atomic standard on hand
than to use atomic standards in the various national laboratories in conjunction with the present time-and-frequency dissemination techniques.

Table 1. Summary of uses of atomic clocks

Application	Clock type
Communications, television carrier stabilization	Cesium, rubidium
Computer data links, secure communication systems	Cesium, rubidium
Navigation systems Loran C Omega	Cesium, rubidium
Satellite tracking	Cesium, rubidium
Deep-space tracking	Cesium, hydrogen
Satellite-borne systems (in preparation)	Rubidium, cesium, hydrogen
Time services	Cesium, hydrogen
Very long-baseline interferometry	Hydrogen, rubidium
Relativity tests	Cesium, hydrogen
General laboratory standards	Cesium, rubidium
Aircraft-collision avoidance (proposed)	Cesium, rubidium

TERRESTRIAL CLOCKS FOR INTERFEROMETRY AND SPACE TRACKING

During the past decade, there has been substantial activity in very long-baseline interferometry (VLBI). Moran [5] recently reported geodetic and astrometric results of VLBI measurements of natural radio sources. A pictorial description of the VLBI technique is shown in Figure 1. The key to the success of this technique is the ability to provide a "common local oscillator" for widely separated stations by using atomic frequency standards at each station — usually hydrogen masers. The signal frequency at a mean wavelength λ is heterodyned to base band and recorded on digital or video tape, along with accurate time information. The noise signals on the magnetic tapes from various locations are cross correlated and integrated to give a fringe pattern whose amplitude depends on the signal-to-noise ratio and the degree of resolution of the object. The phase of the pattern changes owing to the earth's rotation and depends on $\theta(t)$, the angle between the direction of the source and the baseline L separating the stations. The period of the fringe is $\lambda/L \sin \theta(t)$.

A vivid illustration of the accuracy of this technique is given in Figure 2, showing the fractional frequency stability of fringe data taken on galactic H_2O sources at 1.35-cm wavelength by the Haystack Observatory at Westford, Massachusetts, and the National Radio Astronomy Observatory at Greenbank, West Virginia [6]. The figure also contains data taken

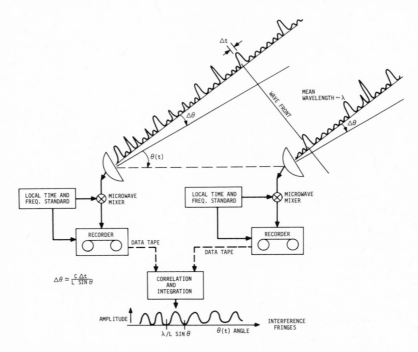

Figure 1. Use of atomic frequency standards in VLBI.

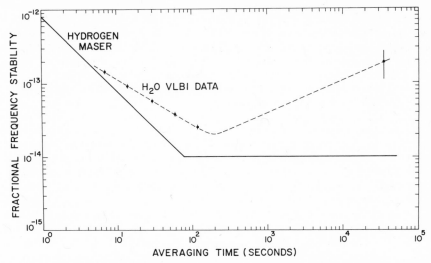

Figure 2.   Fractional frequency deviations ($\Delta f/f$) calculated from the rms
deviations of fringe phase data versus integration time (dashed
line).   The solid line shows 1970 maser performance data [6].

from the same type of hydrogen maser in the laboratory in 1970.   For
averaging intervals up to 100 sec, the data more or less follow those
from the masers.   Beyond 100 sec, large, systematic variations appear,
which are probably due to the frequency multipliers, the antenna system,
and differences in signal propagation at the widely separated sites [7].

The use of VLBI techniques for geodetic monitoring has been successfully demonstrated by scientists at the Jet Propulsion Laboratory (JPL) [8]
with a transportable antenna terminal.   With 28 hours of data, they confirmed a conventionally surveyed 16-km baseline to 3-cm formal uncertainty in all dimensions.   The current state of this technique has been
summarized by Moran [5]:   "The baseline vector between various radio
telescopes has been measured to an accuracy of about 1 m (the accuracy
of the measurement of the length of the baseline vector is better than 50
cm).   Variations in UT1 have been measured to an accuracy of 1 m sec,
polar motion to 500 cm, and the coordinates of radio sources to about 0.1
arcsec.   Improvement of about 1 order of magnitude is expected in the next
five years."

It is clear from such data that this technique has many valuable
applications to various disciplines of science, such as geodesy, geoastron-
omy, and earth physics.

It is natural to ask what the future holds for VLBI and the role that
improved clocks will play.   Swenson and Kellerman [9] and others have

proposed the construction of an array of VLBI stations over the continental USA. This array would obtain the spatial brightness distribution of resolved objects by determining the terms of a two-dimensional Fourier series, each term obtained from the fringe patterns of a pair of stations. This image-forming system would have a resolution better than 10^{-3} arc-sec and would consist of 8 to 10 radio telescopes, each equipped with hydrogen-maser frequency standards having a stability in the 10^{-14} to 10^{-15} region for averaging times from 10^2 to 10^5 sec.

During these rather austere times, it may seem a bit foolhardy, but nevertheless, it has been suggested that we extend the station-to-station baseline of VLBI terminals into space. Such an extension has many desirable features, among them being the absence of problems of propagation through the atmosphere and ionosphere, thus permitting a free choice of wavelengths. The antennas would be unaffected by gravitational forces and can be made large and relatively free of distortion. The orbits can be chosen to provide both a wide range of coverage and a form of aperture synthesis. Communications links, such as those discussed later in Figure 4, can be used to monitor the baseline and to provide accurate data links between the satellite terminals for cross correlating the signals.

CLOCKS FOR USE IN SPACE

Atomic clocks with a stability of 1 part in 10^{14} have been developed for applications in space. With the recent advances in space technology, we can now enlarge our laboratory to span the entire solar system and use massive bodies and large distances to measure relativistic effects. Communication by phase-coherent microwave systems is now possible over enormous distances, and we can realistically consider performing the "gedanken," or thought, experiments discussed in the literature on gravity and relativity.

Traditionally, relativity has been described in terms of systems moving with respect to one another, each containing rods and clocks. Pulsed-light signals connect the systems observationally and provide the basis for comparisons. To make experimental measurements, we can, in fact, use rods and clocks. However, the rod lengths are related to the clocks by the velocity of light, and we can describe distances in terms of wavelengths of the clock frequency if we postulate that the velocity of light is constant in space-time. Thus, we can design relativity experiments that require clocks only.

To illustrate what may be the forerunner of this type of experiment, I should like to describe very briefly an experiment now in preparation [10] by the Smithsonian Astrophysical Observatory and the National Aeronautics and Space Administration to test the equivalence principle, the cornerstone of Einstein's general theory of relativity, as well as a great many more recent theories.

Early in 1976, a rocket probe carrying an atomic hydrogen maser will be sent into a nearly vertical trajectory to a distance of 3 earth radii from the earth's center. During this 2.5-hour mission, the rate of the clock will be compared via a phase-coherent microwave link with a clock system on the ground, and the variation in the apparent rate of the "proper" clock in the probe will be related to the gravitational potential seen by the clock. After corrections for the second-order doppler effect are accounted for by our knowledge of the trajectory, the probe-clock signal frequency is expected to shift upward by 5 parts in 10^{10} when the probe is at apogee. The behavior of frequency with gravitational potential will be measured continuously over the trajectory with an expected precision of 2×10^{-14}, yielding a sensitivity of 40 parts per million for the test.

This test requires extraordinary stability in all aspects of the system propagation, particularly in the removal of all motion-induced doppler shifts and tropospheric and ionospheric effects [11]. Since this type of system may well find use in other future applications, some space will be devoted here to describing it.

Figure 3 shows the concept of the scheme. Basically, it is a phase-coherent microwave system connecting the probe and the ground stations. A transponder aboard the probe will return a phase-coherent signal to the earth in links 1 and 2. If one-half the probe's doppler shift is subtracted from the frequency of downlink 3 from the clock, the first-order doppler shift is removed from its output frequency. Since this is a continuous-wave system, different frequencies must be used in each link. The present system operates at S band (2 GHz), and the transponder input-to-output frequency ratio is 240/221. The different frequencies cause no problems in doppler cancellation when the propagation medium, such as the troposphere, is nondispersive. However, because the three link frequencies

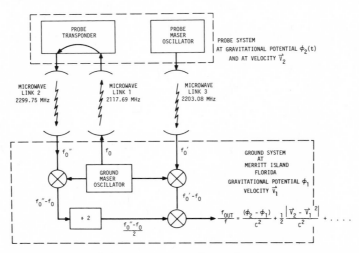

Figure 3. Concept of doppler-canceling clock-comparison system.

are all different, variations of the columnar ionospheric electrodensity during the flight can cause an appreciable effect in the output frequency. The ionospheric effect, or any dispersive effect proportional to frequency^{-2}, can be eliminated in this system by choosing the frequency of the clock signal (link 3) to lie properly between the transponder uplink and downlink frequencies according to the relation

$$R^2/S^2 = (1/2) \ (P^2/Q^2) \ [1 + (N^2/M^2)] \quad ,$$

where P/Q, R/S, and M/N are the frequency ratios shown in Figure 4. To provide a symmetrical data-comparison system that can search for anisotropic propagation or even for "preferred frame effects" [12], the system described above can be extended to allow data acquisition both at the space probe and on the earth.

The symmetrical frequency-comparison system with doppler and ionospheric cancellation is shown in Figure 4. The equation for developing the correct frequencies for the additional link is

$$(P^2/Q^2) = (1/2) \ (R^2/S^2) = [1 + (U^2/T^2)] \quad .$$

Deep-space missions involving clocks have been under consideration for some time [13-16]. The European Space Research Organization

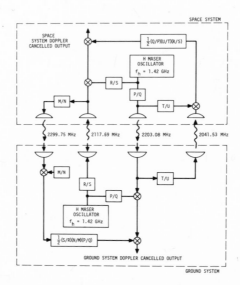

Figure 4. Dual, symmetrical, clock-comparison system with $1/f^2$ dispersion cancellation.

solar relativity mission SOREL studied during 1969-1973 involves several new areas of technology in very advanced states, including two-dimensional spacecraft-drag compensation, a laser, and atomic clocks. The experiment was planned to give improved values of β and γ in the Robertson metric and of J_2, the oblateness of the sun. Other proposals under consideration would "anchor" a satellite about the planet Mercury. By use of a transponder and clock, better data on the planet's motion would be obtained. In addition to this, a redshift test would be made as the planet's eccentric orbit scans the sun's gravity field.

Recently, G. Colombo suggested an earth–Jupiter–sun mission with the spacecraft moving straight into the sun after swinging by Jupiter. The spacecraft, spinning in the direction of fall, would use a one-dimensional drag-compensation system and would fall as a free particle. From the tracking data, Colombo hopes to obtain J_2, the solar dynamical oblateness, to 1×10^{-8}. (Note that R. Dicke's estimate for J_2 is 3×10^{-5}, and for rigid-body solar models, J_2 is about 2×10^{-7}.) The required tracking accuracy of about 10^{-2} cm/sec is now feasible by using atomic frequency standards.

A very preliminary study of earth–Jupiter swingby-injected solar orbital missions has been made by Anderson of JPL [16] to investigate the following scientifically interesting parameters:
1) A measurement of columnar electron density.
2) A measurement of the integrated magnetic field between the earth and the spacecraft.
3) A measurement of the redshift near the sun.
4) From spacecraft dynamics, a determination of the relativistic parameters β and γ to high accuracy.
5) A measurement of the solar oblateness J_2.
6) A determination of the astronomical unit with about 10-cm accuracy.

To give an idea of the dynamics of this type of mission, let us consider dropping a clock in free fall into the sun from a point at rest at the earth's distance from the sun. Figure 5 shows the radial distance versus time, and plotted on the same time scale is the redshift between the probe and the earth. We see that from 60 days out to 64.7 days, the probe has fallen from 0.28 A.U. to about 0.02 A.U., and a change in the redshift of 4.6×10^{-7} occurs. By using today's hydrogen masers, with their stability of better than 1 part in 10^{14}, a test of the redshift at the 10^{-7} level appears feasible if we can tracking the probe velocity to better than 0.2 cm/sec and know its position to within less than 60 m.

C. M. Will has proposed an experiment to test for the existence of nonmetric theories of gravitation by placing clocks using different types of atomic interaction in a spacecraft that scans the sun's gravity field and looking for changes in the relative rates of the clocks as the gravity field changes.

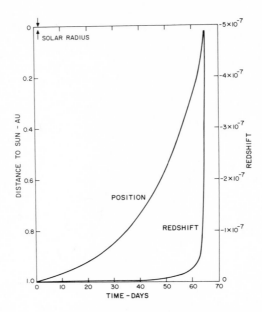

Figure 5.   Illustration of second-order redshift experiment using a clock
in a probe falling toward the sun.

There has long been a question about whether or not the "fundamental constants of nature" are, in fact, constant with time. Typical of these
are the constant of universal gravitation, G, and α, the fine-structure
constant. Both of these will be discussed at later sessions of this conference. Here again, we find that the atomic clock offers a very sensitive
basis for comparing a time scale defined by a hyperfine resonance in an
atom to a time scale related to other processes. So far, I expect we have
succeeded in setting smaller and smaller limits to "null" experiments and
can only ask at what level some consistent, nonzero result will occur.
Experimenters search for means of making increasingly precise measurements, and atomic clocks continue to improve. At the current rate of
progress, within the next few decades we can expect to see hydrogen-
maser and cesium-beam resonator devices with stabilities well into the
10^{-16} domain and with accuracies at the 10^{-14} level. New types of
devices are being investigated by using other kinds of atomic, molecular,
and even solid-state resonance that may well exceed these limits.

Even at the present level of performance, atomic clocks provide
science with an extremely powerful form of metrology. We can expect to
see rapidly growing requirements for atomic clocks and frequency standards in the next decade and must press for continuing development and
more widespread availability to meet this demand.

ACKNOWLEDGMENTS

This work was supported in part by contract NAS-8-27969 from the National Aeronautics and Space Administration and contract N00014-71A-0110 from the Naval Research Laboratories.

I should like to thank all my associates at the Center for Astrophysics for many helpful discussions. In particular, Drs. G. Colombo, J. Moran, and M. Levine were most helpful. I should like also to thank Dr. J. Anderson of Jet Propulsion Laboratory and Dr. H. Hellwig of the National Bureau of Standards.

REFERENCES

[1] Hellwig, H., Proc. IEEE, 63, 212, 1975.
[2] Field, B. F., Finnegan, T. F., and Toots, J., Metrologia, 9, 155, 1973.
[3] Evenson, K. M., and Peterson, F. R., Topics in Applied Physics, vol. 2, Springer-Verlag, Berlin, 1974.
[4] Vanier, J., Can. J. Phys., 38, 1397, 1960.
[5] Moran, J. M., Presented at the 17th International COSPAR Meeting, São Paulo, Brazil, June 1974.
[6] Moran, J. M., Papadopoulos, G. D., Burke, B. F., Lo, K. Y., Schwartz, P. R., Thacker, D. L., Johnson, K., Knowles, S. H., Reisz, A. C., and Shapiro, I. I., Ap. J., 185, 535, 1973.
[7] Counselman, C. C., Proc. IEEE, 61, 1225, 1973.
[8] Thomas, J. B., Fanselow, J. L., MacDoran, P. F., Spitzmesser, D. J., Skjerve, L., and Fliegel, H. F., J. Geophys. Res., in press.
[9] Swenson, G. W., and Kellerman, K. I., Science, in press.
[10] Vessot, R. F. C., in Experimental Gravitation, ed. by B. Bertotti, Academic Press, New York, p. 111, 1973.
[11] Vessot, R. F. C., and Levine, M. W., in Proceedings of the 28th Annual Symposium on Frequency Control, U.S. Army Electronics Command, Ft. Monmouth, N. J., p. 408, 1974.
[12] Will, C. M., in Experimental Gravitation, ed. by B. Bertotti, Academic Press, New York, p. 1, 1973.
[13] Jaffe, J., and Vessot, R. F. C., General Relativity and Gravitation J., 6, 55, 1975.
[14] Israel, G. M., in Proceedings of the Conference on Experimental Tests of Gravitation Theories, ed. by R. W. Davies, Cal. Inst. Tech., Pasadena, Cal., p. 236, 1970.
[15] Will, C. M., Phys. Rev. D, 10, 2330, 1974.
[16] Anderson, J. D., Private communication, 1975.

STABILIZED LASERS AND THE SPEED OF LIGHT

J. L. Hall

Joint Institute for Laboratory Astrophysics[*]

Boulder, Colorado 80302

The speed of light, being perhaps Nature's most universal physical constant, has attracted to its precise measurement the best imaginations and efforts of physicists from many generations. The contemporary set of new techniques -- introduced in pursuit of the "ultimate" speed of light measurement -- includes lasers frequency-stabilized to molecular absorbers and point-contact rectifiers for optical frequencies, to name only two. In this report we briefly summarize these and related developments and will try to project the motivational framework in which these inventions were conceived, developed and refined.

The modern era of direct, high precision speed of light measurements may be said to begin with K. D. Froome's work with a stabilized 36 GHz microwave source. Uncertainty in the wavelength determination dominated the experimental error budget, even though the frequency-doubled radiation of 4 mm wavelength was measured with an ingenious differential microwave interferometer. Froome's 1958 value,[1] c = (299 792.5±0.1)km/sec remained the definitive result in the field for a decade and a half.[2] Still, the limiting problem of diffraction would be much less serious if the wavelength were shorter. Also the wavelength measurement should ideally be made in vacuum.

The gas lasers -- announced less than 3 years after publication of Froome's result -- certainly had sufficiently short wavelengths and were sufficiently monochromatic. But the factor of 10^4 increase in frequency was far too much for existing frequency

[*]Operated jointly by National Bureau of Standards and University of Colorado.

measurement technology. Thus programs[3,4] leading toward a differential speed of light determination were undertaken: small diffraction corrections would be ensured by measuring two nearby optical
wavelengths in a long interferometer, and the frequency metrology
would be made tractable by using the difference frequency[5] between
the two laser lines. The need to control laser frequencies and to
precisely scan interferometer cavities led directly to development
of laser offset-frequency-locking techniques.[6] In Boulder we felt
an urgent need to have an optical wavelength source of sufficient
stability and reproducibility to calibrate in the lab and then carry
to the stable 30 m interferometer in the Poorman's Relief goldmine a
few kilometers away. Following the demonstration of saturated absorption using atomic neon,[7] it was inevitable that long-lived
sharp-line, molecular absorbers such as methane and iodine would
find application in stabilizing lasers.[8-10] Since these molecularly-
stabilized devices quickly surpassed in wavelength reproducibility
the legally-defined International Meter based on the 605.7 nm
krypton line, it became clear that some interesting new metrological possibilities were at hand, including – but far surpassing –
the intended application as interferometric secondary wavelength
standards. The possibilities and interest are strengthened further
with the evolution of techniques which can directly measure these
optical frequencies -- we will return to this subject momentarily.

The iodine- and methane-stabilized lasers have developed very
quickly to an impressive level of performance.[11-13] Thus when the
several (national) standards laboratories reported their wavelength
values to the Comité Consultatif pour la Définition du Mètre (CCDM)
in 1973, the mutual (in-)coherence was less than 4 parts in 10^9 and
was essentially limited by the properties of the krypton 605.7 nm
line defining the International Meter.[14] [In fact this intercomparison also focused attention on a possible ambiguity regarding
treatment of a small intrinsic asymmetry of the krypton line: the
concensus of the CCDM experts was that the optimum invariant reference point lies midway between the maximum and the center of
gravity of the krypton line's profile.[15] It is the rationalization
to this canonical wavelength basis -- rather than change of the experimental results -- which accounts for the small offsets to be
noted between early and post-CCDM reports of the wavelength and
speed of light values.]

With the existence of highly-stabilized laser frequency
sources in the near infrared, it became increasingly interesting to
try to bridge the four-orders-of-magnitude frequency gap between
lasers and the conventional sources and standards in the microwave
region. Such mixing and harmonic generation work with point-
contact-diodes began and flourished at MIT, reaching first to the
HCN laser at 337 μm (890 GHz).[16] Then harmonics of the HCN laser
were used to reach the D_2O laser line at 84 μm,[17] and later the
water vapor line at 28 μm.[18] The 9.3 μm line, R(10) of the CO_2

laser, was measured with 3 harmonics of the 28 μm laser plus a klystron.[19]   In the NBS determination of the 3.39 μm laser frequency,[20]
the tungsten-point-on-nickel diode had currents due to the 3.39 μm
laser, three harmonics of the CO_2 R(30) laser line at 10.2 μm and a
49 GHz klystron. Mixing orders of 12 are now routinely used in the
337 → 28 μm comparison. Using Josephson-tunnel diodes, a harmonic
mixing order of 400 has been demonstrated[21] and a factor 50 is routinely used at NPL (18 GHz → 337 μm). Precise, stabilized CO_2 laser difference frequencies have been measured,[22] and this heterodyne method has been applied to isotopically-substituted $^{13}CO_2$ and
$C^{18}O_2$ lasers as well.[23]   New developments in this area include very
stable,[24] optically-pumped alcohol lasers[25] in the terahertz region,
and extension of the high frequency limit to 150 THz (2 μm).[26]

    While these heterodyne frequency methods produce the highest
precision, they have at present a tendency toward complexity and
expense. It is important to note that these coherent laser frequency sources allow frequency mixing methods to be used, even if
the ultimate readout is based on wavelength -- rather than frequency -- metrology. For example, the wavelengths of the CO_2 laser were determined at NRC by measuring them in the visible after
frequency-up-conversion in a non-linear crystal.[27]   The wavelength
of the 633 nm optical "carrier" was simultaneously determined to
allow extraction of the infrared wavelength. Later the wavelength
of CH_4 was also determined with related techniques.[28]   Very precise
measurements of the 9.3 μm CO_2 line have been made at NPL using the
up-conversion method along with frequency-controlled interferometry.[29]   Such frequency-offset interferometry was also used at NBS.[30]

    Table I displays the available data to be used in determining
the speed of light by the direct   c = f·λ   method. We may make the
following observations:

1) Only the CH_4- and CO_2-stabilized lasers have had both precise
    frequency and wavelength measurements. There is excellent
    agreement between NPL[31] and NBS[20] on the CH_4 frequency, $\Delta\nu/\nu$ =
    $(2\pm6)\times10^{-10}$ and on the CO_2 R(12) frequency, $(2\pm7)\times10^{-10}$.

2) At present the best available value of λ_{CO_2} is the NPL determi-
    nation;[29] the stated error is $\pm2.5\times10^{-9}$. This value is in
    agreement with and about 10 times more precise than the ear-
    lier NRC determination.[27]

3) None of the values of λ_{CH_4} submitted to the CCDM (73) for its
    consideration, (after rationalization to the same krypton con-
    vention) lay beyond 3×10^{-9} from the mean value.[14]   Probably the
    mean value is the most reliable value.

4) An ingenious interferometric scheme allowed measurement of the
    frequency of the 0.633 laser via microwave sideband generation.[32]
    Although the available accuracy was limited to ~6×10^{-8}, it is
    included in Table I as being the only determination of c with
    the visible laser.

Table I. The Speed of Light via the $c = f \cdot \lambda$ Method

Expt	λ (pm)	f (MHz)	c (m/sec)
Froome[a]	4 mm	2 × 36 000	299 792 500 ±100
HeNe[b]	632 991.47 ±0.01	473 613 166. ±29.	299 792 468 ∓018
HeNe:CH_4 P(7)	3 392 231.400[c] ±0.013	88 376 181.627[d] ±0.050	299 792 458.3[e] ±1.2
CO_2:CO_2 R(12)	9 317 246.348[f] ±0.041	32 176 079.482[g] ±0.020	299 792 459.0 ±1.2

[a] see ref. 1,2. [b] see text and ref. 32.

[c] CCDM recommended value and uncertainty; see ref. 15. [d] see ref. 20

[e] CCDM recommended value,.....458. [f] see ref. 29. [g] see ref. 31.

The CCDM (73) notes that some uses of the numerical value of c, notably in astronomy and in lunar ranging, have intrinsic experimental precision capability surpassing the 4×10^{-9} indeterminacy of the present Meter. They expressed the hope that future developments would allow the number c = 299 792 458 m/sec to be preserved (with the expected very small adjustments to be absorbed into the Meter when the definition is refined[14]).

And what about a new definition of the Meter? Unfortunately the known stabilized lasers tend to be complementary in such factors as cost, reliability, precision, ease of use, wavelength diversity, etc.: the "ideal" candidate for a future standard is perhaps not yet known. Thus the CCDM felt it was not yet appropriate to discuss a redefinition of the Meter. Still, the iodine- and methane-stabilized laser wavelengths had been interferometrically intercompared with at least one decade higher precision than feasible in a krypton comparison, so the CCDM could comfortably recommend wavelength values for these two stabilized lasers as secondary wavelength standards. Thus at present we can best preserve the precision and potential accuracy of a measured physical quantity (containing dimensions of length) by referencing the measurement to a suitable stabilized laser for which the CCDM has recommended wavelength values. In still higher precision work, the reference laser's frequency can be measured and dimensionally converted using a conventional value for the speed of light, namely the one recommended by CCDM (1973). Clearly it would be a valuable contribution to identify and develop a metrologically more ideal laser: visible, multicolor, reliable and superstable.

It is not too hard to name a few laser/absorber systems that

might be steps toward the ideal laser for metrology: careful evaluation of their performance potential is quite another matter.
For example, some very interesting results have been obtained with
iodine and Ar^+ lasers. Both molecular beams[33] and cell techniques[34]
have been used. Selenium ion lasers, with so many visible laser
lines, had seemed very promising: unfortunately the plasma noise
problems have not yielded as readily as hoped.[35] Use of an improved
hollow cathode geometry has resulted in 5-color "stable" operation
of the $He:Cd^+$ laser.[36] Another interesting, multicolor laser is
the He:I system with 9 lines between 541 and 880 nm.[37] Certainly
stabilized dye lasers[38] will be attractive in spite of the obvious
problems of complexity, expense and absorption-line identification.
As for the absorber molecule, experience suggests that I_2 will be a
good candidate. Since its spectrum is ideally rich in the visible,
an overlap within a gas laser's tuning range is highly probable.
However for the dye-laser-excitation case, the I_2 spectral richness
will necessitate some kind of <<guide Michelin>>. Perhaps dyelaser pumped I_2 lasers will be possible on a cw basis.[39] If transverse pumping would work, very nice narrow Lamb dips should result.

    The other natural direction is the exploration of the funda-
mental limitations of optical frequency standards. Working with
the methane line at 0.88×10^{14} Hz (3.39 μm), we have recently
achieved an optical resolution better than 1.2 kHz.[40] This value
is sufficient to clearly resolve the recoil-induced spectral doub-
ling,[41],[42] $\Delta\nu/\nu = h\nu/Mc^2$ (=2.163 kHz). Study of the recoil doublet
height ratio vs. pressure confirms the expected[43] pressure-induced
shift of intensity into the lower frequency peak. Our preliminary
result for the intensity ratio $R(P) \equiv I(L)/I(H)$ as a function of
pressure P (in μTorr) is R(0) = 0.98±0.01, R(88±20) = 1.0 and
R(∞) = 1.05±0.04. There will thus be a "magic" pressure near 1
mTorr which will be interesting for experiments in which the recoil
splitting is unresolved: it corresponds to cancellation of the intrinsic pressure-induced blue frequency shift by changing the intensity partition between the two peaks. Since collisional selectivity effects are also occurring,[44],[45] further work will be needed
to make this point clear. Such a regime may be especially interesting for a "transfer-standard" laser stabilized to the CH_4 Coriolis
"E"-line since it is free of hyperfine structure and shifts.[42]

    The high resolution data from our 32 cm by 13 m absorption cell
are not at present being obtained with anything like an optimal
scanning program. Even so, the implied frequency standards possibilities are quite encouraging: Knowledge of the recoil doublet height
ratio to ~1% is obtained in ~300 sec signal averaging time and corresponds to a frequency resetting capability of about 3×10^{-13}, with
random noise presumed the major contribution at this level. For
comparison, the accuracy of the Cs primary frequency standard is
about 1×10^{-13}.

    A new lineshape theory[46] appropriate to the low-pressure,

free-flight regime of standards interest has been developed and applied[47] to the problem of a priori correction of the second-order
Doppler shift. An accuracy of 5–10% of the basic 2×10^{-12} effect is
thought to be achieved. Thus an accuracy capability $\sim 2\times10^{-13}$ should
be attainable with the present techniques. Dramatic progress appears possible with new systems now being analyzed. Thus granting
future progress in infrared frequency synthesis, some unconven-
tional future opportunities and choices may develop.

References

1. K.D. Froome, Proc. Roy. Soc. Lond. A 247, 109 (1958).

2. Confirming experiments as well as earlier results are discussed
   in K.D. Froome and L. Essen, The Velocity of Light and Radio
   Waves (Academic, New York, 1969).

3. J.L. Hall, R.L. Barger, P.L. Bender, H.S. Boyne, J.E. Faller,
   and J. Ward, report to 1968 URSI Laser Measurement Conference
   (Warsaw), in Electron Technology (Warsaw) 2, 53 (1969).

4. V.P. Chebotayev, Radiotech. i Electron. 11, 1712 (1966).

5. J.L. Hall and W.W. Morey, Appl. Phys. Lett. 10, 152 (1967).

6. J.L. Hall, IEEE J. Quant. Electron. QE4, 638 (1968).

7. P.H. Lee and M.L. Skolnick, Appl. Phys. Lett. 10, 303 (1967);
   and V.N. Lisitsyn and V.P. Chebotayev, JETP 27, 227 (1968).

8. R.L. Barger and J.L. Hall, Phys. Rev. Lett. 22, 4 (1969).

9. G.R. Hanes and C.E. Dahlstrom, Appl. Phys. Lett. 14, 362 (1969).

10. S.N. Bagaev, Yu.D. Kolomnikov, V.N. Lisitsyn, and V.P.
    Chebotayev, IEEE J. Quant. Electron. QE4, 868 (1968).

11. CH_4: J.L. Hall, in Fundamental and Applied Laser Physics, pro-
    ceedings of Esfahan Symposium, August 1971, M.S. Feld, A.
    Javan, and N. Kurnitt, Eds. (Wiley, New York, 1973), p. 466ff;
    S.N. Bagaev, E.V. Baklanov, and V.P. Chebotayev, JETP Lett.
    16, 243 (1972); M. Ohi, Y. Akimoto and T. Tako, "Methane-
    stabilized He-Ne lasers with the method of saturated absorp-
    tion," CCDM (1973) Appendix M-10.

12. I_2: G.R. Hanes, K.M. Baird, and J. DeRemigis, Appl. Opt. 12,
    1600 (1973); W.G. Schweitzer, E.G. Kessler, Jr., R.D. Deslattes,
    H.P. Layer, and J.R. Whetstone, Appl. Opt. 12, 2927 (1973);
    J. Helmcke and F. Bayer-Helms, Metrologia 10, 69 (1974);
    W.R.C. Rowley and A.J. Wallard, "Performance studies of He-Ne
    lasers stabilized by $^{127}I_2$ saturated absorption," CCDM (1973)
    Appendix M5; K. Tanaka, T. Sakurai and T. Kurosawa, "Study of
    stabilization of He-Ne lasers using saturated absorption in
    Iodine $^{127}I_2$," CCDM (1973) Appendix M-11.

13. A. Brillet, P. Cerez, and H. Clergeot, IEEE J. Quant. Electron.
    QE-10, 526 (1974).

14.   J. Terrien, Metrologia 10, 9 (1974).

15.   J. Terrien, Nouv. Revue Opt. 4, 215 (1973).

16.   L.O. Hocker, A. Javan, D. Ramachandra Rao, L. Frenkel, and
      T. Sullivan, Appl. Phys. Lett. 10, 5 (1967).

17.   L.O. Hocker, J.G. Small, and A. Javan, Phys. Lett. 29A, 321
      (1969).

18.   K.M. Evenson, J.S. Wells, L.M. Matarrese, and L.B. Elwell,
      Appl. Phys. Lett. 16, 159 (1970); T.G. Blaney, C.C. Bradley,
      G.J. Edwards, and D.J.E. Knight, Phys. Lett. 43A, 471 (1973).

19.   K.M. Evenson, J.S. Wells, and L.M. Matarrese, Appl. Phys. Lett.
      16, 251 (1970).

20.   K.M. Evenson, J.S. Wells, F.R. Petersen, B.L. Danielson, and
      G.W. Day, Appl. Phys. Lett. 22, 192 (1973).

21.   D.G. McDonald, A.S. Risley, J.D. Cupp, K.M. Evenson, and
      J.R. Ashley, Appl. Phys. Lett. 20, 296 (1972).

22.   F.R. Petersen, D.G. McDonald, J.D. Cupp, and B.L. Danielson,
      Phys. Rev. Lett. 31, 573 (1973); and in Laser Spectroscopy,
      R.G. Brewer and A. Mooradian, eds. (Plenum, New York, 1974),
      p. 555.

23.   A.H.M. Ross, R.S. Eng, H. Kildal, Opt. Commun. 12, 433 (1974).

24.   Beat spectral widths well under 1 kHz were readily obtained,
      limited mainly by vibrations. D.A. Jennings, private comm.

25.   A list of ~100 IR laser lines and references to earlier work
      may be found in H.E. Radford, IEEE J. Quant. Electron. QE11,
      213 (1975).

26.   D.A. Jennings, F.R. Petersen, and K.M. Evenson, Appl. Phys.
      Lett. 26, 510 (1975).

27.   K.M. Baird, H.D. Riccius, and K.J. Siemsen, Opt. Commun. 6, 91
      (1972).

28.   K.M. Baird, D.S. Smith and W.E. Berger, Opt. Commun. 7, 107
      (1973).

29.   B.W. Jolliffe, W.R.C. Rowley, K.C. Shotton, A.J. Wallard, and
      P.T. Woods, Nature 251, 46 (1974). In Table I a quadrature ad-
      dition of 4×10^{-9} has been made to their errors since they did
      not consider the uncertainty of the meter definition.

30.   R.L. Barger and J.L. Hall, Appl. Phys. Lett. 22, 196 (1973).

31.   T.G. Blaney, C.C. Bradley, G.J. Edwards, B.W. Jolliffe, D.J.E.
      Knight, W.R.C. Rowley, K.C. Shotton and P.T. Woods, Nature 251,
      46 (1974). I am indebted to Dr. D.J.E. Knight of NPL for the
      CH_4 result quoted in his preprint "Laser Frequency Measurement
      and the Speed of Light."

32. Z. Bay, G.G. Luther, and J.A. White, Phys. Rev. Lett. $\underline{29}$, 189 (1972).

33. L.A. Hackel, D.G. Youmans, and S. Ezekial, Proc., 27th Symposium on Frequency Control, Fort Monmouth, N.J. (1973).

34. F. Bayer-Helms, private communication.

35. A.E. Siegman, private communication.

36. K. Fuji, T. Takahashi, and Y. Asami, IEEE J. Quant. Electron. $\underline{QE-11}$, 111 (1975).

37. G. Collins, private communication.

38. R.L. Barger, J.B. West and T.C. English, Appl. Phys. Lett., to appear July 1 (1975).

39. R.L. Byer, R.L. Herbst, and M. Kildal, Appl. Phys. Lett. $\underline{20}$, 463 (1972).

40. J.L. Hall, K. Uehara and Ch. Bordé, Bull. Am. Phys. Soc. Series II, $\underline{19}$, 448 (1974), and to be published.

41. A.P. Kol'chenko. S.G. Rautian, and R.I. Sokolovskii, JETP $\underline{28}$, 986 (1969).

42. Ch. Bordé and J.L. Hall in Laser Spectroscopy, R.G. Brewer and A. Mooradian, eds. (Plenum, New York, 1974), p. 125ff.

43. The upper level decay rate includes spontaneous emission as well as transit broadening. It is satisfying that the height ratio/pressure broadening data even lead to a reasonable differential, 20 Hz HWHM basis. Also we deduce that the lower level pressure broadens about 5% faster than the upper level, perhaps due to the ~10-fold wider splittings of the upper level as induced by the Coriolis coupling.

44. J.L. Hall, in Lectures in Theoretical Physics, 1969, ed. by W.E. Brittin and K.T. Mahathappa (Gordon & Breach, New York, 1974), and Sixth International Conference in Electronic and Atomic Collisions: Abstracts of Papers (MIT Press, Cambridge, Mass., 1969), pp. 994-6. This effect has brought the pressure-broadening rate down to 4.2 Hz/μTorr in these high resolution experiments (HWHM basis).

45. S.N. Bagaev, E.V. Baklanov, and V.P. Chebotayev, JETP Lett. $\underline{16}$, 243 (1972).

46. Ch. Bordé, D.G. Hummer, C.V. Kunasz, and J.L. Hall "Saturated Absorption Line Shape I: Calculation of the Transit Time Broadening by a Perturbation Approach," in preparation.

47. J.L. Hall, Ch. Bordé, and C.V. Kunasz, Bull. Am. Phys. Soc. Series II $\underline{19}$, 448 (1974).

THE REALIZATION OF THE SECOND

Helmut Hellwig, David W. Allan, Stephen Jarvis, Jr.,
David J. Glaze
National Bureau of Standards, Frequency & Time Standards
Section
Boulder, Colorado USA

INTRODUCTION

A primary cesium beam frequency standard serves to realize the unit of time, the Second, in accordance with the international definition as formulated at the XIII General Conference of Weights and Measures in 1967. The basic design of a cesium standard is shown in Fig. 1. The cesium beam emerges from an oven into a vacuum, passes a first state selecting magnet, traverses a Ramsey type cavity where it interacts with a microwave signal derived from a slave oscillator. The microwave signal changes the distribution of states in the atomic beam which is then analyzed and detected by means of the second state selector magnet and the atom detector. The detector signal is used in a feedback loop to automatically keep the slave oscillator tuned. The line-Q is determined by the interaction time between the atoms and the microwave cavity. Thus a beam of slow atoms and a long cavity leads to a high line-Q. Commercial devices which for obvious reasons are restricted in total size have line-Q's of a few 10^7, whereas high performance laboratory standards with an overall device length of up to 6-m feature line-Q's of up to 3×10^8.

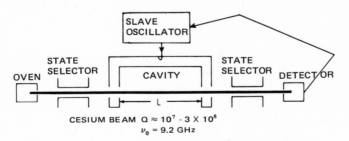

Fig. 1. *Schematic of a Cesium Beam Standard*

330

The National Bureau of Standards has two primary standards for the unit of time. They are both cesium devices and are designated NBS-4 and NBS-6. NBS-4 and NBS-5 (predecessor of NBS-6) have been used since January 1973 for a total of 22 calibrations of the NBS Atomic Time Scale. The independently evaluated accuracies (1) for NBS-4 and NBS-5 are 3.1 x 10^{-13} and 1.8 x 10^{-13}, respectively. NBS-5 was removed from service in February 1974. Major revisions in its oven/detector and vacuum system were carried out and the new system is now designated NBS-6. It has been operating since February 1975 and preliminary data indicate that the accuracy of NBS-6 will exceed the above quoted values.

ACCURACY OF A TIME SCALE

The accuracy of the rate (frequency) of an atomic time scale is the degree to which its Second agrees with the definition of the Sl-second. Primary frequency standards are used to calibrate the rate of an atomic time scale. If these calibration errors are in-dependent from one calibration to any other, and if the calibrations could be compared perfectly in time, then the error of the average would reduce as $n^{-\frac{1}{2}}$, where n is the number of calibrations. How-ever, the calibrations cannot be averaged perfectly: there exists no perfect reference; also, for a given primary standard and even for a set of primary standards the errors of one calibration may well be correlated (in space or time) with some other calibration A more general model for the errors involved in any given calibra-tion (say the ℓth calibration) may be written (2):

$$\sigma_s^2 (\ell) = \sigma_{ruc}^2 (\ell) + \sigma_{\overline{ruc}}^2 (\ell) \tag{1}$$

where $\sigma_s(\ell)$ is the overall accuracy for the calibration, $\sigma_{ruc}(\ell)$ is an estimate of the random uncorrelated errors and $\sigma_{\overline{ruc}}(\ell)$ is an esti-mate of errors that are correlated with some of the past calibrations or with some other primary standard due to similarity of design or evaluation procedure (the bar over "ruc" denotes the logical "not").

An accuracy algorithm can be developed which incorporates the contributions of a series of calibrations by a primary standard. Based on comparions via International Atomic Time (TAI) we find that NBS-4, NBS-5 and the two other operating primary standards at the Physikalisch Technische Bundesanstalt in Germany and the National Research Council in Canada agree within about 1 x 10^{-13}(3). Therefore we are assessing $\sigma_{\overline{ruc}}$ = 0.5 x 10^{-13} (2) and obtain through an accuracy algorithm the filtered calibrations against AT_0(NBS) shown in Fig. 2 (the lower dashed line in Fig. 2 connects the un-filtered original calibration points). The frequency AT_0(NBS) which is derived from an ensemble of commercial cesium standards has not been changed as a result of any of the listed calibration data, but has been maintained, as nearly as possible, as an independent continuing stable frequency reference. Also shown in Fig. 2 are the $\frac{1}{2}$ year average

frequencies of TAI with respect to the NBS best estimate of cesium
frequency obtained from the above mentioned accuracy algorithm. The
comparisons were made via Loran-C data taken at the end of February and
of August (BIH Circular D). The + 1.8 x 10^{-13}gravitational "blue
shift" at Boulder has been accounted for in the data. We would conclude, therefore, that the TAI second is too short by 8 ± 2 parts in
10^{13} at the beginning of 1975. In the fall of 1973 the NBS primary
standards and the German and Canadian primary standards agreed
that the frequency of TAI was about 10 x 10^{-13} too high, i.e., the
length of the TAI second was too short by this amount (3). It may be
anticipated that future, regular input from all available primary
standards will be incorporated into a TAI accuracy algorithm which
would keep the TAI second bounded to within at least 1 x 10^{-13} of the
best realization of the S1-Second. This would significantly decrease
the time errors in TAI with regard to ideal "absolute" time and
correspondingly increase the utility of TAI for fundamental physical
and astrophysical measurements of extended duration.

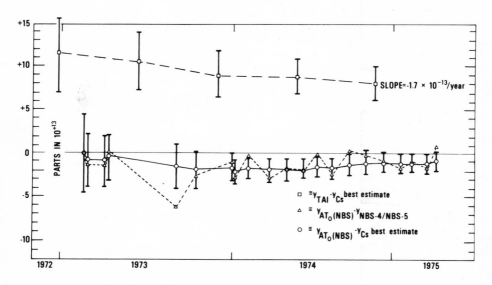

Fig. 2. Plot of the Rates of TAI and AT_0(NBS) With Respect to
the NBS Primary Standards from November 1973 to the Present

STABILITY LIMITATIONS

A prerequisite for achieving high accuracy is high stability.
The ultimate stability limitation of all frequency standards, including cesium, is properly described by a process with a flicker of
frequency spectral density (flicker floor). The physical causes for
this noise are not yet well understood; however, a better understanding and corresponding improvements in the flicker floor are a
prerequisite for significant increases in accuracy. The following
is an attempt to briefly sketch out likely physical mechanisms for
stability limitations.

(a) magnetic fields.

From the well known relationship (4) for the frequency shift
in a magnetic field H in cesium we calculate that a field change
ΔH = 0.18 μG causes a fractional frequency change $\Delta\nu/\nu = 1 \times 10^{-15}$
if a nominal C-field on H = 60 mG is applied in the standard. Be-
cause the applied C-field will always be substantially larger than
the residual magnetic fields in the shielded region, we need not
be concerned about magnetic field changes in directions other than
the applied C-field. In a controlled laboratory environment, we may
find changes of the environmental field of at most 10^{-3} G. Thus,in
order to assure 10^{-15} frequency stability we need shielding factors
of almost 10^4. This should not be difficult to achieve; however,this
is only a valid statement for beam tubes with C-fields applied in a
radial direction as in most tubes including all NBS tubes. If a
cavity configuration is chosen where the C-field must be applied in
the axial direction, the effectiveness of the shielding is reduced
in relation to the length/diameter ratio L/D. In long beam tubes
this could become critical since this geometric effect (5) reduces
the longitudinal shielding factor, for example, by more than a factor of 10 as compared to the shielding of the radial component in
the case of L/D = 10, and will approach zero for L/D → ∞.

Of special concern, beyond the nominal shielding factor, may
be the stability in time of the residual magnetization of the
shields. Little is known about such effects. We have some experimental evidence suggesting that time-varying mechanical stress may
cause field changes in NBS-6 (4 m length of the shield package).
We measure field changes of the order of a few hertz (a fraction of
a hertz in NBS-4) in the Zeeman frequency over days which is equivalent to a stability of about 10^{-14} in the clock frequency; this is
the flicker noise level observed in stability comparisons between
NBS-4 and NBS-6 and, before, between NBS-4 and NBS-5.

(b) oven temperature.

The oven temperature acts via the velocity dependent effects:
cavity phase shift caused by a phase difference δ between the two
sections of the cavity, and the second-order Doppler effect: The
frequency shift due to velocity dependent effects can be written:

$$\Delta\nu = - V_p \frac{\delta}{2\pi L} - V_D^2 \frac{\nu_0}{2c^2} \tag{2}$$

where L is the separation between the two cavity sections, c is the
speed of light, ν_0 is the cesium resonance frequency and V_p and V_D
are some mean velocities of the atomic beam (6). Generally, the beam
optics strongly influences the mean velocity. However, as a worst
condition we may assume that a temperature fluctuation δT of the cesium oven directly influences the mean velocity via

$$\frac{\delta V}{V} = \frac{\delta T}{2T} \tag{3}$$

The mean velocity in cesium tubes is of the order of 100 m/s and oven
temperatures of about 360°K are typical. The fundamental frequency
bias due to velocity dependent effects rarely exceed a value of 10^{-11}.
For these conditions we calculate a needed oven temperature stabil-

ity of slightly better than 0.1°K for 10^{-15} fractional frequency
stability. This is well within technical possibilities.
(c) cavity temperature variations.
     A temperature difference δT between the two cavity arms
causes a change in L of $\Delta L = \alpha\ L\ \Delta T$. α is the coefficient of thermal expansion. This causes a change in the cavity phase difference of

$$\delta(\delta) = \Delta L\ (\partial\delta/\partial\chi) \tag{4}$$

where $(\partial\delta/\partial\chi)$ is the maximum phase gradient across the cavity window.
and relates to the cavity Q ($\partial\delta/\partial\chi = 0$ in an ideal,lossless cavity).
Eq. 4 combined with the phase term in Eq. (2) yields

$$\delta\Delta\nu/\nu = (\alpha\ V_p/\ 2\pi\nu)(\partial\delta/\partial\chi)\ \delta T \tag{5}$$

which interestingly, is independent of L. We calculate for a copper
cavity, estimating $\partial\delta/\partial\chi < 3 \times 10^{-4}$ rad/cm (9); a needed temperature
difference stability of 0.5 degrees to assure a frequency stability
of 10^{-15}.
(d) microwave power.
     The effective mean velocities V_p and V_D in Eq. (2) are depen-
dent on the interrogating microwave power. This is because the trans-
sition probability has the following proportionality (6).

$$P \sim \sin^2 2\ b\tau \tag{6}$$

where b^2 is proportional to the microwave power, and τ is the transit
time for an atom in one cavity section. If the velocity distribution
is known, V_p and V_D can be calculated exactly (7). This, in fact, has
been used for the determination of δ in accuracy evaluations of cesium standards at the National Bureau of Standards (1). In the presence
of a non-vanishing cavity phase difference δ, fluctuations of the microwave power will cause frequency fluctuations. In order to estimate the sensitivity we simplify Eq. (2) by setting $V_p = V_D$ and obtain

$$\frac{\delta\Delta\nu}{\nu} = \left[\frac{V}{c^2} + \left(\frac{\Delta\ \nu}{\nu}\right)_P \frac{1}{V}\right] \delta V \tag{7}$$

where $(\Delta\nu/\nu)_P$ corresponds to the phase shift term in Eq. (2).
For V = 100 m/s and $(\Delta\nu/\nu)_P = 10^{-11}$ we obtain $\delta V \simeq 10^{-2}$ m/s in order
to limit $\delta\Delta\nu/\nu$ to 10^{-15}. This is a very stringent requirement; from
typical velocity distributions (1,6) it may be estimated that this
corresponds to a microwave power stability of about 10^{-2} dB. Of
course, smaller values for ($\Delta\nu/\nu)_P$ or δ are realizable and thus
relax this requirement.

FUNDAMENTAL ACCURACY LIMITATIONS

     Unfortunately, we cannot assume that δ is a unique value for a
given cavity and fully determined by its geometry. We have to consider that the cavity is lossy and that openings for passage of the
beam are of finite size. This leads to the concept of distributed
cavity phase variations which make δ dependent on the particular tra-

jectory location in the cavity (8). All known accuracy evaluation techniques for the determination of δ affect the atomic trajectories, whether it is beam reversal, power shifts (via the velocity dispersive nature of the beam optics), etc. (3). Therefore, a full determination of δ and the associated frequency bias is an elusive goal, especially since a modeling of the distributed phase variations in the cavity appears not to be practical. The effect obviously is reduced by the choice of long cavities; e.g., we estimated a fundamental limitation in the determination of δ for NBS-5, i.e., its accuracy, of somewhat better than an equivalent non-evaluable frequency bias of 1×10^{-13} (8). The question arises of how to increase further the available accuracy under this condition. If no radical departure from a conventional beam tube design is considered, then the following design recommendations could be made: Geometrically narrow beams with stable trajectories, (comp. next section for NBS-6) reduction of velocity dispersive character of the beam optics, and better controlled cavities with reduced electrical losses. Narrow cylindrical or sheet type beams could be obtained by hexapole or special dipole optics, respectively. A more fundamental approach appears possible: significantly slower atomic velocities. This would reduce all velocity dependent effects correspondingly and would have the added benefit of a spectrally narrower atomic resonance. This approach is experimentally a challenge; however, it appears to be the only one promising substantial improvement. If slow beams of adequate intensity could be realized, then accuracy values of much better than 10^{-13} appear possible. Cesium may or may not be the atom of choice in this case.

ACCURATE CLOCKS

It is not only desirable to increase the accuracy of the primary standard, but also to make use of this accuracy. The present requirement of interrupting the operation of the standard in order to evaluate its full accuracy is deficient in that it disallows the use of the standard's superior stability performance for the generation of time. Conversely, the use of the primary standard as a clock leads to a deterioration of its realized accuracy because some parameters affecting its frequency may slowly change, unnoticed.

An analysis of the effects which are most likely to be slowly time dependent and which have significant impact on the accuracy shows the following possibilities:

(1) microwave power.

The power could be monitored independent of the beam tube and corrected (by a servo) if necessary.

(2) magnetic field.

The field can be measured via the magnetic field dependent transitions ($m_F \neq 0$). A measurement during clock operation could be executed by interrogating symmetrically with respect to the $m_F = 0$ transition. The regular servo could easily discriminate against such additional (likely, much slower) modulation. This technique could be realized with no effect on the frequency other than a small deterioration of the shot-noise limited stability of the standard. Magnetic field inhomogeneities, however, may limit the usefulness of this approach.

(3) cavity phase difference.

Here, advantage may be taken of a spatial velocity dispersion of the beam optics which forces different velocities on different trajectories. In the deflection plane of dipole optics, we thus have a velocity scale, and different detector positions will detect different mean velocities. Two detectors installed at an offset from each other in the direction of deflection essentially examine two beams with different mean velocities. Simultaneous use of these detectors will give information on changes in the cavity phase difference. For example, one detector may be used for the main frequency servo, the other to generate an error signal which depends on δ (comp. Eq. (2)). This information could be used continually to apply bias corrections or to drive a servo system. Other uses are possible such as the creation of a mixed signal which is in first-order independent of δ. We have installed such a detector in NBS-6 in order to establish the feasibility of such a technique. NBS-6 with its on-line beam geometry making use of both atomic levels in two separate beams (1) would also allow the use of the dual detector to optimize trajectory symmetry and thus to evaluate more accurately the distributed cavity phase (see above). Thus, NBS-6 may surpass significantly the accuracy of NBS-5.

All of the above measures are no full substitute for a true accuracy evaluation; however, once such an evaluation is executed these techniques would maintain this accuracy in the operating standard; their use would allow a long-term clock operation of the primary standard with no sacrifice in the realized accuracy.

REFERENCES

(1) D. J. Glaze, et al., IEEE Trans. on I&M, Vol. IM-23,
       No. 4, pp. 489-501, December 1974.
(2) D. W. Allan, et al., Metrologia, 1975 (to be published).
(3) H. Hellwig, Proc. of the IEEE, Vol. 63, No. 2,
       pp. 212-229, February 1975.
(4) R. C. Mockler, Advances in Electronics & Electron Physics,
       Vol. 15, pp. 1-71, 1961.
(5) A. J. Mager, IEEE Trans. on MAG., Vol. MAG-6, p. 67-75 (1970).
(6) H. Hellwig, et al., Metrologia, pp. 107-112, September 1973.
(7) S. Jarvis, Jr., Metrologia, pp. 87-98, October 1974.
(8) R. F. Lacey, Proc. of the 22nd Annual Symp. on Freq. Contr.,
       pp. 545-558, April 1968.
(9) S. Jarvis, Jr., NBS Tech Note #660, January 1975.

STABILITY AND REPRODUCIBILITY OF He-Ne-LASERS STABILIZED BY SATURATED ABSORPTION IN IODINE

Jürgen Helmcke

Physikalisch-Technische Bundesanstalt

D-33 Braunschweig, Bundesallee 100, Fed.Rep. of Germany

1. INTRODUCTION

The iodine stabilized He-Ne laser is of particular interest as a wavelength standard in the visible. It is now possible to build transportable lasers and comparisons have been performed recently in the BIPM and NPL between lasers of the BIPM, NPL and PTB /1/. In our laboratory lasers of the two systems $^{129}I_2$, ^{22}Ne /2/ and $^{127}I_2$, ^{20}Ne /3/ have been built. Most of the work has been done on the system $^{129}I_2$, ^{22}Ne as there exists a characteristic triplet of strong saturation dips with peak sizes up to one percent of the output power.

2. EXPERIMENTAL SETUP

Fig. 1 shows a block diagram of the experimental setup. The laser heads have been designed identically for the two systems /5/ with following exceptions: the lengths of the iodine cells are 8 cm for $^{129}I_2$ and 12 cm for $^{127}I_2$. The corresponding resonator lengths are 30 cm and 35 cm, respectively. The temperature of the cooling fingers can be servocontrolled within $\pm$ 0,2 $^{\circ}$C in the range of 0 $^{\circ}$C to 20 $^{\circ}$C. Furthermore this temperature is monitored using an additional thermocouple. Commercial cold cathode laser tubes with special gas fillings are used. The excitation at 3,39 μm is suppressed by means of a glass plate which is cemented to one of the brewster windows.

The frequency of the lasers is stabilized in the usual way to the center of the saturation dip using the zero crossing of the third harmonic of the modulation frequency /4/. The contents of higher harmonics of the modulation frequency generator is

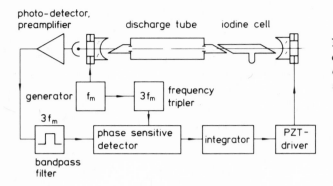

Fig. 1: Block
diagram of the
experimental
setup.

−70dB. The corresponding distortions in the frequency modulation
are measured to be smaller than −65dB in the case of 5 MHz$_{pp}$ scan.

3. FREQUENCY STABILITY

The stability of the lasers was investigated by observing the
beat frequency fluctuations of two identical lasers which are
stabilized independently to different absorption lines. The beat
frequency was measured by means of a computing counter which can
be programmed to calculate the Allan-variance. Fig. 2 shows the
measured fractional frequency fluctuations $\sigma(2,\tau)$ in dependence
of the counting time for the different systems $^{129}I_2$, ^{22}Ne and
$^{127}I_2$, ^{20}Ne. The measured curves follow in both cases with good
approximation the $\tau^{-1/2}$-dependence at integration times between
1s and 1000s. They go down to about 2.10^{-13} within the measured
range at 1000s integration time. The better short term stability

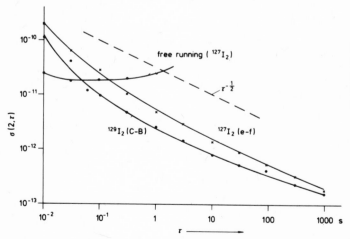

Fig.2: Fractional
frequency fluctua-
tions $\sigma(2,\tau)$ in
dependence of the
counting time τ.

of $^{129}I_2$, ^{22}Ne as compared to the $^{127}I_2$, ^{20}Ne-system can be explained by the larger peak size of the $^{129}I_2$-saturation dips.

4. REPRODUCIBILITY

The reproducibility of the lasers was investigated by varying several parameters and observing the corresponding frequency shifts. This can easily be performed by measuring the beat frequency changes of two stabilized lasers (one fixed, the other varied). Until now, only the reproducibility of the $^{129}I_2$-stabilized lasers has been investigated in some detail.

The frequency offsets produced by the electronic circuit are estimated to be less than about 5 kHz: No significant frequency changes were observed if the ac- or the dc-gain of the phase sensitive detector were varied by a factor of ten or if the laser radiation on the photodetector was attenuated by the same order of magnitude. The frequency offsets produced by a zero offset of the integrator can be examined in the following way: The "rise time" τ_1 of the integrator output voltage with no modulation at the laser is compared to "rise time" τ_2 for a known zero offset at the integrator. The frequency offset produced by the integrator follows as the product of the ratio τ_2/τ_1 and the frequency offset produced by the known zero offset. For our lasers this offset is less than 3 kHz. Furthermore all of the electronic components, including the frequency generator, frequency tripler, and the phase sensitive detector have been exchanged successively and no significant frequency shift was obtained.

The influence of the laser power was investigated by slightly mis-aligning the laser tube and by using various laser mirrors with different reflectivities. Part of this work was done together with Dr. Wallard during the laser comparison at the NPL /1/.

We have found three small additional saturation dips near the $^{129}I_2$ components A, B and C at high power levels inside the resonator/6/.Two of them are clear to see in the upper curve of Fig. 3. Furthermore with increasing power there occurs some irregularity in the wing of component A confirmed by repeated recordings. The new dips nearly vanish at low power levels as can be seen in the lower curve of Fig. 3. The origin of these lines is not known till now.

Frequency shifts depending on radiation power inside the laser and modulation amplitude have been investigated. The power has been varied by misaligning the laser tube and replacement of mirrors with different reflectivities (99,6 %, 99,2 %,and 98 %). At high power level and with high modulation amplitude the frequency shift can be as high as 100 kHz for a change in power by a factor of three. For small modulation amplitudes and low power

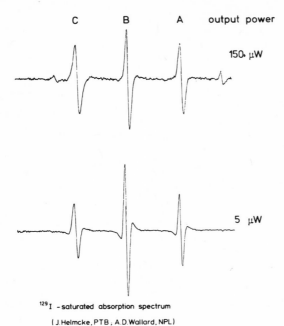

Fig.3: Saturation
spectrum of ^{129}I
for two different
power levels.

inside the cavity,however, this dependence is much less. Fig. 4
shows a typical dependence, observed for a small modulation
amplitude of 1,5 MHz$_{pp}$ and for mirror reflectivities of 99,2 %
and 98 %, respectively. In this example the shift is only 20 kHz
for a power change from 20 μW to 130 μW at the high reflecting
mirror end. The corresponding dependence on the modulation
amplitude is shown in Fig. 5 for different power levels as
parameters. From Fig. 5 is to follow that under the conditions of
reduced radiation power corresponding to less than about 100 μW
output power and reduced modulation amplitude, that means less
than about 5 MHz$_{pp}$, the extreme frequency spread is $\pm$ 15 kHz
corresponding to $\pm$ 3.10^{-11} λ. The iodine pressure was 13 N/m^2
at cooling finger temperature of 11,7 $^{\circ}$C. Frequency differencies
of two ^{129}I$_2$-lasers stabilized under reduced conditions e.g.
modulation amplitude 2 MHz$_{pp}$ were always less than 15 kHz,
which is consistent with the values given in Fig. 5.

The dependence of the frequency f on the iodine pressure
was measured under the conditions of reduced power and of a
small modulation amplitude of 1.5MHz$_{pp}$. For this purpose the
temperature of the cooling finger was varied in the range from
7°C to 17°C. The frequencies decrease linearly with pressure to

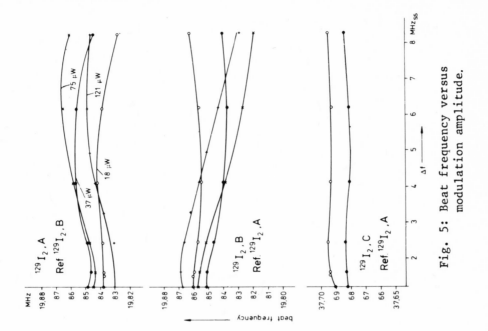

Fig. 5: Beat frequency versus modulation amplitude.

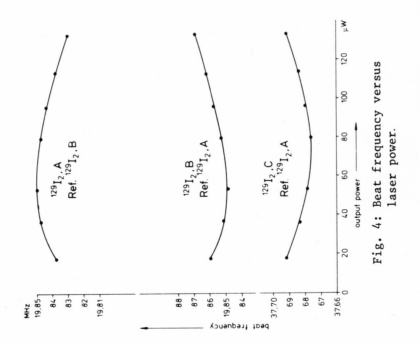

Fig. 4: Beat frequency versus laser power.

describe by the coefficients $-6,02$KHz/Nm^{-2} (-800KHz/Torr) for the components ^{129}I$_2$,A and B and $-6,54$KHz/Nm^{-2} (-870KHz/Torr) for the component ^{2,129}I$_2$,C$_9$ These results are in good agreement to those obtained for ^{2,129}I$_2$,k by Schweitzer et al. /2/.

For a given iodine pressure the frequency variations of the used iodine stabilized lasers by a certain power dependence combined with a modulation dependence are limited to $\pm 3 \cdot 10^{-11}$f. Taking into account possible further systematic effects which might shift the laser frequency, the reproducibility of our lasers operated under the discussed conditions may be estimated to be within $\pm 5 \cdot 10^{-11}$f.

REFERENCES

1. CHARTIER, J.-M.; HELMCKE, J.; WALLARD, A.D.: (to be published)

2. SCHWEITZER, W.G., Jr.; KESSLER, R.D., Jr.; DESLATTES, R.D.; LAYER, H.P.; WHETSTONE, J.R.: Appl. Opt. 12, 2927 (1973)

3. HANES, G.R.; BAIRD, K.M.; DEREMIGIS,J.; Appl. Opt. 12, 1600 (1973)

4. WALLARD, A.D.: J.Phys. E 5, 927 (1972)

5. HELMCKE,J.; BAYER-HELMS, F.: IEEE Trans. IM -23, 592 (1974)

6. HELMCKE, J.; WALLARD, A.D.:(to be published)

PERFORMANCE AND LIMITATIONS OF IODINE STABILIZED LASERS

A.J. WALLARD

Division of Quantum Metrology

National Physical Laboratory, Teddington, Middx., England

INTRODUCTION

The $^{127}I_2$ stabilized laser has now reached the stage of development where a reproducibility of better than 20 kHz ($\Delta\nu/\nu = 4 \times 10^{-11}$) may consistently be achieved between lasers which operate at the same iodine temperature. The purpose of this paper is to report the present reproducibility and performance of the NPL lasers, to suggest the likely limitations of the present systems, and to indicate possibilities for further improvement. Both 127 and 129 iodine isotope lasers have been built using third harmonic (3f) servocontrols, which have high Q rejection filters for f and 2f signals and a low Q 3f band-pass filter between the photoamplifier detector and phase sensitive detector (psd). This arrangement avoids problems in the electronic servosystem due to the large fundamental and second harmonic components which are present in the amplifier signal and which might become rectified in the psd to give rise to a frequency offset (1). The 127 lasers operate with an iodine temperature of 18 $\pm$ 0.05 °C which results in maximum height saturated absorption features on the laser power output for these lasers. At temperatures both above and below this, the peak size either decreases or remains constant.

$^{127}I_2$ STABILIZED LASERS

During the past year, the sensitivity of the stabilized frequency to electronic shifts (3f distortions in the frequency tripler, psd shifts due to zero offsets and f and 2f signals, and the variation and control of integrator settings for zero input signal) has probably been reduced to less than 5 kHz and the present reproducibility is limited by the remaining electronic shifts, nonlinearities in the PZT length modulator and uncertainties in the control and measurement of the iodine pressure.

Reproducibility has, as before (2) been measured by beat frequency experiments in which two lasers are locked firstly to two nearby although not usually adjacent peaks in the defg group. The lasers are then reversed by exchanging the peak to which each is locked. Half the difference between the averaged beat frequencies which are measured in the two cases is taken to be the reproducibility. Recently, 30 such measurements, each over 100s have been made in this way and show an average reproducibility of 6.0 kHz about the mean of each measurement. The spread of the measured values for a component separation with the two lasers always locked to the same two components had a standard deviation of 7 kHz and the distribution of the means of the reversals had a standard deviation of 3.7 kHz. Two other independent estimates of the reproducibility of these lasers are available from recent international comparisons between the NPL and the PTB (November 1974) and the BIPM (May 1975). During both these comparisons, two NPL lasers were available for beat measurements against the laser from the other laboratory. During the PTB measurements during which beats with both NPL lasers were made simultaneously with the PTB $^{129}I_2$ laser, 88 measurements under a variety of conditions gave a mean frequency difference between the two NPL lasers of 10.9 kHz with a spread which had a standard deviation of 8 kHz. During the BIPM intercomparisons in which individual sets of results for the BIPM laser and one of the two NPL lasers were made at different times, a reproducibility of 5.1 kHz was found with a standard deviation of 6.1 kHz from 25 measurements. This figure, however, includes day to day variations in the BIPM laser reproducibility.

It appears, then, that a reproducibility which approaches $1x10^{-11}$ may be achieved if the limiting frequency shifts are carefully studied. To make such studies, a frequency stability of better than $1x10^{-11}$ is desirable over convenient integration times of about 10s. The present stability of the beat between two lasers is shown in Figure 1 and the flattening and rise of the curve is probably due to slow electronic shifts and fluctuations in iodine pressure (a stability of 0.05 °C is required for the 1 kHz stability around 18 °C). A new, more rigid cavity is now available which should improve the short term stability and enable systematic studies of frequency shifts at the $1x10^{-11}$ level to be made. These new designs should also give information about the factors which at present limit the stability to some 400 Hz at times of 800s.

$^{129}I_2$ STABILIZED LASERS

This type of laser has recently been studied following the visit of Dr Helmcke of the PTB last year during which power shifts of the laser frequency were first observed. These shifts are probably due to the effects of new saturated absorption features which appear around the ABC triplet at high laser powers. These features, we suspect, are attributable to a mixed 127-129 iodine molecule, which becomes more important at the higher iodine operating pressures. The lasers used to study these features had narrow bore ^{22}Ne gain tubes and high reflectivity mirrors to obtain high intracavity laser powers. The iodine cells were filled at the PTB and were operated at about 11 °C. Figure 2 shows the third derivative spectrum in the region of the ABC triplet for three different powers and shows the clear emergence of the new features to the low frequency side of component A. The separation of A from the peak nearest to A is 15.1 MHz. At lower iodine temperatures, the appearance and effect of the new peaks is far less marked and, in particular, for the NBS $^{129}I_2$ laser, (3) which operates with a high mirror transmission and at an iodine temperature of 2 °C such spectra are difficult to observe. Figure 3 shows the A-e separation as a function both of laser power and scan amplitude for the A peak in $^{129}I_2$. It seems likely that this power shift, which has not been observed with $^{127}I_2$ lasers for a similar power variation, is closely linked with the appearance of the new structure, although the clearly visible peaks are some 15 MHz away. With both the PTB and NPL lasers, relative shape and height variations of the ABC peaks have also been observed and this suggests the possibility of additional structure within them. However, no careful experiments have been made to investigate this, nor have the effects of different purity iodine samples or cell filling techniques been studied. It may also still be possible to exploit the inherent appeal of the high signal to noise ABC triplet in a low gain, low temperature system although probably not with the same degree of reproducibility as the $^{127}I_2$ lasers.

INTERNATIONAL COMPARISONS

Some of the results of various international comparisons have been mentioned already. The recent series of experiments using 127 lasers between the BIPM and NPL has not yet been carefully analysed, but the average reproducibility over 96 individual measurements between the lasers was 6.8 kHz with a standard deviation of 4 kHz. The typical standard deviation of the mean of a component reversal experiment was about 4 kHz, and a maximum of eight component reversals was made for each separation. For the 129 laser studies, a direct comparison of the results of PTB and BIPM on the same laser is difficult because of the power shift of some 50 kHz. However, for the low power situation in which the laser usually worked and after a correction for different iodine pressures, the PTB $^{129}I_2$ laser may be compared with the 129 laser donated by the NBS to the BIPM. Under these conditions, the reproducibility of these lasers then also falls below 20 kHz. Similarly, measurements made at the BIPM and NPL with 127 lasers from each laboratory against the NBS–BIPM 129 laser also agree, as would be expected, at this level.

REFERENCES

1.   K C Shotton and W R C Rowley.   NPL Quantum Metrology Report Number 28.

2.   A J Wallard.   IEEE Trans Inst. Meas. IM23 532–535, 1974.

3.   W G Schweitzer, E G Kessler, R D Deslattes, H P Layer, J R Whetstone.   Appl. Opt. 12 2927–2938, 1973.

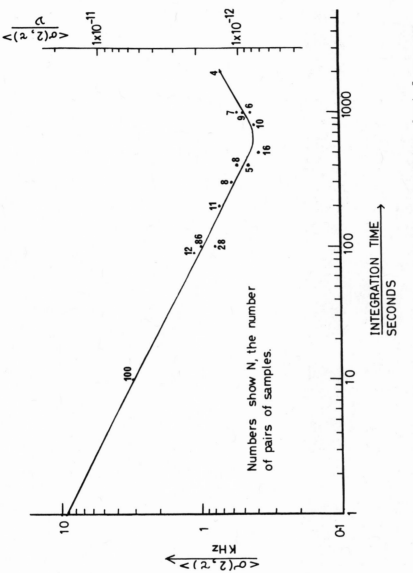

Fig. 1. Allan variance and relative frequency stability, beat frequency
between two $^{127}I_2$ lasers.

Frequency

A B C

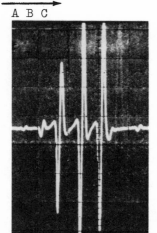

135 μW

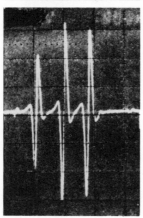

75 μW

Figure 2

Third derivative
spectra showing
new features to
the low frequency
side of A. Scale is
20 MHz/division and
the mirror
transmission is 99.8%

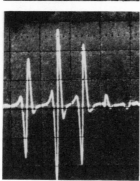

10 μW

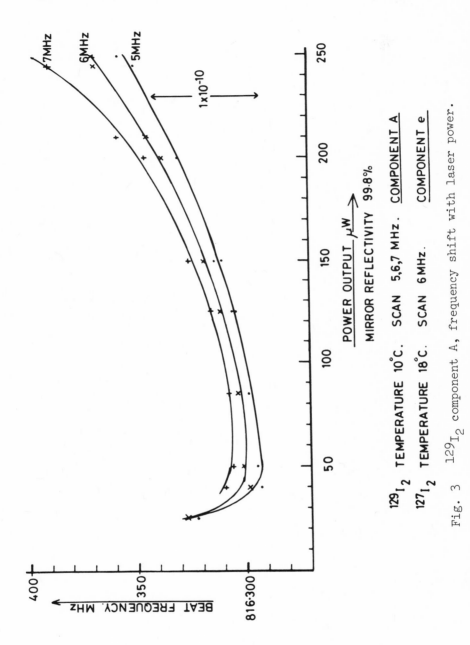

Fig. 3 129_{I_2} component A, frequency shift with laser power.

REPRODUCIBILITY OF METHANE- AND IODINE-STABILIZED OPTICAL FREQUENCY STANDARDS[x]

A. BRILLET, P. CEREZ, S. HAJDUKOVIC[xx] and F. HARTMANN

Laboratoire de l'Horloge Atomique

91405 ORSAY, France

INTRODUCTION

Helium-neon lasers stabilized to the saturated absorption peaks of CH_4 (3.39 µm) or I_2 (0.633µm)exhibit a very high frequency stability, reaching 5 parts in 10^{14} for the former and 8 parts in 10^{13} for the latter over observation times of a few tens of a second[1]. Their reproducibility, on the other hand, is presently only about one part in 10^{11}. Although this value is quite adequate for the contemplated use of these devices as length standards, it is far poorer than the reproducibility of the cesium beam primary frequency standard (10^{-13}). Several laboratories have thus been led to undertake a systematic study of all possible limitations and associated physical problems. The present paper reports our recent results in this direction.

IODINE-STABILIZED HELIUM-NEON LASER

As is well known[2], the R(127) line of $^{127}I_2$ presents 21 hyperfine components, 10 of which (customarily labeled a,b,...i,j)[3,4] falling within the Doppler emission profile of commercial 0.633 µm He-Ne lasers. In our short, internal cell devices, the defg quartet appears at the top of this profile whereas the abc and hij triplets appear on the edges. This last situation is undesirable since it results in a very high sensitivity to electronic distorsions within the servo loop[5], a typical value being a 24 kHz (5×10^{-11}) frequency offset for 0.1 percent third harmo-

(x) Work supported by BNM and CNRS
(xx) Permanent address : Institute "Boris Kidric",Belgrade,Yugoslavia

nic distortion. In order to achieve a good reproducibility, use of the triplet lines is thus to be avoided.

A second factor influencing the choice of a peak is the modulation amplitude effect due to the wings of neighboring lines [4]. It results in a frequency "pushing" (with third-harmonic locking), which increases with the modulation amplitude. The d and g peaks are thus shifted by 3 kHz (6×10^{-12}) for the value of the frequency excursion (6 MHz peak to peak) giving rise to maximum signal. The e and f peaks, located approximately halfway between two neighbors whose "pushings" compensate each other, are much less sensitive to this effect (400 Hz, or $\simeq 10^{-12}$, in the same conditions) and are therefore to be preferred.

The principal remaining factors to be considered for the e and f lines are then the pressure shift (- 1.12 MHz/torr, or - 3×10^{-11}/°C around 20°C) [4] and electronic offsets of the PSD and integrator. We have thus been led to fix independently for each laser the temperature of the iodine cell cold point, using a platinum resistance thermometer, with a absolute accuracy of 5×10^{-2}°C. The effect of a DC offset, on the other hand, could be reduced to less than 1.6 kHz (3.5×10^{-12}). In these conditions, the frequency difference between two lasers never exceeded 4 kHz (8.4×10^{-12}), irrespective of cavity configuration, laser tube gain and mirror reflectivities, so that this figure can be considered as a good estimate for the overall reproducibility of our devices.

Although this value could of course be improved, let us note that the R(127) peaks of iodine are far from a best choice for a frequency reference because of their large width ($\simeq$ 3 MHZ FWHM) and the small value of their absorption coefficient ($\simeq 3 \times 10^{-5}$ $cm^{-1} \times torr^{-1}$) due itself to a sparsely populated lower state (5×10^{-7} of the total population in each hyperfine level) and a modest oscillator strength for the transition ($\simeq 3 \times 10^{-4}$). With the advent of tunable lasers and the need for an efficient frequency stabilization of these devices, much better lines will certainly be found.

METHANE-STABILIZED HELIUM-NEON LASER

The usual way for stabilizing the He-Ne laser at 3.39 μm is to make use of the $F_2^{(2)}$ line of methane at 2947.912 cm^{-1}, which is in close coincidence with the center of the Ne Doppler emission profile. Using a 60 cm cavity containing a low noise, DC excited gain tube and a 20 cm absorption cell filled with methane at a pressure of 10 millitorr, we observe a 200 kHz FWHM peak in the output power with 2 percent contrast which, when used as a reference, allowed us to reach a frequency stability of $2 \times 10^{-13}/\sqrt{\tau}$ where τ is the observation time[1]. The device is not optimized and stability figures around $10^{-15}/\sqrt{\tau}$ could certainly be obtained with a better matching of the Ne and CH_4 saturation parameters, but this is of little interest as long as reproducibility is not improved upon its present value of 10^{-11}.

It has been known for some time[6,7] that the main factor limiting reproducibility is the power shift (- 2 x 10^{-13} for a one percent increase of oscillation level). This is due to a differential saturation effect of the three main hyperfine components of the $F_2^{(2)}$ line [8] which, because of their close spacing ($\simeq$ 10 kHz) are not resolved in conventional devices. Rather than attempting to build a high resolution apparatus enabling observation of an isolated hyperfine component, we investigated the possibility of using the methane E line at 2947.811 cm^{-1}, which is free from hyperfine structure. The line can be excited by Zeeman tuning the He-Ne laser, the - 3 GHz offset being obtained on the $\sigma-$ component with a magnetic field of about 1800 gauss. We have built a 2 meters long laser cavity with a one meter internal absorption cell (methane pressure 10 millitorr) and a one meter gain tube filled with a high pressure (5 torr) of the He-Ne mixture in order to increase the homogeneous linewidth and to ensure single mode operation. The transverse magnetic field was created by 5 permanent magnets which could be translated for tuning. The gain and absorption cells had Brewster windows oriented as to select the σ Zeeman components ; however, this did not completely prevent the unshifted π component from oscillating. As a result, the $F_2^{(2)}$ and E lines could be observed simultaneously but, due to strong coupling and hysteresis effects, no accurate measurement of the $F_2^{(2)}$ - E frequency interval could be obtained in these conditions. We have therefore inserted within the cavity a 10 cm cell filled with 30 torr of methyl bromide, which presents two absorption lines in the vicinity of 2947.912 and 2948.012 cm^{-1} [9] ; the resulting 20 or 30 percent absorption prevents the π and $\sigma+$ components from oscillating without affecting the $\sigma-$ component. The laser output power was then 600 µW and exhibited a 200 kHz FWHM peak with one percent contrast ; on the other hand, due to the detrimental effect of the magnetic field on the DC discharge, amplitude noise was much higher than with $F_2^{(2)}$ devices.

The experimental set up enabling observation of the beat note between this laser and a $F_2^{(2)}$ line-stabilized local oscillator is shown on Fig. 1. The beat detector was either a cooled InSb or a selected room temperature InAs photovoltaic reverse polarized diode. The nominal bandpass of these detectors was only 400 MHz ; however, with an E laser power of 600 µW and a $F_2^{(2)}$ laser power of 500 µW, we observed after careful matching a 3 GHz beat note with a fairly high signal to noise ratio (Fig. 2) but a rather low level (- 89 dBm) which prevented a direct counting of its frequency. Such a measurement was performed by mixing this signal with the output of a klystron signal generator and counting alternatively the klystron and mixer output frequencies with a computing

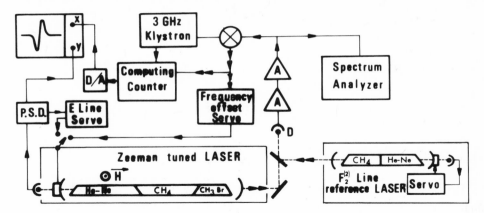

Fig. 1 - Block diagram of the frequency comparison between E and $F^{(2)}$ lines of methane at 3.39 μm. D : matched fast photovoltaic detector ; A : 25 dB gain tunnel diode amplifier.

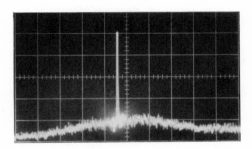

Fig. 2 - Spectrum analyzer display of the 3.032 GHz beat note between E and $F_2^{(2)}$ lasers. Detector output power is -89 dBm. Horizontal scale 20 MHz/div. Vertical scale 10 dB/div. IF bandwidth 10 kHz.

counter programmed to restore the signal frequency.

The E laser could either be locked to the E line or frequency-offset locked to the $F_2^{(2)}$ laser in order to record the E line shape. Both methods have been used to determine the $F_2^{(2)}$ - E frequency interval. The precision of the measurements was limited by the klystron frequency instabilities (± 2 kHz over one second). Taking into account the 1 kHz reproducibility of the $F_2^{(2)}$ laser and a slight (± 2 kHz) asymmetry of the observed E line probably due to inadequate magnetic shielding, the value of this frequency interval is :

$$\nu \ (F_2^2 - E) = 3 \ 032 \ 560 \pm 5 \ \text{kHz}$$

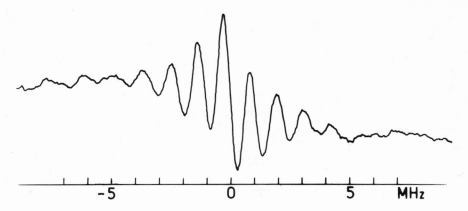

Fig. 3 - Stark splitting of the methane E line in derivative
form. Electric field is 1500 V/cm.

which is in agreement with, but more precise than the value given
by Magyar (ν = 3 032 565 $\pm$ 100 kHz)[10].

   We have performed some preliminary studies of the E - laser
reproducibility. In contrast to the $F_2^{(2)}$ case, the oscillation
level of a laser locked to the E line could be varied by more than
a factor of 2 without inducing a power shift larger than 2 kHz
(2.3×10^{-11}). We also investigated the influence of an electric
field. It is known[11,12], that the E line exhibits a first-order
Stark splitting into 13 components with a frequency separation of
8.21 Hz/V x m^{-1}. However, this splitting is symmetrical and the
second order Stark effect vanishes[11] so that the E line center
is fairly insensitive to electric fields. We have observed with
our device the splitting of the line in a field of 1500 V/cm
(Fig. 3) and the constancy of the center frequency within our 5
kHz precision.

   It thus appears that the reproducibility of the E - line device should presently be limited by more fundamental effects we
are now to discuss.

FUNDAMENTAL REPRODUCIBILITY LIMITATIONS

   Optical frequency standards, and among them saturated absorp-
tion-stabilized lasers, are subject to fundamental physical phenomena which can limit their ultimate reproducibility and accuracy.
The most important are Doppler (1st and 2nd order) and recoil effects.

First order Doppler broadening is essentially eliminated by the saturated absorption technique, which basically amounts to observing only those atoms traveling perpendicularly to the light beam. The main effect of molecular thermal motion is then to yield a transit time contribution to the linewidth, linked to the finite beam diameter and wavefront curvature. However, this last effect can also result in a shift of the saturated absorption peak, associated with the travel of the saturation-created population hole across the light beam, as has been shown by Bordé et al.[13]. We have evaluated the importance of this shift in the case of our experiments : it is found to be smaller than 3×10^{-12} in the methane device and smaller than 10^{-12} in the iodine device.

The recoil effect gives rise to a splitting of the saturated absorption into two peaks separated by a frequency interval $h\nu^2/mc^2$ [14]. This corresponds to a separation of 2.15 kHz (2.4×10^{-11}) for methane and 3.9 kHz (8.2×10^{-12}) for iodine. These peaks can be resolved with a very high resolution apparatus, as has been shown for methane by Hall et al. [7,15]. In such case, the recoil effect is not a factor of inaccuracy since it can be fully accounted for. However the two peaks will not be resolved in more conventional devices, especially with iodine. The position of the signal maximum will in this case be dependent on several factors such as the ratio of the relaxation rates in the upper and lower levels [14] and the degree of saturation [16,17]. In such situation it will probably be difficult, unless a careful evaluation of above factors could be made by independent means, to reach a reproducibility and accuracy better than one part in 10^{12}.

Finally, we must consider the second order Doppler effect. In the simple case when the linewidth is not limited by transit time effects and is not too small, this effect gives rise to a relative frequency shift equal to $-kT/mc^2$. This gives a room temperature shift of -152 Hz (-1.72×10^{-12}) [18] for methane and -51 Hz (-1.08×10^{-13}) for iodine. A simple temperature control of the absorption cell should then ensure a reproducibility higher than a few parts in 10^{14}, provided the absorber is not too much heated by the laser beam. In very high resolution experiments, where the transit time brings an important contribution to the linewidth, an accurate evaluation of the second order Doppler correction requires a more careful analysis [7,13].

CONCLUSION

Methane- and iodine- stabilized lasers are now reaching the point where their reproducibility (parts in 10^{12}) is limited by fundamental effects such as Doppler (1st and 2nd order) and recoil phenomena. In the present state of our knowledge, it seems that iodine devices will probably never reach the 10^{-12} level. Methane E-line devices are more promising in this respect, as probably are cw tunable (dye) lasers used in conjunction with other atomic or molecular absorbers. Let us finally point out that, together with saturated absorption, other possibilities for observing narrow resonance lines exist, such as use of collimated beams [19] or two-photon absorption [20]. The relative importance of the above mentioned fundamental limitations can be different in each case : let us for example recall that recoil effects are absent in two-photon absorption [21]. The potentialities of these techniques in the optical frequency standard domain should then be more thoroughly investigated.

REFERENCES

(1) A. Brillet, P. Cérez and H. Clergeot, IEEE J. Quantum Electron. QE-10, 526 (1974)
(2) M. Kroll, Phys. Rev. Lett. 23, 631 (1969)
(3) G.R. Hanes and C.E. Dahlstrom, Appl. Phys. Lett. 14, 362 (1969)
(4) P. Cérez, A. Brillet and F. Hartmann, IEEE Trans. Instr. Meas. IM-23, 526 (1974)
(5) G.R. Hanes, K.M.Baird and J. de Remigis, Appl. Opt. 12, 1600 (1973)
(6) N.B. Koshelyavskii, V.M. Tatarenkov and A.N. Titov, ZhETF Pis. Red. 15, 461 (1972) (JETP Lett. 15, 326 (1972))
(7) J.L. Hall, Proceedings of the CNRS International Colloquium on Sub-Doppler Spectroscopy, (Editions du CNRS, Paris, 1974) p. 105
(8) J.L. Hall and C. Bordé, Phys. Rev. Lett. 30, 1101 (1973)
(9) M. Betrencourt, M. Morillon-Chapey, C. Amiot and G. Guelach-vili, J. Mol. Spectrosc.., To be published
(10) J.A. Magyar, Thesis (University of Colorado, 1974) unpublished
(11) K. Uehara, K. Sakurai and K. Shimoda, J. Phys. Soc. Japan, 26 1018 (1969)
(12) A.C. Luntz and R.G. Brewer, J. Chem. Phys. 54, 3641 (1971)
(13) C. Bordé, C.V. Kunasz, D.G. Hummer and J.L. Hall, Phys. Rev. To be published
(14) A.P. Kol'chenko, S.G. Rautian and R.I. Sokolovskii, ZhETF 55, 1864 (1968) (JETP 28, 986 (1969))
(15) J.L. Hall, K. Uehara and C. Bordé, To be published
(16) C.G. Aminoff and S. Stenholm, Phys. Lett. 48A, 483 (1974)
(17) E.V. Baklanov, Opt. Comm. 13, 54 (1975)
(18) S.N. Bagaev and V.P. Chebotaev, ZhETF Pis. Red. 16, 614 (1972) (JETP Lett. 16, 433 (1972))
(19) T.J. Ryan, D.G. Youmans, L.A. Hackel and S. Ezekiel, Appl. Phys. Lett. 21, 320 (1972)
(20) K.C. Harvey, R.T. Hawkins, G. Meisel and A.L. Schawlow, Phys. Rev. Lett. 34, 1073 (1975)
(21) B. Cagnac, G. Grynberg and F. Biraben, J. Phys. 34, 845 (1973)

SOME EXPERIMENTAL FACTORS AFFECTING THE REPRODUCIBILITY AND STABILITY OF METHANE-STABILIZED HELIUM-NEON LASERS

B.W. Jolliffe, W.R.C. Rowley, K.C. Shotton, P.T. Woods

National Physical Laboratory

Teddington, England

Introduction

The frequency of the methane-stabilized helium-neon laser was first measured by Evenson et al (1). In June 1973, their result was combined by the Comité Consultatif pour la Définition du Mètre (2,3), with the results of four independent measurements of its wavelength, to give a recommended value for the speed of light (c = 299 792 458 ms⁻¹) with an uncertainty of 4 parts in 10^9 .

Measurements are currently in progress at the National Physical Laboratory to determine independently both the frequency and the wavelength of this methane-stabilized He-Ne laser radiation. In order that the laser stability should not limit the accuracy of the results, a frequency reproducibility of better than 1 part in 10^{10} (9 kHz) is required, with a similar frequency stability for averaging times of a few seconds.

A preliminary experiment to determine the frequency of this radiation (4) suffered from a large uncertainty in the correction applied for the frequency offset due to the steep background slope of the helium-neon power curve in the vicinity of the methane saturated-absorption feature. The use of a commercially available gain tube, filled with the naturally occurring mixture of neon isotopes, resulted in a correction of 60 ± 30 kHz. A further aspect of the preliminary measurement was the high pressure of methane used (5 Pa, 40 mtorr) in the intra-cavity absorption cell, resulting in a wide absorption feature of small amplitude. The full width at half maximum was greater than 2 MHz and the amplitude only 0.2% of the maximum power output.

In order to improve these, and other experimental factors affecting the properties of the methane-stabilized lasers at NPL, four aspects of their performance have been studied in some detail. These are:

a)   The short term stability, over averaging times of 1 ms to
     100 s.
b)   The long term reproducibility.
c)   The frequency offsets due to gain curve background slope.
d)   The effect of variations of the methane pressure in the
     intra-cavity absorption cell.

The arrangement for the series of experiments to be described is shown in Figure 1. The lasers now being used represent a considerable improvement over those used in the earlier frequency measurement. Laser 1 includes a new absorption cell 200 mm in length, whereas laser 2 uses a 100-mm cell. Both cavities use dc-excited gain tubes 260 mm in length, filled with 27 Pa (0.2 torr) of Ne-22 and 270 Pa (2 torr) of He-3. The peaks of the output power curves are frequency shifted by both pressure and by the choice of isotope mixture, into near coincidence with the methane saturated-absorption feature. The troublesome 'background slope' correction which must be applied to a first derivative locked laser is thus reduced (by a factor of forty for laser 1) with a corresponding reduction in its uncertainty. The servocontrol systems are usually of the first-derivative type, but third derivative techniques are used for some studies.

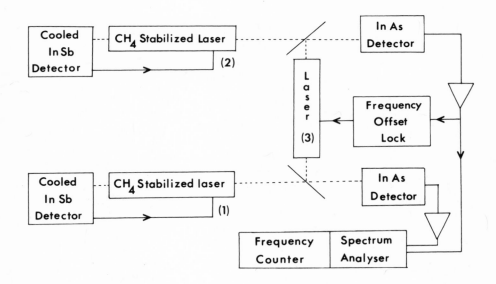

Fig.1. The experimental arrangement for beat frequency measurements.

To avoid the difficulties associated with the measurement of a beat frequency near zero, a third laser (laser 3) is used. This laser is locked with a frequency offset of 4 MHz from laser 2, and the beat frequency between lasers 1 and 3 is monitored.

(a) Short-Term Stability

As a preliminary to obtaining frequency stability measurements between lasers 1 and 2, it was necessary to assess the performance of the system used to offset-lock laser 3. Stability measurements of 70 Hz and 10 Hz for 1 s and 10 s averaging times respectively, expressed as the Allan variance (5) of the beat frequency, were obtained for the performance of this offset-lock, using the indium arsenide detector D2. As these values are.more than a factor of ten better than the laser stabilities expected, the offset-lock system is regarded as satisfactory.

The free-running stability between lasers 1 and 2 has been determined for averaging times between 1 ms and 10 s. No modulation was applied to either laser, and a beat width of 30 kHz was observed on a spectrum analyser with a bandwidth of 1 kHz and a scan rate of 1 kHz/ms. Thermal drift of the cavity structure becomes evident at averaging times longer than 100 ms, the drift rate being about 10 kHz per second. The 100 ms free-running stability is 3.7 kHz, ie. better than 5 parts in 10^{11}.

With both lasers locked to their methane saturated-absorption features, the performance of laser 1 has been investigated for variations of the servocontrol system parameters. The effect of varying the dc gain and the use of a 6 db or 9 db per octave servosystem response were studied. It was clear that the 9 db frequency response was preferable and that the greater the dc gain the better the stability, especially over averaging times between 1 s and 60 s. Under optimum gain conditions, the stability between the two stabilized lasers was 600 Hz (7 parts in 10^{12}) over 1 s, and 45 Hz (5 parts in 10^{13}) over 60 s averaging times. Of this 600 Hz, 70 Hz could be attributed to the offset lock, so that the true 1 s stability is 6 parts in 10^{12}. The stability results are summarised in Figure 2.

It should be noted at this point that these lasers were operating in conditions similar to those to be encountered during the measurement of their frequencies, ie. enclosed by an acoustically isolating box, but with fairly rudimentary antivibration mounts and in the proximity of operating vacuum systems.

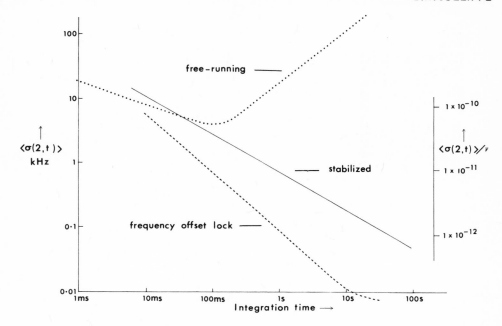

Fig.2.   The Allan variance and relative frequency stability of the
methane-stabilized lasers.

(b) Reproducibility

The long term reproducibility was measured by disconnecting
the servo-loop and retuning the optics of one laser, while the
other remained under servocontrol. The standard deviation of an
individual result, in a sample of 5 observations, was 2 kHz.

(c) Background Slope Effect

Two techniques were used to measure the frequency offset
caused by the slope of the helium-neon power curve against
frequency in the vicinity of the methane saturated-absorption
feature. In the first, the first-derivative curve was plotted for
both lasers, being the output voltage of the PSD used for the
servocontrol system. Figure 3 shows the form of this trace for
laser 2 with a scan of 2 MHz/cm, indicating a peak-to-peak width of
the first derivative curve of approximately 500kHz. The small
saturated-absorption feature obtained with the 100-mm cell in this
laser made the effect of background slope more important. Laser 1,
however, used a new 200-mm absorption cell which produced a feature

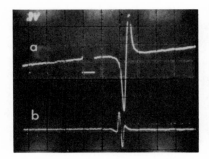 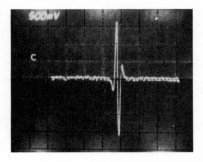

Fig.3. 1st and 3rd derivatives of the methane saturated-absorption
feature recorded at 2 MHz/cm, with a scan amplitude of 600 kHz peak
to peak. (a) 1st derivative, 2V/cm, zero shown; (b) 3rd derivative,
2V/cm; (c) 3rd derivative, zero on centre line, 500 mV/cm.

some 20 times the size. The offset due to the background slope for
this laser was calculated from the the recorded trace as
1.47 ± 0.1 kHz for the modulation amplitude used.

The second method of assessing the effect of background slope,
used as a check, was to relock each laser using a third-derivative
technique. As Figure 3 indicates, this method virtually eliminates
the frequency offset produced by the large background slope.

With both lasers locked by the first-derivative technique, a
beat frequency of 25.8 kHz was measured between lasers 1 and 2, as
shown in Figure 4. Laser 1 was then relocked using a third
derivative servocontrol, and the beat frequency remeasured as
28.4 kHz. Hence the offset due to background slope for laser 1,
together with any electronic offsets present in the servosystems,
was 2.6 kHz. This value is subject to an uncertainty of 600 Hz
introduced by relocking laser 1. The procedure was repeated for
laser 1 with first-derivative, and laser 2 with third-derivative
locks. The result indicated a background slope offset for laser 2 ,
with the weak saturated-absorption feature, of about 24 kHz. At the
time of test, only one third-derivative system was available so
that the beating of both lasers with third-derivative servocontrols
was not possible. The greater uncertainty in the first/third
derivative comparison makes it useful as a check only, and the
result was considered to be in satisfactory agreement with the
value of 1.5 kHz determined by the first method.

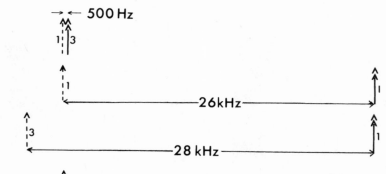

Fig.4.Comparison of lasers locked to 1st and 3rd derivatives.

(d) Methane Pressure Effect

The saturated-absorption feature width has been measured at pressures of 1.5 Pa (10 millitorr) and 6 Pa (40 millitorr). From the peak separation of the first-derivative discriminant, the full widths at half maximum were determined as 900 ± 30 kHz and 2.08 ± 0.07 MHz respectively. The feature amplitude is more than five times greater at 1.5 Pa than 6 Pa.

The laser designated laser 1, to which the above experiments relate particularly, is being used in an experiment to measure the absolute frequency of the methane-stabilized laser, the principle of the experiment having been described earlier (4). Preliminary work is also in progress towards an intercomparison of the wavelengths of the iodine and methane-stabilized helium-neon lasers, using a recently constructed servocontrolled Fabry-Perot interferometer.

We wish to thank A.J. Wallard for useful discussions, particularly in connection with the third derivative locking technique.

References

1.   EVENSON, K.M., et al, Appl. Phys. Lett., 22, 192, (1973).

2.   TERRIEN, J., Nouv. Rev. Optique, 4, 215, (1973).

3.   Com. Consult. Définition Mètre, (Com. Int. Poids. Mesures), 5th Session, (1973).

4.   BLANEY, T.G., et al, Nature, $\underline{254}$, 584, (1975).

5.   BARNES, J.A., et al, IEEE Trans. Inst. Meas., $\underline{IM20}$, 105, (1971).

MOLECULAR BEAM STABILIZED ARGON LASER

L.A. Hackel, D.G. Youmans, and S. Ezekiel

Research Laboratory of Electronics
Massachusetts Institute of Technology
Cambridge, Massachusetts 02139

ABSTRACT

A single frequency 5145Å argon ion laser has been stabilized
to a hyperfine component in a molecular beam of I_2. A long term
stability of 1 x 10^{-13} was achieved for an integration time of
200 seconds. Reproducibility of the laser frequency was better
than 7 x 10^{-12}. An upper limit on intensity shift was 2 x 10^{-14}
for a 1% change in laser intensity. Frequency shift caused by
external magnetic fields was less than 2 x 10^{-13} per gauss.

INTRODUCTION

The use of a resonance transition in a molecular (or atomic)
beam as a stable reference for long term laser frequency stabilization was first discussed by Ezekiel and Weiss (1) in 1966. However, it was not until late in 1967 that the feasibility of this
scheme was experimentally demonstrated by the observation of laser
induced fluorescence in a molecular beam(2). Other novel applications of atomic beams such as atomic beam lasers have been suggested as early as 1965 by Basov and Letokhov(3). A comprehensive discussion of the many possibilities of using atomic beam
techniques to achieve a stable coherent radiation source is found
in a review article by Basov and Letokhov(4).

In this paper we present recent data on the performance of a
5145Å argon ion laser stabilized to a hyperfine resonance in a
molecular beam of $I_2{}^{127}$.

STABILIZATION SYSTEM

The setup used for stabilization is shown in Figure 1. A

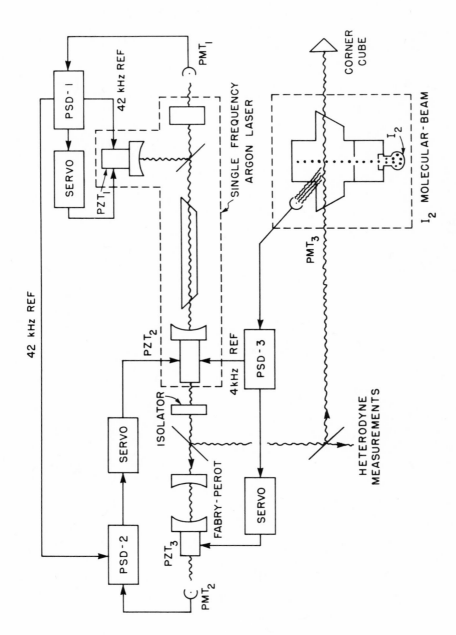

Figure 1. Stabilization Scheme.

single frequency 5145Å argon ion laser is short term stabilized
by locking the laser frequency to an external Fabry-Perot. Long
term stabilization is accomplished by locking the external Fabry-
Perot to an I_2 hyperfine component excited by the laser in a
molecular beam.

The argon laser used is similar to that described earlier.(5)
The laser cavity is made with granite end blocks and invar spacer
rods 98 cm in length. The discharge tube is designed to operate
free from plasma oscillations and the discharge power supplies
are highly filtered. Single frequency operation is obtained with
a 2.0 cm long intracavity Fox-Smith mode selector as shown in
Figure 1. A low bandwidth servo is used to enable the mode selector to track the laser frequency. The frequency discriminant for
this loop is obtained by modulating the length of the mode selector at 42 kHz and synchronously detecting the phase of the resulting laser intensity modulation in a phase sensitive demodula-
tor (PSD-1). The error signal available at the output of PSD-1
is then fed through a servo amplifier to the piezo-electric
crystal (PZT_1) controlling the length of the mode selector. In
this way the mode selector automatically tracks the laser frequency over the 5 GHz tuning range of the argon laser without any
mode hopping.

In order to improve the short term stability the laser frequency is locked to an external cavity by means of a second feedback loop. A Tropel model 216 Fabry-Perot with a 1 MHz width
(FWHH) and a 300 MHz free spectral range is used as the short term
external reference. The frequency discriminant for this second
loop is obtained by utilizing the small 10 kHz frequency modulation of the laser frequency resulting from the length modulation
of the mode selector discussed earlier. The appropriate discriminant is therefore obtained by synchronously detecting the
output of the Fabry-Perot at 42 kHz in PSD_2. The discriminant
signal is then fed through a servo amplifier to the piezo electric crystal (PZT_2) controlling the length of the primary laser
cavity. With a bandwidth of 1 kHz, the laser jitter is reduced
to about 15 kHz.

The short-term stabilized laser is then used to induce
resonance fluorescence in a molecular beam of I_2. The laser
frequency can be smoothly tuned across the I_2 spectrum by
varying the length of the Fabry-Perot reference cavity.

The beam apparatus is a high vacuum chamber maintained at a
pressure of less than 1×10^{-7} torr. A glass oven contains
reagent grade I_2 which effuses through a narrow primary slit. A
second slit some distance away determines the degree of beam
collimation. I_2 molecules which are not part of the collimated
beam are pumped out by liquid nitrogen traps (surface area about

1200 cm^2) that surround the interslit region. With the oven at
room temperature, the I_2 vapor pressure is about 0.3 torr. The
source slit is typically 0.05 mm wide and 2 cm high, allowing
5 x 10^{17} molecules per second to emerge from the oven.

In the experiments described in this paper the separation
between slits is 18 cm and the width of the second slit is set
at 0.18 mm, thus forming a beam with a geometrical width of 1
m radian which corresponds to a Doppler width of 500 kHz for I_2.
This particular slit geometry is chosen for convenience. In our
experiments on I_2 lineshape studies, the Doppler broadening contribution from the slit geometry is reduced to less than 20 kHz.

The laser beam is expanded to a diameter of 1 cm before it
enters the interaction region where it orthogonally excites the
I_2 beam. The laser beam is then reflected back through the I_2
beam by a precision corner cube. The induced I_2 fluorescence is
collected with an externally mounted 38 mm diameter fl.0 lens
system and focused onto an EMI9554B dry ice cooled photomultiplier
with an extended S-20 photocathode. The photomultiplier is
mounted directly below the interaction region. A Corning-type
3-68 sharp cut filter is placed between the lens system and
phototube to reduce scattered 5145Å light. The photomultiplier
has a dark count of 40 per second and an additional 40 counts per
second are due to scattered 5145Å light.

The observed I_2 hyperfine lines are the $\Delta F = \Delta J$ components
of the P(13) (43-0) and R(15) (43-0) transitions as discussed
earlier.(6) Figure 2 shows a typical spectrum. The hyperfine
structure in I_2 is primarily the result of nuclear electric
quadrupole interactions with smaller contributions from spin-
rotation, spin-spin and magnetic octupole interactions(7). The
derivative of the I_2 lines is obtained by modulating the laser
frequency with a 4 kHz sinusoidal signal applied to PZT$_2$, which
results in a laser frequency excursion of approximately 300 kHz
peak to peak. The observed signal to noise in a typical derivative at the output of PSD-3 is approximately 120 for τ = 30 msec.
The output of PSD-3 is then fed through a servo amplifier to the
piezo-electric crystal (PZT$_3$) controlling the length of the
reference Fabry-Perot. In this way, the laser frequency is long
term stabilized to a selected transition in the I_2 molecular beam.

LONG TERM STABILITY MEASUREMENTS

To evaluate the long term laser frequency stability, two
identical lasers were independently stabilized and the outputs
heterodyned. One laser was locked to the second line from the
left in the I_2 spectrum in Figure 2 and the other laser to the

third line. These particular lines were chosen for the stabilization experiments because they appear isolated from other hyperfine structure transitions. A complete identification of the
weaker $\Delta F = 0$ transitions confirmed that there are no $\Delta F = 0$
transitions overlapping the second and third lines. The beat
frequency of 59.8 MHz corresponding to the separation between
the lines was detected on a high speed (TIXL59) photodiode. The
photodiode output was amplified and fed simultaneously to a frequency counter, a frequency discriminator and a spectrum analyser.

An analysis of the frequency counter output using the two
sample Allan variance gave the results shown in Figure 3. For
averaging times of 200 seconds a long term stability of
1.2×10^{-13} was achieved. The plot in Figure 3 shows that the
stability improved approximately with $\tau^{-1/2}$ as is the case for
a shot noise limited stability. In addition, the long term
stability did not show any bottoming out for large integration
times indicating the absence of time dependent shifts with a
period shorter than 200 seconds.

REPRODUCIBILITY MEASUREMENTS

The reproducibility of the frequency of a molecular beam
stabilized laser depends on several factors, the most important
of which is the orthogonality between the laser and molecular
beam. Any deviation from orthogonal excitation of the I_2 beam
results in a shift of the center frequency of the I_2 transition
as well as a nonsymmetrical broadening of the line due to the
velocity distribution of molecules in the beam. Frequency shifts
due to nonorthogonality can be reduced considerably by reflecting
the laser beam back through the I_2 beam by means of a precision
corner reflector. In this way two iodine transitions are observed which are symmetrically displaced with respect to the true
line center. For a small misalignment angle, the two I_2 transitions overlap such that the overall effect of retroreflection is
to convert a shifted line to an unshifted but broadened line as
long as the intensities of the two laser beams are equal. Clearly, the smaller the misalignment angle, the less the contribution
due to unequal intensities.

In our reproducibility experiments we determined the orthogonal alignment between laser and molecular beam by deflecting
the laser beam and observing the width of the I_2 transition as a
function of angle. In this way we were able to achieve orthogonal
alignment within 5×10^{-5} radians which corresponds to an error
in the line center of 24 kHz. This misalignment error was reduced further by reflecting the laser beam back through the I_2
beam with a one second of arc corner cube reflector. The re-

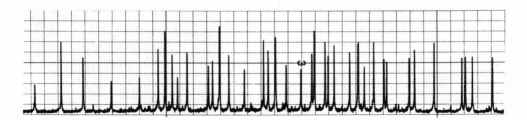

Figure 2. I_2^{127} Hyperfine Structure Excited by 5145 Å Argon
Ion Laser.   Frequency Scale: 30 MHz/box.

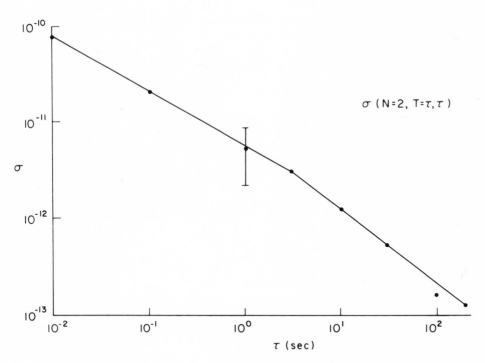

Figure 3.   Allan Variance (**σ/τ**) Plot.

flected beam, which was 10% less intense than the incident beam,
was prevented from reentering the laser by making the laser beam
strike the corner cube slightly off center, thus inducing a
slight translation in the reflected beam. The procedure for
reproducing the laser frequency was to unlock one of the stabilized lasers and remove the alignment mirror and corner reflector, then to realign the laser beam as described above and
measure the new beat frequency with respect to the fixed laser.
The results of 5 attempts are -1.10, 6.85, -4.70, 2.31, 1.80 kHz
with respect to an arbitrary zero. This gives a standard deviation in the reproducibility of 7.0×10^{-12}.

In addition, we measured the shift of the stabilized laser
frequency as a function of laser intensity. We found the shift
to be less than 2×10^{-14} for a 1% change in a laser intensity
at 1 mW. This value is an upper limit at present pending a more
detailed study.

The effect of external magnetic fields was investigated by
applying a 50 gauss field at various angles with respect to the
I_2 interaction region. A shift of 2×10^{-13} per gauss was
measured. This type of shift of course can be reduced significantly by suitable magnetic shielding.

LIMITATIONS OF MOLECULAR BEAM STABILIZED LASER

The performance achieved so far in stabilizing the 5145Å
argon laser to an I_2 molecular beam is by no means the best that
can be accomplished with this scheme. Stability measurements
can be improved by increasing the signal-to-noise in the induced
fluorescence. The expected limit on signal to noise in our system is about 10^3 for $\tau = 1$ sec. With an I_2 linewidth of 100
kHz, it should be possible to measure a stability of 1×10^{-13}
for $\tau = 1$ second. By using multiple beams the S/N could be increased by another order of magnitude which would enable a
stability of 1×10^{-14} for $\tau = 1$ second. Nozzle sources may
also prove useful.

The fundamental limitation on reproducibility will depend
on level shifts due to the radiation field, second order Doppler
and recoil shifts.

We have estimated the level shift to be 1.3×10^{-18} for a
1% change in laser intensity of 1 mW/cm^2. This estimate is based
on energy shift calculations by Mizushima(8). The second order
Doppler shift for I_2 at room temperature is $10^{-15}/°C$ and it can
be kept small by careful control of the oven temperature. Molecular recoil causes an intensity dependent line shift which is
estimated to be less than 10^{-14} for a 1% change in laser intensity. This could become the major limitation on reproducibility.

Other causes of line shift include the shift due to overlapping Lorentzian tails from neighboring hyperfine transitions which is calculated to be less than 2×10^{-17} for a 1% change in intensity; shift due to the slope of the laser power curve as a function of frequency which is calculated to be less than 2×10^{-15} for a 1% change in intensity.

The present reproducibility performance of 7×10^{-12} is only preliminary. Careful alignment procedures with low loss optics are yet to be carried out.

This work was supported by the Air Force Office of Scientific Research.

REFERENCES

1. S. Ezekiel and R. Weiss, Research Laboratory of Electronics, Report No. 80, M.I.T. January 15, 1966.
2. S. Ezekiel and R. Weiss, Phys. Rev. Letters 20, 91 (1968).
3. N.G. Basov and V.S. Letokhov, JEPT Letters, 49, 3, (1965).
4. N.G. Basov and V.S. Letokhov, Electron Technology, 2, 2/3 pp 15-20, Institute of Electron Technology P.A. Sci, Warsaw 1969.
5. D.G. Youmans, L.A. Hackel and S. Ezekiel, J.App. Phys. 44, 2319 (1973).
   T.J. Ryan, D.G. Youmans, L.A. Hackel and S. Ezekiel, App. Phys. Letters 21, 320 (1972).
6. D.J. Ruben, S.G. Kukolich, L.A. Hackel, D.G. Youmans and S. Ezekiel, Chem. Phys. Letters 22, 326 (1973).
7. L.A. Hackel, K.H. Casleton and S. Ezekiel (to be published).
8. M. Mizushima, Phys. Rev. 133, A414 (1964).

FREQUENCY STABILIZATION OF AN AR^+ LASER WITH MOLECULAR IODINE

Frank Spieweck

Physikalisch-Technische Bundesanstalt

D-33 Braunschweig, Bundesallee 100, Fed.Rep. of Germany

The frequency of the green argon ion laser line at 515 nm can be stabilized by various methods. In the preceding report by Hackel, Youmans, and Ezekiel, stabilization with an external molecular beam has been described (1). Hohimer, Kelly, and Tittel used an external heated iodine absorption cell and stabilized the laser frequency to one point of half the intensity of the absorption profile (2). Bordé, Camy, Vialle, and Decomps reported on a stabilization scheme containing an iodine cell within an external ring interferometer (3). Stabilization should be possible, too, by means of the saturation fluorescence spectroscopy developed by Hänsch, Levenson, and Schawlow (4,5).

All these methods require two or even more servo loops resulting in complex stabilization systems. It will be shown below, however, that it is possible to build up simple and compact stabilization devices with only one servo loop.

1. FREQUENCY STABILIZATION WITH AN EXTERNAL $^{129}I_2$ ABSORPTION CELL

Frequency stabilization of a short cavity argon ion laser with an external $^{129}I_2$ absorption cell has already been described in a previous paper (6). Figure 1 shows an improved version of the stabilization system. An air-cooled laser (Spectra-Physics, model 162) has been fitted with a piezoelectric drive and a modified suspension of the fan in order to reduce vibrations. The laser is operated at a discharge current of 5 A close to the threshold yielding single-mode radiation with a power of about 300 μW. The mode purity is monitored with a small optical spectrum analyzer and a scope. Part of the laser beam passes a folded absorption path in a iodine cell of 40 cm length placed in parallel to the

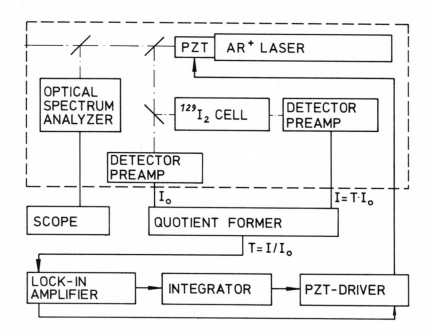

Figure 1. Schematic diagram of Ar^+ laser frequency stabilization
with an external $^{129}I_2$ absorption cell.

path in an iodine cell of 40 cm length placed in parallel to the
laser. Within the whole laser frequency tuning range of c/2L =
460 MHz, there exists only one transmission minimum showing a
depth of 30 %. As the laser gain profile is not perfectly flat,
the laser frequency is not stabilized directly to the minimum of
the transmitted intensity I_{min} but to the minimum of transmission

$$T_{min} = I_{min} / I_o$$

by means of an electronic quotient former placed between the
detectors of the intensities I and I_o and the lock-in amplifier
(see figure 1). The modulation frequency of 800 Hz has an amplitude of 10 MHz peak-to-peak. Figure 2 shows the output of the
lock-in amplifier yielding a frequency instability of less than
$10^{-9} \nu_o$.

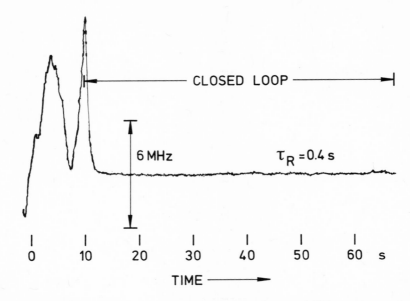

Figure 2. Output voltage of the lock-in amplifier shown in
figure 1, τ_R: time constant of the recorder.

2. FREQUENCY STABILIZATION WITH AN INTERNAL I_2 ABSORPTION CELL

A schematic diagram of the second stabilization system is
given in figure 3. A commercially available laser with cavity
extension (Coherent Radiation, model CR-3) has been used. The
rear mirror has been coated for selective reflectivity at 515 nm,
the output mirror has a reduced transmission of 3 %. Both reflec-
tors have been mounted on piezoelectric drives. Single-mode
operation has been obtained by an internal oven stabilized etalon.
Different internal iodine absorption cells were used, a $^{127}I_2$ cell
of 8 cm length and a $^{129}I_2$ cell of 12 cm length. In the latter
case, mode selection can be achieved also with a second internal
I_2 cell containing $^{127}I_2$, instead of the etalon (7); then, only a
few $^{129}I_2$ hyperfine components lie within the laser frequency
tuning range (8).

The modulation frequency of 330 Hz has an amplitude of 9 MHz
peak-to-peak. Saturated absorption peaks can be detected by
observing the third harmonic of the modulation frequency contained
in the laser radiation. In this case, the etalon tilt angle need
not be stabilized by an additional servo loop, as discussed in a
separate paper (9). Figure 4 shows some $^{127}I_2$ hyperfine components
obtained at a vapour pressure of 9 N/m^2. The laser frequency can

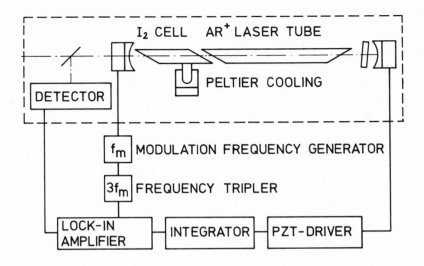

Figure 3. Schematic diagram of Ar$^+$ laser frequency stabilization
with an internal I$_2$ absorption cell, f$_m$ = 330 Hz.

be locked to any of the stronger components already observed by
Youmans, Hackel, and Ezekiel (10) as well as by Sorem and Schawlow (4). The output power is about 20 mW, at a discharge current
of 17 A. If the laser frequency is locked to the strongest
^{127}I$_2$ hyperfine component, the resulting instability is less
than 10^{-10} ν_0, as can be estimated from figure 5.

3. WAVELENGTH AND BEAT FREQUENCY MEASUREMENTS

The wavelength of the air-cooled laser stabilized with an
external ^{129}I$_2$ absorption cell has been measured in comparison
with the primary standard of length yielding the value

$$\lambda_{vac}^{(1)} = (514.674206 \pm 0.00001) \text{ nm.}$$

Beat frequency measurements have been performed between the air-
cooled laser and the long laser stabilized to one of the I$_2$ hyperfine components, revealing an uncertainty of the air-cooled laser
of about 1 MHz, i.e. 2.10^{-9} ν_0. According to a measured beat frequency of 775 MHz, the second of the stronger ^{127}I$_2$ hyperfine
components (at 140 MHz in the arbitrary scale of figure 4) has a
wavelength of

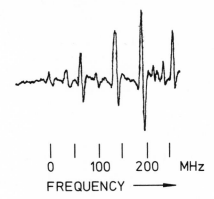

Figure 4. Output voltage of the lock-in amplifier shown in
figure 3 (third harmonic of the modulation frequency),
obtained with an internal $^{127}I_2$ absorption cell.

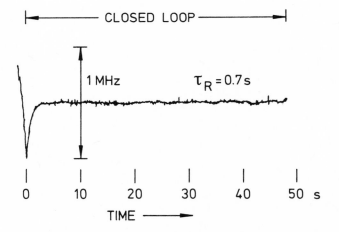

Figure 5. Output voltage of the lock-in amplifier shown in
figure 3, τ_R: time constant of the recorder.

$$\lambda_{vac}^{(2)} = (514.673521 \pm 0.00001) \text{ nm}.$$

Due to a corresponding beat frequency of 836 MHz, the strongest $^{127}I_2$ hyperfine component shown in figure 4 has a wavelength of

$$\lambda_{vac}^{(3)} = (514.673468 \pm 0.00001) \text{ nm}.$$

4. DISCUSSION

The frequency stabilized green argon ion laser radiation is an interesting wavelength standard for length metrology, especially as the wavelength is relatively short. Combined with the red helium-neon laser line, the interference order number can be determined by coincidence methods.

Stabilization of the air-cooled laser is unambiguous because after an eventual mode hop the laser is locked to the same frequency again. The second method with an internal iodine absorption cell is, in principle, as simple as in the case of an I_2 stabilized helium-neon laser. Its uncertainty should be less than $10^{-10} \nu_o$.

REFERENCES

1. HACKEL, L.A., YOUMANS, D.G., and EZEKIEL, S., AMCO-5 (1975).

2. HOHIMER, J.P., KELLY, R.C., and TITTEL, F.K.,
   Appl.Opt. 11, 626 (1972).

3. BORDÉ, C., CAMY, G., VIALLE, J.L., and DECOMPS, B.,
   CPEM 1974, IEE Conf.Pub.No. 113.

4. HÄNSCH, T.W., LEVENSON, M.D., and SCHAWLOW, A.L.,
   Phys.Rev.Lett. 26, 946 (1971).

5. SOREM, M.S., and SCHAWLOW, A.L., Opt.Commun. 5, 148 (1972).

6. SPIEWECK, F., Appl.Phys. 3, 429 (1974).

7. SPIEWECK, F., Appl.Phys. 1, 233 (1973).

8. SPIEWECK, F., Thesis, TU Hannover (1974).

9. SPIEWECK, F., submitted to Metrologia.

10. YOUMANS, D.G., HACKEL, L.A., and EZEKIEL, S.,
    J.Appl.Phys. 44, 2319 (1973).

SATURATED-ABSORPTION-STABILIZED N_2O LASER FOR NEW FREQUENCY STANDARDS

B.G. Whitford, K.J. Siemsen and H.D. Riccius

National Research Council of Canada

Division of Physics, Ottawa, Ontario, Canada K1A OR6

The N_2O laser operates in approximately the same spectral region, around 10.6 μm, as the well-known CO_2 laser. The two molecules resemble each other in being linear triatomic and having roughly the same energy levels. One would expect therefore that the technique of observing the standing-wave saturation in CO_2 lasers via the fluorescence (1) could be applied to the N_2O laser. Such is indeed the case. The 4.5 μm fluorescence from room temperature N_2O in an absorber cell placed in the laser cavity has been observed and has a strength comparable to that of the 4.3 μm CO_2 fluorescence. Because the centre frequencies of the N_2O laser transitions can be determined to within 1 part in 10^{10} by this technique, and because there are twice as many N_2O lasing transitions in the 00^01-10^00 band, compared to CO_2 case, the N_2O laser should be useful as a secondary frequency standard in the 10 μm region. It is the purpose of this paper to study some of the characteristics of a Lamb-dip stabilized N_2O laser which has been used for absolute frequency measurements of N_2O laser transitions.

The experimental set-up used was similar to that reported previously (2). The N_2O and CO_2 lasers had discharge tubes 1.4 m and 0.4 m long, and resonant cavities 1.96 m and 0.95 m long, respectively.   The use of two identical gratings blazed at 8 μm at one end of each cavity provided single line operation of the N_2O laser on P-branch lines from J = 3 to 35 and on R-branch lines from J = 1 to 40, and on CO_2 laser lines from P(4) to P(38). Typical power output for each of the two lasers was 200 mW. The N_2O laser was operated with a flowing 15% N_2 - 16% N_2O - 69% He gas mixture giving a total pressure of 5 Torr in the discharge tube, whereas the CO_2 laser had a commercial sealed-off discharge

tube.   Power enhancement in the N_2O laser by the addition of CO
has been reported previously (3).   In order to avoid CO_2 laser
action no carbon monoxide was added to our gas mixture.

The N_2O laser requires higher flow rates than an identical
flowing CO_2 laser.   Figure 1 shows the laser power output as a
function of the power input with different flow rates (cm^3/s S.T.P.)
as parameter.   The dashed lines indicate that the discharge
becomes unstable.   The maxima particularly at higher flow rates
arise because of the competing effects of better population inversion and rapid dissociation of N_2O molecules with increasing power
input (4).   The dissociation of the N_2O molecules is a severe
problem for making a sealed-off N_2O laser.

Narrow fluorescence resonances are observed when the absorption
of the N_2O molecules is saturated at low pressures by the N_2O laser
radiation.   The fluorescence of these molecules from the upper
level of the laser transition is observed in the region of 4.5 μm.
In the experiment, a short absorption cell was placed inside the
laser cavity. A sapphire window on top of the absorption cell
allowed the observation of the fluorescence signal with a liquid-
nitrogen cooled gold-doped germanium detector.   The interior of
the cell was aluminium-coated.   As shown in Fig. 2, the intensity
of the spontaneous emission at 4.5 μm from the 00^01—10^00 band

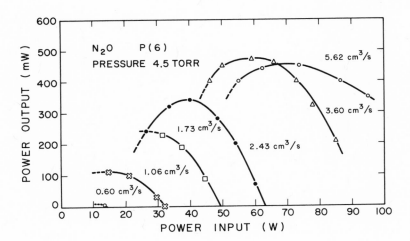

Fig. 1. N_2O laser power output as a function of power input with
        different flow rates (cm^3/s S.T.P.) as parameter.

of N_2O is comparable to that observed in CO_2.   The recorder
tracings in Figure 2 were obtained by using the same laser and
absorption cell under identical experimental conditions.   Only the
gas mixture in the laser discharge tube and the absorber gas in the
cell were changed. Laser frequency tuning was achieved by means of
a piezoelectric tuner causing a small displacement of the laser
mirror.   The pressure in both absorption cells was kept at 0.04
Torr.   The flow rate of the absorber gas was very small, just
sufficient to exceed any leaks in the system.

Figure 3 shows the fluorescence power output of 70 N_2O laser
transitions as a function of rotational quantum number J.   The
laser power was kept constant at 50 mW by adjusting flow rate or
discharge current accordingly. The solid line represents the
thermal equilibrium population among the rotational levels of the
10^00 vibrational level, calculated for T = 300 K and taking all
lower vibrational levels into account (5).   From the rather good
agreement between the experimental and theoretical curves it is

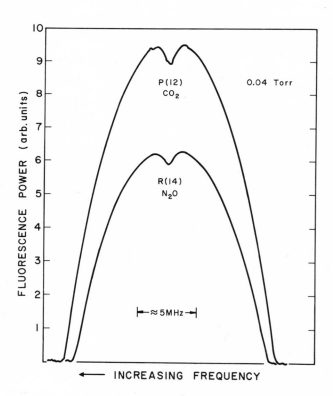

Fig. 2.   Typical recorder tracings of the 4.3 μm and the 4.5 μm
          fluorescence signals for CO_2 and N_2O, respectively.

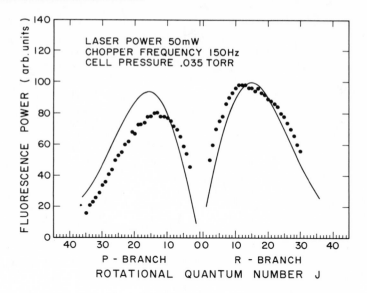

Fig. 3.   Fluorescence power output of N_2O laser transitions as a
function of rotational quantum number J.

evident that the fluorescence power measured at the centre of the
Lamb dip represents essentially the thermal population of the 10^00
level.

     The following procedure was used to measure the absolute
frequencies of the N_2O, 00^01—10^00 laser band by heterodyning with
known CO_2 laser frequencies of the 00^01-10^00 band in a tungsten-
nickel diode.   The radiant energy of approximately 50 mW from
each laser, with parallel polarization of the electric field vectors,
was focussed with a 3 cm focal length IRTRAN-2 lens onto the diode
junction.   The output of the diode was fed into a spectrum analyzer
which had been specially calibrated against NRC[133]Cs standard to
permit determination of beat frequencies ranging from 0.6 to 35 GHz
to within ± 5kHz. Difference frequency signals from the tungsten-
nickel diode up to 4 GHz were observed directly on a Tektronix
1L20 spectrum analyzer.   Those signals above 4 GHz were first down-
converted to a frequency below 4 GHz by injecting one of the microwave frequencies 8.255, 10.205, or 11.735 GHz into the tungsten-
nickel diode, together with the N_2O and CO_2 frequencies.   The microwave frequencies were phase-locked to a 5 MHz signal from the
NRC[133]Cs standard.   The fundamental microwave frequency, or its
second or third harmonic was used, as required, to obtain a diode
output below 4 GHz.

The N_2O absolute frequencies were deduced from the measured beat frequencies and the CO_2 transition frequencies derived by the use of the absolute value for R(30) given by Evenson et al. (6) and rotational constants recently determined by Petersen et at (7). By using a stabilized CO_2 laser as the frequency reference, 33 N_2O laser frequencies have been measured with a precision of 2 parts in 10^9 or better.   The values of the measured beat frequencies and the calculated N_2O frequencies with their statistical errors have been published elsewhere (8), together with more precise values for the band centre and for the rotational constants.

References

1.   FREED, C. and JAVAN, A., Appl. Phys. Lett. 17, 53 (1970).

2.   WHITFORD, B.G., SIEMSEN, K.J. and RICCIUS, H.D., Optics Commun. 10, 288 (1974).

3.   DJEU, N., Kan, T. and WOLGA, G., IEEE J. Quantum Electronics QE-4, 783 (1968).

4.   CHEO, P.K., IEEE Quantum Electronics QE-3, 683 (1967).

5.   HERZBERG, G., Spectra of Diatomic Molecules, D. Van Nostrand (1950).

6.   EVENSON, K.M., WELLS, J.S., PETERSEN, F.R., DANIELSON, B.L. and DAY, G.W., Appl. Phys. Lett. 22, 192 (1973).

7.   PETERSEN, F.R., MCDONALD, D.G., CUPP, J.D. and DANIELSON, B.L., Phys. Rev. Lett. 31, 573 (1973).

8.   WHITFORD, B.G., SIEMSEN, K.J., RICCIUS, H.D. and HANES, G.R., Optics Commun. in press.

Optically pumped far-infrared laser for infrared frequency measurements

C.O. Weiss and G. Kramer

Physikalisch-Technische Bundesanstalt

D-33 Braunschweig, Bundesallee 100, Fed.Rep. Germany

1) Introduction

Optically pumped far-infrared (FIR) lasers /1/ are much better
suited to infrared frequency measurements than the discharge lasers
because: i) The absence of free electrons yields high frequency
stability. ii) About 400 laser transitions are available in the
30 μm to 2 mm region. iii) Dimensions of these lasers can be made
much smaller than those of discharge lasers.
The mechanism of these lasers is shown in Fig. 1: CO_2-laser radiation is absorbed on a rot.-vibr. transition of a polar molecule.
The lower rotational levels of the vibrational ground state are
thermally populated, while the first exited vibrational level is
practically empty. Absorption of the CO_2-laser radiation leads to
population inversion of the 2-3 and possibly of the 4-1 rotational
transition.

2) Experimental set-up

An optically pumped laser device is shown in Fig. 2. The FIR laser
consists of 2 gold-coated mirrors of 10 cm diameter, 80 cm spacing,
with a near-folded-confocal resonator geometry. Pump radiation
enters the resonator through the 3 mm coupling hole in the flat
mirror. The FIR radiation emerges through the same hole. Due to
diffraction, the FIR beam expands outside the cavity and is deflected by a third mirror. The resonator is placed inside a vacuum
chamber, filled with the laser gas. Windows of ZnSe and TPX transmit the pump- and FIR-radiation. Sealed-off operation is possible
because the optical pumping is nondestructive.
Our first experiments with this laser showed, that the FIR output
power stability strongly depends on the pump frequency stability.
The relative spectral position of the 9,2 μm CO_2 R(20) line and of
the HCOOH absorption line is shown in Fig. 3. Pumping will be most

efficient at absorption-line center and changes of the pump frequency consequently affect the FIR output power. In addition, light
reflected back from the FIR laser influences the pump laser frequency so that a stable FIR output is difficult to achieve.
It is therefore necessary to stabilize the frequency of the pump
laser. To pump various gases it is desirable to stabilize the frequency at any desired value within any CO_2-laser line. A frequency
stabilizing system was set up as shown in Fig. 4. The reference
laser is stabilized to the center of a CO_2 line. Stabilization is
accomplished by observation of the dip in the 4.3 μm fluorescence
of CO_2 gas when excited by a standing-wave field. /2/ The laser frequency is modulated and the output of an InSb detector, that measures the fluorescence, is detected at the modulation frequency, thus
generating an error signal, which controls the resonator length.
Part of the reference radiation is mixed with a small part of the
pump laser radiation. The beams are combined on a fast detector
which generates the beat frequency between the lasers. This frequency is controlled to a value corresponding to the desired offset
from line center by a frequency control loop which acts on the pump
laser resonator. With this set-up the pump frequency can be stabilized at any frequency within ± 30 MHz from the center of any
CO_2-line. Alternatively with the 4.2 K Ge: Cu detector shown a
Wo-Ni whisker diode was used. The improvement in FIR power stability can be seen from Fig. 5 where the output is recorded
without and with pump frequency stabilization. The FIR power was
monitored by a Golay-cell, pressure in the FIR laser was measured
by a capacitive pressure gauge.

3) Measurements
As with this pumping system stable and reproducible pumping is
achieved, the dependence of the FIR output power on the system
parameters could be measured. Fig. 6 shows the FIR power as a
function of pressure and pumping power. FIR power increases approximately linearly with pump power. The optimum pressure is 100 mTorr,
nearly independent of pump power. For a pump power of 10W, 10mW FIR
power were measured, corresponding to a 4 % quantum efficiency.
Observation of the FIR laser output while the resonator length is
scanned allows measurement of the FIR wavelength and of the FIR
oscillation bandwidth. Fig. 7 shows the 432 μm HCOOH line profile.
The band width is 8.4 MHz, four times the Doppler width, and hence
gives an estimate of the rotational relaxation rate. Fig. 8 shows
the output power as a function of pump frequency for the 147 μm
CH_3NH_2 line and the 432 μm HCOOH line. It can be seen that the
HCOOH line may be pumped by a pump laser stabilized to the CO_2-line
center while in the case of the CH_3NH_2 line the spacing of pump and
absorption line is larger so that an offset pump frequency is
necessary.
The frequency of the 432 μm HCOOH line-center was measured against
the 13th harmonic of a stabilized 50 GHz klystron. The frequency
is: 692 951.4 ± 0.1 MHz.

It seems possible to measure CO_2 laser frequencies using only one
intermediate FIR laser of this type. For this purpose it is however
necessary to increase the FIR output power. To improve on the
quantum efficiency, beside a knowledge of the spectroscopic data
(energy levels, Einstein coefficients) of the laser molecule, a
quantitative understanding of the collision processes is required.
We have therefore started measurements in this direction.

4) Vibrational relaxation
Two main relaxation rates are of importance in a FIR laser:
i) Rotational relaxation between the laser levels. ii) Vibrational
relaxation of the first exited vibrational level to the ground
state. In addition velocity cross-relaxation within any level may
also play some role. Rotational relaxation is much faster than the
vibrational, the latter may therefore represent the limiting "bottle
neck" of the laser cycle /3/. To gain information about the vibrational relaxation, measurements of the linear as well as the saturated absorption of CO_2 laser radiation in HCOOH were carried out.
In the evaluation of these measurements no distinction was made
between the rotational sublevels of a vibrational state, since
about 10 rotational levels lie within kT and the rotational relaxation is fast.
In the case of inhomogeneous broadening, which is a reasonable
assumption for the absorbing transition, the absorption coefficient
is:

(1) $\mathcal{H}(q) = \mathcal{H}_0 (1 + q/q_s)^{-\frac{1}{2}}$, where $\mathcal{H}_0$ is the linear absorption
coefficient $\mathcal{H}_0 = n_\ell \cdot B_{\nu 1,2} \cdot h \cdot \nu/c$ (n_ℓ: density of molecules,
B_ν: Einstein coefficient, q: photon density, q_s: saturation photon
density). A simple rate equation model yields

(2) $q_s = \pi \cdot \Delta\nu_H \cdot R/(4 \cdot B_{1,2} \cdot h \cdot \nu)$

where $\Delta\nu_H$ is the homogeneous line-width, R is the vibrational
relaxation rate and collisional exitation is neglected.
$\Delta\nu_H$ is estimated from the FIR line width, the density of the lower
absorbing level is roughly estimated to 1/10 of the total density.
The uncertainty of this estimate affects the value of the Einstein
coefficient but nearly cancels out for the vibrational relaxation
rate.
The absorption measurements were carried out with a CO_2-laser stabilized for simplicity to the CO_2 line center. The absorption cell
length was 120 cm. At a fixed intensity the transmission through
the absorption cell was measured as a function of HCOOH pressure.
The linear absorption measurements (high pressure) yield an Einstein coefficient of

$$B_{1,2} = \int B_\nu d\nu = 4 \cdot 10^{15} \; m^3 s^{-2} J^{-1}$$

The saturation densities q_s calculated from the saturation measurements are shown in Fig. 9 as a function of pressure. Evaluation of

the measurement data was done by iteratively solving the radiation
transport equation, integrated over the cell length for q_s, using
the absorption coefficient (1). We expect the saturation density to
vary quadratically with pressure, since both the homogeneous line
width $\Delta_{\nu H}$ and R are proportional to pressure. This is indeed the
case as can be seen from Fig. 9. Using the measured q_s values we
find the vibrational relaxation rate from (2):

$$R = 2 \cdot 10^4 \text{ s}^{-1} \text{ at } 100 \text{ mTorr.}$$

This value may be uncertain to some extent because of the assumptions involved. Nevertheless we can draw some conclusions from this
value. Assuming that rotational deexitation of the upper FIR laser
level is caused only by stimulated emission, we can calculate the
limiting FIR power for the case that equal populations are maintained by the pump radiation on the absorbing transition i.e. for
complete pump saturation. For the excited volume in our laser we
obtain a limiting power of 0.6 W at 100 mTorr. With the available
pump power obviously we do not reach this vibrational relaxation
limit. The small quantum efficiency must consequently be attributed to the collisional deexitation of the laser transition. It
can therefore be expected that the quantum efficiency will increase
with pump power and resonator Q. FIR gain measurements are however
needed to assure these qualitative conclusions and to give quantitative values of FIR gain coefficient, Einstein coefficient, and
rotational relaxation rates.
Knowledge of these data will enable one to find the optimum design
and operating conditions of FIR lasers. The availability of stably
operating FIR lasers will also be valuable for these measurements.

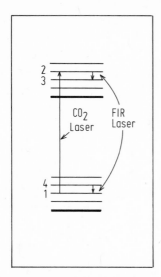

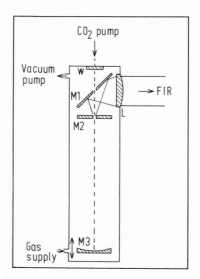

Fig. 1: Energy levels and
         transitions in a FIR
         laser.

Fig. 2: FIR laser schematic, M_1,
         M_2, M_3 gold mirrors
         W: ZnSe-window, L: TPX
         lens, M_3 is movable
         to tune the resonator.

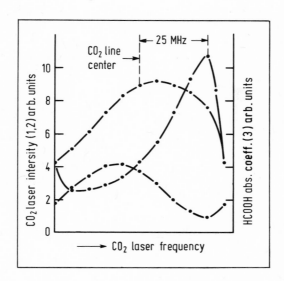

Fig. 3: CO_2-laser power (1), power transmitted through HCOOH
         absorption cell (2), HCOOH absorption coefficient (3),
         frequency scale is non-linear.

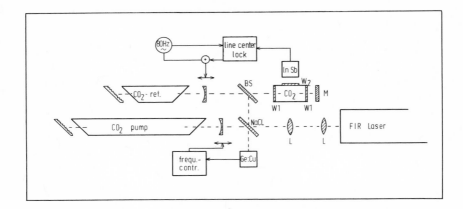

Fig. 4: Frequency-stabilized FIR laser pumping system.
BS: beam splitter, NaCl: NaCl plate, W_1: Ge-window,
W_2: Al_2O_3-window
InSb: detector 77 K, Ge: Cu: detector 4.2 K.

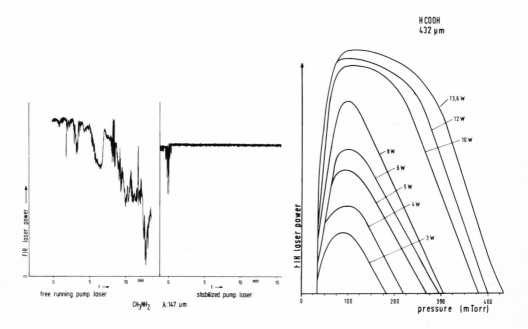

Fig. 5: FIR power output using
free running and
stabilized pump laser.

Fig. 6: FIR power output as
function of pressure and
pump power.

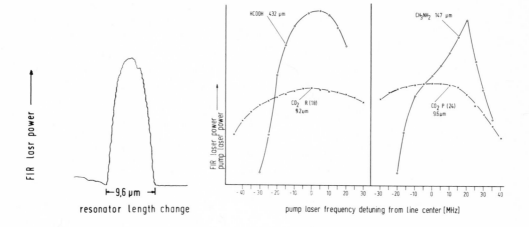

Fig. 7: FIR laser emission line   Fig. 8: FIR power and pump power
        profile, total width is   as function of pump
        8.4 MHz.   frequency for the 432 μm
                                              HCOOH-and the 147 μm
                                              CH_3NH_2-laser.

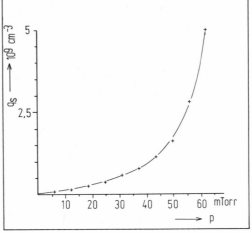

Fig. 9: Measured saturation
        photon density q_s as
        function of HCOOH
        pressure.

References:
1) Chang, T.Y.; Bridges, T.J., Opt.comm. 1 no 9 423 (1970)
2) Freed, C.; Javan, A., Appl.phys.lett. 17 no 2 53 (1970)
3) Tucker, J.R., Conf. on submillimeter waves, Atlanta 1974
   technical program p 17

TWO-PHOTON LASERS

D. E. Roberts and E. N. Fortson

University of Washington

Seattle, Washington 98195 U.S.A.

INTRODUCTION

In low-pressure gas lasers the gain linewidth for a single
transition is determined by the Doppler width of the transition.
Such a width is typically 10^{-5} or 10^{-6} times the transition frequency. Various methods have been used to stabilize laser frequencies to tolerances far better than the Doppler width.[1]
Most important among these are the Lamb dip and saturated absorption methods.

Here we wish to discuss the feasibility of another technique
that could avoid Doppler broadening in gas laser output. We discuss the possibility of operating a gas laser in such a way that
oscillation occurs simultaneously on two transitions in cascade.
Then, provided the homogeneous widths of the two transitions are
smaller than the Doppler width, there should be sharp features on
a plot of power output versus cavity tuning. These features would
have widths corresponding to the homogeneous widths, and the sum
of the two output frequencies would be defined only by atomic level
separations and not by Doppler broadening.

LINE NARROWING

Line-narrowing effects in coupled Doppler-broadened transitions
have been discussed before in various contexts.[2]-[7] Also, the
possibility of constructing a laser operating on a two-transition
cascade has been suggested.[5] Here we will make calculations that
bear on the feasibility of such a laser and discuss some of the
factors affecting the ultimate linewidth.

The explanation of the cancellation of Doppler shifts in such a laser is similar to the explanation of Doppler cancellation in the two-photon spectroscopy method that was recently suggested [8] demonstrated.[9] We consider transitions among three approximately equally spaced levels 2, 1, 0 (in order of descending energy) in which 2 → 1 and 1 → 0 will be the cascade laser transitions. Let us assume that the species having this level system is placed in an optical cavity and that the inversion between the states 2 and 1 is sufficient to maintain laser oscillation at the 2 → 1 transition frequency. This oscillation frequency would then be tunable over a frequency range limited by the Doppler width.

Now we consider the gain on the 1 → 0 transition. Even without considering line-narrowing effects, it is clear that the presence of the 2 → 1 saturating field will increase the gain on the 1 → 0 transition by increasing the population in the level 1. To simplify the discussion, it is convenient to suppose that the 2 → 1 laser field propagates only in one direction in the cavity, as could be the case in a ring laser. Then a sharp feature appears on the gain profile for a field with a frequency within a Doppler width of the 1 → 0 transition frequency, and propagating anti-parallel to the laser field. The gain feature will have a width determined by the homogeneous widths of the states 0 and 2, and will occur at a frequency such that the sum of the laser frequency and the 1 → 0 frequency add up to the total level separation 2 → 0. A primary result of the calculation below is that for typical inversion conditions and for a typical (saturating) 2 → 1 laser field, the narrow gain feature on the 1 → 0 gain profile will be comparable in size to the total 1 → 0 gain that would exist without any 2 → 1 field at all.

GAIN CALCULATION

The procedure in this calculation will be first to get an expression for the power delivered to the 1 → 0 field for atoms of one velocity class, and then to integrate over velocity to get the total power delivered to the field.

We use units in which $\hbar = 1$. The equation of motion describing our three-level system is

$$\dot{\rho} = i[\rho, H] - \frac{1}{2}\{\alpha, \rho\} + R \qquad (1)$$

$$\rho = \begin{pmatrix} \rho_{22} & \rho_{21} & \rho_{20} \\ \rho_{12} & \rho_{11} & \rho_{10} \\ \rho_{02} & \rho_{01} & \rho_{00} \end{pmatrix} \qquad H = \begin{pmatrix} 0 & H_{21}exp\,i\Delta w_{21}t & 0 \\ H_{12}exp\text{-}i\Delta w_{21}t & 0 & H_{10}exp\,i\Delta w_{10}t \\ 0 & H_{01}exp\text{-}i\Delta w_{10}t & 0 \end{pmatrix}$$

$$R = \begin{pmatrix} R_2 & & \\ & R_1 & \\ & & R_0 \end{pmatrix} \qquad \alpha = \begin{pmatrix} \gamma_2 & & \\ & \gamma_1 & \\ & & \gamma_0 \end{pmatrix}$$

Here ρ is the density matrix describing the three-level system; H is the perturbation Hamiltanian due to the laser field (H_{12} and H_{21}) and to the weak field for which gain is being calculated (H_{10} and H_{01}). ρ and H satisfy $\rho_{ij} = \rho_{ji}{}^*$ and $H_{ij} = H_{ji}{}^*$. R is the matrix giving rates at which the levels are populated and α gives the decay rates out of the states; γ_2, γ_1, γ_0 are the homogeneous widths of the levels 2, 1, 0. $\Delta w_{21} = E_2 - E_1 - \omega_{21}$ and $\Delta w_{10} = E_1 - E_0 - \omega_{10}$ are the detunings from the transitions $2 \to 1$ and $1 \to 0$; here E_2, E_1, E_0 are the level energies and ω_{21}, ω_{10} are the applied field frequencies, after Doppler shifting. For $H = 0$, the solution of (1) is $\rho_{ij} - \delta_{ij} R_j/\gamma_j = N_j$, the zero-field population densities. We choose for the dimensions of ρ and R volume $^{-1}$ frequency $^{-1}$; $2\pi R_j/\lambda$ is the number of atoms per volume per time per velocity along the beam. Here λ is the transition wavelength. The assumption that the R_j are independent of the Δw_{ij} is equivalent to the assumption that the Doppler width is much greater than the homogeneous widths.

After removing oscillating factors in the usual fashion [2],[4] from (1) we have

$$\left. \begin{aligned}
\dot{\rho}_{21} &= (\rho_{22}-\rho_{11})\,i\,H_{21} + \rho_{20}i\,H_{01} - \rho_{21}A_{21} \\
\dot{\rho}_{10} &= (\rho_{11}-\rho_{00})\,i\,H_{10} - \rho_{20}i\,H_{12} - \rho_{10}A_{10} \\
\dot{\rho}_{20} &= (\rho_{21}i\,H_{10} - \rho_{10}i\,H_{21}) - \rho_{20}A_{20} \\
\dot{\rho}_{22} &= (\rho_{21}i\,H_{12} - \rho_{12}i\,H_{21}) - \rho_{22}\gamma_2 + R_2 \\
\dot{\rho}_{11} &= -(\rho_{21}i\,H_{12} - \rho_{12}i\,H_{21}) + (\rho_{10}i\,H_{01} - \rho_{01}i\,H_{10}) - \rho_{11}\gamma_1 + R_1 \\
\dot{\rho}_{00} &= -(\rho_{10}i\,H_{01} - \rho_{01}i\,H_{10}) - \rho_{00}\gamma_0 + R_0
\end{aligned} \right\} \quad (2)$$

Equations (2) are first solved for the simple case in which $H_{10} = 0$, H_{12} arbitrary. The result for the steady state is the usual two-level result.

$$
\left.
\begin{aligned}
\underline{\rho}_{22} &= N_2 - (N_2 - N_1)\,\frac{w/\gamma_2}{1 + 2w\gamma_{12}/\gamma_1\gamma_2} \\[2mm]
\underline{\rho}_{11} &= N_1 + (N_2 - N_1)\,\frac{w/\gamma_1}{1 + 2w\gamma_{12}/\gamma_1\gamma_2} \\[2mm]
\underline{\rho}_{00} &= N_0 \qquad\qquad \underline{\rho}_{21} = \frac{iH_{21}}{A_{21}}\left(\underline{\rho}_{22} - \underline{\rho}_{11}\right)
\end{aligned}
\right\} \quad (3)
$$

$$\underline{\rho}_{10} = \underline{\rho}_{20} = 0$$

$$W \equiv Z_{21}\cdot 2\,\mathrm{Re}\,\frac{1}{A_{21}} \qquad\qquad Z_{21} \equiv |H_{21}|^2$$

$$A_{10} \equiv \gamma_{10} + i\Delta\omega_{10} \qquad\qquad A_{21} \equiv \gamma_{21} + i\Delta\omega_{21}$$

$$A_{20} \equiv \gamma_{20} + i(\Delta\omega_{10} + \Delta\omega_{21}) \equiv \gamma_{20} + i\Delta\omega_{20}$$

$$\gamma_{ij} = \tfrac{1}{2}\left(\gamma_i + \gamma_j\right)$$

Now we use the solution above to solve equations (2) approximately for the case $|H_{10}| \ll \gamma_2, \gamma_1, \gamma_0$. Defining $\Delta\rho_{ij} = \rho_{ij} - \underline{\rho}_{ij}$ to be the change in the density matrix component ρ_{ij} caused by the application of the field H_{10}, we find by substituting (3) into (2) that for small H_{10}, in the steady state:

$$
\left.
\begin{aligned}
0 &= (\Delta\rho_{22} - \Delta\rho_{11})iH_{21} + \Delta\rho_{20}iH_{01} - \Delta\rho_{21}A_{21} \\[1mm]
0 &= (\underline{\rho}_{11} - \underline{\rho}_{00})iH_{10} - \Delta\rho_{20}iH_{12} - \Delta\rho_{10}A_{10} \\[1mm]
0 &= \underline{\rho}_{21}iH_{10} - \Delta\rho_{10}iH_{21} - \Delta\rho_{20}A_{20} \\[1mm]
0 &= (\Delta\rho_{21}iH_{12} - \Delta\rho_{12}iH_{21}) - \Delta\rho_{22}\gamma_2 \\[1mm]
0 &= -(\Delta\rho_{21}iH_{12} - \Delta\rho_{12}iH_{21}) + (\Delta\rho_{10}iH_{01} - \Delta\rho_{01}iH_{10}) - \Delta\rho_{11}\gamma_1 \\[1mm]
0 &= -(\Delta\rho_{10}iH_{01} - \Delta\rho_{01}iH_{10}) - \Delta\rho_{00}\gamma_0
\end{aligned}
\right\} (4)
$$

To find the power delivered to the $1 \to 0$ field we need only solve for ρ_{00}:

$$\frac{\Delta\rho_{00}\gamma_0}{Z_{10}} = 2\,Re\;\frac{(\underline{\rho}_{11} - \underline{\rho}_{00})A_{20} - \underline{\rho}_{21}\,iH_{12}}{Z_{21} + A_{10}A_{20}} \quad (5)$$

$$= N_{10}\cdot 2\,Re\;\frac{A_{20}}{Z_{21}+A_{10}A_{20}} + N_{21}\left[1 + 2w\,\frac{\gamma_{12}}{\gamma_1\gamma_2}\right]^{-1}$$

$$\times\left[\frac{w}{\gamma_1}\cdot 2\,Re\frac{A_{20}}{Z_{21}+A_{10}A_{20}} + Z_{21}\cdot 2\,Re\left(\frac{1}{A_{21}}\cdot\frac{1}{Z_{21}+A_{10}A_{20}}\right)\right]$$

Here $Z_{10} = |H_{10}|^2$. Equation (5) agrees with more general results of reference 2. $\Delta\rho_{00}\gamma_0$ is interpreted as the number of photons per unit time delivered to the $1\to 0$ field per volume of medium per Doppler frequency interval. To get the total photon rate per volume, expression (5) must be integrated over frequency. Due to the way we have chosen the propagation directions for the fields, the detuning $\Delta\omega_{20}$ is the same for atoms of all velocities, so in the integration only $\Delta\omega_{10}$ and $\Delta\omega_{21} = \Delta\omega_{20} - \Delta\omega_{10}$ change.

We have been assuming that the Doppler width is very large so N_{21} and N_{10} are independent of the detunings. We find for the total photon rate P per volume

$$\frac{P}{Z_{10}} \equiv \frac{\gamma_0}{Z_{10}}\int_{-\infty}^{\infty}\rho_{00}\,d\Delta\omega_{10} = N_{10}\cdot 2\pi$$

$$+ N_{21}\cdot 2\pi\cdot\frac{2Z_{21}}{\gamma_1 Q}\cdot\frac{\dfrac{\gamma_0 + \gamma_2 Q}{2}}{\Delta\omega_{20}^2 + \left(\dfrac{\gamma_0 + \gamma_2 Q}{2}\right)^2}$$

where $Q = (1 + 4Z_{21}/\gamma_1\gamma_2)^{1/2}$. Equation (6) agrees with reference 2.

The above expression for P shows that the gain due to the inversion N_{10} is independent of cavity tuning regardless of the field intensity Z_{21}. (Of course the limited Doppler linewidth would eventually limit this gain for large detuning.) The gain due to the inversion N_{21} on the other hand is sharply frequency dependent. Its size is also of great interest: for mildly saturating Z_{21} ($Z_{21} \simeq \gamma_1\gamma_2$) and $N_{21} \simeq N_{10}$, the gain for $\Delta\omega_{20} = 0$ due to the N_{21} inversion is comparable to the gain due to the N_{10} inversion.

DISCUSSION

While it seems clear that a laser could be constructed that operated according to the above prescription, it is likely that such a laser would not approach the linewidth standards of saturated absorption methods. This is for the same reason that the conventional Lamb dip usually does not provide a good frequency standard [1]: the rather high pressures needed to get laser oscillation cause the Lamb dip to be broad compared with the gain features due to saturated absorption in low-pressure absorption cells. In CO, for example, the Lamb dip is not much narrower than the Doppler-broadened gain curve. [10] CO would otherwise be a good candidate for a two-photon laser medium, having as it does many cascade laser transitions of nearly equal frequency. One possibility for narrowing linewidths would be to have two discharges of the same gas in the same laser cavity: one high-pressure discharge to provide sufficient gain to maintain oscillation, and one at low pressure to provide the unshifted and unbroadened sum frequency resonance. A similar idea has been applied successfully to achieving a Lamb-dip in the He-Ne laser which is no broader than the radiative width of the laser transition [1].

It is possible also to use a second laser to supply an external field at, say, the $2 \rightarrow 1$ frequency, and then adjust the cavity tuning to achieve oscillation only on the $1 \rightarrow 0$ transition. The above calculations apply in this case as well, and there may be some advantages since this method avoids the necessity of obtaining oscillation simultaneously at two frequencies with one cavity setting.

REFERENCES

(1)   For a review, see V.S.Letokhov and V. P. Chebotaev, Kvant. Elektron. (Moscow) 1, 241 (1974). (Sov. J. Quant. Elect. 4 137 (1974).)

(2)   M. S. Feld and A. Javan, Phys. Rev. 177, 540 (1969).

(3)   B. J. Feldman and M. S. Feld, Phys. Rev. A5, 899 (1972).

(4)   T. Hansch and P. Toschek, Z. Phys. 23b, 213 (1970).

(5)   N. Skribanowitz, et al, Appl. Phys. Lett. 19, 161 (1971).

(6)   N. Skribanowitz, S. P. Herman, and M. S. Feld, Appl Phys. Lett. 21, 466 (1972).

(7)   I. M. Beterov and R.I. Sokolovskii, Usp. Fiz. Nauk. 110, 169 (1973) (Sov. Phys. - Usp. 16, 339 (1973)).

(8)   L. S. Vasilenko, V. P. Chebotaev, and A. V. Shishaev, ZhETF Pis. Red. 12, 161 (1970) (JETP Lett. 12, 113 (1970).)

(9)   M. D. Levenson and N. Bloembergen, Phys. Rev. Lett. 32, 645 (1974); F. Birahen, B. Cagnac, and G. Grynberg, Phys. Rev. Lett. 32, 643 (1974); T. W. Hansch, et al, Opt.Comm.11, 50 (1974).

(10) C. Freed and H. A. Haus, IEEE J. Q. Elec. QE-9, 219 (1973).

LIGHT-SHIFTS PRECISION MEASUREMENTS OF THE 0-0 TRANSITION IN A ^{133}Cs GAS CELL

M. ARDITI and J.L. PICQUÉ

Institut d'Electronique Fondamentale
and Laboratoire Aimé Cotton
Université Paris-Sud et CNRS, 91405. ORSAY - France

I. INTRODUCTION

Light-shifts in ground states have been previously studied both theoretically[1], and experimentally with the use of spectral lamps[2]. The recent developement of a CW, narrow band, tunable GaAs laser[3] has enabled us to obtain a check of the theory in the case of coherent monochromatic radiation, and to make accurate measurements of the light-shift of the 0-0 transition in a cesium gas cell as a function of the radiation frequency and intensity.

II. THEORY

For the light-shift $\Delta\omega$ and width γ of the hfs transition between the 0-0 levels of the ground state of the alkali atoms, expressions having the same form than that of Ref. (1) can be derived in the case of a monochromatic coherent radiation weak enough so that $/I/ \ll \Gamma$:

$$\Delta\omega = /I/^2 (\Omega - \Omega_0) \left[(\Omega - \Omega_0)^2 + (\Gamma/2)^2 \right]^{-1} \tag{1}$$

$$\gamma = /I/^2 \Gamma \left[(\Omega - \Omega_0)^2 + (\Gamma/2)^2 \right]^{-1} \tag{2}$$

where Γ = life time of excited state, Ω_0 = optical resonance frequency, Ω = radiation frequency
I = optical analogous of the Rabi nutation frequency.

In our experiment, the laser was tuned to one hyperfine component ($^2S_{1/2}$, F = 3 or F = 4 $\longleftrightarrow$ $^2P_{3/2}$) of the 8521 Å, D_2 resonance line in a Cs gas cell. Due to the velocity distribution in the vapor (Doppler width $\sim$ 400 MHz), the hyperfine structure of the $P_{3/2}$

excited state (splittings $\sim$ 200 MHz) was not resolved. In that
case, the theoretical expression for the light-shift of the 0-0
hfs ground state transition[4] is derived by simply adding shifts
of the type (1) for different values of the central frequency Ω_o
corresponding to the Doppler profile around each unresolved hyperfine component.

III. EXPERIMENTAL

The experiments were carried out on a resonance cell containing Cs vapor and a buffer gas. The CW GaAs laser[3] served both
to provide efficient hyperfine pumping[5] and to produce the light-
shift. The power available from the laser in single-mode operation
was of the order of 1 mW in a beam of 0.1 cm^2 cross-section ; the
lifetime of the excited state being $\Gamma^{-1} \simeq$ 30 ns.,the experimental
conditions were always such that $/\bar{1}/ < \Gamma$. The laser frequency was
set at different values around the frequency of one or the other of
the two hyperfine components issuing from the level F = 3 or F = 4
of the ground state ; for each laser frequency, the shift of this
level was probed by measuring the microwave frequency of the transition F = 3, m_F = 0 $\longleftrightarrow$ F = 4, m_F = 0.

Apparatus :   The laser single-mode frequency was locked to a
transmission resonance in an external Fabry-Perot cavity, resulting in a linewidth of a few MHz[6]. The laser frequency displacement was achieved by varying the length of the reference interferometer and measured by means of a second interferometer.

The resonance cell was placed in a D.C. magnetic field of
1 G parallel to the laser beam, to remove the degeneracy of the hfs
transition frequencies. The cell was irradiated with microwave power
from a horn fed by a klystron. The klystron was phase-locked to a
microwave signal at 9192.63... MHz produced by mixing a 9180 MHz
signal generated by harmonic multiplication from a 5 MHz crystal
oscillator and a 12.63... MHz signal from a stable and tunable
crystal oscillator. The 5 MHz crystal oscillator frequency was
synchronized with a radio-signal at 2.5 MHz generated by station
FFH, which is monitored by an atomic frequency standard. By measuring the 12.63... MHz frequency within a 0.1 Hz, an accuracy of
a few parts in 10^{11} could be obtained for the frequency of the
microwaves. The 0-0 transition was detected by monitoring the fluo-
rescent light from the cell, using synchronous detection as in optically pumped gas cell frequency standards.

Results :   Fig. 1 shows experimental data of the 0-0 frequen-
cy shift $\Delta\omega$ obtained by varying the laser frequency Ω step by step
across the absorption profile of the F = 4 component. In that case
the Cs cell contained 6 Torr neon as a buffer gas and was kept at
room temperature (20° C). The resonance linewidth was typically of
the order of 1.5 KHz. The theoretical curve, plotted as a solid
line in Fig. 1, is seen in good agreement with the experimental

points. The symmetrical dotted-line curves represent the contribution of each hyperfine component to the F = 4 level shift. From a
comparison between the theory and the experiment in that example,
we deduce a laser beam intensity of 30 µW/cm². By using calibrated
optical attenuators, we obtained other curves (Fig. 2) that well
verified the light-shift proportionality to the light intensity
and crossed at a zero-shift point ($\omega = \bar{\omega}_0$, $\Omega = \bar{\Omega}_0$) to within the
experimental uncertainties. From the determination of this point
for several buffer gas pressures, we could make accurate measurements of pressure-shifts of the 0-0 hfs transition. We obtained
in this manner a linear pressure-shift coefficient of 555 ($\pm$ 25)
Hz/t for neon buffer gas and 800 ($\pm$ 40) Hz/t for molecular nitrogen.

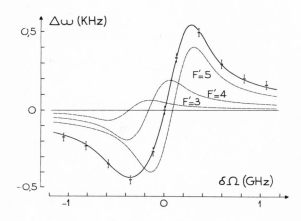

Fig. 1
Experimental data and
theoretical curves for
the light-shift $\Delta\omega$ =
$\omega - \bar{\omega}_0$ of the 0-0 transition vs laser detuning
$\delta\Omega = \Omega - \bar{\Omega}_0$ in a cesium
gas cell (6 T Ne)
(Dotted lines : contri-
bution of unresolved
hyperfine transitions).

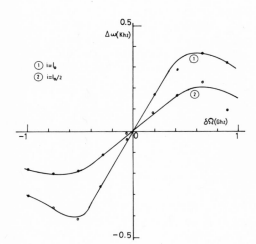

Fig. 2

Experimental light-shift curves
for laser light intensities I_0
and $I_0/2$ in a cesium gas cell
(5 T N_2).

    The previous results were typical ones obtained with light
intensities of the order of those used in conventional optical
pumping experiments. With increasing laser beam intensities, we
observed a dissymmetry and a broadening of the microwave resonance
curves. This phenomenon occured in an approximately symmetrical
way for opposite laser detunings $\Delta\Omega = \Omega - \bar{\Omega}_0$ and increased with
increasing detunings. Some resonance signals are reproduced in
Fig. 3 in the derivative form in which they were obtained. Fig. 4
shows the linewidth and the resonance signal (peak to peak of the
derivative curves) as a function of the laser detuning. The deformation of the resonance curves can arise from several effects[7].
Further experiments are now in progress to study the phenomena
occuring at high intensities.

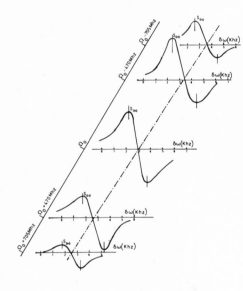

Fig. 3

Derivative of the magnetic
resonance signal S_{oo} vs microwave frequency for various
laser detunings.

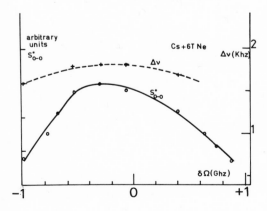

Fig. 4

Line-width and microwave
resonance signal S_{oo} vs
laser detuning.

IV. <u>APPLICATIONS</u>

Light-shifts in ground states may be used for some practical applications. For example, in the case shown in Fig. 1, the slope of the central portion of the S curve is about 3 Hz/MHz. We took advantage of this high sensitivity of the 0-0 frequency to laser frequency variations and of the high resolution of the microwave equipement to check the stabilization of the GaAs laser. To this effect, the microwave excitation frequency was locked to the center of the 0-0 resonance, as in optically pumped gas cell frequency standards. We deduced from its variations a stability of the laser frequency of about ± 2 MHz over a period of 1 minute.

A reverse process could also be used to stabilize the laser wavelength by maintaining the microwave frequency at a given value. A first attempt was made using the set-up sketched in Fig. 5. Part 1 of the set-up represents an optically pumped gas cell frequency standard using a laser light source. The frequency of the laser was then locked to the frequency of a stable variable oscillator by the error signal coming from a frequency-phase discriminator, as shown in Part 2 of the sketch. Preliminary results indicated a long-term stability of ± 1 MHz for the laser frequency. Frequency scanning of the laser along the central portion of the S-curve of Fig. 1 was achieved by varying the frequency of the variable oscillator. Stabilization at the zero-shift point gives a laser frequency independent of the optical radiation intensity (one could also

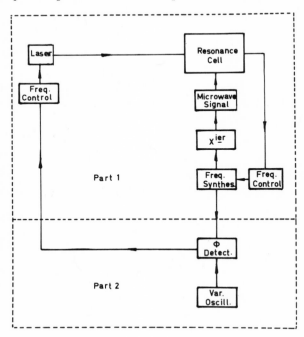

Fig. 5

Block diagram of laser frequency stabilization system.

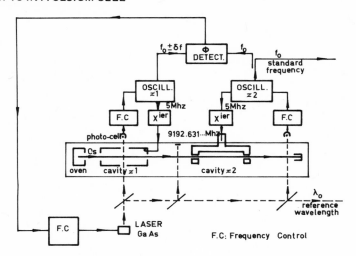

Fig. 6

Proposed apparatus for simultaneous optically pumped cesium
beam frequency standard and cesium wavelength reference.

obtain this point, without the reference oscillator, when the
frequency of the standard does not change with a modulation of the
laser beam intensity). However, with a system of the type of Fig.
5, the laser frequency could depend too much on the parameters
which affect a gas cell frequency standard (such as buffer gas
pressure, temperature, aging, etc.). A more absolute laser frequency could be obtained by replacing the gas cell in Part 1 of
this system by a cesium atomic beam feeding a microwave cavity in
vacuum. Also in this case, one could use in Part 2 a cesium beam
frequency standard, thus obtaining an optical frequency reference
directly connected with the time standard. Finally, we could combine two microwave resonators in series with a cesium beam (Fig.
6) to obtain simultaneously a wavelength reference and a frequency
standard using optical pumping[8] (or the latter could also use
the Rabi's method of magnetic deflection).

V. REFERENCES

(1) J.P. Barrat and C. Cohen-Tannoudji, J. Phys. Rad. 22, 329,
    443 (1961)
(2) M. Arditi and T.R. Carver, Phys. Rev. 124, 800 (1961)
    C. Cohen-Tannoudji, Ann. Phys. (Paris) 7, 423 (1962)
    P. Davidovits and R. Novick, Proc. IEEE 54, 155 (1966)
    E.N. Bazarov and V.P. Gubin, Radio Eng. Electron. Phys. 13
    1324 (1968)
    G. Busca, M. Têtu and J. Vanier, Appl. Phys. Lett. 23, 395
    (1973)

(3) J.L. Picqué, S. Roizen, H.H. Stroke and O. Testard, Appl.
    Phys. 6, 373 (1975)
(4) B.S. Mathur, H. Tang and W. Happer, Phys. Rev. 171, 11 (1968)
(5) J.L. Picqué, IEEE J. Quant. Electron. QE-10, 892 (1974)
(6) J.L. Picqué and S. Roizen, submitted to Appl. Phys. Lett.
(7) M. Arditi and J.L. Picqué, to be published
(8) M. Arditi and P. Cérez, IEEE IM-21, 391 (1972)

COMPARAISON DE LONGUEURS D'ONDE À L'AIDE D'UN INTERFÉROMÈTRE DE MICHELSON

Klaus Dorenwendt

Physikalisch-Technische Bundesanstalt

D-33 Braunschweig, Bundesallee 100, Rép.Féd. d'Allemagne

Au cours des dernier mois des comparaisons entre les longueurs d'onde d'un laser asservi sur le méthane et la radiation standard du krypton 86 ont été effectuées à la PTB. Après un bref apperçu de l'appareil et de la méthode de mesure les résultats et les erreurs systématiques seront discutés.

1. APPAREIL

Un interféromètre à deux ondes du type de Michelson, ressemblant à celiu du BIPM, a été utilisé pour les mesures (Fig. 1). La partie essentielle de l'interféromètre se trouve dans un caisson à vide V; les sources lumineuses L, K avec le monchromateur M, les récepteurs E_1, E_2 et l'optique d'adaptation sont montés sur des tables à l'extérieur du caisson. Pour diminuer la dérive thermique le caisson est bien isolé et les différentes pièces de l'interféromètre sont fixées sur un banc en invar. Les réflexions parasites sur les séparatrices T et les fenetres du caisson sont évitées par l'emploi de lumière polarisée sous l'angle de Brewster.

2. MÉTHODE DE MESURE

Le miroir mobile S_1 effectue une translation de $-$ 200 mm à + 200 mm par rapport à la différence de marche nulle. On mesure les variations d'ordre d'interférence pour les deux radiations et on en déduit le rapport des longueurs d'onde.

Les ordres d'interférences correspondant à la translation du miroir sont déterminés en faisant ce déplacement égal à la longueur connue d'un calibre à bouts A en zerodur. Pour cela on utilise un interféromètre auxiliaire, non représenté dans le dessin, devant

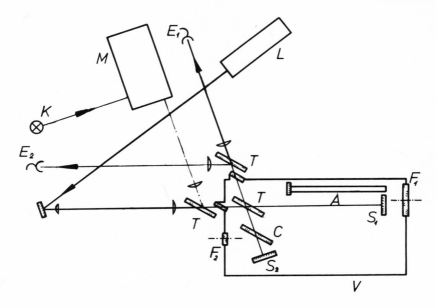

Fig. 1.: Principe de l'interféromètre.

la fenêtre F_1 qui permet d'observer des interférences en lumière
blanche entre les faces de mesure du cabibre et le revers traité
du miroir mobile.

    Les parties fractionaires des interférences sont déterminées
par la méthode dite "des quartre pointés" [1,2,3]. Une lame compensatrice rotative C varie le chemin optique autour de la valeur
cherchée. On mesure le flux émergent de l'interféromètre pour
quatre positions de la lame et un calcul simple donne la partie
fractionaire. La précision de cette méthode dépend de l'exactitude
avec laquelle les positions de C peuvent etre ajustées. Dans un
deuxième interféromètre auxiliaire, non représenté dans le dessin,
on repère à travers la fenêtre F_2 des Interférences à ondes mul-
tiples entre un miroir fixe et un miroir lié à la lame compensatrice. Les franges étroites, pointées à l'aide d'un réticule,
contrôlent la position de C suffisamment bien pour réaliser des
changements du chemin optique dans l'interféromètre principal avec
une exactitude qui est meilleure que $10^{-3} \lambda$ de la lumière visible.

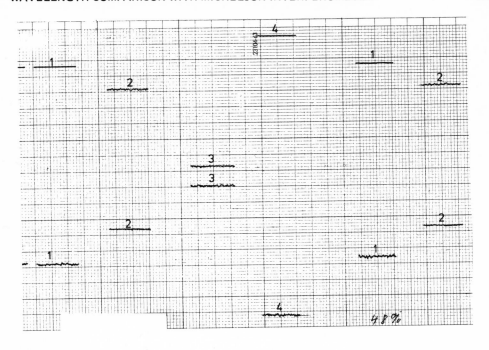

Fig. 2.: Enregistrement du flux sortant de l'interféromètre
pour quatre positions de C.

La figure 2 montre l'enregistrement du flux emergent de
l'interféromètre. Pour gagner du temps les mesures sont faites
simultanément pour les deux radiations avec les memes positions de
la lame compensatrice. On reconnaît les traits nets appartenent au
laser et les traits couverts d'un léger bruit de la radiation
krypton.

La figure 3 montre huit mesures successives des parties
fractionaires. Pour égaliser l'effet des longueurs d'onde
différentes léchelle varie d'un facteur six dans les deux
radiations. On reconnaît la dérive de l'interféromètre. Les
écarts des points de la droite permettent d'évaluer la précision
de la détermination des parties fractionaires.

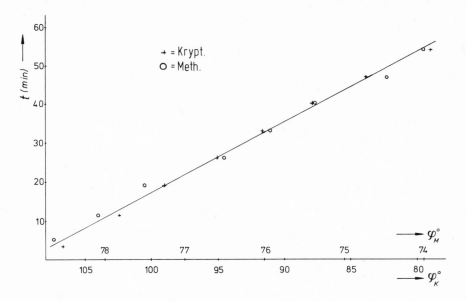

Fig. 3: Mesures des parties fractionaires $\mathcal{S}_K$; $\mathcal{S}_M$ en degré pour une différence de marche de 400 mm en fonction du temps t.

3. RÉSULTATS

Huit séries de mesures (Fig. 4) ont été effectuées. Chaque série comporte à peu près huit déterminations de l'ordre d'inter-férence ou huit valeurs pour la longueur d'onde cherchée du laser asservi sur le méthane.

Entre les différentes séries le réglage de l'interféromètre a été refait. L'écart-type des séries varie selon les conditions de mesure de $1.10^{-9}\ \lambda$ à $4.10^{-9}\ \lambda$. La valeur moyenne de toutes les mesures a donné le résultat $\lambda_M = 3,392231426$ µm. La lampe à krypton a été utilisée sous les conditions recommandées et aucune correc-tion due à un défaut éventuel de son profil spectral n'a été apportée a la valeur ci-dessus.

4. ERREURS SYSTÉMATIQUES, EXACTITUDE

Les différences apparentes entre les séries montrent l'exi-stence d'erreurs systématiques, dépendant du réglage de l'interféro-mètre. La première cause possible pour de telles erreurs est une inclinaison différente des deux faisceaux par rapport à l'axe optique. Nous croyons que son influence n'excède pas $2.10^{-9}\ \lambda$.

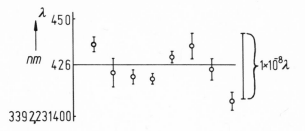

Fig. 4: Valeurs de la longeur d'onde du laser asservi sur le méthane.

La deuxième cause est plus complexe. Les centres de gravité des deux faisceaux ne tombent pas sur le meme endroit des miroirs. Avec les défauts de planéité et de réglage de parallélisme des miroirs il en résulte une autre erreur systématique plus difficile à évaluer. Nous estimons cependant que son influence aussi ne dépasse pas 2.10^{-9} λ. Le réglage de l'interféromètre introduit donc une incertitude de 4.10^{-9} λ.

La diffraction de la lumière, notamment dans l'infrarouge, altère les résultats d'une mesure interférentielle. La figure 5 montre le chemin que nous avons choisi pour tenir compte de ce phénomène $[4]$. Une onde plane à l'entrée de l'interféromètre est brouillée par la diffraction aux bords du diaphragme E. A la sortie les ondes Σ_1 et Σ_2, ayant parcouru des chemins inégaux dans les deux bras de l'interféromètre, se présentent sous une forme différente. Dans le plan focal de la lentille O_2 elles peuvent etre décrites d'une manière simple grâce à l'aspect peu compliqué de la diffraction à l'infini $[voir p.e. 5]$. L'expression de chaque onde contient une fonction $G(u,v)$ qui est à des constantes près la transformée de Fourier de la répartition d'amplitude $g(x,y)$ dans le diaphragme E. Les fonctions exponentielles suivantes représentent une courbure des surfaces d'onde, dépendant des chemins d_1, d_2 parcourus par les ondes entre E et O_2. La dernière fonction exponentielle de Σ_2 décrit le déphasage des ondes, dû au déplacement z du miroir M_2. Les signaux S d'un récepteur à la sortie résultent d'une intégration de la répartition d'intensité dans ce plan. En donnant S la forme de l'avant dernière ligne Φ (z) exprime le déphasage des interférences, provoqué par la diffraction. Nous avons calculé Φ (z)

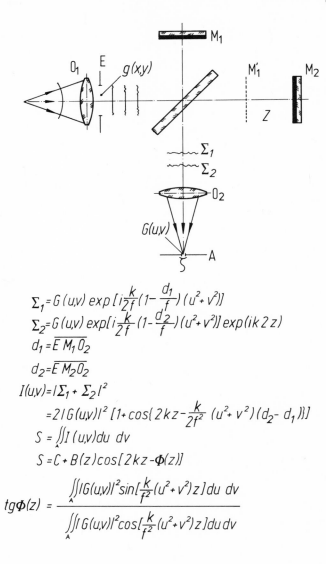

$$\Sigma_1 = G\,(u,v)\,exp\,[\,i\frac{k}{2f}(1-\frac{d_1}{f})\,(u^2+v^2)]$$

$$\Sigma_2 = G\,(u,v)\,exp[\,i\frac{k}{2f}\,(1-\frac{d_2}{f})(u^2+v^2)]\,exp\,(ik\,2\,z)$$

$$d_1 = \overline{E\,M_1\,O_2}$$

$$d_2 = \overline{E\,M_2\,O_2}$$

$$I(u,v) = |\Sigma_1 + \Sigma_2|^2$$

$$= 2\,|G(u,v)|^2\,[1 + cos\,\{2\,k\,z - \frac{k}{2f^2}\,(u^2+v^2)\,(d_2 - d_1)\}]$$

$$S = \iint_A I\,(u,v)\,du\ dv$$

$$S = C + B(z)\,cos\,[\,2\,k\,z - \Phi(z)]$$

$$tg\Phi(z) = \frac{\displaystyle\iint_A |G(u,v)|^2 sin\,[\frac{k}{f^2}(u^2+v^2)\,z\,]\,du\ dv}{\displaystyle\iint_A |G(u,v)|^2 cos\,[\frac{k}{f^2}(u^2+v^2)\,z\,]\,du\,dv}$$

Fig. 5: Influence de la diffraction sur une mesure interférentielle.

pour notre interféromètre avec un diamètre de 25 mm du diaphragme E, une focale de 1000 mm de la lentille O_2 et un diamètre de 0,5 mm du trou de sortie. Il en résulte une correction de $-5.10^{-9}\lambda$ qui doit etre apportée au résultat du chapitre précédent. A cause de la surface non uniforme de notre récepteur nous croyons cependant que cette correction n'est sûre qu' à 2.10^{-9} λ.

Nous avons donc trouvé pour la longueur d'onde du laser asservi sur le méthane

$$\lambda_M = 3,392231409 \ \mu m.$$

Nous estimons que le résultat ci-dessus devrait etre exact à 6.10^{-9} λ près, tenant compte des incertitudes du réglage et de la diffraction.

REFERENCES

1. TERRIEN, J., Optica Acta (Paris), 6, 301 (1959)

2. ROWLEY, W.R.C., HAMON, J., Rev.Opt., 42, 519 (1963)

3. CARRÉ, P., Metrologia, 2, 13 (1966)

4. DORENWENDT, K., BÖNSCH, G., à paraître, Metrologia

5. GOODMAN, J.W., Introduction to Fourier Optics, p. 86,
           Mc. Graw-Hill, San Francisco (1968)

WAVELENGTH INTERCOMPARISON OF LASER RADIATIONS USING SERVO-LOCK INTERFEROMETRY

W.R.C. Rowley, K.C. Shotton, P.T. Woods

National Physical Laboratory

Teddington, England

Introduction

The general level of accuracy in wavelength intercomparison has improved steadily over the last twenty years, largely as a result of the development of frequency stabilized lasers. In 1957, five laboratories reported intercomparisons of the wavelengths of cadmium and krypton-86 radiations (1), which were in mutual agreement within a scatter of ± 3 parts in 10^8. By 1973, the measurement techniques had improved such that six reporting laboratories (2) agreed to ± 2 parts in 10^9 on the wavelength value of the iodine-stabilized helium-neon laser radiation at 633 nm with respect to the krypton-86 primary standard. The reproducibilities of lasers stabilized by saturated absorption techniques are, however, far in advance of these figures. Such lasers have stimulated developments in interferometry to utilize their potential for measurement purposes, such as in the determination of those fundamental constants which are related to the length unit.

One such development made possible by the narrow-band tunability of single mode gas lasers is that which may be described as servo-lock interferometry. It may be used with Michelson and Fabry-Perot interferometers as well as with resonant cavity interferometers (3). With this method of operation, a tunable 'slave' laser makes it possible for the interferometer to transmit both the radiations under intercomparison without any length or path difference scanning. The principal advantage of this technique is that it avoids many of the systematic errors which may arise in scanning and interpolation systems, such as any non-linearity of scan, the dispersion of refractive index or temperature changes resulting from gas pressure scanning, the distortion of recorded interference patterns due to amplifier bandwidth limitations, or loss of parallelism occurring during length scanning. Compared with continuous scanning systems, the technique also has an advantage for weak sources due to the more efficient use of the observation time.

A servo-locked Fabry-Perot interferometer system has been developed at NPL for measurements on stabilized laser radiations. A version of this system, based on a 20-cm Fabry-Perot etalon has been used for wavelength measurements of CO_2 laser radiation, via up-conversion, with respect to a stabilized 633 nm laser, as part of a determination of the speed of light (4). A measurement uncertainty of $\pm$ 2.5 parts in 10^9 was achieved for the 9.3 μm infrared wavelength, corresponding to an intercomparison of the 633 nm laser and 679 nm up-converted radiations with an accuracy of $\pm$ 1.7 parts in 10^{10}. A new interferometer system using a 1-metre Fabry-Perot etalon has since been constructed, and as described below, it has demonstrated the capability of making interferometric wavelength intercomparisons to an accuracy of one part in 10^{11}.

1-Metre Fabry-Perot Interferometer

The optical system associated with the interferometer is shown in Fig 1. The Fabry-Perot etalon is enclosed in a tubular stainless steel vessel, with fused silica windows, which is maintained at a pressure below 10^{-8} Pa (10^{-6} torr) by an ionisation pump. The vessel is thermally insulated and maintained at a constant temperature, a few degrees above ambient, by a thermistor bridge and heater winding in a feedback loop. Mirror optics are used for collimating and focussing, with additional folding mirrors to form a fairly compact arrangement, and to minimise the off-axis angle of the 1-metre focussing system.

The system is illuminated through a rotating diffuser, which destroys the spatial coherence of the laser radiations, enabling substantially uniform illumination to be achieved over the exit pinhole (0.3 mm diameter) and across the area of the interferometer plates (30 mm diameter). The exit pinhole is focussed onto the photodetector to minimise the effects of non-uniform sensitivity, which would otherwise be equivalent to non-uniform illumination in the plane of the etalon.

The Fabry-Perot etalons have interchangeable reflecting plates, flat to $\lambda/100$ or better, and spacers with lengths between 20 mm and 1 metre. The design, shown in Fig 2, uses silica tubes of 45 mm internal diameter. At each end, piezoelectric plates (in pairs for increased sensitivity) are attached at three positions using epoxy resin, and to each piezoelectric pair an invar stud is

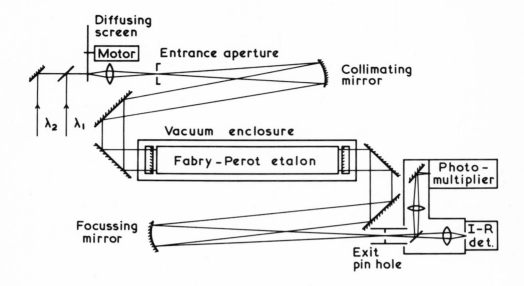

Fig.1. Interferometer optical system.

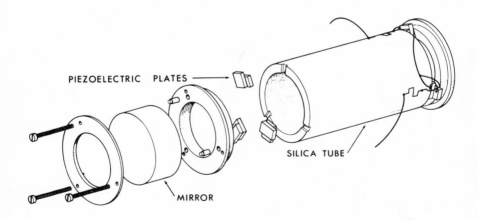

Fig.2. Details of the Fabry-Perot etalon construction.

similarly attached. The studs are given lateral support by being
cemented into an aluminium ring. This enables them to be ground and
polished coplanar with the two sets parallel to about one visible
wavelength. Light retaining rings are used to press the
interferometer flats against the invar studs. In use, the
piezoelectric adjustments at one end are used only for angular
adjustment and those at the other end for length adjustment. For
servocontrol purposes, a continuous length modulation is applied
with a peak to peak amplitude of $\lambda/140$ at 716 Hz.

Servocontrol System

The method of operation under servocontrol may be explained
with the aid of Fig. 3. The interferometer length is adjusted under
servocontrol to a transmission peak for light from a visible He–Ne
laser, which has its frequency locked 20 MHz offset from the 633 nm
iodine-127 stabilized He–Ne laser serving as the wavelength
standard. The frequency offset laser has inherently good short term
stability and the long term reproducibility of the
iodine-stabilized laser is transferred to it by using a slow
response lock system (τ = 30 ms). The average offset frequency is
controlled digitally from the beat between the two lasers.

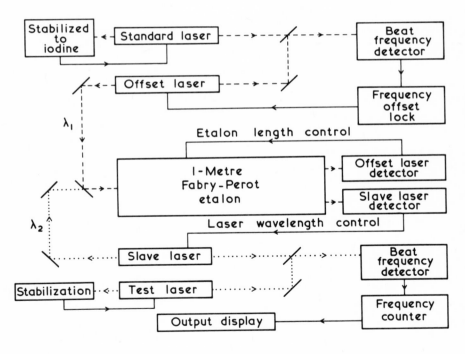

Fig. 3. System of laser and interferometer servocontrols.

Light from the tunable 'slave' laser, having roughly the same wavelength as the 'test' stabilized laser, also enters the interferometer. The slave is tuned and servo-locked to a transmission peak of the interferometer. It will then, like the offset laser, have an integral number of wavelengths in the etalon path difference. The ratio of the wavelengths of the two lasers is given by the ratio of their order numbers, although more than one etalon length is necessary to eliminate the effects of dispersion in the phase changes occuring in the reflecting surfaces. The small wavelength difference between the radiations of the slave and the test lasers is then determined by counting their beat frequency. When the standard and test radiations are of greatly different wavelength it is possible to use separate detectors for each and to have both radiations passing through the etalon simultaneously. When the wavelengths are similar, however, the radiations are used alternately, and the etalon must be stable enough for its length to remain constant during the (15 second) periods when it is not under active control.

The servo-lock system electronic design (5) is too detailed for description here, but certain critical aspects can be identified. The basic principle is to detect any signal in the interferometer photoelectric output which is synchronous with the 716 Hz length modulation, and to derive from this a rectified error signal which is integrated and fed back to the piezoelectric adjustment so as to eliminate any 716 Hz component. Systematic errors may arise from effects which generate false error signals. Voltage offsets in the electronic system after rectification are clearly of fundamental importance and are minimised by having as much amplification as possible before the rectification circuit. The limit to the allowable signal at the rectifier may, however, be lower than anticipated because even slight non-linear behaviour can cause a rectification effect for non-synchronous signals. The rejection of the rectifier to the harmonic of the modulation frequency is also very important.

Optical Adjustments

The two most critical aspects governing the absolute accuracy of the instrument are the deviations of the servo-lock systems from ideal performance, and the adjustment of the two beams of radiation onto the common axis of the interferometer.

The optical adjustment is critical because of flatness errors in the etalon plates. These are of the order of magnitude $\lambda/100$ or less, corresponding to roughly 5 parts in 10^9 of a one-metre etalon. The extent to which the instrument performs better than this is directly related to the adjustment precision. Accurate average parallelism of the etalon is readily achieved with the instrument under servo-lock to the offset laser. When maintained on an intensity maximum in this way, the transmitted intensity through the small exit pinhole is sensitive to the instrument finesse, and hence on its average parallelism, enabling the adjustment to be made simply and objectively. Similarly the exit pinhole can be centred on the interference ring pattern with excellent precision.

The choice of image size at the diffuser and its focussing onto the entrance aperture are important. The diffuser is imaged through the interferometer onto the exit pinhole with a magnification of 2.7, so that any intensity non-uniformity corresponds to a change of the average optical path difference. The magnitude of this for the Gaussian intensity distribuion of a laser beam may be calculated, and also verified experimentally as described below. The choice of image size and also the angular scattering uniformity of the diffuser are necessarily compromises between optical transmission efficiency of the instrument and the magnitude of systematic effects. In practice it has been found useful to improve the illumination uniformity by using asynchronous symmetrical sawtooth scanning of the beam positions on the diffuser, with amplitudes comparable to the beam diameters.

Experimental Verification of Performance

Tests for systematic effects arising from defects of optical adjustment or electronic control have been made using a 1-metre Fabry-Perot etalon, with a 633 nm laser as the slave and the iodine-stabilized laser acting as both test laser and standard. Initially observations were made without using an offset laser, so that the slave/test beat frequency should have been centred about zero frequency. It was measured using a bi-directional counting technique (6). A discrepancy of 100 kHz (2 parts in 10^{10}) was observed, due to the frequency modulation of the standard laser coupled with the non-perfect symmetry of the interference pattern. When using the frequency offset laser however, the beat is centred about the offset frequency (20 MHz $\pm$ 20 Hz) and should ideally be identical to it. In practice it is found that under good adjustment conditions the average discrepancy is less than 5 kHz (1 part in 10^{11}).

Because the standard and test wavelengths are the same, they are passed through the etalon alternately. The etalon length is servocontrolled from the offset laser for 10 seconds, then held constant for 15 seconds while the slave is under control and an 8-second count of the slave/test beat is made. The time constant of the etalon servosystem is 4 seconds, while that of the slave laser to the etalon is 20 ms. The drift rate of the etalon (typically 4 parts in 10^{12} per second) is measured from the steady change of piezoelectric voltage when it is under servocontrol, and an appropriate correction applied.

Observations are usually made in sets of ten counts, taking about 5 minutes to complete. The standard error of the mean of a set is 3 parts in 10^{11}. Adjustments of etalon parallelism made between sets do not alter the mean value outside this uncertainty if the two optical beams are well aligned onto a common axis. If the alignment is noticeably maladjusted, the means of sets can change by up to one part in 10^{10} when the etalon parallelism is readjusted. For nominally good optical adjustments, 700 individual observations (70 sets of 10) corresponding to 18 realignments of the two laser beams and of the interferometer internal optics, over a period of a month, had a mean value correct to 1 part in 10^{11} with a standard error of the mean of 4 parts in 10^{12}.

Possible systematic errors arising in the servocontrol systems (voltage drifts, harmonic response, voltage offsets, incoherent rectification, etc) have been checked regularly, and can individually be kept below the equivalent of an error of 4 parts in 10^{12}.

It is concluded that the instrument has demonstrated an accuracy capability of one part in 10^{11} for the intercomparison of similar wavelengths. It is currently being used to remeasure, using up-conversion, the wavelength of R(12) CO_2 laser radiation at 9.3 μm, and will shortly be used for a direct wavelength intercomparison of the 633 nm and 3.39 μm stabilized lasers.

References

1.   Com.Consult.Définition Mètre,(Com.Int.Poids Mesures),
      2nd Session, M15, (1957).

2.   Com.Consult.Définition Mètre,(Com.Int.Poids Mesures),
      5th Session, M15, (1973).

3.   MIELENZ, K.D.,Proceedings of the conference on Precision
      Measurement and Fundamental Constants, held at Gaithersburg,
      Maryland, August 1970. Edited by D.N.Langenberg & B.N.Taylor,
      NBS Special Publication 343, 53, (1971).

4.   BLANEY,T.G., et al, Nature, 251, 46, (1974).

5.   SHOTTON, K.C. and ROWLEY, W.R.C.,
      NPL Quantum Metrology Report No. 28, (1975).

6.   ROWLEY,W.R.C., J. Phys. E., 8, 223, (1975).

A FIELD COMPENSATED INTERFEROMETER FOR WAVELENGTH COMPARISON

P. BOUCHAREINE

INSTITUT NATIONAL DE METROLOGIE
Conservatoire National des Arts et Métiers
292 rue Saint-Martin, F-75141 PARIS CEDEX 03

We have built at the Institut National de Métrologie (INM) a two beam interferometer, of Michelson type, for optical wavelength comparison. It is in vacuum and the separating plate works at Brewster's angle. Its main quality is to give a flat field of interference at a path difference chosen for measurements (temporarily 250 mm), hence providing a large flux. The standard line of krypton provides a signal-to-noise ratio which allows us to work with a continuous scanning of path difference and a time constant less than 10^{-1} second.

PRINCIPLE OF FIELD COMPENSATION

Proposed by P. Connes (1) the compensation of interference field is obtained by fitting some geometrical properties of the two interfering beams. If two rays interfere at a point M with an angle i, the path difference between the rays is related to the position of M by

$$\frac{\partial \Delta}{\partial x} = i \, ,$$

where x is the amplitude of a displacement perpendicular to the fringes. The larger the angle i, the thinner are the fringes. If the rays are interfering at infinity, they are parallel, and i = 0. Fringes are infinitely wide; but in that case, we are interested by the variation of Δ with the angle i of the two rays with a reference direction:

$$\frac{\partial \Delta}{\partial i} = x \, ,$$

417

where x is the distance between the planes containing the rays and parallel to the fringes at infinity. A stationary path difference is given if the rays are superposed. This is achieved in the Michelson interferometer when $\Delta = 0$, or when $i = 0$, at the center of rings. If the path difference is not null, it decreases as the angle of incidence increases. For high values of Δ a fine pinhole is needed to isolate rays with a definite order of interference. If N fringes have been scanned since zero path difference, the solid angle thus allowed is $2\pi/N$.

Several optical devices (1 to 6) have been proposed to keep the rays superposed at any path difference, for any angle of incidence. An extensive review has been given by J. Ring (4). In 1966, M. Cuisenier and J. Pinard (3) described the cat's eye interferometer used in Fourier spectroscopy. They mentioned the possibility, based on an idea of P. Connes, of compensating the interference field by varying the curvature of a mirror. We have built such an interferometer and, to our knowledge, it is the first attempt to realize that proposal.

A cat's eye is a retro-reflector made of a primary concave mirror, with a secondary one at its focus. Easier to make than a cube-corner, a cat's eye has the same property: to give a reflected beam symmetrical to the incident one with respect to a point O. The center of symmetry is the image given by the primary mirror of the center of curvature of the secondary one. With such a device, fringes are always located at infinity, independent of any tilting of the reflector, because reflected beams are always parallel. To compensate the interference pattern, it is necessary to superpose the images, in the separating plate, of the two centers of symmetry O and O' of the cat's eye in the two arms of the interferometer. This is achieved, when one cat's eye is translated, by changing the radius of curvature of the secondary mirror.

THE INTERFEROMETER

An optical drawing is given in fig. 1. A beam enters one half of a cat's eye (actually a quarter of it, since two independent channels are side by side in the same interferometer) and gives two output beams interferometrically opposite in phase; these two beams, separated from the incident one, can be simultaneously measured; thus fluctuations of the source can be corrected.

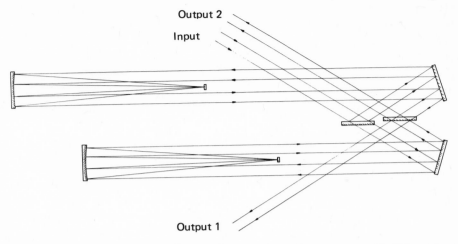

Figure 1 - Ray tracing in the cat's eye interferometer. Collimating mirrors, which fold the beams, are not reproduced. Field widening is given by the small secondary mirrors.

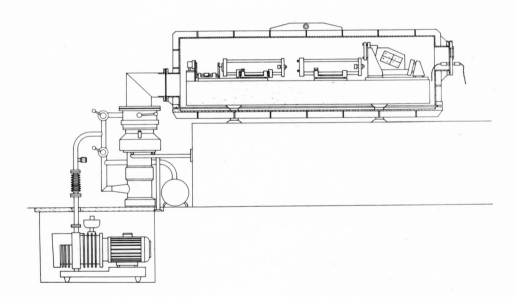

Figure 2 - The interferometer in its vacuum tank. Light enters the tank at the upper right, is folded down on the separating plate which is seen leaning, then folded again in the arms of the interferometer. The mechanical pump is under the ground level.

The optical path and the phase shift in the separating plate are well compensated in the mixing plate. The very good symmetry of the interferometer is seen with white light fringes at zero path difference. Phase measurements of fringes for different wavelengths at zero path difference have shown no significant phase shift, except for wavelengths shorter than 450 nm. This slight difference is due to some defects in the separating and mixing plates.

Experimental Arrangement (fig. 2)

The interferometer is built on a diabase basement (200 x 500 x 2 000 mm) weighing 600 kg and laying in a vacuum tank of stainless steel with damping rubber feet. The tank itself is laying on vibration-damping material on a concrete pier, isolated from the building. The interferometer is thus well protected against the town environment (four underground railways are travelling near the Conservatoire National des Arts et Métiers).

Interferometric Adjustments

There is no fine adjustment in this interferometer. The visibility of the fringes depends on the stability of the block on which are mounted the separating and mixing plates. It is made of melted aluminum alloy stabilized by a special heat treatment. The plates are supported against brass knobs that were ground with polishing tape. This alignment of the plates was long and tedious, but the instrument is very stable. At zero path difference, a fringe visibility of 90% could be achieved; this visibility has been kept above 85% for two months in the evacuated tank.

Varying Path Difference

It is not possible in this interferometer to modify the path difference by rotating the compensating plate, which is also a mixing plate and works by reflection. We have chosen to translate one cat's eye with an elastic parallelogram. A good translation, over 50 μm, reproducible within a few ångströms, was achieved with broad and thin steel blades on which the cat's eye was suspended. This device gives a good stability, particularly when the tank is being closed and the interferometer seriously shaken.

The translation is produced with a differential screw bending
a lever to reduce by 10 the amplitude of the movement. The cat's eye
can thus easily be moved at a speed of 300 Å · s^{-1}. One difficulty
was to avoid lateral efforts due to the rotation of the screw. A fairly
good decoupling was obtained with a clock pivot and its two sapphire
bearings. So a severe irreversibility in movements due to slipping of
steel balls on glass planes was completely removed.

RESULTS

Path differences are defined by the fringes given by a frequency
stabilized laser beam travelling in one channel of the interferometer.
We need not know the wavelength; we need only know that it is steady.
In the second channel are successively transmitted the standard line
of krypton, the line to be measured and the standard line again.
Fringe phase is measured at each pulse given by the laser fringes; as
the wavelength ratio is known with high accuracy, each phase angle
(or fractional excess) gives a value of the phase angle at the starting
point. On fig. 3 are plotted such values measured on 20 steps. Crosses
are measured with a forward movement of the cat's eye and circles with
a backward movement. The standard deviation for one measurement
is $4 \cdot 10^{-3}$ fringe, when the points indicated by large crosses have
been removed. These points are given by laser pulses that are near
the top or bottom of the fringes, where a small error in amplitude
is responsible for a large uncertainty in phase angle. A mean phase
is computed with points where cos φ is less than 0.96.

These first measurements were made with a continuous translation of the cat's eye and a few points were sampled with a time
constant less than 0.1 s. Of course, it will be necessary to use a step-
by-step motion to increase the time constant and avoid systematic
errors due to time delay in sampling operation.

Phase fringes measured for several wavelengths at a given path
difference allow computation of integer number of fringes for that
path difference. After correction for chromatic phase shift, experimentally measured at zero path difference, the wavelength of each
spectral line is thus measured by comparison with the primary standard
of length. Table I gives some results obtained with the first records
we were able to make at the beginning of 1975. They are compared
to measurements published in 1962 by some national laboratories and
to computed values of Humphreys and Kaufman (7) noted 'hu'.

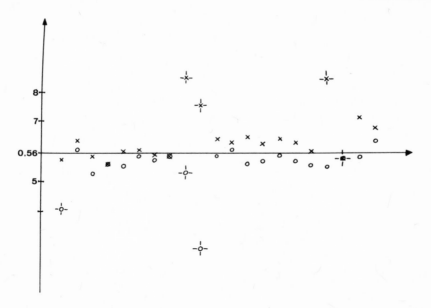

Figure 3 – Phase angle of an interferogram measured for a fixed path difference at several laser fringes. Crosses and circles refer to forward and backward motion of the cat's eye. Scale is in fractional excess (phase angle divided by 360°).

ACKNOWLEDGMENTS

I should like to thank some of the people who helped us in building this instrument. L. Henry (Paris VI University) assisted us in the vacuum tank manufacturing; M. Clinard (Institut d'Optique, Orsay) polished the mirrors and the separating plates; E. Pelletier (Marseille) evaporated the multiple dielectric coatings. Computer routines of P. Carré (BIPM) were helpful in writing ours. A. Janest (INM) took a large part in the technical realization. I am indebted to J. Hamon (BIPM), J. Vergès and G. Guelachvili (Laboratoire Aimé Cotton, Orsay) for helpful discussions.

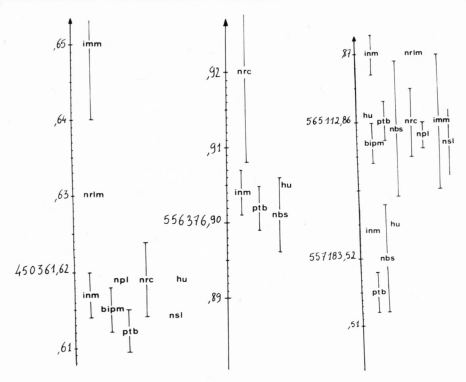

Table I - First and temporary results of wavelengths of krypton-86
lines, expressed in picometers.

REFERENCES

1.   P. Connes, Rev. Opt. 35, 375 (1956)
2.   P. Bouchareine, P. Connes, J. Phys. Rad. 24, 134 (1963)
3.   M. Cuisenier, J. Pinard, J. Phys. 28, C2, 97 (1967)
4.   J. Ring, J.W. Schoffield, Appl. Opt. 11, 507 (1972)
5.   M.J.E. Golay, J. Opt. Soc. Am. 63, 1217 (1974)
6.   J.G. Hirschberg, Appl. Opt. 13, 233 (1974)
7.   V. Kaufman, C.J. Humphreys, J. Opt. Soc. Am. 59, 1614
       (1969)

DETERMINATION OF NEW WAVELENGTH STANDARDS IN THE

INFRARED BY VACUUM FOURIER SPECTROSCOPY

Guy Guelachvili

Laboratoire Aime Cotton
C.N.R.S. II, Bat. 505
91405, Orsay, France

Abstract : 59 absolute wavenumbers of the 1-0 band of $^{12}C^{16}O$ are given with an uncertainty of ± 2 MHz.

Fourier Transform Spectroscopy using a Michelson Interferometer has the advantage over other traditional methods that one can simultaneously measure the wavelengths of a great number of transitions to high precision. It is therefore well suited to the provision of wavelength Standards. The absorption spectra of free molecules possess properties required for such Standards i. e. reproductibility, negligible shifts and easy use. Particularly suitable are the vibration-rotation spectra of diatomic or linear molecules characterized by rotational structures regularly and discretely distributed over large spectral ranges.

The extension towards the Infrared of the working spectral range of the molecular Fourier Spectrometer at Lab. Aimé Cotton has allowed a new measurement, under vacuum, of the 1-0 band of the $^{12}C^{16}O$ molecule. The 2-0 band of the same molecule was used as reference ; this has been measured previously on the same interferometer, the reference wavelength then being the "raie orangée du Krypton".

I - FOURIER SPECTROSCOPY AND ABSOLUTE WAVELENGTH MEASUREMENT.

To analyze a spectrum by Fourier Spectroscopy using a Michelson interferometer one essentially counts fringes as the path difference varies. These fringes result from the combination of a great number of monochromatic fringes, each having an intensity

and a frequency dependent upon the spectrum to be analyzed. For
a given spectrum, the resolution will be higher the greater the
number of fringes. This method is identical to those used in the
metrologic determination of length. Such work, however, is generally limited to quasi-monochromatic light, and unlike Fourier
Spectroscopy (F.S.) harmonic analysis is not performed to recons-
titute the spectrum. The quality of the wavenumber measurement by
F.S. will depend upon the quality of the counting and then upon
first, the path difference, secondly, the fringe intensity measurements. The Fourier spectrometer has been described elsewhere [1]
[2][3]. Taking into account the complexity of its functioning, and
the automatic nature of its operation, it is difficult to take proper account of all the possible systematic errors ; we therefore
draw a distinction between "metrological" experiments and "spectroscopic" experiments. The quality of the latter has been illustrated by the increase of the precision of numerous studies ; they can
normally furnish wavenumbers with relative error of $\pm 5 \cdot 10^{-7}$. A
detailed study of the various systematic errors which could affect
metrological work has been given in a chapter of [4]. It turns out
to be essential to measure the actual path difference Δ . The
exact superposition in space and time of the measured beam and the
measuring beam, from the source to the detector and from the beginning to the end of the experiment, minimizes the systematic errors.
The different sources of error then affect as similarly as possible
the measured wavelength and the standard of wavelength used for
measuring Δ . Let σ_{REF} be the wavenumber of the standard, and
$\sigma_{REF,EXP}$ its experimental value. Each wavenumber σ , the measured value of which is σ_{EXP} can be deduced from the formula :

$$\sigma = (1+\alpha) \, \sigma_{EXP} = (1 + \frac{\sigma_{REF} - \sigma_{REF,EXP}}{\sigma_{REF}}) \, \sigma_{EXP} \quad . \quad (1)$$

II – <u>ABSOLUTE WAVENUMBER MEASUREMENT OF THE 1-0 V.R. BAND OF $^{12}C^{16}O$</u>

A) <u>Experimental ; Description of spectra.</u>

Seven independent spectra of about 1 500 000 samples have been
recorded, each, over a period of about 20 hours, over a spectral
range of 5 000 cm^{-1}. <u>Figure 1</u> represents one of these spectra with
very low resolution. The combined effects of the detector (InSb ,
PV) and the white source (Globar) which would give an intensity of
the background considerably stronger at 2 000 cm^{-1} than at
4 000 cm^{-1} are balanced by the filter which enhances the second region. On the other hand the opacity of the filter in the 3 000 cm^{-1}
region, allows a higher source intensity, and therefore a better
signal to noise (S/N) ratio which is about 500. Only one detector,
at the symmetrical output of the interferometer was used in order
to avoid chromatic effects due to the beam splitters. The resolu-

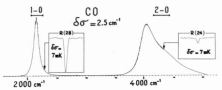

Fig. 1. Low resolution individual spectrum, with two windows on
         high resolution.

tion, apodized by the same function as in [5] is $7 \ 10^{-3}$ cm^{-1}. It
is midway between the respective Doppler width of the 2-0 ($11 \ 10^{-3}$
cm^{-1} at 300°K) and the 1-0 ($5.5 \ 10^{-3}$ cm^{-1} at 300°K). The absorption cell is a White type cell of 1 meter. The absorption length
is 40 m, the CO pressure 50 μ , giving no appreciable pressure
shifts (According to [6] the pressure shift is $5 \ 10^{-7}$ cm^{-1}). The
observed rotational transitions are consequently located, for the
2-0 between P(26) at 4 138 cm^{-1} and R(26) at 4 337 cm^{-1}, and
for the 1-0 between P(26) at 2 004 cm^{-1} and R(33) at 2 252cm^{-1}.
The measured halfwidth of the non-saturated lines are for each band
respectively $12 \ 10^{-3}$ cm^{-1} and $9.5 \ 10^{-3}$ cm^{-1}. In figure 1 are also
shown two small portions of the spectrum computed with maximum resolution in the R(28) region of the 1-0 and R(24) region of the
2-0 . The wavenumber scale is expanded enough to show the width of
the lines. A general presentation of the same spectrum is given in
figure 2 which represents only the 2-0 and 1-0 spectral region.
The linewidth in this case is too small to appear in the figure. A
localized normalisation is used to make the background constant and
yields a better appreciation of the relative intensities of the lines without reducing the noise amplitude. In the 2 000 cm^{-1} region
other vibrational transitions, under investigation at the present
time, are clearly present. They are due to isotopic species
$^{13}C^{16}O$, $^{12}C^{18}O$, $^{12}C^{17}O$. The spectra have been calculated by means
of programs written by H. Delouis [7]. The wavenumbers of the ob-
served transitions are measured in the same conditions as in [5].

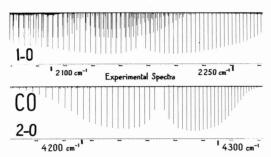

Fig. 2. High resolution plots of the spectrum shown in Fig. 1.

B) Results and Precision.

Each spectrum labelled I $(I=1,\ldots,7)$ gives two simultaneous sets k,ℓ of values denoted by $\sigma_{EXP}(I,K)$, $\sigma_{EXP}(I,L)$. The first with $K=1,\ldots,49$ corresponds to the experimental wavenumbers of the 2-0 band (25 R lines, 24 P lines), the second with $L=1,\ldots,59$ corresponds to the experimental wavenumbers of the 1-0 band (32 R lines, 27 P lines). Only lines with a S/N ratio higher than 20 appear in each set. The $P(4)$ and $P(24)$ transitions have been eliminated from the ℓ set since they coïncide with $R(8)$ and $P(14)$ respectively of the 1-0 band of $^{13}C^{16}O$. The determination of the $\alpha(I)$ (see formula 1) is performed starting from the $\sigma_{EXP}(I,K)$ by using the band center notion. Each set $\sigma_{EXP}(I,K)$, (or the set derived from them by application of $\alpha(I)$) has to sa-tisfy the following relation :

$$\sigma(m) = C_0 + \sum_n C_n\, m^n \tag{2}$$

where $m=-J$ for the P branch, and $m=J+1$ for the R branch. J is the rotational quantum number, C_0 the band center, and $C_1\ldots C_n$ parameters linearly depending upon the molecular constants B, D, H, ... of the 2 vibrational levels of the transition. The correct value of the wavenumber C_0 corresponds for each set to only one value of α . These values are then determined in such a way that the $C_{0,k}(I)$ (calculated band center of the corrected set k numbered I) obtained by a least squares analysis, coïncides with $C_{0,REF}$ band center of the standard 2-0 band given previously in [5] equal to 4 260.06264 cm^{-1}. The computation is performed in double precision with $n_{max} = 4$. Knowing the seven $\alpha(I)$ one can derive absolute experimental wavenumbers $\sigma_{ABS,EXP}(I,L)$ from the $\sigma_{EXP}(I,L)$ by application of formula 1.

The validity and the consistency of this correction can be il-lustrated by the following. Consider

$$A(I) = \frac{1}{59} \sum_{L=1}^{59} (\sigma_{EXP}(1,L) - \sigma_{EXP}(I,L)) \quad \text{and}$$

$$B(I) = \frac{1}{59} \sum_{L=1}^{59} (\sigma_{ABS,EXP}(1,L) - \sigma_{ABS,EXP}(I,L)) ,$$

which represent respectively the mean value of the differences bet-ween experimental wavenumbers of the set $\ell,1$ and of the set ℓ,I , and the mean value of the difference between these same wavenumbers but corrected by $\alpha(1)$ and by $\alpha(I)$. In figure 3 $A(I)$ and $B(I)$ are plotted versus I . The dispersion of the results is reduced by a factor of about 4 by application of the $\alpha(I)$. The distribu-tion $A(I)$ gives an idea of the reproductibility of the experiments but does not involve any information on the actual divergences between absolute positions, which have to be found using the $\alpha(I)$.

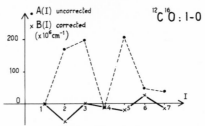

Fig. 3. Dispersion of mean wavenumbers for uncorrected and corrected
         spectra.

A second assessment of the consistency of the correction procedure
can be obtained without using the previous $\alpha(I)$ and taking into
account only the seven spectra.

$$\text{Consider} \quad D(I) = \frac{1}{49} \sum_{K=1}^{49} \left(\sigma_{EXP}(1,K) - \sigma_{EXP}(I,K) \right)$$

and $\alpha_k'(I)$ a correction factor identical to the $\alpha(I)$ and determined for $I=2,\ldots,7$, by the minimization of $D(I)$. Consider
also $\alpha_\ell'(I)$, minimizing the previously defined $A(I)$. In other
words, the wavenumbers of the 2-0 and the 1-0 values of the
first spectrum are used as standards for respectively the 2-0 and
the 1-0 of the other spectra which are then corrected by $\alpha_k'(I)$
and $\alpha_\ell'(I)$. Ideally one should obtain $\alpha_\ell'(I) - \alpha_k'(I) = 0$ for each
I. Table 1 gives what is actually obtained.

I	1	2	3	4	5	6	7
$\left(\alpha_\ell'(I) - \alpha_k'(I)\right) \times 10^8$	0	2.64	-0.90	0.93	1.41	0.72	-0.18

The mean value of these differences $\Delta\alpha = \frac{1}{6} \sum_{I=2}^{7} \left(\alpha_\ell'(I) - \alpha_k'(I) \right)$
is equal to $5.3 \ 10^{-9}$. It represents the discrepancy between the
actual correction and the ideal correction law, which corresponds
to a shift of $5.3 \ 10^{-9} \times 2\,000 \ cm^{-1} \simeq 10^{-5} \ cm^{-1}$ or 300 KHz for the

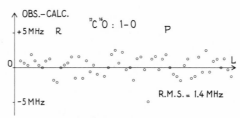

Fig. 4. Difference between observed and calculated wavenumbers given
         in Table 3.

1-0 wavenumber scale determined from the 2-0 region. The absolute experimental wavenumber of the 1-0 , $\sigma_{ABS,EXP}(L)$, are finally obtained from the mean value $\sigma_{ABS,EXP}(L) = \dfrac{1}{7} \sum\limits_{I=1}^{7} \sigma_{ABS,EXP}(I,L)$. A least squares analysis leads to calculated absolute wavenumbers $\sigma_{ABS}(L)$ which are the proposed standards given in Table 2 ; we give also the differences $[O-C](L) = \sigma_{ABS,EXP}(L) - \sigma_{ABS}(L)$ between observed and calculated wavenumbers. This $[O-C](L)$ distribution is shown in figure 4, and do not exceed $+76 \times 10^{-6}$ cm^{-1} ; in other words 2.3 MHz and −2.1 MHz.

J	L	$\sigma_{ABS}(L)$ P(J) cm^{-1}	O-C $\times 10^{3}$ cm^{-1}	L	$\sigma_{ABS}(L)$ R(J) cm^{-1}	O-C $\times 10^{3}$ cm^{-1}
0				28	2147.08158	0.02
1	1	2139.42652	0.04	29	2150.85645	0.02
2	2	2135.54663	0.05	30	2154.59603	0.00
3	3	2131.63202	-0.02	31	2158.30016	0.05
4		2127.68285		32	2161.96870	0.02
5	4	2123.69926	-0.17	33	2165.60149	0.00
6	5	2119.68140	0.03	34	2169.19840	-0.01
7	6	2115.62942	0.05	35	2172.75928	0.02
8	7	2111.54346	-0.02	36	2176.28397	0.04
9	8	2107.42367	0.04	37	2179.77234	-0.06
10	9	2103.27019	0.06	38	2183.22423	-0.07
11	10	2099.08318	-0.00	39	2186.63951	-0.02
12	11	2094.86278	-0.03	40	2190.01802	0.01
13	12	2090.60913	-0.04	41	2193.35961	0.02
14	13	2086.32238	-0.06	42	2196.66415	0.05
15	14	2082.00269	0.00	43	2199.93148	-0.05
16	15	2077.65019	0.04	44	2203.16145	0.04
17	16	2073.26504	-0.05	45	2206.35393	-0.05
18	17	2068.84737	-0.05	46	2209.50876	0.03
19	18	2064.39734	0.04	47	2212.62581	-0.02
20	19	2059.91509	-0.02	48	2215.70491	0.00
21	20	2055.40077	0.07	49	2218.74594	-0.04
22	21	2050.85452	-0.02	50	2221.74873	-0.05
23	22	2046.27650	0.03	51	2224.71316	-0.04
24		2041.66684		52	2227.63908	-0.00
25	23	2037.02569	0.04	53	2230.52630	0.06
26	24	2032.35321	-0.00	54	2233.37472	0.08
27	25	2027.64952	0.04	55	2236.18420	0.02
28	26	2022.91479	-0.01	56	2238.95457	0.07
29	27	2018.14915	-0.05	57	2241.68569	-0.02
30				58	2244.37743	-0.00
31				59	2247.02963	-0.06

Table 2 - Calculated absolute wavenumbers of the 1-0 band of $^{12}C^{16}O$. The uncertainty is ±2 MHz (or ±70 μcm⁻¹). Observed-calculated wavenumbers differences are given when the observed line has been used in the computation.

The estimation of the uncertainty has to be the statistical combination of a) the uncertainty affecting the reference wavenumbers (or $C_{0,REF}$), b) the probable systematic error due to the instrument and the method, c) the random error. The two first terms have been estimated elsewhere [4][5] and are respectively equal to 3 and 1×10^{-8}. The third is given by

$$\frac{1}{C_{0,ABS}} \times \left[\sum_{I=1}^{7} \frac{(C_{0,ABS} - C_{0,ABS}(I))^2}{7} \right]^{\frac{1}{2}} \quad (3)$$

and is equal to $\dfrac{39 \times 10^{-6} \text{ cm}^{-1}}{C_{0,ABS}} = 1.8 \times 10^{-8}$. Finally the absolute wavenumbers of the Table 3 are given with a relative error $\dfrac{\Delta\sigma_{ABS}}{\sigma_{ABS}} = 3.5 \times 10^{-8}$. In other words, the uncertainty is equal to

$\pm 70\ 10^{-6}$ cm^{-1} or ± 2 MHz. Let remark that the main error is coming from the reference. The molecular constants are given below with an uncertainty obtained by formula identical to (3).

$C_0 = 2\ 143.27157 \pm 0.00007$	
$B_0 = 1.92252908 \pm 30\ 10^{-8}$	$B_1 = 1.90502618 \pm 30\ 10^{-8}$
$D_0 = 6.1206\ 10^{-6} \pm 0.0007\ 10^{-6}$	$D_1 = 6.1203\ 10^{-6} \pm 0.0007\ 10^{-6}$
$H_0 = 7.44\ 10^{-12} \pm 1\ 10^{-12}$	$H_1 = 6.85\ 10^{-12} \pm 1\ 10^{-12}$

To our knowledge reference [8] gives the best previously published values of the 1-0 , $^{12}C^{16}O$ wavenumbers proposed as wavelength standards in the infrared. The comparison between these values and those of Table 4, shows a discrepancy the general form of which, parabolic versus J , is similar to the one found for 2-0 ([5] Figure 4). However in the present case the discrepancy is more accentuated ($3.4\ 10^{-3}$ cm^{-1}) and corresponds to a larger relative error. Our precision is then increased by a factor of 30 when compared to the precision of the best previous measurements. The values B_0 D_0 have already been given in [5]. We note the excellent agreement between the B_0 values (1.9225291 and 1.92252908). This corroborates the value of the velocity of light which was given in the same paper equal to 299792.46 m/s. The small shift between the D_0 comes from the new determination of H_0 which was previously assumed to be zero. Taking into account the opposite signs of D_v and H_v (formula (3)), introduction of H_0 must reduce D_0 slightly. This is in agreement with experiment.

The author is grateful to C. Amiot and F. Gautier for their helpful assistance in the computation part of this work.

References

[1] J. Connes et al. Nouv. Rev. Opt. Appl. I, 1, (1970).
[2] G Guelachvili Nouv. Rev. Opt. Appl. III, 6, (1972), 317.
[3] C Amiot, G Guelachvili J. Mol. Spect., to be published.
[4] G Guelachvili, Thèse d'Etat, Orsay 1973.
[5] G Guelachvili Opt. Comm. 8, 3, (1973).
[6] J P Bouanich, C Brodbeck Revue Phys. Appl. 9, (1974), 475.
[7] H Delouis, Thèse d'Etat, Orsay 1973.
[8] K.N Rao et al. Wavel. Stand. in the I. R. Ac. Press.
                                              N. Y. London, (1966).

PRELIMINARY WAVELENGTH MEASUREMENTS

OF A $^{127}I_2$ STABILIZED LASER AT IMGC

F. Bertinetto and A. Sacconi

Istituto di Metrologia "G.Colonnetti" (IMGC)

Torino, Italy

1. IODINE STABILIZED He-Ne LASERS

Iodine stabilized lasers will be applied at IMGC as secondary wavelength standards in laser frequency calibrations, measurements of "g" and silicon lattice parameter determination.

In developing our system we gave special attention to problems encountered by general users, such as the identification of the peak actually selected for stabilization when only one laser is available.

The cavity structure consists in two flanges supporting the mirror mounts, while a push-pull technique provides alignment of the mirrors; three invar rods, 39 cm long, are used as spacers, the thermal expansion of which is compensated by the mirror mounts; two separate PZT transducers provide modulation and stabilization; the $^{127}I_2$ cell, 10 cm long, is placed inside the cavity and a cold cathode tube is used with natural Ne and He filling.

The third harmonic technique is used to provide stabilization (Fig.1). A quartz stabilized oscillator generates a frequency of about 55 kHz, which is divided by 2^4; a circuit similar to that described by Wallard [1] performs a further division by 6. At this point,

the first to the third harmonic ratio is 40 dB, and becomes at least 75 dB after filtering. Hence, a satisfactory modulation signal is obtained at f_o = 573 Hz. A
second chain of flip-flops creates a reference frequency $3f_o$ = 1719 Hz. In our system it is possible to change
the logic state before and after each flip-flop by means
of ex-OR. Thus the phase of the $3f_o$ signal is changed,
with respect to that of the f_o one, with a setting of
approximately 6°.

After the detector and pre-amplifier, four filters
are used to reject the first, second and fourth harmonics and obtain only a good third harmonic signal, the
ratio of the third to undesirable harmonics being more
that 80 dB. The band pass filter has a 3 dB width of
about 100 Hz, thus limiting the stabilization band to
50 Hz.

We have recently developed a PSD, which can accept
a 10 V p-p signal at its input. In this way, it reduces
possible subsequent offset errors, with a definite advantage over the commercial instrument previously employed that can accept only a 3 V p-p signal. The phase compensated integrator, similar to those used elsewhere /2/,
is suitable for operating up to 1 kHz. The stabilizing
signal to PZT is fed by a high voltage amplifier with an
operational module providing a negative variable output
from 0 to 250 V. The amplifier bandwidth is 5 kHz and
its output ripple, less than 1 mV p-p, is mainly caused
by the power supply. Its spectrum approaches that of
white noise, so the third harmonic component is practically absent.

The detector and pre-amplifier are connected also to
a first harmonic band-pass filter, followed by an analog
phase shifter; a logic signal of the laser modulation
frequency is derived from the tripler circuit. The smoothed output of a second PSD is then used to avoid ambiguities concerning the peak whereon the servosystem is actually locking the laser.

We found this technique of parallel detection of the
first harmonic quite useful and reliable, requiring only
a previous calibration of the PSD output and a rough sta-

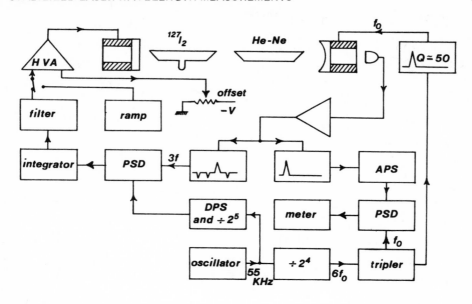

Fig.1. Schematic diagram of the control system.
PSD phase sensitive detector; DPS digital phase
shifter; APS analog phase shifter.

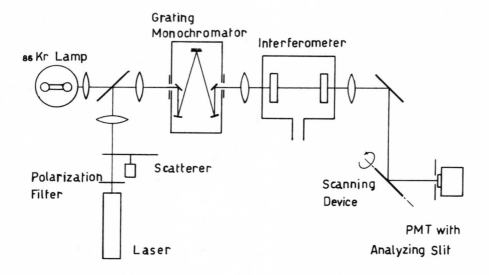

Fig.2. Experimental arrangement for wavelength
measurements.

bilization of the temperature of the iodine cell.

Up to now the cell has been operated at room temperature in our laboratory, namely (20 ±0.5) °C over long periods. We are planning to place the cavity in an isolated chamber and further stabilize the cell temperature within ±0.01 °C.

2. WAVELENGTH MEASUREMENTS

The absolute vacuum wavelengths of the h and i components have been measured by direct interferometric comparison with the ^{86}Kr primary standard line emitted as recommended /3/.

For the measurement being described we used an evacuated Fabry-Pérot interferometer with photoelectric scanning detection developed from the experimental arrangement (Fig.2), which proved to yield results in excellent agreement with precision measurements of other laboratories /4/.

To get further confirmation of the reliability of our results, we compared the 606 nm line with that of another lamp recently acquired. The two wavelengths resulted equal within an observed statistical uncertainty (standard deviation of the mean for 88 individual determinations) of 1.3 fm, i.e., well within the accuracy limits for these lamps.

A data acquisition system has been recently incorporated, so that the intensities of the interferential patterns can be digitally recorded for computer analysis. Properly amplified signals from the photomultiplier were displayed on a strip-chart recorder, as well as digitized and stored in a small computer. Each fringe of the interferential pattern consists of approximately 100 points.

The usual least square method enabled us to evaluate the fractional part of the order number from measurement of the diameters of six consecutive rings. The fractional parts of interference number having thus been obtained, the integer order numbers were determined with the method of exact fractions for five ^{86}Kr lines and laser line.   Finally, the value of the laser wavelength was calculated from the nominal wavelength value of $605\ 780.211 \cdot 10^{-12}$ m , adopted as standard value without any correction for spectral profile asymmetry or for extrapolation to unperturbed conditions.

The dispersion of phase change on reflection was taken into account using three different plate separations: 187 mm, 113 mm and 33 mm. For the three single distances we performed five or six sets of measurements, each of them requiring only 20 min and comprising ten individual wavelength determinations. Each determination is derived through the sequential scan laser – standard – laser. Between one measurement set and the next the usual precautions were taken to randomize all possible systematic errors.

The repeatability of a set of measurements was characterized by a relative standard deviation of the mean of $(4 \div 7) \cdot 10^{-9}$.

Our preliminary vacuum wavelength values are

$$632\ 991\ 397.7 \quad fm \quad (i \text{ component})$$
$$632\ 991\ 370.2 \quad fm \quad (h \text{ component})$$

the statistical variation corresponds to a standard deviation of the mean $\sigma_m = 1.6$ fm .

Further experiments and measurements are in progress to improve the laser system and the measurement apparatus, in order to evidence possible systematic errors.

REFERENCES

/1/ Wallard, A.J., Wilson, D.C. : J. Phys.E, 7, 161 (1974)
/2/ Wallard, A.J., NPL, private communication; Cérez,F., L.H.A., private communication
/3/ CIPM, Procès-verbaux des Séances, 2e série, 31, 76 (1963)
/4/ Sacconi, A., Fontana, S. : Metrologia 11, 33 (1975)

WAVELENGTH MEASUREMENTS IN THE UV AND VACUUM UV

G.H.C. Freeman and Vera P. Matthews

National Physical Laboratory

Teddington, Middlesex, U.K.

INTRODUCTION

The NPL is using a Michelson Interferometer for the measurement of wavelengths in the UV and VUV. The interferogram produced by a particular line is examined at a number of widely separated path differences by scanning over a short distance (1 to 4 fringes), so as to obtain the visibility and also the phase relative to that of a reference wavelength interferogram. These directly yield the shape of the line and its wavelength.

APPARATUS

All the optical components of the apparatus (1) are mirrors except for the beam splitter, thus allowing a large wavelength range to be observed. Each line is selected by a concave grating and the intensity at the centre of the ring pattern is measured using a photomultiplier. The interferogram (variation of the central intensity with path difference) is sampled by scanning one of the interferometer mirrors in 60 steps of 5 nm each (total 300 nm). This produces a change in optical path of about one visible fringe, or more at shorter wavelengths. The other mirror can be moved manually to set the path difference at which the interferogram is sampled. A computer fits the equation $I = A + B \cos (CM + 2 \pi \Delta)$, where I is the intensity and M the step number. The visibility $V = B/A$, and the phase Δ is in fractions of a fringe.

MEASUREMENT

The method of exact fractions is used to determine the order
of interference for each line at each of the path differences.
The fractional order of interference Δ is obtained from the
cosine fit. One of the lines must be either the primary standard
(λ 606 nm Kr^{86}I) or a secondary standard that is known with
sufficient accuracy.

ERRORS

When using unpolarized radiation, the largest source of error
is caused by the beam that is reflected at the uncoated surface
of the parallel sided beam splitter (2). This produces a second
set of fringes superimposed on the main ones. The effect of these
can be minimized by choosing a beam dividing coating that reflects
more than it transmits, but this is not practicable if a large
wavelength range has to be covered, as the visible transmission
would be too low, the only suitable coating being aluminium with
an overcoat of magnesium fluoride. Figure 1 shows the reflection
and transmission coefficients for the coated fused silica beam
splitter that was used for the present measurements. With this
beam splitter, the unwanted reflected beam produce a maximum
variation in the phase of the observed interferograms of $\pm$ 0.0081,
$\pm$ 0.028 and $\pm$ 0.032 fringes at λ 546 nm, λ 253 nm and λ 185 nm
respectively. The complementary visibility variations are 10%,
30% and 33% of the visibility value. The amount by which the
observed values are altered from one set of measurements to the
next depends on the relative phases of the interferograms from the
wanted and unwanted beam, and this changes with the beam splitter
temperature and tilt, 0.1 kelvin or 3×10^{-6} radians of tilt
being sufficient to produce the maximum change indicated above.

SOURCES

Two commercial (GEC) sealed-off Hg^{198} lamps were investigated.
Each had 2 mg of mercury isotope, one with 1 torr argon and the
other 3 torr argon as carrier gas. The envelope and water jacket
were of spectrosil. Deionized water was circulated through the
water jacket and was controlled to $\pm$ 0.03 °C. The optical depth
of water was about 2 mm and provided the water was changed each
month, the λ 185 nm line was transmitted with little attenuation.
A 300 MHz, coil coupled, power supply consuming about 20 watts of
power was used to excite the lamps.

Both lamps were measured at 4 °C and the 3 torr one at 20 °C
as well. The line shapes were also studied at a series of
temperatures up to 30 °C.

REFERENCE WAVELENGTH

The green and blue lines were used as internal wavelength standards. Rowley (3) measured these on an air scanned Fabry-Perot interferometer against the primary standard, Table 1. Previous work by Rowley (4) had shown that the wavelength of the green line cannot be assumed if accuracies of better than ± 20 fm are required. Also shown in table 1 are the CCDM recommendations (4).

For the measurements at 20 °C, the blue line was used with the minor wavelength correction measured by Emara (5). Rowley (6) has shown that the green line becomes asymmetrically self-absorbed as the temperature is increased, and Emara (5) has shown that the wavelength shifts by 0.85 fm degree^{-1} (13.6 fm for 4 to 20 °C). Terrien has reported similar changes (7).

RESULTS

Figure 2 shows the results for the two lamps at 4 °C. Plotted are the visibilities, and the difference between the measured phase and that calculated from an assumed wavelength for the line (we use Kaufman's values (8)). By fitting a straight line to the phase differences by a least squares method, a small correction can be calculated to convert the assumed wavelength to the true one. These are given in Table 2 and the errors are those from the least squares fit (1 standard deviation).

The table also shows the temperature and pressure shifts.

The scatter of points is explained entirely by the above effect of the unwanted beam reflected at the uncoated surface. For the 1 torr lamp, the interferometer temperature was only controlled to ± 0.1 °C and the larger number of points is sufficient to show the total range of values that the unwanted beam can produce. For the 3 torr lamp results, the temperature was repeatable to ± 0.01 °C from one set of measurements to the next and so the scatter is much less. However there could be a hidden systematic shift and this could amount to ± 20 fm for all the lines.

The phase curves for the λ185 nm line undergo a change of 0.5 fringe where the visibility curve has its minimum at about 60 mm path difference. Here the visibility of the interferogram from the wanted beam from the coated surface of the beam splitter is small and that of the unwanted beam from the uncoated surface is dominant and is the one that was measured, hence the larger scatter.

In Figure 2 the solid curves in the visibility results are a least squares fit assuming the widths are a convolution of a Lorentzian (includes natural) and Doppler broadening, after

Terrien (9). At 4 °C the lines from the 1 torr lamp all have widths (full width at half height) that are 1 x 10^{-6} (± 10%) of their wavelength, e.g. the λ 365 nm line is 3.0 m^{-1} wide. The 3 torr lamp produces lines that are 20% wider, this being mainly an increase in the Lorentz width. At 20 °C the 3 torr lamp lines are not appreciably wider except for the λ 253 nm and λ 185 nm lines. The λ 253 nm shows self-absorption, producing a minimum in the visibility at about 120 mm path difference, and is self-reversed at 30 °C. The λ 185 nm line is reversed under all conditions.

CONCLUSION

We have confirmed Kaufman's wavelengths (8) to within our combined errors. Our pressure shifts are slightly larger than his. Because the temperature shifts are of a comparable value, the lamps should be controlled in temperature to ± 1 °C if accuracies of 3 fm are required. If the λ 253 and 185 nm lines are to be used then the temperature should be as low as possible.

Improved accuracy of measurement could be obtained if the reflection from the uncoated surface of the beam splitter were removed, either by using a polarizer or perhaps an unsupported semi-reflecting film.

REFERENCES

1.   FREEMAN G.H.C.,   1971, Optics Communications, 4, 66-68.

2.   FREEMAN G.H.C.,   1971, National Physical Laboratory, Quantum
        Metrology Report Qu19.

3.   ROWLEY W.R.C.,   1972, unpublished.

4.   CCDM. Comité Consultatif pour la définition du mètre. Comité
        international des poids et mesures, Paris, Gauthier-
        Villars, 1962.

5.   EMARA S.H.,   1961, J. Res. NBS, 65A, 473-474.

6.   ROWLEY W.R.C.,   1963, unpublished.

7.   TERRIEN J.,   1954, National Physical Laboratory Symposium
        No 11 on Interferometry, 437-456.

8.   KAUFMAN V.,   1962, J. Opt. Soc. Amer., 52, 866-870.

9.   TERRIEN J.,   1967, J. de Physique, 28, C2/3-C2/10.

10. Average values for lamps filled with 1 and 3 torr gas taken
        from ROWLEY W.R.C.,   1962, reference (4), pages 99-103.

<u>TABLE 1</u>   Wavelength of Hg[198] reference lines

(a) Green line (5462Å)   Units are fm

	Argon carrier gas pressure		Pressure shift
	1 torr or less	3 torr	
Our lamp Rowley (3)	546 227 083 ± 6	122 ± 6	39
NPL 1962 (10)	072	090	18
CCDM (4)	05-	---	--
Kaufman (8)	046 ± 10	066 ± 10	20

(b) Blue line (4359Å)

	Argon carrier gas pressure		Pressure shift
	1 torr or less	3 torr	
Our lamp Rowley (3)	435 956 260 ± 10	285 ± 10	25
NPL 1962 (10)	252	263	11
CCDM (4)	24-	---	--
Kaufman (8)	225 ± 10	249 ± 10	24

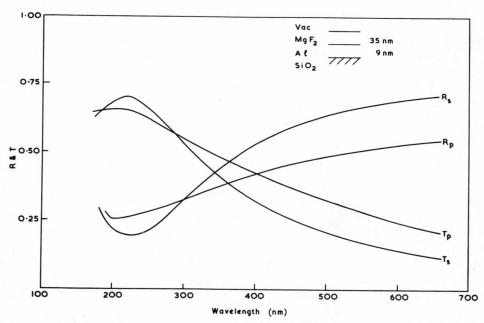

Fig. 1.   The Reflection and Transmission Coefficients of Fused Silica Beam Splitter at 45°.

TABLE 2

Wavelength values and shifts obtained on the Michelson Interferometer

Kaufman $\frac{1}{4}$ torr observed (8)	1 torr lamp 4 °C	3 torr lamp 4 °C	3 torr lamp 20 °C	1-3 torr NPL	Pressure shift $\frac{1}{4}$-3 torr Kaufman (calculated)	Temperature shift 3 torr lamp 4-20 °C
579 226 834 ± 10	827 ± 49	951 ± 30	940 ± 66	124	- 14	- 11
577 119 829 ± 10	894 ± 9	917 ± 30	967 ± 35	23	11	50
546 227 046 ± 10	083 reference wavelength	122 ± 11	142 ± 11	39	27	20
435 956 225 ± 10	235 ± 9	272 ± 11	290 reference	37	21	18
404 771 455 ± 10	451 ± 13	455 ± 23	499 ± 23	4	12	44
365 119 666 ± 10	660 ± 8	725 ± 12	716 ± 18	65	15	- 9
253 726 877 ± 10	867 ± 9	885 ± 8	890 ± 20	18	9	5
184 949 195 (calculated)	210	170	--	- 40	7	--

Units are fm (100 fm = 1 mÅ)

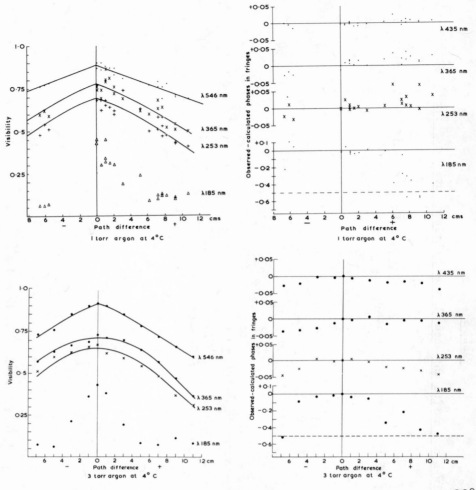

Fig. 2.   The Visibility and Phase Measurements for the Two Hg[198] Lamps.

APPLICATIONS OF THE JOSEPHSON EFFECT

J. E. Mercereau

California Institute of Technology, Low Temperature

Physics, 63-37, Pasadena, California   91125

In just a little more than a decade the Josephson effect[1] has evolved from the status of a concept in theoretical physics to an officially adopted technique for maintaining international voltage standards.[2] This rapid development reflects the very fundamental importance of Josephson's discovery for which he shared the Nobel Prize in Physics in 1973. Many other applications of Josephson's discovery have been carried out in the general area of unique and sensitive quantum-electronic devices. The intent of this paper is to present a summary of some of this development, with special emphasis on the Standards application to experts unfamiliar with superconducting phenomena.

The fact that any purely solid state electronic phenomena can be operationally related to the fundamental quantum constants of nature is itself unusual. A more normal circumstance is to expect that the inevidable thermal chaos which accompanies large collections of particles will reduce the fundamental precision of quantum effects to statistical uncertainty. Another version of the same circumstance is to say that in solids the motion of electrons is in general uncorrelated except for very short distances—usually on the order of Angstroms. However, in a superconductor, electrons engage in a pairing interaction which leads to a long range correlation in the electronic motion which relates directly to the Josephson effect and its fundamental precision and sensitivity.

This view of superconductivity (originated by F. London in 1935) is that it is a peculiar __macroscopic__ quantum electronic state of ordered electronic __momentum__ extending throughout the material. This macroscopically ordered quantum state Ψ can be described symbolically in quantum language quite simply as having an amplitude ψ whose square is equal to the density of pairs, and a phase φ so that $\Psi = \psi e^{i\varphi}$. Usually, the amplitude is considered to be a constant throughout the superconductor while φ may be a function of space and time. Josephson's prediction concerns two superconductors (or two such macroscopic quantum systems) between which it is possible to transfer electrons. This circumstance sets the domain of what has becomes known as "Weak Superconductivity"—the two macroscopic quantum systems must be coupled together so that electron transfer between them is possible, but at the same time coupled weakly enough that the two parts can be considered independently. He predicted that a supercurrent I_J would flow between the two coupled superconductors such that this pair current depended on the sine of the quantum phase difference ($\varphi = \varphi_1 - \varphi_2$) between the two superconductors and that this phase difference evolved in time at a rate proportional to the difference in the chemical potential of a pair, $\Delta\mu_p$, between superconductors, or: $I_J = I_0 \sin\varphi$ and $\dot{\varphi} = \hbar^{-1}\Delta\mu_p$.

In equilibrium, $\Delta\mu_p$ is determined by the voltage V which is maintained between the superconductors ($\Delta\mu_p = 2eV$) and a voltage-frequency relationship can be defined (where $\omega \equiv \dot{\varphi}$) $V = (\hbar/2e)\omega$. Thus a steady voltage maintained between the superconductors results in an oscillating supercurrent at a frequency $2eV/\hbar$. The time dependence of the current is determined solely by the ratio of fundamental constants, with no "materials parameters" entering the relationship. Essentially, this is because of the macrosopic nature of the quantum states between which the current flows. This particular point has been examined extensively[4] and no deviations have been found or are expected to, at most, parts in 10^8. However, since the precision of the phenomena depends on equilibrium conditions, measurements must be carried out sufficiently far from the weak coupling region that equilibrium prevails. If this is not done, considerable deviation in the Josephson Voltage-frequency relation ship possible.[5]

Weak coupling has been achieved by a number of junction techniques: electron tunneling, proximity effect, point contacts, etc., however, the quantum aspects of the supercurrent in all cases has been found to be the same. From an operational point of view, the Josephson Junction device can be considered as a Voltage to Frequency converte $V = (\hbar/2e)\omega$ whose conversation ratio is set by fundamental constants and is known to be precise to about one part in 10^8. Additionally, the Current-Voltage relationship $I_J = I_0 \sin(2e/\hbar)(\int Vdt + \delta)$ is highly non linear and lends itself to many unique instrumentation possibilities—some of which are discussed in the next section.

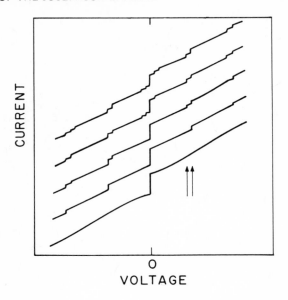

Fig. 1. Radiation induced phenomena in Josephson junction I-V curves (see text).

STANDARDS APPLICATIONS

Voltage. Perhaps the most well developed Standards application of the Josephson effect is its use to maintain the Volt. This application comes from a phenomena first observed by Shapiro[6] and involves a frequency modulation of the Josephson oscillation by means of an applied microwave field. The modulation results in distorting the dc current-Voltage (I-V) curve of a Josephson device by introducing current "steps" at constant voltages $V = n(\hbar\omega/2e)$, where ω is the frequency of the microwave field. This phenomena is shown in Fig. 1, the lowest I-V curve is with no radiation applied, the next curve above shows the development of "steps" induced by an applied radiation field. By precise measurement of voltage and frequency and a determination of the integers n, a direct evaluation of $\hbar/2e = V/n\omega$ can be made. This procedure was first reported by Parke et. al.[7] in 1967. The difficulty in this application is that convenient frequencies are typically 10 GHz which means that the step separation is only about $20\mu V$, and so a large number of steps must be counted to achieve precision of measurement. A practical limit on step number is a few hundred so that a direct

measurement of the interval between steps is limited to voltages
in the millivolt range. Thus a sensitive D.C. bridge is required
to achieve accuracy with this technique. However, such a direct
measurement of h/2e has been done and the results were used in a
critical evaluation of quantum electrodynamics and a reevaluation
of the "best" values of the fundamental constants.[8]

More recently a different method has been developed to relate
the junction step voltage to voltages more nearly equal to that of
the "standard cells" ($\sim$ 1 volt). This involves a transfer technique utilizing a Hamon resistor network[9] which can be accurately
switched from a resistance R to that of N^2R. In practice, this
technique works as follows: a voltage V_J (defined in terms of h/2e
and ω) is established by applying radiation to a junction $V_J=n\hbar\omega/2e$,
(see Fig. 2a). This voltage V_J then used to calibrate a current
source supplying the resistor R, (Fig. 2b). Then by switching R
precisely to a resistance N^2R, the circuit becomes a calibrated
voltage in the range of N^2V_J.

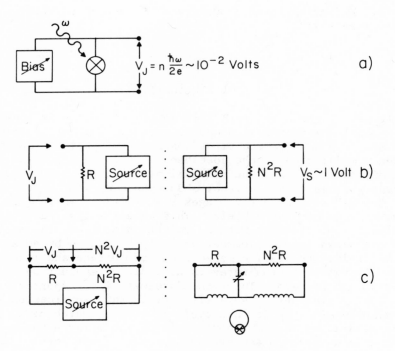

Fig. 2. Schematic drawing of various resistance bridge networks
for comparing the Josephson voltage V_J to the primary standard.
(See text).

This technique has been developed to a very sophisticated level by B. T. Field, at. al.[10] at NBS, Washington. In that application, two PbO tunnel junctions in series are stripline coupled to a radiation source at 10 GHz. This coupling results in well defined steps out to n = several hundred so that V_J for these junction pairs can be reliably about 10 mV. Hamon transfer networks have also been developed with a ratio in the range of 100 to a precision of about 2 parts in 10^8. On 1 July 1972, V_{NBS} was defined in terms of the Josephson effect such that $2e/h = 483593.420$ GHz/V_{NBS}. By means of this definition and using stripline coupled junctions with the precision Hamon networks, the U. S. volt can now be maintained to an overall precision of about 3 parts in 10^8.

Attenuation. Kamper[11] has suggested that Josephson techniaues can be usefully applied to the precise determination and calibration of radio frequency (rf) attenuation. This occurs because it can be shown that if a Josephson device is inserted into a superconducting loop, the impedance of that loop becomes periodic in the magnitude of magnetic flux enclosed by the loop. For example, the data to the right of Fig. 3 shows the rf impedance of the Josephson loop structure at the left of Fig. 3. The general characteristic of the impedance is that it varies as a Bessel Function of the rf flux. In Kamper's application, the loop is pumped at a frequency of about 1.2 GHz and is used to monitor rf flux in the 30 MHz range. For this mode of operation, the microwave impedance Z depends periodically on the rf flux φ as $Z \propto J_o(2e\varphi/\hbar)$. A very precise determination of relative values of rf flux can be made due to the small size of the flux quantum, $h/2e \sim 207 \times 10^{-15}$ Tm^2, and the accurately known periodicity of the Bessel Function. This technique has been developed for calibration of rf attenuators and has achieved agreement with the NBS Calibration Service to within $\pm$ 0.002 dB. In addition to precision, this application has the additional advantage of very wide dynamic and frequency range. Thus, it would seem to be a strong future contender for the position as a primary standard of attenuation.

Time and Length. Because of the extreme non-linearity of the Josephson current-voltage relationship, a Josephson device also acts as a very effective mixer of high frequency radiation—the mixing efficiency in general being a product of Bessel Functions of argument $(2ev/\hbar\omega)$, where v is the voltage amplitude of radiation at frequency ω. Thus the mixing efficiency does not necessarily decrease monitonically with increasing frequency. This aspect of the Josephson phenomena has been pursued by McDonald[12] and co-workers for Standards application. An indication of this effect can be seen in the upper curves of Fig. 1, where radiation at the two frequencies (indicated by the arrows) show beat frequency "steps" at low voltage (frequency). McDonald et. al. have shown efficient mixing up to 31.5 THz and direct harmonic mixing of 9 GHz microwave radiation with a 3.8 THz infra-red laser. It should be noted that

these high frequencies are far above (nearly 50 times) the gap
frequency of the superconductor; nevertheless, the Josephson phenomena still prevails—although with somewhat diminished amplitude.

One of the goals of this work is to provide a simplified technique for directly determining the frequency (relative to the primary standard) of a high frequency source which is sufficiently
stable to be used for interferometry in length metrology. Such a
source might be the 88-THz radiation from a He-Ne laser, but since
the primary frequency standard ($^{133}C_s$ atomic beam) is at approximately 9.2 GHz, comparison between these two frequencies is at present very difficult. If it could be achieved, however, a unified
practical standard of length and time might be possible. The hope
is that the Josephson phenomena can help to simplify this comparison because of its precise nature and efficiency at high frequency
mixing. The limits of the Josephson technique in this regard have
not yet been reached and work is continuing.

Current Development. This discussion is space limited to only
a very few items which bear particularly strongly on the Standards
application. Recent work by Palmer[14] with large series arrays of
junctions has indicated conditions under which self coupling of the
Josephson oscillation can occur. The array thus acts as if it were

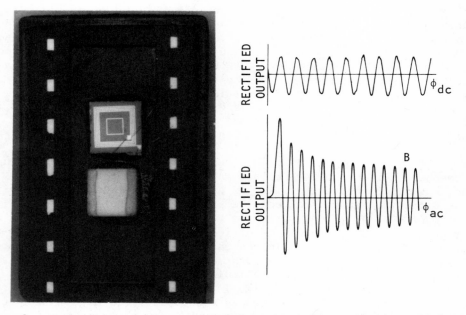

Fig. 3. Left; micro photograph of Josephson loop circuit. Right;
Impedance of this circuit as a function of magnetic flux; upper dc
flux, lower rf flux.

quantized in units of N(h/2e) where N is the number of series elements. An ingenious technique has been developed by Sullivan[15] for accurately determining resistance ratios at cryogenic temperatures. By this technique, current ratios can be determined to an accuracy of a part in 10^9. A completely cryogenic voltage comparator system has been constructed[16] around these methods and is now undergoing tests. Development of a field portable Josephson voltage standard is being pursued by a collaborative effort between NBS Washington, and the ScT Corporation[17]. This device will have the capability to maintain the standard volt to one part in 10^7, an improvement of about a factor of 10 over the existing Standard Cell techniques. A prototype of this device will be tested in about six months and marketable instruments should be available within the next year.[17]

ACKNOWLEDGEMENTS

The author wishes to express his gratitude to many colleagues who generously provided information and background materials, in particular Drs. T. Finnegan, W. Goree, R. Kamper and B. N. Taylor.

REFERENCES

1. B. D. Josephson, Phys. Lett. 1, 251 (1962).
2. Nat. Bur. Std. Tech. New Bull. 56, 159 (1972).
3. J. Bardeen, L. N. Cooper, and J. R. Schrieffer, Phys. Rev. 106, 162 (1957).
4. For references see: B. F. Field, T. F. Finnegan and J. Toots, Metrologia 9, 155 (1973).
5. M. L. Yu and J. E. Mercereau, Bull. Am. Phys. Soc. Ser. II 17, 1196 (1972).
6. S. Shapiro, Phys. Rev. Lett. 11, 80 (1963).
7. W. H. Parker, B. N. Taylor and D. N. Langenberg, Phys. Rev. Lett. 18, 287 (1967).
8. B. N. Taylor, W. H. Parker and D. N. Langenberg, Rev. Mod. Phys. 41, 375 (1969).
9. T. F. Finnegan, A. Denenstein and D. N. Langenberg, Phys. Rev. B 4, 1487 (1971).
10. B. J. Field, T. F. Finnegan and J. Toots, Metrologia 9, 155 (1973).
11. Robert A. Adair, et. al., IEEE Trans. IM-23, 375 (1974).
12. D. G. McDonald, et. al., Low Temp. Phys. LT-13, Vol. 4, 592 (1974). Plenum Press, Appl. Phys. Lett. 24, 335 (1974).
13. R. A. Kamper and J. E. Zimmerman, J. Appl. Phys. 42, 132 (1971).
14. D. Palmer and J. E. Mercereau, Appl. Phys. Lett. 25, 467 (1974).
15. D. B. Sullivan and Ronald F. Dziuba, IEEE Trans. IM-23, 256 (1974).
16. Ronald F. Dziuba, Bruce F. Field and Thomas F. Finnegan, IEEE Trans. IM-23, 264 (1974).
17. Superconducting Technology Corporation (ScT), Mountain View, California   94040.

A BRIEF REVIEW OF THE A.C. JOSEPHSON EFFECT DETERMINATION OF 2e/h

B.W. Petley

Division of Quantum Metrology

National Physical Laboratory, Teddington, England

In the decade between 1962 and 1972 the Josephson effects were developed from a slightly controversial theoretical beginning to the use in national standards laboratories of the a.c. Josephson effect as a quantum method of maintaining the volt. In this brief review we concentrate on the methods used to measure 2e/h, rather than a theoretical discussion of whether it really is 2e/h that is being measured.

In his original paper on superconductive electron tunnelling, Josephson showed that in the presence of electromagnetic radiation of frequency f, in addition to an alternating supercurrent a direct supercurrent could pass between the superconductors whenever the potential difference V between them satisfied the simple equation

$$2eV = n\,hf \qquad \ldots (1)$$

where e was the electronic charge, h the Planck constant and n was an integer. Experimental confirmation of the Josephson effects followed within the year and in particular Shapiro[2] and others demonstrated the reality of the effects shown by the above equation. It became apparent that the Josephson effects were observable in a number of configurations. Thus the two superconductors could be separated by a thin neck, either as a point-contact or an evaporated film (Dayem link), or the oxide barrier could be semi-conducting or photoconducting or even be replaced by a normal metal. Generally however the 2e/h measurements have used tunnel or point-contact barriers.

A group of workers at the University of Pennsylvania in a

450

series of experiments of increasing precision[3-7] demonstrated that
equation (1) was valid to the highest precision. They also
performed the classic 1969 review[8] of the best values of the atomic
constants in order to sort out difficulties concerning the fine-
structure constant determinations, for if one used the 1963 Cohen
and DuMond[9] evaluation of the atomic constants one might well
conclude that equation (1) was only obeyed to $\sim$ 20 ppm. However
the 1969 evaluation showed that it was 2e/h to 2.5 $\pm$ 6.4 ppm
while Petley and Morris[10],[11] confirmed the Pennsylvania value for
2e/h using different junctions, apparatus and methods. Thus
2e/h measurements were used to deduce a value for α^{-1} via the
equation

$$\alpha^{-1} = \left[\frac{c}{4R_\infty} \frac{\Omega_{abs}}{\Omega_\ell} \frac{\mu'_p}{\mu_B} \cdot \frac{2e/h}{\gamma'_p}\right]^{\frac{1}{2}} \qquad \ldots (2)$$

Since 1969 the reviews of the constants have tended to assume
that it is, for present purposes, precisely 2e/h in equation (1)
and one may expect the reviewers to re-examine the validity of
equation (1) each time that new experimental data become available.
Certainly there is internal evidence that equation (1) is a
constant of superconductivity which is independent of the experimental parameters to precisions of better than a part in
10^8.[7],[12],[13]   (A very good discussion is given in references
(7&14).)   It is salutary to recall however that the detailed
behaviour of a barrier, when all of its parameters are included,
is not amenable to calculation to high precision and recourse must
be made to analogues to solve the highly non-linear differential
equations. Thus it is only really in the one aspect - the voltages
at which the supercurrent steps occur - that the behaviour of the
barriers is understood to high precision.

It is important to note that it is not equation (1) that is
verified experimentally and the measured value of 2e/h is actually

$$\left(\frac{2e}{h}\right) = \left(\frac{2e}{h}\right)\left(\frac{V_n}{V_1}\right)\left(\frac{V_{BI69}}{V_n}\right)\left(\frac{V_{69BI}}{V_{BI69}}\right)\left(\frac{A_{ABS}}{A_{BI69}}\right)\left(\frac{\Omega_{abs}}{\Omega_n}\right) \qquad \ldots (3)$$

The suffixes 1 and n refer to the local and nationally maintained
units, 69BI refers to the current value of the BIPM unit and
BI69 is the BIPM unit at the time of the 1st January 1969 readjustment of the maintained volts. The other symbols have the same
meaning as used in current reviews by Cohen and Taylor[15].

It was only in the very early measurements that the quantities
on the right hand side of equation (3) were negligible in compari-

son with the experimental error. As soon as the precision became better than a part in ten million they became progressively more important. Indeed it can be argued that since 1972 the experiments have not been measuring 2e/h at all, but are merely giving information about changes in the appropriate nationally maintained voltage unit since 1969.

The experimental measurements are made by biassing the junction onto a high order supercurrent step to give a potential difference of between 1 and 10 mV and this potential must be compared with a standard cell emf. This comparison is usually effected by passing the same current through two resistors in series and balancing the two emf's simultaneously, by adjusting the current to balance the cell emf and changing the incident frequency to balance the junction galvanometer. With tunnel junctions the frequency change is limited by the need to stay close to the junction self-resonant frequency in order to couple the incident microwave radiation into it efficiently. With room temperature potentiometric systems there is inevitably a small thermal emf, $\sim$ 50 nV, and so a small emf is injected into the low voltage circuit. This emf $\sim 10^{-4}$ V_J can also be used to balance the junction galvanometer since it is only required to a part in 10^4 to be $\sim 10^{-8}$ V_J. The thermal emf's are eliminated by current reversal and, while this method can lead to problems with the repeatability of the switch thermals in the low voltage circuit, these can be avoided, as in the NPL equipment, by shunting the galvanometer with a shorting switch while the current is reversed.

It was found at the University of Pennsylvania[7] that it was possible with one or two tunnel junctions in series to achieve measurable steps to at least 10 mV and, although this technology has been transferred to NBS and recently to BIPM and elsewhere, most laboratories have tended for their initial work to rely on getting measurable steps to between 1 and 5 mV and the values of the resistance ratio are determined accordingly. Although the resistance ratio has been measured by a.c. methods[11] the tendency has been to rely on Hamon[16] resistors. Binary build-up resistors have also been used[7,17] and give similar results.

It is perhaps important to remember that as recently as 1969 professional metrologists hesitated to claim to measure $\sim$ mV direct potentials with an absolute precision of much better than a few tenths of a part in a million and this remains to some extent true today outside the specifically designed 2e/h experiments.

THE DIFFICULTY OF COMPARING THE RESULTS

It rapidly became apparent that the 2e/h determinations were in remarkably good agreement.[5,10] However, just as the University of Pennsylvania group had experienced difficulty in establishing

V_1/V_n, it became apparent that there were equal difficulties in attempting to compare the determinations at NBS[18], NPL[19], NSL[20],[21] (now NML) and the PTB[22],[23],[24]. Usually the differences between the nationally maintained volts were established by means of the triennial comparisons with the BIPM maintained volt, but attempts were made by the laboratories concerned with the 2e/h measurements to effect a better voltage intercomparison. The results of these and the 2e/h measurements were considered and tabulated by Eickie and Taylor 1972[25]. Denton at the NPL[26] showed however that if one took all of the results for 2e/h and the voltage comparisons then one could, assuming a linear drift, extrapolate them back to obtain a value of 2e/h = 483 594.0 GHz/V_{BI69} where V_{BI69} was the value of the BIPM volt at 1st January 1969. Of course once the drifts were demonstrated the voltage drifts of all countries were affected by this new information in a way that is hard to determine.

Thus the work on the Josephson effects had enabled the absolute drifts of the various nationally maintained voltage standards to be seen for the first time. Denton's paper was considered by the Comité Consultatif d'Electricite (CCE) of the Comité International des Poids et Mesures (CIPM), at its 13th meeting held in October 1972, and they adopted a resolution (Declaration E-72) which was subsequently approved by the CIPM at its 61st meeting

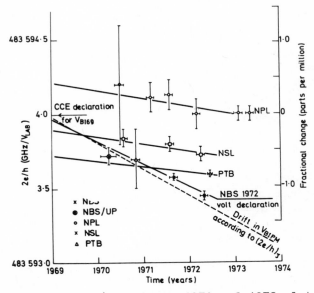

Fig 1   The values of 2e/h between 1970 and 1973 plotted in terms of their appropriate nationally maintained volt. Note: There is no common basis for the error assignments. The lines are least squares fitted, taking into account voltage intercomparisons between 1970 and 1973 as well[15],[16].

(also October 1972) which reads in part: 'Therefore, the CCE
judges that, according to these results, V_{BI69} is equal, to
within one half of one part per million, to the voltage step that
would be produced by the Josephson effect in a junction irradiated
at a frequency of 483 594.0 GHz.'

Cohen and Taylor[15] reanalysed the data considered by Denton
and came to substantially the same result. The significance of
1 January 1969 is that most countries changed their voltage units
at about that time in order to be in close agreement with the
BIPM volt and with the best estimate of the absolute volt. (Any
electrical measurements before 1969 should strictly be reconsidered
in the light of the absolute drifts of a few tenths of a ppm per
year that have now been revealed). If subsequently it is decided
to make a further change, in the light of a better knowledge of
the absolute ampere, it can be made quite simply. The present
situation is that PTB, NPL, NSL(NML), VNIIM and BIPM appear to have
stabilised their volt according to the CCE resolution, while the
NBS have so far stabilised their volt at the value that it had
drifted to by July 1972, corresponding to 483 593.42 GHz.

Although it was necessary to make the initial measurements
with the potentiometric apparatus at room temperature, there are
considerable advantages to be gained in putting at least the low
voltage side of the potentiometry into the cryostat. The first
such system was described by Sullivan (1972)[27] and used a SQUID
as galvanometer, although it was not capable of sub-ppm precision
without further modification. Gallop and Petley[28] described a
successful cryogenic system shortly afterwards which was capable
of sub-ppm precision, using a Clarke[29] type d.c. operated SLUG as
the galvanometer. Other systems have been described since by
Dziuba et al.[30] Kose et al.[24] and Krasnov et al.[17]. Some of these
suggest a performance which is at least as good as the room
temperature systems. Gallop and Petley[19b] have argued that from
considerations of Johnson noise limitations a 1 mV junction will
suffice to calibrate a standard cell and that the major advantage
of the higher junction voltages $\sim$ 10 mV cannot be fully exploited.
However such arguments depend very much on the performance of the
particular apparatus that is adopted. Most systems now operate to
a few parts in 10^8 and it is almost impossible to assess their
relative precision. It should be posssible to produce a
commercially engineered system without too much difficulty.

It has been seen that the Josephson effects have made the
transition from slightly controversial theory to being a quantum
method of maintaining the volt in ten exciting years. Those who
measure 2e/h are probably unique among experimenters in that it
is the unit that is changed in order to realise the CCE value and
the answer is known before they begin! It still seems incredible
that it apparently is 2e/h in equation (1) to at least a part

in a million but it appears that all is well to better than a part
in 10^8 at present. Space has not permitted a discussion of the
theoretical aspects of 2e/h and this would not be appropriate at
an experimentally biased conference. However most of the theoret-
etical attacks came earlier and these were discussed by Clarke[14]
and Finnegan et al.[7]. Further comments have been made by
Langenberg and Schrieffer[31], Bracken and Hamilton[32] and Fulton[33].
Discussions will no doubt continue and further experimental evidence will come within a year or so. Improved precision in our
knowledge of the fundamental constants will play a key part in
this fascinating saga.

REFERENCES

1. JOSEPHSON, B.D., Phys. Lett. 1, 251 (1962)

2. SHAPIRO, S., Phys. Rev. Lett. 11, 80 (1963)

3. PARKER, W.H., TAYLOR, B.N. and LANGENBERG, D.N., Phys. Rev. Letters 18, 287 (1967)

4. TAYLOR, B.N., PARKER, W.H., LANGENBERG, D.N. and DENENSTEIN, A., Metrologia 3, 89-98 (1967)

5. PARKER, W.H., LANGENBERG, D.N., DENENSTEIN, A. and TAYLOR, B.N. Phys. Rev. 177, 639-664 (1969)

6. FINNEGAN, T.F., DENENSTEIN, A. and LANGENBERG, D.N., Phys. Rev. Letters 24, 738-742 (1970)

7. FINNEGAN, T.F., DENENSTEIN, A. and LANGENBERG, D.N., Phys. Rev. B, 6, 1487-1521 (1971)

8. TAYLOR, B.N., PARKER, W.H. and LANGENBERG, D.N., Rev. Mod. Phys. 41, 375 (1969)

9. COHEN, E.R. and DuMOND, J.W.M., Rev. Mod. Phys. 37, 537 (1965)

10. PETLEY, B.W. and MORRIS, K., Physics Letters 29A, 289-90 (1969)

11. PETLEY, B.W. and MORRIS, K., Metrologia 6, 46 (1970)

12. MACFARLANE, J.C., App. Phys. Lett. 22, 549-550 (1973)

13. CLARKE, J., Phys. Rev. Lett. 21, 1566-1569 (1968)

14. CLARKE, J., Amer. J. Phys. 38, 1071-95 (1970)

15. COHEN, E.R. and TAYLOR, B.N., J. Phys. & Chem. Ref Data 2, 663 (1973)

16. HAMON, B.V., J. Sci. Instr. 31, 450 (1954)

17. KRASNOV, K., MASOUROV, V. and FRANTSOUZ, E. (to be published) VNIMM work.

18. FIELD, B.F., FINNEGAN, T.F. and TOOTS, J., Metrologia, 9, 155 (1973)

19a.   GALLOP, J.C. and PETLEY, B.W.,   IEEE Trans on Instrum & Meas
       IM-21, 310-316 (1972)

19b.   GALLOP, J.C. and PETLEY, B.W.,   IEEE Trans on Instrum & Meas
       IM-23, 267 (1974)

20.   HARVEY, I.K., MACFARLANE, J.C. and FRENKEL, J.C. Phys. Rev.
       Lett, 25, 853 (1970)

21.   HARVEY, I.K., MACFARLANE, J.C. and FRENKEL, R.B.,   Metrologia
       8, 116-124 (1972)

22.   KOSE, V., MELCHERT, F., FACK, H. and SCHRADER, H.J.,   PTB
       Mitt. 81, 8 (1971)

23.   KOSE, V., MELCHERT, F., FACK, H. and HTEZEL, W.,   IEEE Trans
       on Instrum & Meas IM-21, 316-315 (1972)

24.   KOSE, V., MELCHERT, F., ENGELARD, FACK, H., FUHRMANN, B.,
       GUTMANN, P. and WARNECKE, P.,   IEEE Trans. on Instrum & Meas
       IM-23, 271 (1974)

25.   EICKIE, Jr. W.E. and TAYLOR, B.N.,   IEEE Trans on Instrum &
       Meas IM-21, 316 (1972)

26.   DENTON, J.D.,   Unpublished submission to October 1972 CCE
       meeting

27.   SULLIVAN, D.B.,   Rev. Sci Instr 43, 499-505 (1972)

28.   GALLOP, J.C. and PETLEY, B.W. Electron Lett. 9, 488 (1973)

29.   CLARKE, J.,   Phil. Mag. 13, 115 (1966)

30.   DZIUBA, R.F., FIELD, B.F. and FINNEGAN, T.F.,   IEEE Trans. on
       Instrum & Meas IM-23, 256 (1974)

31.   LANGENBERG, D.N. and SCHRIEFFER, J.R.,   Phys. Rev. B 3,
       1776-1778 (1971)

32.   BRACKEN, T.D. and HAMILTON, W.D.,   Phys. Rev. B 6, 2603 (1972)

33.   FULTON, T.A.,   Phys. Rev. B 7, 981-2 (1973)

DETERMINATION OF 2e/h AT THE B.I.P.M.

T. J. Witt and D. Reymann

Bureau International des Poids et Mesures

Pavillon de Breteuil, 92310 Sèvres-France

INTRODUCTION

Determinations of 2e/h have been made at the Bureau International des Poids et Mesures (B.I.P.M.) since February 1974 by means of the ac Josephson effect. In order to compare our results with those of another laboratory with an accuracy surpassing the several parts in 10^7 obtainable by transporting standard cells and in order to check for systematic errors, we transported the critical components of our measuring system to the Physicalisch-Technische Bundesanstalt (P.T.B.), Braunschweig and made 2e/h measurements there.

MEASUREMENTS AT B.I.P.M.

The measuring system at the B.I.P.M. is similar to that described by Finnegan, Denenstein, and Langenberg [1]. Our Josephson junctions are lead-lead oxide-lead tunnel junctions of about 200 mΩ normal resistance having resonant frequencies near 9 GHz. We obtain a 10 mV output by connecting two junctions on the same substrate in series. The radiation source is an X-band frequency-stabilized klystron.

The 10 mV output from the junctions is compared to a standard cell emf by means of a series-parallel comparator (SPC) [2] which produces a 100/1 voltage ratio by using two networks of ten 100 Ω resistors which are connected alternatively in series and parallel. Special precautions were taken in the building of the SPC in order to minimize self-heating effects in the main resistors. These precautions include 1) use of open, card-wound resistors, 2) close matching of the temperature coefficients of the two networks, 3) mounting both strings in the same oil-filled can, 4) selecting resistors with a resistance versus temperature maximum at 26 °C, the maintained SPC temperature and 5) controlling the temperature

457

to ± 0.01 K. The power coefficient was measured under operating conditions by the bridge within-a-bridge method [3]. The 1 σ systematic error due to the self-heating is estimated to be about 5 x 10⁻⁹.

The SPC weighs 22 kilograms and is portable. The errors due to mismatches among the main resistors and among the current and voltage fans [4] are easily controlled to parts in 10^{10} by means of a self-checking procedure which can be accomplished with a microvolt meter, an important feature for the transportable instrument.

The null detectors are commercial galvanometer-amplifiers which were modified by the addition of a decoupled recorder output. The recorded deflections allow us to interpolate drifts in the current supply and in the thermal emfs in the leads from the junctions to the SPC.

The working group of 6 standard cells, V_w, against which the 2e/h measurements are actually made, is mounted in a commercial temperature-regulated standard cell enclosure located in the screened room in which the 2e/h measurements are made.

A second group V_R of 4 standard cells in a similar enclosure located outside the screened room in a temperature controlled environment is used as a reference group. This reference group is necessary for controlling the working group. Its mean value in terms of V_{69-BI}* was determined at approximately 2-month intervals. At present the value of V_{69-BI} is defined as the mean of a group of 42 saturated standard cells in an oil bath located about 3 meters away from the screened room. Approximately two weeks are required to compare the reference group to V_{69-BI}.

One day's run consists of about six separate 2e/h measurements. Standard cell comparisons between groups V_R and V_W are made before and after each run and the temperature of V_W is measured repeatedly throughout the run. Also, for each run, leakage resistance measurements in the junction leads and checks for non-vertical steps are made. Less often we measure thermal emfs, check the main and fan resistor mismatches, recalibrate the frequency counter time base and measure all leakage resistances.

Results

The values of 2e/h are calculated in the following manner : 1) calculate a value of the "equivalent standard cell frequency,"

$$F_s = \beta n f = \frac{2e}{h} V_s$$ (where V_s is the standard cell voltage, β is the SPC ratio, n is the step number, and f the frequency) for the cell used in the Josephson effect measurement at the operating

---

* By V_{69-BI} we mean the actual as-maintained volt at B.I.P.M. It is thus a quantity which varies in time. The number in the subscript recalls the 1 January 1969 readjustment of the B.I.P.M. volt.

temperature ; 2) apply a temperature correction to normalize all
results to the same standard cell temperature ; 3) calculate
$F_s(\bar{V}_w)$, where $\bar{V}_w$ is the average of the working group, using the
results of the cell measurements made before and after the run ;
4) from these same data calculate $F_s(\bar{V}_R)$, i.e., F_s for the average
of the reference group ; 5) from a least squares fit to the
standard cell comparison data for $\bar{V}_R$ in terms of V_{69-BI}, calculate
V_R in units of V_{69-BI} ; 6) calculate 2e/h.

The resulting values of 2e/h in terms of V_{69-BI} are shown in
Fig. 1. The first four points, the solid circles, denote
measurements having rather high uncertainties (about 2×10^{-7}) due
to the fact that no standard cell measurements were made on these
days. Each remaining point was weighted by the reciprocal of the
square of its total standard deviation. The total standard
deviation was taken to be the quadrature sum of the observed random
standard deviation of the mean and a constant standard cell
uncertainty of 4×10^{-8}. A linear least squares fit was made to

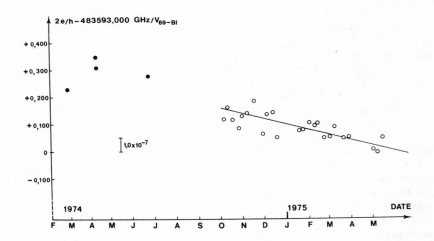

Figure 1.- Measured values of 2e/h in V_{69-BI} as a function of time
for the period of February, 1974 to May, 1975. The first four
points, represented by solid circles, are results of preliminary
measurements for which the uncertainties were rather high since
standard cell comparisons of the working group were not made on
these days. The remaining points were used to determine a linear
least squares fit, shown in the figure, which indicates a drift
in V_{69-BI} of - 0.48 µV/year. Error bars have been omitted for
clarity. The total standard deviation, including the observed
random uncertainty of the mean plus the uncertainty due to standard
cells and other sources is about 6.3×10^{-8} for the points after
January, 1975.

these points and the resulting line, representing the drift in
V_{69-BI}, is included in the figure. The value of 2e/h <u>for</u>
<u>1 January, 1975</u> was

$$2e/h = 483\ 593.100 \pm 0.030\ \text{GHz}/V_{69-BI}.$$

The 1 σ uncertainty[*] of the mean value is 6.3×10^{-8}. The value of
the drift of V_{69-BI} obtained from the least squares fit is

$$\frac{d}{dt}\ V_{69-BI} = -\ 0.48 \pm 0.14\ \mu V/\text{year}.$$

The uncertainty in the drift is the sum of the uncertainty of the
least squares fit to the 2e/h versus time data plus the uncertainty
of the least squares fit of the $\bar{V}_R$ data.

Comments

We divide the uncertainty assignments, which are preliminary
estimates, into two main sections, those due to the SPC and those
due to standard cell measurements, as given in tables I and II.
Table III presents the sum of all uncertainty terms.
      The uncertainty associated with the SPC is, neglecting random
effects which are included in table III, about 1.7×10^{-8}. If we
assume a random contribution of about 1.0×10^{-8}, we arrive at
2.0×10^{-8}, approximately the value obtained by Finnegan and
Denenstein [2]. The uncertainty in 2e/h associated with the
standard cells is dominated by that due to temperature changes in
the thermoregulated enclosure for V_w which, in the worst case,
attained 2 mK during a day's run. The variations in measured
2e/h values were correlated with these drifts, implying that the
temperature changes occurred rather uniformily without creating
gradients across the standard cells which would result in
differences in the emfs. We observed that the temperature of the
thermoregulated enclosure consistantly increased during 2e/h
measurements. This was not strongly related to temperature changes
in the screened room because even after reducing the ambient
temperature variations by a factor of about 6, the enclosure
temperature variations remained rather unchanged. Nevertheless,
the values of the random standard deviation of the mean of the
2e/h measurements decreased from about 2 to slightly less than
1×10^{-8} with improved control of the ambient temperature.
      The sum of all uncertainties appears in table III. We lack
the space here to explain how all of the individual uncertainties
were estimated and the values given here are preliminary estimates.
Most of those associated with the SPC were calculated from direct
measurements made either periodically (for those expected to change
in time) or during construction.

---

[*] All uncertainties are given as one standard deviation
estimates.

TABLE I

Estimated 1 σ uncertainties due to SPC	
Deviations among main resistors	0.01×10^{-8}
Deviations among I and V fan resistors	0.1
Transfer resistance of tetrahedral junctions	0.6
main resistor self-heating	0.5
temperature stability of SPC	0.3
current stability	0.3
calibrating signal	0.8
leakage resistance	0.7
thermal emf's	0.9
Root sum square total	1.7×10^{-8}

TABLE II

Estimated 1 σ uncertainties in standard cell measurements	
thermal emf's	1.5×10^{-8}
leakage resistance	1.0
temperature stability of working group	4.
temperature stability of reference group	2.2
Root sum square total	4.9×10^{-8}

TABLE III

Estimated 1 σ uncertainties : TOTALS	
uncertainties due to SPC	1.7×10^{-8}
uncertainties in standard cell comparison	4.9
random uncertainty of the mean	3.2
frequency measurement and stability	1.0
spectral purity of radiation	1.
thermal emf's in junction leads	0.7
leakage resistance in junction leads	0.05
possible non-vertical steps	0.4
Root sum square total	6.3×10^{-8}

The uncertainties associated with the standard cells are results of direct measurements for the case of thermal emf's and leakage resistance. That due to the temperature stability is estimated from the variations of $\bar{V}_w - \bar{V}_R$ before and after the runs and from the resolution (about 0.04 μV) of the standard cell comparisons.

Our value of 2e/h can be compared with the mean $\overline{(2e/h)}$ of the values reported by four national standards labs by means of the February, 1973 International Comparison [5] if we suppose that V_{69-BI} has drifted linearly since that time. The results are

$$\frac{(2e/h)_{BI} - \overline{(2e/h)}}{(2e/h)_{BI}} = 2.4 \times 10^{-7} \pm 3.3 \times 10^{-7} \text{ (February 1973)}.$$

The calculated uncertainty in our value for the drift of V_{69-BI} combined even with the rather optimistic estimate of 2×10^{-7} for the uncertainty in the International Comparison yields an uncertainty of about 3.3×10^{-7} in the difference.

COMPARISON WITH P.T.B.

In April of this year, the first international comparison of Josephson effect voltage systems was made by transporting by automobile the SPC, Josephson junctions, galvanometers and miscellaneous equipment from B.I.P.M. to P.T.B.

After about one week of set up time we were able to make measurements of 2e/h versus U, the mean of the group of standard cells which are used to conserve the unit of emf at P.T.B. between Josephson effect measurements [6]. In all we made a total of five runs. Two of the runs were consecutive with measurements made by the P.T.B. using the same standard cell.

The mean of our five determinations of U was about 50 nV higher than that assumed by the P.T.B. at the time of the measurement. For the two days on which consecutive measurements were made our value was higher by about 45 nV.

The total uncertainty of our measurements is about 4×10^{-8} and is nearly equal to that claimed by P.T.B. [6] and the results lie within the range of the combined uncertainties of the two measuring systems. Thus

$$\frac{(2e/h)_{BI}^{PTB} - (2e/h)_{PTB}}{(2e/h)} = 5.0 \pm 5.6 \times 10^{-8} \text{ in April, 1975.}$$

The rather smaller uncertainty in the B.I.P.M. measurements at P.T.B. as compared to those made at B.I.P.M. is due to better temperature stability of the standard cell enclosure used in the 2e/h runs at P.T.B. and to the capability at P.T.B. of comparing directly to U before and after Josephson effect measurements thus eliminating the need for a reference group with its associated uncertainty.

We should like to note that a number of experimental conditions differ between the 2e/h measuring systems at B.I.P.M. and P.T.B. Not only were the voltage comparators quite different, but also the frequencies (9 GHz for B.I.P.M. versus 70 GHz for P.T.B.), temperatures(1.5 K versus 4.2 K), junction output voltages (10 mV versus 3 mV), and types of junction (tunnel versus point contact).

CONCLUSIONS

We conclude that, to within an uncertainty of 6.3×10^{-8}, our value of 2e/h appears to be correct and that the present value of V_{69-BI} is decreasing at a rate of about 0.48 $\mu V/_{year}$.

Direct Josephson effect comparison measurements are practical and make it possible to compare national voltage units with accuracies surpassing those obtainable from standard cell measurements. They can thus be used to assure international continuity among voltage standards.

Transportable Josephson effect apparatus can also serve to check, as in the measurements described above, on the estimates of the systematic errors in Josephson effect voltage systems. One can do such experiments by comparing two voltage comparators at the one volt level and then performing the measurements of the junction output voltage from each device thus completely eliminating the intervention of standard cells. Also it appears to be quite possible to test, at accuracies of nearly 1×10^{-8}, the possibility of dependence of measured 2e/h values on the experimental parameters of temperature, frequency, material, etc. in a single measurement which would be free of all standard cell uncertainties.

ACKNOWLEDGEMENTS

One of us (T.J.W.) wishes to thank the National Bureau of Standards for providing a special grant for buying the components of the SPC. He also wishes to thank Dr. T.F. Finnegan of N.B.S. and Dr. A. Denenstein of the University of Pennsylvania for many useful discussions and suggestions.

We gratefully acknowledge our colleague Mr. G. Leclerc for patient and careful standard cell measurements and Dr. V. Kose and Dr. F. Melchert of P.T.B. for their help and hospitality during our work there.

REFERENCES

[1] Finnegan, T. F., Denenstein, A., and Langenberg, D. N., Phys. Rev. B4, 1487 (1971).

[2] Finnegan, T. F. and Denenstein, A., Rev. Sci. Instr., 44, 944 (1973).

[3] Wenner, F., J. Res. Nat. Bur. Std., 25, 253 (1940).

[4] Page, C. H., J. Res. Nat. Bur. Std., 69C, 181 (1965).

[5] Leclerc, G., Rapport sur la 13e Comparaison des Etalons Nationaux de Force Electromotrice (janvier-avril 1973) (Bureau International des Poids et Mesures, Pavillon de Breteuil, Sèvres, France), [1975].

[6] Kose, V., Melchert, F., Engelland, W., Fack, H., Fuhrmann, B., Gutmann, P., and Warnecke, P., IEEE Trans. Instr. Meas., IM-23, 271, 1974.

2e/h DETERMINATION BY JOSEPHSON EFFECT IN ETL

T.Endo, M.Koyanagi, K.Shimazaki, G.Yonezaki, and

A. Nakamura

Electrotechnical Laboratory

5-4-1, Mukodai-machi, Tanashi, Tokyo, JAPAN

1. INTRODUCTION

Precise determination of 2e/h by the AC Josephson effect would be important not only from the physical point of view but also from the establishment of a better voltage standards.

In 1974, experimental status in ETL reached to 0.1 ppm level with 1 mV Josephson voltage(1). Recently, the voltage has reached to 10 mV, and a new determination of 2e/h has been done with the national voltage standard unit of Japan in April, 1975. Before this experiment, the laboratory was moved to the next room of another laboratory where the national voltage standard is maintained. Standard cells used in the experiment are continuously calibrated with the national standard via cables as long as 10 m.

Reports will be given on the recent experiment along with the results obtained before.

2. EXPERIMENTS

Two Pb junctions are used in series for obtaining 10 mV. The junction has the area of 0.1x0.9 mm^2, and the normal resistance of 40 mΩ. Microwave at 9.3 GHz is fed through 50 Ω strip line(2) made of evaporated Pb film and a copper plate attached to the waveguide wall as shown in Fig.1. Incident power of about 100 mW is coupled to the strip line from the waveguide by a E-coupling probe. The junction produces 5 mV each with the current width of 100 to 200 μ A.

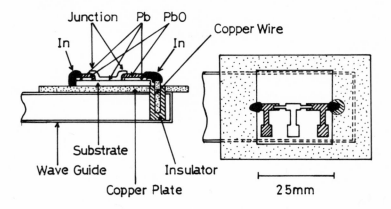

Fig.1   Two Pb Josephson junctions are connected in series
         on a glass substrate. Microwave is fed through the
         50 Ω strip line composed by lead film and a copper
         plate attached to waveguide wall.

For the oxidation of Pb film, glow discharge method is used in dry
oxygen atmosphere of about 16 Pa. Discharge continues 20 min. at
400 V and 7 mA before the oxide reaches to the thickness corresponding to 40 mΩ normal resistance of the junction(3).

A double series-parallel exchange comparator(4) is used in this
experiment as shown in Fig.2. In each Hamon type resistor set H1 and
H2, 15 main resistors of 45 Ω are installed. This provides 225 to 1
ratio which can be used for 4.5 mV Josephson voltage. The set has
extra fan resistors at the position of the 10-th main resistor, so
that one can obtain 100 to 1 ratio corresponding to 10 mV. Scattering in the resistance of the main resistors is kept within 10 ppm,
1 Ω voltage fan resistors within 100 ppm, and 10 mΩ current fan
resistors within 1 percent. Four terminal junctions are made of
copper cube with the side lengths of 15 mm, and their cross resistance is no more than 20 nΩ.

Whole comparator set is placed in an oil bath which is temperature regulated within 0.005°C. Temperature coefficient of two Hamon
resistor sets is 1 ppm/°C and 0.8 ppm/°C respectively. Self heating
effect of the main resistors is below 0.002 ppm. Current supply to
the comparator is similar to that of Finnegan and Denenstein

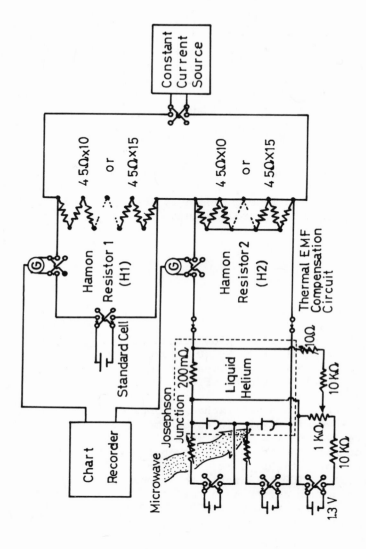

Fig. 2   DC circuit for the determination of 2e/h. Double series–parallel exchange comparator is used as a voltage divider.

(4), and has drift of 1 ppm/h or less. Series-parallel exchange of the comparator can be performed in one action from the outside of the oil bath, and the movement of copper shorting bars of 5x10x35 mm^3 makes new connection between mercury cup switches.

Working standard cells are contained in a temperature regulated air bath which has temperature stability of 0.001°C as far as the room temperature is kept at 25+0.5°C. Before and after the experiment, the standard cells are calibrated against the national voltage standard. From the long term calibration data(3 months), relatively large uncertainty(1σ=0.14 ppm) is forced to be assigned to the calibration.

In the experiment, the Hamon set H1 is used in series connection(H1s) and H2 in parallel(H2p) at first. Voltage drop across H1s is to be compared with the working cell, and voltage across H2p with the Josephson voltage. Reference signal is introduced in the circuit for determining the amount of small deviation when the comparison is made between the H1s voltage and the cell. On the other hand, microwave frequency is slightly changed to attain a balance between the H2p voltage and the Josephson voltage. The exact value of frequency can be read on the frequency counter calibrated against the national frequency standard JJY. Resolution of the latter voltage comparison reflects the uncertainty of the frequency determination.

Four comparisons are usually made with the current direction through the comparator + - - +. Then, the similar comparisons are performed for H2s and H1p combination. One run of the experiment consists of eight comparisons, four with the first combination and four with the second. Observed frequency and voltage drop across the Hamon set with series connection may be written as f1, Vs1 for the first and f2, Vs2 for the second combination respectively. Then, the value of 2e/h will be obtained by

$$\frac{2e}{h} = n \left[\left(\frac{f1}{Vs1} - \frac{f2}{Vs2}\right) \frac{H1s}{H2p} - \frac{H2s}{H1p} \right]^{1/2} , \qquad \ldots \ldots (1)$$

where n is the number of steps used (528), and H2p/H1s, H1p/H2s are the resistance ratios in the first and the second combinations of the comparator. It is easily seen that informations are required only for H1p/H1s and H2p/H2s but for H2p/H1s or H1p/H2s.

In the Hamon type resistor sets used in this experiment, H1p/H1s and H2p/H2s are both equal to 1/100 within relative uncertainty of 10^{-8}. Using this value, the ratios are also derived as

$$\frac{H2p}{H1s} = \frac{1}{100} \sqrt{f1/f2} , \quad \frac{H1p}{H2s} = \frac{1}{100} \sqrt{f2/f1} , \qquad \ldots(2)$$

providing Vs1=Vs2 which is satisfied as far as the standard cell

Table 1

Sources of uncertainty in 2e/h determination

Sources	Relative uncertainties (one standard deviation)

1) Voltage measurement

    Voltage divider

Main resistor mismatch	0.003 $\times 10^{-8}$
Four terminal junctions	0.1
Fan resistor mismatch	0.002
Self heating effect of main resistor	0.2

Leakage resistance	0.2
Resolution of standard cell balance	0.4
Calibration signal of standard cell balance	1.7
Thermal EMF of switches and connections in the standard cell circuit	1.4

2) Microwave frequency measurement

Frequency counter resolution and JJY calibration	0.5
Resolution of junction balance	0.5

3) Random uncertainty of the mean value (4 runs)

    1.4

Subtotal (RSS) 2.7 $\times 10^{-8}$

4) Working standard cell

Calibration with national standard voltage	14.0
Thermal EMF of switching system in the calibration circuit	1.4

Total (RSS) 14.3 $\times 10^{-8}$

voltage is kept constant during the experiment. Thus, the ratios of
the comparator could also be checked in each run though they are not
necessary for 2e/h determination.

3. RESULTS

With four runs, obtained result is

$2e/h$ = 483 593.674 $\pm$ 0.069 GHz/V(ETL,75,4,8) ($\pm$ 0.143 ppm),

and

$H2p/H1s$ = 0.999 999 545 x 10^{-2} ($\pm$ 0.027 ppm) ,

$H1p/H2s$ = 1.000 000 455 x 10^{-2} ($\pm$ 0.027 ppm) .

Assigned uncertainties are one standard deviation, and the sources
are listed in Table 1. The largest uncertainty arises from the
mutual comparisons between the working standard cells and the national standard cells. Uncertainty of 2.7×10^{-8} is assigned for the

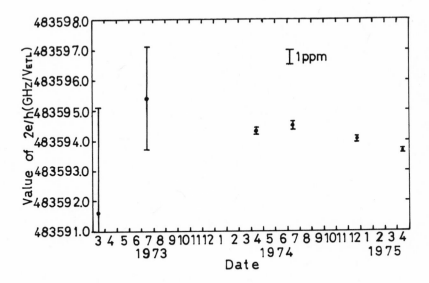

Fig.3   Time dependent variation of the observed 2e/h
in the voltage unit of ETL.

comparison of the Josephson voltage with the working standard cell. During the experiment, the Josephson junctions are kept at 1.788 K.

In Fig.3, values of 2e/h in the unit of ETL are plotted since March, 1973. Except the value of April in 1975, all values are determined with 1 mV of the Josephson voltage. Two values after October, 1974 are obtained in the new laboratory. During recent one year, there is constant decrease of the observed value of 2e/h. This would be originated from the decrease in the voltage unit maintained in ETL. However, it was difficult to confirm this variation of the national voltage unit by other means than the Josephson method. In ETL, a trial will be started soon for monitoring the national voltage unit with the Josephson voltage.

Authors are grateful to R.Ishige, H.Hirayama and K.Murakami for their kind help and discussions.

REFERENCES

1) T.Endo, M.Koyanagi, S.Koga, G.Yonezaki, and A.Nakamura: Fifth International Cryogenic Engineering Conference, (May, 1974 at Kyoto).

2) T.F.Finnegan, J.Wilson, and J.Toots: IEEE Trans. Mag. vol. MAG-11 (1974) p.821.

3) M.Koyanagi, S.Koga, T.Endo, G.Yonezaki, and A.Nakamura: Japanese J. Appl. Phys. vol.44, Suppl. (1975) p.135.

4) T.F.Finnegan, and A.Denenstein: Rev. Sci. Instr. vol.44 (1973) p.944.

Etape d'une détermination absolue du coefficient 2e/h au L.C.I.E.

F. DELAHAYE, N. ELNEKAVE, A. FAU

L.C.I.E. - B.N.M. - France

B.P. 8   92260 Fontenay-aux-Roses - France

Le choix d'un système cohérent d'unités dérivées des grandeurs mécaniques nous contraint à effectuer un certain nombre d'opérations expérimentales constituant, selon Maxwell, des déterminations absolues. Ces déterminations permettent de situer dans le système choisi la valeur des étalons qui représentent les unités secondaires. C'est le cas de toutes les unités électriques.

Les opérations conduisant aux déterminations absolues sont en général longues et compliquées. Elles comprennent en effet la conception et la réalisation des instruments avec lesquels seront effectuées les mesures. Les précisions qu'on peut en attendre seront moins bonnes que celles généralement obtenues dans la comparaison de grandeurs de même nature. La pratique métrologique courante exigera donc des références directes sous forme d'étalons stables pour toutes les unités.

Ce type de référence est remarquablement réalisé pour le volt par l'effet Josephson alternatif. Cet effet permet d'obtenir une différence de potentiel continue, proportionnelle à une fréquence et au coefficient quantique $2e/h$. La perennité de l'unité de f.e.m. se trouve ainsi assurée.

Le coefficient $2e/h$ peut être évalué à partir de constantes physiques déjà déterminées, telles que la vitesse de la lumière, le rapport des masses de l'électron et du proton, la masse atomique du proton, le nombre d'Avogadro, la constante de structure fine et le nombre de Rydberg$\underline{/1/}$ ou par d'autres combinaisons équivalentes. Il peut s'exprimer à travers les déterminations absolues de l'ampère et de l'ohm. La procédure la plus directe consiste cependant à effectuer une détermination absolue du volt à l'aide d'un électromètre et à y rapporter le coefficient 2e/h.

Les étapes de cette détermination absolue comprennent des mesures de tension élevées (5 à 10kV) à l'aide d'un électromètre à constante calculée et le transfert des résultats à des niveaux très faibles (5 à 10mV) obtenus dans les expériences cryogéniques; une étape intermédiaire au niveau du volt permet d'alléger le travail expérimental en utilisant comme échelon provisoire la f.e.m. d'un groupe de piles Weston.

L'électromètre absolu

Cet électromètre est du type Kelvin et comprend essentiellement un condensateur à armatures planes circulaires. L'électrode gardée est suspendue à la balance et se trouve, comme l'anneau de garde, portée au potentiel des terres. L'électrode non gardée est à un potentiel de 5 à 10kV relativement aux précédents.

Les électrodes sont des disques massifs en silice fondue, rodés et polis optiquement. Elles sont revêtues d'un dépôt d'or de 0,3 µm d'épaisseur, assurant à la fois une bonne conductibilité électrique et un bon pouvoir réfléchissant. La planéité contrôlée interférométriquement, est de 0,06 µm pour l'anneau de garde et de 0,03 µm pour l'électrode mobile et l'électrode haute tension.

La distance des électrodes est définie par 6 cales en silice optiquement adhérentes situées entre les bords extérieurs de l'anneau de garde et de l'électrode haute tension. L'électrode gardée, suspendue au fléau est retenue par une butée triple, solidaire de l'anneau de garde. Cette butée possède des dispositifs de réglage originaux permettant d'obtenir la coplanéité des électrodes gardée et de garde, et d'assurer en même temps la simultanéité du décollement des trois points d'appui de l'électrode mobile.

Le réglage de la coplanéité de l'électrode mobile et de l'anneau de garde s'effectue interférométriquement. La méthode utilisée est la méthode classique du coin d'air dans un interféromètre de Michelson. L'électrode inférieure étant retirée, on dispose une lame séparatrice sous l'anneau de garde et on projette sur cette lame un faisceau laser hélium-néon élargi (20mm de diamètre environ) qui se divise en deux parties, une partie est transmise vers un miroir plan de référence, l'autre partie est réfléchie verticalement vers les électrodes. La lame séparatrice est disposée de telle façon que le faisceau qui atteint les électrodes est centré sur l'entrefer, une partie du faisceau est réfléchie par l'électrode mobile, l'autre partie par l'anneau de garde. Après recombinaison à la sortie de l'interféromètre avec le faisceau de référence, on obtient donc deux systèmes de franges noires rectilignes, parallèles et équidistantes correspondant à chacune des électrodes. Les franges ne sont pas localisées et peuvent être observées facilement sur un écran. Le réglage consiste donc à rendre ces deux systèmes identiques. Mais ceci ne permet d'obtenir que le parallélisme des deux électrodes et pour arriver à la coplanéité, il faut passer en lumière blanche (non monochromatique) et éclairer l'interféromètre par une source étendue. On observera à ce moment là à l'aide d'une lunette des franges colorées localisées sur les électrodes et en particulier dans chacun des deux systèmes l'unique frange noire correspond à l'arête du coin d'air. Il suffira alors de faire coïncider les deux franges noires pour être sûr de la coplanéité.

Fonctionnement de l'électromètre

Dans l'utilisation normale de l'électromètre, l'électrode mobile est retenue sur ses butées par l'action simultanée de son poids et de la force d'attraction électrique. La somme de ces forces est équilibrée en disposant une masse sur l'extrémité opposée du fléau . Dans ces conditions une diminution légère de la tension appliquée entraîne le décollement de l'électrode mobile. La valeur de la tension pour laquelle le décollement a lieu est alors notée.

La reproductibilité du décollement, qui peut être mise en question, a été vérifiée expérimentalement. Elle est déterminée par une variation de charge pratiquement égale à la sensibilité de la balance soit 25 μg. La force d'attraction de l'électromètre, pour un champ de 1kV/mm, correspond à une masse de 3,6g. Dans ces conditions, la sensibilité relative en tension est de $4 \cdot 10^{-6}$.

La constante de l'électromètre

La "constante" de l'électromètre est le rapport de la force d'attraction des électrodes au carré de la différence de potentiel. Elle s'obtient pour un déplacement suivant une coordonnée x en dérivant l'expression de l'énergie du condensateur

$$F = \frac{1}{2} U^2 \frac{dC}{dx} \tag{1}$$

$$K(x) = \frac{F}{U^2} = \frac{1}{2} \frac{dC}{dx} \tag{2}$$

Pour un condensateur idéal à électrodes planes et parallèles

$$K(x) = -\frac{1}{2} \cdot \frac{C}{x} \tag{3}$$

On notera que pour un condensateur gardé le déplacement virtuel considéré doit être compatible avec la coplanéité électrode de garde, électrode mobile.

Les deux défauts de l'électromètre qui empêchent d'appliquer la formule simple précédente sont les conditions aux limites de l'anneau de garde et l'existence d'un entrefer d'isolement entre anneau de garde et électrode mobile. Un répartiteur de tension permet d'éliminer la première cause d'erreur, en uniformisant le champ dans toute la région comprise entre les électrodes. Quant au rôle de l'entrefer isolant, son effet peut être inclus dans la formule ci-dessus légèrement modifiée.

On considère que pour un condensateur plan idéal le produit xC de la capacité par la distance des électrodes reste constant lorsque x varie. Pour le condensateur réel avec entrefer, ce produit varie légèrement. Nous écrirons donc

$$y = xC = A\left[1 + \delta(x)\right] \tag{4}$$

où

$$|\delta(x)| \ll 1$$

et A est une constante.

On en tire

$$\frac{dC}{dx} = -\frac{C}{x} + \frac{A\,\delta'(x)}{x} \# -\frac{C}{x}\left[1 - \alpha\delta'(x)\right]$$

et

$$F \# -\frac{U^2}{2}\cdot\frac{C}{x}\left[1 - \alpha\delta'(x)\right] \qquad (5)$$

La dérivée $A\delta'(x)$ est déterminée avec bonne approximation par des mesures, pour 4 valeurs de x, de la capacité de l'électromètre.

Jonctions Josephson à couches minces

Nous utilisons des jonctions à couches minces du type Niobium-oxyde de Niobium-Plomb. Celles-ci ont l'avantage de se conserver à température ordinaire, contrairement aux jonctions Pb, oxyde de plomb, Pb qui doivent être maintenues de façon permanente à basse température (T $\angle$ 77 K).

Pour obtenir le maximum de précision lors de la comparaison de la f.e.m. d'une pile étalon et de la tension Josephson, celle-ci doit être d'au moins 5mV, avec une amplitude en courant d'au moins 20 μA. De plus, la puissance hyperfréquence nécessaire pour obtenir ce résultat doit être aussi faible que possible.

L'obtention de ces caractéristiques nécessite l'optimisation de la jonction en ce qui concerne :

- ses dimensions géométriques qui fixent la fréquence de résonance
- son couplage avec le rayonnement H.F.

Dans les conditions optimales d'adaptation [2], nous avons obtenu des marches au niveau de 5mV, ayant une amplitude en courant de 20 μA avec une puissance H.F. de 5mW et des marches au niveau de 10 mV, d'une amplitude de 10mA avec une puissance de 22mV. La jonction utilisée avait un courant critique de 2mA et une résistance tunnel normale de 0,25 Ω .

Réducteurs de tension

Deux réducteurs de tension sont utilisés dans la détermination absolue de 2e/h. Le premier associé à l'électromètre
comprend une résistance haute tension de 10MΩ composée de
100 éléments de 100kΩ ; la mise en parallèle de ces 100 éléments conduit à une résistance de 1000Ω que nous comparons
par une procédure de substitution à la résistance basse tension.
La tension appliquée normalement à l'électromètre soit 10000 V
environ, peut ainsi être comparée à celle d'une pile étalon de
type Weston saturée. Un bain d'huile à température régulée à
0,001 degré près et un système d'écrans à potentiels répartis
complètent l'équipement de ce réducteur dont la précision est
estimée à 5.10^{-7}.

Le deuxième réducteur comprend deux ensembles de
10 résistances de 100Ω que l'on peut relier en série ou en parallèle suivant le système de Hamon. Les connexions des résistances sont du type symétrique axial, comportant une liaison
centrale et trois liaisons périphériques à 120° sur un disque de
cuivre; ces liaisons introduisent des erreurs inférieures à 10$^{-7}\Omega$
lors du passage série-parallèle.

Les deux ensembles de 10 résistance peuvent succes-
sivement être utilisés en configuration série parallèle, fournis-
sant deux rapports très voisins de 100. Une procédure où la
fréquence d'irradiation f des jonctions Josephson sert de para-
mètre permet un auto-étalonnement des rapports, la force élec-
tromotrice d'une même pile s'exprimant par

$$E = 10^2 \, af \frac{R_1}{R_2} = 10^2 af' \frac{R_2}{R_1}$$

$$\Rightarrow \quad E = 10^2 \, a \sqrt{ff'} \tag{6}$$

R_1 et R_2 étant respectivement les valeurs moyennes des résistances de chaque ensemble et a une constante .

La précision du réducteur, compte tenu de la procédure d'emploi, est évaluée à 2 10^{-8}.

Résultats

Les mesures sur la base nationale française de f.e.m. ont permis d'attribuer à 2e/h la valeur suivante au 1er janvier 1973 :

$$2 \ e/h = 483 \ 594,64 \ GHz/V \qquad (7)$$

Depuis cette date, la procédure de conservation a été inversée, et le volt français est maintenu par effet Josephson avec le coefficient de transfert précédent. La reproductibilité des mesures, estimée à mieux que 1.10^{-7} en 1973 et 1974 a été portée à $5 \ 10^{-8}$ par l'emploi du nouveau réducteur de tension.

Les premières mesures à l'électromètre confirment, dans leurs limites actuelles de précision(soit 3.10^{-5}) la valeur de l'unité française de f.e.m. . Avec le perfectionnement prochain de l'appareil nous pensons arriver dans un délai de quelques mois à une précision de l'ordre de quelques millionièmes ce qui permettra de confirmer ou d'infirmer les valeurs de 2e/h actuellement admises à partir des mesures absolues de l'ampère et de l'ohm. L'objectif final de l'électromètre est toutefois une exactitude de 10^{-6} ou mieux, ce qui, nous l'espérons permettra de vérifier les résultats du calcul de 2 e/h à partir d'autres constantes physiques déjà déterminées.

Bibliographie

/1/   V.0. Aroutunov, SV Gorbatsevitch, K.A. Krasnov
       "Comment augmenter la précision des étalons électriques
       et magnétiques"
       Colloque International sur l'Electronique et la Mesure
       Paris 26 au 30 mai 1975

/2/   T. Pech, J. Saint-Michel
       "Design of stable thin film Josephson Tunnel Junctions
       for the maintenance of voltage standards"
       I.E.E.E., Trans. on Magnetics MAG-II, 817 (1975)

CRYOGENIC VOLTAGE STANDARD AT PTB

V. Kose, B. Fuhrmann[*], P. Warnecke, F. Melchert

[*]Physikalisch-Technische Bundesanstalt and
Technische Universität Braunschweig
Braunschweig, West Germany

INTRODUCTION

For several years precise 2e/h measurements via
the ac Josephson effect have been routinely carried out
at the Physikalisch-Technische Bundesanstalt (PTB)
[1,2] and other standard laboratories throughout the
world. The total (1σ) uncertainty obtained for 2e/h is
4 parts in 10^8. This accuracy should also be guaranteed
between the different measurement apparatus for 2e/h
determinations in the various standard laboratories.
This comparison can be realized by two conceivable procedures. In one case electrochemical standard cells are
transported from one laboratory to another and the
EMF's of the cells are compared with the local
Josephson reference voltages. During the 1973 intercomparisons at BIPM the results of NBS, NML, NPL, and PTB
agreed to within $\pm$ 2 parts in 10^7, yielding an accuracy
which was evidently limited by the use of thermo-regulated electrochemical transport standards. On the other
hand improvement of the consistency of the data can be
achieved by transporting Josephson apparatus as was
done for the first time between BIPM and PTB. Fig. 1
and Fig. 2 show the results of this intercomparison
between the mean EMF of 39 saturated standard cells and
the 1-volt Josephson reference voltage of BIPM (cross
points) and PTB (dots) carried out in April and May
1975. There is a close agreement within a few parts in
10^8, although very different Josephson apparatus and
dc techniques were involved. These results should en-
courage further developments of portable Josephson

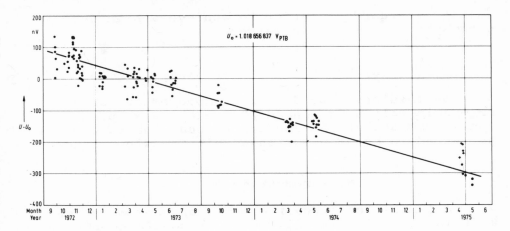

<u>Fig.1</u> Voltage comparison on 1-volt level between the
mean EMF of 39 saturated standard cells and the
Josephson dc voltages of PTB (dots) and BIPM (crosses).
Based on 2e/h=483 594 GHz/V $_{BI69}$ U means the EMF of
the cells converted to 20 °C and U_0 is the value of the
EMF in February 1973. Each point presents a mean value
of a measurement series comprising usually 10 independent measurements. The mean EMF drifts almost linearly
with time at a rate of -1.46×10^{-7} V per year (not including the BIPM data of April 1975).

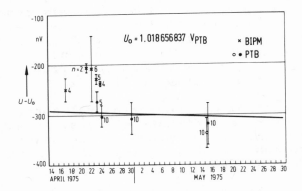

<u>Fig.2</u> Expanded time scale of Fig. 1. n means the num-
ber of the measurements. The error bars indicate the
standard deviation. All measurements were obtained via
cell no.4 located in a thermo-regulated enclosure,
except the one (see circle on May 75) which was measured via cell no.3.

voltage standards in order to achieve a high degree of
consistency in the 2e/h values.

CRYOGENIC VOLTAGE STANDARD

In this paper we describe improvements of our
cryogenic-voltage standard [2] which is at present
under test and can be used as a local laboratory voltage standard or as a portable volt transfer standard.

Measurement Mode

A simplified circuit diagram of the measurement
mode is shown in Fig. 3. The Josephson reference voltage of $\approx$ 3 mV is produced as described earlier [1].
While feeding a constant dc current (from battery B2)
through the resistive divider, the voltage drop across
the 0.1-Ω resistor can be compared with the dc voltage
by means of the null detector. The null detector system
consists of a two-hole SQUID used as a picovoltmeter
which has a sensitivity of less than 10 pV/$\sqrt{Hz}$ for an
input resistance of 0.1Ω, so that a balance of 3 parts
in 10^9 can be achieved in the 3-mV circuit. A dc voltage change across the 0.1Ω resistor causes an output
voltage of the SQUID system which generates a negative
feedback voltage in series with the input circuit by
means of the potential divider formed by 0.1Ω and the
feedback resistor $R_F + 32\Omega$. The extremely constant voltage of interest at the 1-V level appears across the
total resistance of 32Ω. The thermal EMF's due to
temperature gradients on the 1-V potential leads are
compensated by adjusting the current from battery B1
by means of r as described earlier [2].

Size reduction of the peripheral apparatus is
achieved by using solid state microwave components
like Gunn and IMPATT oscillators, the power supplies
of which are considerably smaller than those of
klystrons used so far.

In the following we report on a cryogenic resistive divider which establishes a ratio of 320:1 and
consists of only two resistors having an extremely low
relative temperature coefficient of resistivity (RTCR)
of $\frac{1}{\varrho} \frac{d\varrho}{dT}$ = 2 parts in 10^7/K, where ϱ means the resistivity.   Due to this small RTCR no temperature con-
trol is necessary by operating at 4.2K. The resistance ratio can be determined by one simple measurement

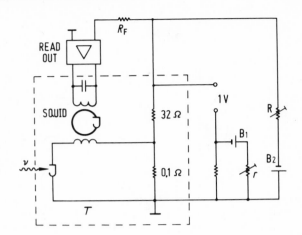

Fig.3 Simplified circuit diagram of the cryogenic voltage standard operating at 2 K or 4.2 K. The battery B2 feeds a current through the divider (comprising a 32-Ω and a 0.1-Ω resistor) which is automatically controlled by the readout voltage of the SQUID null detector. Compensation of the thermal EMF's on the potential leads are accomplished by adjusting r.

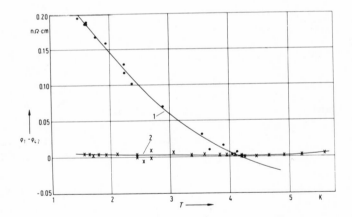

Fig.4 Temperature dependent resistivity of two cryogenic resistance alloys.
Curve 1: Copper germanium alloy having a mass content Ge of 6% and residual magnetic impurities of a few ppm. $\rho_{4.2}=16$ $\mu\Omega$cm.

Curve 2: Aluminium magnesium alloy having a mass content Mg of 4.9%, Mn of 0.1%, Fe of 0.15%, and Cr of 0.07%. $\rho_{4.2}=3.6$ $\mu\Omega$cm.

using a specially designed inductive voltage comparator (IVC).

Voltage Divider

Stimulated by the precise electrical metrology
at low temperatures progress has been made in the last
two years by producing low temperature alloys with a
very low RTCR of a few parts in $10^6/K$ $[3,4]$. Curve 1
in Fig. 4 shows the temperature dependence of the resistivity ϱ for a CuGe alloy having a mass content Cu
of 94% and Ge of 6%. As original materials we used
99.9998% pure Cu and 99.999% Ge. The negative slope
of the temperature dependent resistance is caused by
a few ppm residual magnetic impurities of Cr, Fe, Mn,
which were present in the original copper material
(Kondo effect). Reducing the content of magnetic impurities from a level of a few ppm to zero would
change the RTCR from negative values via zero to positive ones. However, this experiment has not been realized with copper alloys up to now.

The temperature coefficient of resistivity $\frac{d\varrho}{dT}$
(TCR) of aluminium alloys $[5,6]$ is much less influenced by magnetic impurities than that of alloys
based on noble metals. Curve 2 in Fig. 4 shows the
temperature dependent resistivity for an AlMg alloy,
having an RTCR smaller than 2 parts in $10^7/K$ between
2 K and 4 K, where $\varrho_{4.2\ K} = 3.6\ \mu\Omega cm$. This special
aluminium alloy is commercially available as DIN-alloy
AlMg5, containing a mass content of 5% magnesium and
magnetic impurities of about 3000 ppm, i.e. about
1000 times more than that of the above mentioned CuGe
alloy. Indeed, we measured that the RTCR changes its
sign depending on the mass content of the magnetic
admixture.

The 32-Ω and the 0.1-Ω resistors of the divider
were constructed by bifilar winding the AlMg5 alloy
wire on a rectangular thread which was cut into a poly-
tetrafluorethylene rod. The diameter of the wire of
both resistors was 0.15 mm. The power coefficient for
a 1-Ω resistor was estimated to - $1 \cdot 10^{-9}/mW$, assuming
a heat transfer coefficient of 0.1 $W/cm^2 K$.

CALIBRATION OF THE DIVIDER

The cryogenic resistive divider has been calibra-

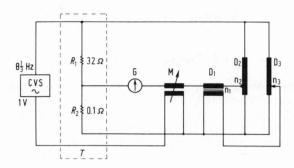

<u>Fig.5</u> Calibration of the cryogenic resistive divider
(R_1 and R_2) at temperature T=2K or 4.2K. D_1, D_2, and
D_3 are inductive voltage dividers with enhanced input
resistances, having ratios $n_1=10^{-3}$, $n_2=3\cdot10^{-3}$, and
$0 \leqslant n_3 \leqslant 1$. CVS means a constant voltage source, G a
balance detector, and M a mutual inductance.

ted by means of a low frequency comparator system shown
in Fig. 5. This inductive voltage comparator (IVC) consists of a balance detector G with a very low rms noise
voltage of 10 pV at a bandwidth of 0.1 Hz and three inductive voltage dividers D_1, D_2, D_3, which have electronically increased input resistances of 10^9 ohms.
Divider D_3 is a variable divider with 8 decades whereas
the dividers D_1 and D_2 have fixed ratios of $n_1=10^{-3}$ and
$n_2=3\cdot10^{-3}$ respectively. In order to get a very low output resistance of about 30 mΩ for the voltage comparator and a small error, a special design of one-step-
dividers is used for D_1 and D_2 $\underline{/7,8/}$. The circuit is
energized by the constant voltage source CVS with a
voltage of about 1 V at a frequency of 8 1/3 Hz. The
in-phase part of the voltage drop across R_2 is compensated by the sum of output voltages of D_2 and D_1. Small
quadrature terms (inductive) can be cancelled by a voltage derived from the variable mutual inductance M, which
is fed by the current through R_1 and R_2.

At balance the equation for the resistive ratios
reads

(1) $R_2/(R_1+R_2)=n_2+n_1n_3=3\cdot10^{-3}+10^{-3}n_3,$

where $0\leqslant n_3\leqslant1$. A range of ratios from $3\cdot10^{-3}$ to $4\cdot10^{-3}$
or nearly 1:333 to 1:250 may be calibrated. If divider
D_1 is reversible by a switch (not shown in the figure)
a range from 1:500 to 1:250 will be covered. As the input resistances of the dividers are 10^9 ohms the influ-

ence of the lead resistances on the accuracy is sufficiently small.

DISCUSSION

There are several advantages in using an ac system
for calibration: Thermal EMF's do not affect the comparison. As the power dissipation in the resistive divider (R_1 and R_2) is the same during the measurement and
calibration mode the effects of temperature coefficient
and power coefficient are reduced to second order.

On principle there exist differences between the
ac and the dc resistances of a resistor which, however,
become extremely small at 10-Hz-frequency. The estimated error which is caused by inductances and capaci-
tances, dielectric losses, eddy currents, and skin
effects is smaller than 1 part in 10^{11}. Another possible error is the Peltier effect [9] which could arise
from the use of spot welding the three potential leads
to the resistance wire. This effect as well as the
Thomson effect can be neglected as can be shown by a
detailed analysis.

REFERENCES

[1] KOSE,V.,MELCHERT,F.,FACK,H.,and HETZEL, W.,
    IEEE-IM 21,314(1972); also PTB-Mitteilungen 82,
    230(1972)
[2] KOSE,V.,MELCHERT,F.,ENGELLAND,W.,FACK,H.,
    FUHRMANN,B.,GUTMANN,P.,and WARNECKE,P.,IEEE-IM 23,
    271(1974)
[3] SULLIVAN,D.B.,Rev.Sci.Instrum.42,612(1971)
[4] WARNECKE,P.and KOSE,V.,Rev.Sci.Instrum. to be
    published
[5] CAPLIN,A.D. and RIZZUTO,C.,Phys.Rev.Lett.21,746
    (1968)
[6] BABIC,E.,FORD,P.J.,RIZZUTO,C.,and SALAMONI,E.,
    Solid State Commun.11,519(1972)
[7] FUHRMANN,B.,IEEE-IM 23,352(1974)
[8] EMSCHERMANN,H.H. and FUHRMANN,B., "One-step in-
    ductive voltage divider with ratio up to 1:1000",
    paper presented on EEMTIC 75,May 13 - 15,1975,
    Ottawa
[9] KIRBY,C.G.M. and LAUBITZ,M.J.,Metrologia 9,103
    (1973)

REVIEW OF THE MEASUREMENT OF μ_p/μ_N

B.W. Petley

Division of Quantum Metrology

National Physical Laboratory, Teddington, England

The magnetic moment of the proton, μ_p, is conventionally expressed in terms of the nuclear magneton μ_N (= $\frac{1}{2}\hbar e/M_p$) since the latter is the value predicted on the basis of the simple Dirac theory. The first measurements were made by Frisch and Stern in 1933[1] who found that the measured value was between two and three nuclear magnetons. By 1940 the anomaly had been established to better than 0.1%.[2,3] Progress continued, with the work of Bloch and Jeffries[4] and of Sommer, Thomas and Hipple.[5] The precision is now at the sub-part per million level. The pre-1970 determinations were reviewed by Taylor, Parker and Langenberg (1969)[6] and by Petley.[7]

The post-war determinations all use the suggestion of Alvarez and Bloch (1940)[8] that μ_p/μ_N may be simply measured by taking the ratio of the proton spin precession frequency f_s, to its cyclotron frequency f_c, in sensibly the same magnetic flux. Thus

$$\mu_p/\mu_N = (f_s/f_c) \cdot (B_c/B_s) \qquad \qquad \dots (1)$$

where B_s and B_c are the flux densities in which the two frequencies are measured and $B_s \simeq B_c$. There is little direct theoretical interest in a precise value of μ_p/μ_N and the main contribution to science comes through the combination of μ_p/μ_N with other precise determinations. One contribution concerns the equation linking the Faraday, F, the gyromagnetic ratio of the proton γ_p, u the nuclidic mass of the proton, and μ_p/μ_N which is

$$F = (u \times 10^{-3}) \cdot (\gamma_p)/(\mu_p/\mu_N) \qquad \qquad \dots (2)$$

It is apparent from equation (2) that the dimensions of γ_p are basically C/kg and so it suggests that there may be an alternative method of measuring γ_p which does not necessitate measuring dimensions very precisely.

It is experimentally convenient to measure f_s', the spin precession frequency of protons in a water sample. The frequency however is not entirely an atomic property but is partly a material property. The major part of the correction to that for the free proton is amenable to calculation[9] and experimental verification.[10] There remain however some small corrections which become of major importance below 0.1 ppm precision, these include chemical shifts due to dissolved gases and impurities, temperature dependent chemical shifts, and shielding effects due to the sample container etc. An air bubble or sealing pip on a spherical sample can change the effective shape factor, or, if cylindrical, the sample may be too short. The r.f. tuned circuit may give frequency pulling, while the copper winding may distort the field. If an electromagnet or permanent magnet is used then the insertion or movement of apparatus within the $\sim 5\,cm$ air-gap may distort or change the magnetic field. (Conventional shielding calculations can give a false impression by neglecting the effect that materials in the air-gap have on the reluctance of the magnetic circuit.) If the magnetic field is non-uniform the size of the nmr sample becomes an important consideration, for the field may be averaged by it in a way different from that intended by the experimenter. Beyond these considerations there are others which concern whether γ_p' may vary with magnetic[11] or gravitational fields.[12]

The Sommer, Thomas and Hipple 'omegatron' measurement unquestionably had the greatest impact on the recommended values of μ_p/μ_N up to 1970 and a possibly even greater impact on vacuum technology. Unfortunately, its apparent simplicity is deceptive for the detailed performance depends on all of its dimensions.[13] The early reviewers were faced with the problem of interpreting what Sommer, Thomas and Hipple meant by: 'However we feel that the ratio lies within the range $2.792\ 68_5 \pm 0.000\ 06$ ..... The assigned error is several times the estimated probable error.' Bearden and Watts (1951)[14] in their review treated the error as a standard deviation. DuMond and Cohen[15] however regarded it as two sigma and assessed the standard deviation as 11 ppm. Later, Fystrom, Petley and Taylor[16] argued that it was much more plausible to regard the stated error as a standard deviation. Indeed once this is done the measurements before 1970 look remarkable good, Figure 1. It is interesting that the weighted mean, assuming the correctness of the assigned errors is about 2.792 777 (10),3ppm, which is in remarkable agreement with the more recent work. If anything the experimenters are shown to be slightly pessimistic in their error assignments. Sommer, Thomas and Hipple combined their value with the determination of γ_p[17] and also with μ_p/μ_o[18] in just the way one would expect if at

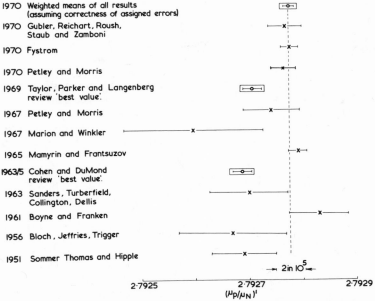

Figure 1. The values obtained for
μ_p/μ_N up to 1970

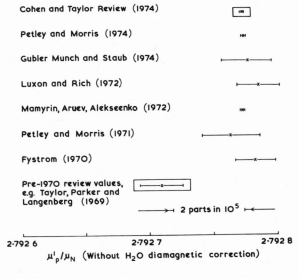

Figure 2. The values obtained for
μ_p/μ_N since the 1969 evaluation

the time of publication their error was intended to be a standard deviation.

The lesson is that experimenters should give as much detail about their error estimate as possible and make it very clear how their final number is intended to be treated.[19]   A further question is whether one is justified in always taking a mean with the results weighted as the inverse square of the variance.   The difficulty is greatest, when it is the most precise determination which has a pseudo precision.   Thus the careful and painstaking work of Boyne and Franken[20] and Mamyrin and Frantsuzov[21] had less impact than they deserved, neither did that of lesser precision by Bloch and Jeffries,[4] Sanders and Turberfield,[22] Petley and Morris[23] and Marion and Winkler [24] (– ten years is a long time to wait in order to be vindicated as a metrologist!)

Outstanding among the post-1970 determinations (Figure 2) is the measurement of Mamyrin et al.[25]   Their apparatus was similar to the mass synchrometer described by Smith,[26] with the ions accelerated by the two successive passages of the rf modulator and detected by deflecting them onto an electron multiplier.   The apparatus was generally similar to that used by Mamyrin and Frantsuzov[21] in their 1965 determination.   As with all determinations the cyclotron frequency was shifted by electrostatic fields or field gradients and the effects of these were eliminated by bringing two different masses of ions successively to resonance. Thus they used $He_4^+ - Ne_{20}^+$, $He_4^+ - Ne_{20}^{++}$ and $He_4^+ - Ar_{40}^+$ ions. It is probable that the electrostatic fields varied widely over the ion orbit.

In their 1974 determination they used a double-ion source which allowed both pairs of ions to be accelerated simultaneously.   As with the omegatron there was a small magnetic field from the heated spiral tungsten cathode in the ion source which was reduced to less than 0.1 ppm by compensating leads and current reversal.   Unlike the omegatron, evidence was found that their filament moved slightly when the current was reversed.   Their resonances were 20 ppm wide at the half height and the asymmetry amounted to 0.2 ppm.   The field at the orbit centre was ± 7 ppm different from the field over the ion orbit but it was considered that by using a spherical water sample $\sim$ 60 mm^3 (perhaps too large considering the field non-uniformity), the field could be averaged over the ion orbit to 0.2 ppm.   The effect on the magnetic field of movement of the magnetic resonance mass spectrometer chamber and cover plate was 2.85 ± 0.18 ppm, although in the error assignments but not in the table of corrections, the 0.16 ppm error appropriate to the chamber without the cover is taken.   Since the ions spend twice as long with the intermediate velocity as with the initial and final velocity it is not immediately obvious what the relativistic correction (0.1 ppm error) should be although one assumes that it

should be the rms velocity.   Their largest correction was that for
the transit time in the modulator.[27]   The rf field was at the
196th harmonic of the cyclotron frequency and so the observed
resonance frequency was

$$f_{res} = f_c (196 + p) \qquad \ldots (3)$$

where p was calculated theoretically to be − 0.002 671 8(98) and
was 0.6 ppm smaller than obtained in their 1964 determination.
The error in this correction was assessed as 0.01 ppm, but there
is the possibility that the electrical separation of the modulator
grids differs from the accurately measured mechanical separation.
Since the amplitude of the rf modulator voltage was of the order
of one third of the ion acceleration voltage, the confinement of
the field within the modulator must have presented extreme difficulty.   The elimination of frequency shifts which vary as the
reciprocal of the ion mass is of course a major worry in all of the
μ_p/μ_N determinations.   The final result was modified by Cohen and
Taylor[28] to allow for the revised nuclidic mass of helium following
the work of L.G. Smith.

The other determination with sub part-per-million precision is
that of Petley and Morris[29] using their quadrupole omegatron.
They used frequency scanning at a constant magnetic field.   The
performance was checked against theory to high precision and the
critical parameters, magnetic field, rf voltage, partial and total
pressure, ion trapping voltage, electron current and energy, were
all investigated over as wide a range as possible.   The ions used
for most of the measurements were H_2^+, HD^+, He^+ and D_2^+ with
some check measurements between masses 15 and 18 (CHD^+, H_2O^+ etc).
Protons were measured and gave a sensitive check of the relativistic
correction.   Dimensionless parameters were also calculated from
the cyclotron frequencies and checked against those derived from
nuclidic mass tables.   The ability to change experimentally the
critical parameters and investigate their effect is of course a
major advantage of the omegatron over fixed-slit devices.

It seems probable that (μ_p/μ_N) has been determined to better
than a ppm although the closeness of the results may be deceptive.
This conclusion is supported by the other high precision measurements of Fystrom[30] (whose result should be decreased if the relativistic correction is revised to allow for orbit drift), Luxon and
Rich[31] and Gubler et al.[32]   These determinations have very little
in common in the method of estimating the cyclotron frequency.
Overall the precision has advanced by roughly an order of magnitude
per decade.   For the future one can expect the precision to continue to advance a little further, using free protons or other ions,
to perhaps 0.1 ppm.   Beyond that it seems likely that other methods
must be employed.   Thus a measurement of the difference between

the Rydberg constants for hydrogen and deuterium by saturable absorption spectroscopy,[33] might lead to a better measurement of m_e/M_p - which is of more direct interest than μ_p/μ_N.

REFERENCES

1.   FRISCH, R. and STERN, O.   Zeits. f. Physik 85 4 (1933)

2.   RABI, I.I.   Phys Rev 51 652 (1937)

3.   RABI, I.I., ZACHARIAS, SMILLMAN, K.R. and KUSCH, P. Phys 55, 526 (1939)

4.   BLOCH, F. and JEFFRIES, C.D.   Phys. Rev. 80, 305 (1950)

5a. SOMMER, H., THOMAS, H.A. and HIPPLE, J.A.   Phys. Rev. 82, 697 (1951)

5b. SOMMER, H. Ph.D. Thesis. Agricultural and Mechanical College of Texas, (unpublished) (1950)

6.   TAYLOR, B.N., PARKER, W.H. and LANGENBERG, D.N. Rev. Mod. Phys. 41, 375 (1969)

7.   PETLEY, B.W.   Precision measurements and fundamental constants (NBS Special Publication 343) ed. D.N. Langenberg and B.N. Taylor (Washington: USGPO) pp 159-167 (1971)

8.   ALVAREZ, L.W. and BLOCH, F.   Phys. Rev. 57, 111 (1940)

9.   RAMSEY, N.F.   Phys. Rev. 78, 699 (1950)

10. MYINT, T., KLEPPNER, D., RAMSEY, N.F. and ROBINSON, H.E. Phys. Rev. Lett. 17, 405 (1966)

11. RAMSEY, N.F.   Phys. Rev. A1, 1320 (1970)

12. YOUNG, B.A.   Phys. Rev. Lett. 22, 1445 (1969)

13. PETLEY, B.W. and MORRIS, K.   J. Phys. E. 1, 417 (1968)

14. BEARDEN, J.A. and WATTS, H.M.   Phys. Rev. 81, 73 (1951)

15. DuMOND, J.W.M. and COHEN, E.R.   Rev. Mod. Phys. 25, 691 (1953)

16. FYSTROM, D., PETLEY, B.W. and TAYLOR, B.N.   ref (7) pp 187-91 (1971)

17. THOMAS, H.A., DRISCOLL, R.L. and HIPPLE, J.A.   Phys. Rev. 78, 787 (1950)

18. GARDNER, J.H. and PERCELL, E.M.   Phys. Rev. 76, 1262 (1949)

19. CODATA Bull. 9, December 1973

20. BOYNE, H.S. and FRANKEN, P.A.   Phys. Rev. 123, 242 (1961)

21a. MAMYRIN, B.A. and FRANTSUZOV, A.A.   Dokl. Akad. Nauk, USSR 159m 777 (1964) Sov. Phys-Doklady 9, 1082 (1965)

21b. MAMYRIN, B.A. and FRANTSUZOV, A.A., Zh. Eksp. : Teor. F12 48, 416 (1065) Sov. Phys - JETP 21, 274 (1965)

21c. MAMYRIN, B.A. and FRANTSUZOV, A.A.   Proceedings of the Third international Conference on Atomic Masses, R.C. Barber Ed. (University of Manitoba Press, Winnipeg), 427 (1968)

22. SANDERS, J.H. and TURBERFIELD, K.C.   Proc. Roy. Soc. A, 272, 79 (1963)

23a. PETLEY, B.W. and MORRIS, K.   Ref (21c), 461 (1968)

23b. PETLEY, B.W. and MORRIS, K.   Nature 213, 586 (1967)

24. MARION, J.B. and WINKLER, H.   Phys. Rev. 156, 1062 (1967)

25. MAMYRIN, B.A., ARUYEV, N.N. and ALEKSENKO, S.A.   Zh. eksp. teor Fiz. 63, 3 (1972) Sov. Phys-JETP 63, 1-9 (1973)

26. SMITH, L.G. and DAMM, C.C.   Rev. Sci. Instrum. 27, 638 (1958)

27. ALEKSEENKO, S.A. ARUEV, N.N., MAMYRIN, B.A., FRANTSUZOV, A.A. and PROSKURA, M.P.   Izmer. Tekh. 16, No 2, pp 7-8 (1973) Eng transl. Meas. Tech, 16, No 2, 162-66 (1973)

28. COHEN, E.R. and TAYLOR, B.N.   J. Phys. & Chem.   Ref Data 2, 663-734 (1973)

29. PETLEY, B.W. and MORRIS, K.   J. Phys. A. 7, 167 (1974)

30. FYSTROM, D.O.   Phys. Rev. Lett, 25, 1469 (1970)

31. LUXON, J.L. and RICH, A.   Phys. Rev. Lett. 29, 665-8 (1974)

32. GRUBLER, H., MUNCH, S. and STAUB, H.H.   Helv. Phys. Acta. 46, 722 (1974)

33. HANSCH, T.W., NAYFEH, M.H., LEE, S.A., CURRY, S.M. and SHALIN, I.S.   Phys. Rev. Lett. 32, 1336 (1974)

MAGNETIC MOMENT OF THE PROTON IN H_2O IN UNITS OF THE BOHR MAGNETON

William D. Phillips, William E. Cooke, and Daniel Kleppner

Res. Lab. of Elec. and Department of Physics

M.I.T., Cambridge, Massachusetts 02139, U.S.A.

We have made a new measurement of the magnetic moment of protons in a spherical sample of water in units of the Bohr magneton. This ratio, $\mu_{p'}/\mu_B$, is needed in many atomic physics experiments for converting into atomic units the value of a magnetic field measured as an NMR frequency. It is used in the determination of α from $2e/h$ experiments and in the evaluation of other constants such as γ_p and m_p/m_e. Also, by comparing $\mu_{p'}/\mu_B$ to μ_p/μ_B, the free proton moment in Bohr magnetons[1], one obtains the absolute value of the diamagnetic shielding factor for protons in water. Shielding factors of different molecules can be intercompared to very high precision, so that the present result permits absolute determination of the shielding factor for any molecule.

We have measured $g_j(H)/g_{p'}$, from which $\mu_{p'}/\mu_B$ is easily calculated, to an estimated precision of 10 parts per billion (ppb); it differs from the currently accepted value by 152 ppb, about twice the latter's estimated uncertainty.

The measurement was made by comparing the NMR frequency of protons in a nearly spherical sample of water with the electron spin flip frequency of ground state atomic hydrogen in a maser[1]. The magnetic field was 0.35T, giving a proton frequency of 15 MHz. The electron transition, characterized by $(m_s=-1/2, m_I=-1/2) \rightarrow (m_s=+1/2, m_I=-1/2)$ in the high field limit, occurred at 9.2 GHz. The two frequencies were compared by repeatedly interchanging the NMR sample and maser storage bulb, and observing free precession decay for each sample. The electron frequency was corrected for the effects of hyperfine struction and nuclear spin using the accurately known values of the hydrogen hyperfine separation[2] and the ratio $g_j(H)/g_p(H)$ [1]. The ratio of the corrected electron

492

frequency to the proton frequency gives $R = g_j(H)/g_{p'}$.

Both samples were contained in spherical quartz bulbs 1.28 cm.
in diameter with entrance apertures approximately 2mm in diameter.
These bulbs were affixed to a quartz spacer which separated them
by 5.08 cm. on centers. The bulb serving as the hydrogen maser
storage bulb was coated with $(CF_3CH_2CH_2)_2SiCl_2$, a substance similar
to dri-film[3], to inhibit wall relaxation. Spin polarized hydrogen
from an atomic beam source entered through the neck of the bulb
and was retained by a .75 mm thick, coated collimator with 25μ
holes. The bulb containing the water sample had a similar collimator at its entrance so that the two sample holders were as nearly
identical as possible. A coating of $(CF_3CH_2CH_2)_2SiCl_2$ on the
water bulb and collimator prevented the water from wetting the
surface of the bulb and made it easier to contain the water in the
bulb. This was the major motivation for using water rather than
one of the organic NMR standards such as $(CH_3)_4Si$. Such liquids
were nearly impossible to contain in the bulb because they would
wet the surface of the bulb and collimator and be drawn up into
the collimator.

When the samples were interchanged, only the bulb assembly was
moved. Since the bulbs were nearly identical each sample was in
very nearly the same magnetic environment when under observation.
The bulb under observation was centered in a cylindrical microwave
cavity, whose cylindrical portion was made of a tightly wound copper
strip which also served as the NMR pickup coil. This arrangement
resulted in a poor NMR filling factor, but assured that the NMR
resonance fields were uniform over the sample volume.

The same processing procedure was used for both the NMR and
maser free precession decay signals. The signal was converted to
an audio frequency of about 1 or 2 kHz by heterodyning it with a
local oscillator offset either above or below the signal frequency.
This audio signal was sampled by an analog to digital converter
at a rate about four times the audio signal frequency. The digitized
signal was stored by computer, and successive signals were averaged
until the desired signal-to-noise ratio (S/N) was obtained. Several
such averages were obtained with the local oscillator offset both
above and below the original signal frequency before the samples
were interchanged and the data transferred to magnetic tape for
later analysis.

The data were fourier analyzed using a fast fourier transform
algorithm. The maser signal was taken to be a simple damped sine
function and its transform was fitted by an absorptive and dispersive lorentzian. The full width at half maximum was typically
600 Hz. A S/N of 30 was typically obtained from about 5 seconds
of signal averaging. The scatter of points (standard deviation)
taken before an interchange was generally 4 ppb. This is con-

sistent with the known magnitude of the magnetic field fluctuations. A typical digitized maser signal and its fourier transform fit to a lorentzian is shown in Fig. 1.

The NMR signal was similarly fourier transformed. Since the
NMR line shape is the convolution of a relaxation broadened line
with the inhomogeneous magnetic field distribution, the NMR line
cannot be fit to any simple function. Because of this, the real
and imaginary parts of the fourier transform were combined to
produce a purely absorptive line shape and the centroid of the
line was calculated. Computer modeling of the signal confirmed
that this procedure was valid for a wide range of possible field
distributions. NMR linewidths were typically 1.5 Hz with a S/N
of about 100 obtained from 30 to 60 seconds of averaging. The
standard deviation of points taken over a period of a few minutes
was about 13 ppb. The scatter is larger than in the case of the
maser data partly because the time between measurements is longer
so that field fluctuations are larger, and partly because the
method of finding the centroid is not as precise as the line fitting procedure used with the maser signal where the lineshape is
known. An NMR signal and its absorptive fourier transform are
shown in Fig. 2.

A set of data consisted of the results of 4 or 5 complete
interchange cycles, each cycle taking typically 6 minutes. The
electron-proton frequency ratio involved quantities measured at
different times and is therefore sensitive to field drift. However,
the drift rate was generally small, 1 or 2 ppb per minute and the
effects of drift were effectively eliminated by using linear interpolation between successive frequency measurements. The standard
deviation for individual ratio measurements in a set of data was
typically 8 ppb.

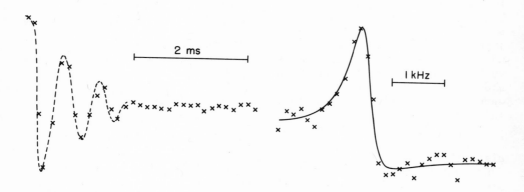

Figure 1: Maser signal and fourier transform with lorentzian fit.

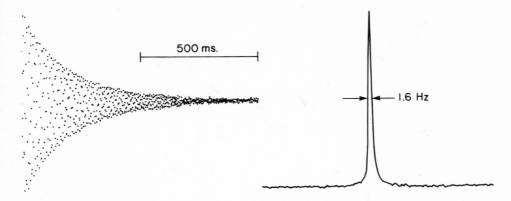

Figure 2: NMR signal and fourier transform.

There are a number of potentially serious sources of error in
an experiment such as this which requires interchange of samples in
a magnetic field. Essentially they come down to the problem of
assuring that both samples experience the same field distribution.
Effects of local disturbances were minimized by making the two
sample holders identical insofar as possible. The inside of each
bulb was spherical to about 0.1% and diameters of different bulbs
were identical to the same precision. To eliminate the effects of
uncontrollable features such as small variations in wall thickness,
the bulbs were interchanged, so that each bulb was used both as an
NMR sample holder and as a maser storage bulb.

Several corrections remained in spite of the bulb interchange
procedure, the largest arising from the meniscus of the water
sample. The meniscus resulted in a roughly hemispherical intru-
sion into the top of the spherical sample. Due to the bulk diamagnetism of the water there was a shift in the average field value
from what would have been obtained with a perfectly spherical
sample. In addition there were corrections due to the presence
of atomic hydrogen in the meniscus region and the collimator,
regions which were inaccessible to the water sample. There were
unavoidable field gradients in these regions due to the diamagnetism of the bulb stem and collimator. The accuracy of the final
result depended to a great extent on the precision with which all
of these corrections could be calculated. To check the correction
procedure, the entire experiment was repeated with a second set
of bulbs, with different meniscus size. There is also a correction
for the effect of the water diamagnetism on the hydrogen sample
which is the same for all configurations. A summary of the
corrections for each of the 4 bulb configurations, identified by
the bulb serving as NMR bulb, is presented in Table I, along with
the uncertainties in parentheses.[4]

TABLE I: Corrections to g_j/g_p, ppb

NMR bulb	#1	#2	#3	#4
Statistical error	(19)	(5)	(4)	(8)
Meniscus corrections for non-sphericity	-47(16)	-41(14)	-15(5)	-9(2)
Gradients across meniscus	-30(10)	-27(9)	-10(5)	-4(2)
Water diamagnetism	-8(2)	-8(2)	-8(2)	-8(2)
Collimator	8(4)	8(4)	3(1)	3(1)
Uncalculated	(12)	(12)	(5)	(5)

The statistical error quoted is the standard deviation of the results obtained from the various sets of data taken with a single bulb configuration. The uncalculated effects which are included as additional uncertainties include such effects as gradients due to bulb asymmetries, errors in positioning the bulbs in the field, and regions of the bulb other than the meniscus which are sampled by hydrogen atoms but not by the water.

Bulbs 1 and 2 were used together as bulb set A, and 3 and 4 were set B. The unweighted average of the two results from each set constitutes the final result for that set. Care must be taken in computing the uncertainty of the final result since the errors for the two bulbs in a set are not independent. This is due to the fact that the bulbs are very similar and the corrections shown in Table I are calculated using the same method for each bulb. As a result, only the statistical error is combined as if it were independent for the two bulbs and the other errors are averaged. A summary of the results is given in Table II.

The difference between the results for the two sets is only 1.5 ppb. Since the corrections for set A were much larger than for set B due to the larger meniscus, the good agreement suggests that the correction procedure was the proper one.

The final result is the weighted mean of the results of the two sets. We have assumed the errors to be independent since they arise mostly from statistics or meniscus related corrections for very different meniscuses. We have added quadradically an uncertainty of 5 ppb to account for the possibility of shifts due to ambient magnetic field gradients. The magnitude of the effect was estimated by measuring the shift when an additional gradient was introduced. The final result with this added uncertainty is $g_j(H)/g_{p'} = 658.216\ 009\ 1(68)$ (10ppb), for a spherical sample of

TABLE II: Corrected values of $g_j(H)/g_p$,

single configuration result   set result

#1: 658.216 026 3(192) (29ppb)
                                          A: 658.216 010 0(176) (27ppb)
#2: 658.215 993 6(139) (21ppb)

#3: 658.215 956 1(59) (9ppb)
                                          B: 658.216 009 0(63) (10ppb)
#4: 658.216 061 9(68) (10ppb)

water at 34.7(1)C. The uncertainty represents the confidence
interval of a standard deviation and includes statistical error as
well as our best estimate of the uncertainty introduced by all
known systematic effects.

From this result we can easily calculate the proton moment, in
Bohr magnetons; using the calculations of Grotch and Hegstrom[5]
for $g_j(H)/g_e = (1-1.7705 \times 10^{-5})$ and the accepted value[6] of
$g_e/2 = 1.0011596567(35)$ we obtain

$$\mu_p'/\mu_B = .001520992983(17)(11ppb)$$

Using the theoretical value[3] for $g_p(H)/g_p$(free) =
$1-1.7733 \times 10^{-5}$ and the experimental result[1] $g_j(H)/g_p(H) =$
658.210706(6) we can obtain the diamagnetic shielding factor
for water:

$$\sigma = 25.790(14) \times 10^{-6}$$

The most accurate previous measurement of $g_j(H)/g_p$, is due
to Lambe and Dicke[7]. In their experiment an electron spin flip
transition in hydrogen was measured by microwave absorption using
a sample of atomic hydrogen in a buffer gas. Their result is
$g_j(H)/g_p$, = 658.215 908 8(436) (66ppb). The quoted error is twice
the statistical error, and takes into account possible frequency
shifts due to the buffer gas. Our method differs from theirs
chiefly in the use of a hydrogen maser which yields a linewidth
30 times narrower and a S/N 10 times larger than the microwave
absorption method, as well as eliminating the buffer gas. In
addition, the NMR linewidth and S/N have each been improved by a
factor of 3. The two results differ by 152 ppb, more than twice
the estimated error in the Lambe-Dicke experiment. Part of the
discrepancy can be attributed to temperature effects. The proton
moment in water decreases with rising temperature at a rate of
about 10 ppb/°C[8]. The temperature of our samples was maintained
accurately at 34.7(1)C by the magnet's thermal control system.
Lambe and Dicke did not measure the temperature of their samples,
which were at the ambient field gap temperature[9]. If we assume
that the ambient temperature was 30C we can account for about 50 ppb

of the discrepancy.

Another possible source of error is the cylindrical neck of
the Lambe–Dicke sample holder. We estimate that the bulk diamagnetic effect of the teflon sample holder would reduce the field in
the neck by roughly 1000 ppb. The neck volume was about 15% of
the total volume which would imply an error in their final result
of the order of the observed discrepancy.

ACKNOWLEDGEMENTS

We wish to thank Francine E. Wright who developed the method
for analyzing the NMR data and Stuart B. Crampton for providing
the bulb coating material.

REFERENCES

*This work was supported by the National Science Foundation Grant
 GP-39061X1

1. P. F. Winkler, D. Kleppner, T. Myint and F. G. Walther, Phys.
   Rev. A5, 83 (1972).
2. H. Hellwig, R. F. C. Vessot, M. W. Lecine, P. W. Zitzewitz,
   D. W. Allen and J. W. Glaze, I.E.E.E. Trans. Instrum. Meas.
   19, 200 (1970).
3. The coating material was supplied by Prof. Stuart B. Crampton
   of Williams College.
4. Such corrections will generally not be of concern to the
   experimenter who uses NMR to measure magnetic fields, since
   most of the errors arise from the necessity in our case of
   using interchangeable sample holders.
5. H. Grotch and R. A. Hegstrom, Phys. Rev. A4, 59 (1971).
6. E. R. Cohen and B. N. Taylor, J. Phys. Chem. Ref. Data 2,
   4 (1973).
7. E. B. D. Lambe, Ph.D. Thesis, Princeton University 1959
   (unpublished).
8. J. C. Hindman, J. Chem. Phys. 44, 4582 (1966). See also
   Schneider et al. J. Chem. Phys. 28, 601 (1958) and Jacobson,
   B.S. Thesis, M.I.T. 1972 (unpublished).
9. E. B. D. Lambe (private communication).

A PROGRESS REPORT ON THE g-2 RESONANCE EXPERIMENTS

Hans G. Dehmelt

University of Washington

Seattle, Washington 98195, U.S.A.

INTRODUCTION

The spin resonance experiments on slow free electrons in vacuum have a long history. In 1953 Bloch proposed to trap electrons in an electric potential well of depth $\sim 10^{-5}$ V and unspecified shape superimposed upon a magnetic field of ~ 1000 G. Cyclically the latter could be made so inhomogeneous that the effective magnetic hill seen by all electrons not in the lowest, unmagnetic Rabi-Landau level would overcompensate the electric well. Thereby, only those in the lowest level would be retained. Spin- or cyclotron-transitions to the next higher levels induced subsequently would be signaled by loss of the electrons. The levels referred to are given by $E = (2n + 1 + g_s m)\mu_0 H$, with $n = 0, 1, 2,\ldots$ and $m = \pm 1/2$, (Rabi, 1928). Also at Stanford in connection with their cyclotron-resonance work in which trapping was carefully avoided Franken & Liebes showed in 1956 that a small external electric field should shift the cyclotron frequency $\omega_c \equiv 2\pi\nu_c$ by $\delta\omega_c \propto \phi_{zz}/\omega_c$. This suggested that the Penning (1937) trap in the form described by Pierce (1949) should be well suited for the simultaneous measurement of ω_c and of the spin resonance frequency $\omega_s \equiv 2\pi\nu_s$ on stored electrons. For such a trap the axial field gradient ϕ_{zz} and $\delta\omega_c$ are constant throughout its volume and $\delta\omega_c = -\omega_m$ may be determined by measuring the axial oscillation frequency $\omega_z \equiv 2\pi\nu_z$ or the magnetron (drift) frequency $\omega_m \equiv 2\pi\nu_m$ on the electrons in situ. This allows the use of well depths which comfortably exceed the contact potential uncertainties of .1 - 1 V commonly encountered in radio tubes. Consequently, work with ~ 2 V deep Penning traps was begun at the University of Washington. Axial resonances at $\nu_z \simeq 2.7$ MHz about 10 kHz wide, and ~ 15 kHz wide cyclotron resonances at $\nu_c \simeq 81$ MHz were observed and efforts to detect the spin resonance of the stored electrons via interaction with a polarized Na-beam were initiated by 1959. In the course of this work the relation $2\omega_c\omega_m \simeq \omega_z^2$ was demonstrated experimentally. Expressions were derived for the thermalization time $\tau_{IT} \simeq (4Mz_0^2)/(e^2R_s)$ of a single ion of mass M and initial energy W_{IO} oscillating inside a trap of cap separation $2z_0$ interacting with a tuned circuit of shunt resistance R_s and also the power signal-to-noise ratio $S/N \simeq W_{IO}/kT$ available in the tuned circuit with an observation time $\sim\tau_{IT}$ was obtained (Dehmelt 1961, 1962). As no spin resonance was observed it was decided to go to smaller, deeper traps, $\nu_z \simeq 60$ MHz, in higher fields, $\nu_c = 3 - 22$ GHz, and to study thermalization and relaxation processes occurring in the cloud, which were feared to interfere with the detection of the spin resonance. Of special interest here were interactions between the electrons, with the tuned circuit and, via Majorana flops, with (modulated) magnetic field gradients, e.g. (Kleppner et al. 1962). The hope was to use relaxation effects to provide a link between spin and cyclotron motion that spin resonance might be detected by heating of the cloud (Dehmelt & Walls 1968), eliminating the need for the Na-beam. Having adsorbed a number of former members of the Washington ion-rf-spectroscopy group the Bonn/Mainz group entered the field very vigorously in 1965, taking up the Penning trap/Na-beam combination after spin resonance of He$^+$ in a Paul trap polarized by spin exchange with a Cs beam had been demonstrated previously (Dehmelt & Major, 1962; Fortson, Major & Dehmelt, 1966). This group was the first to report spin and g-2 resonances (Fortson, Graeff, Major, Roeder and Werth, 1968; Graeff, Klempt and Werth, 1969). The work at the three labs prior to 1972 is the subject of a chapter in the 1972 review article of Rich & Wesley. Part of it is also covered in the 1967 and 1969 review articles on the rf spectroscopy of stored ions by the author. Earlier work has been reviewed by Hughes (1959) and Farago (1965).

THE STANFORD EXPERIMENT

Any experiments with slow electrons are beset with many problems unfamiliar to experimentalists well versed in the handling of "stiff" and much more popular high energy beams. It is all the more surprising that the remarkable data which have been reported for the Stanford electron-positron free-fall apparatus (Fairbank et al., 1973) have not stirred the interest of experimentalists in this field more. It has been proposed (Knight, 1965) to adapt this time-of-flight apparatus for which energy resolutions of $\sim10^{-10}$ eV at 4°K and $\sim10^{-5}$ eV at 300°K have been reported to simultaneous measurement of spin and cyclotron resonances on electrons in the lowest Rabi-Landau level (Rich & Wesley, 1972). Land & Raith (1974) have described experiments with a somewhat similar time-of-flight apparatus developed for collision studies and report a resolution of $\sim5 \times 10^{-3}$ eV at 300°K.

THE BONN/MAINZ EXPERIMENTS

Using the same Penning trap/polarized Na-beam apparatus but noise-instead of ejection-thermometry Church & Mokri (1971) have repeated the g-2 resonance experiments of Graeff, et al. (1969) relying on the spin dependence of the cooling of the electron cloud associated with impact excitation of the Na D-lines. Graeff et al. (1972) have monitored the Na-beam emerging from the trap for spin flips it might have undergone upon interacting with the electron cloud. The spin resonance has been observed in this way. Kienow, et al.(1974) have introduced a superconducting 65 kG Magnet and a new mode of ejection-thermometry relying on particle counting. As the depth of the trapping well is gradually lowered electrons of lower and lower energy leave the trap through a hole in one end cap, are accelerated and counted. The authors have tested their apparatus by determining a sequence of energy distributions as the stored cloud is allowed to cool via spontaneous emission of cyclotron radiation for increasing intervals. Through the heating of the cloud under cyclotron excitation they have also observed cyclotron signals of 1 ppm width. In a separate experiment also using a superconducting magnet and introducing some important modifications Graeff et al. (1975) are reviving the (1953) proposal of Bloch. The authors propose to trap electrons in a Penning trap, thermalize their axial and cyclotron motions at $4^{O}K$, eject them by diabatic reduction of the well depth and pass them through the fringing fields along the axis of the superconducting solenoid. About half of the electrons should be in the lowest unmagnetic level and should therefore experience no acceleration "gliding down" this magnetic "potential hill". Changes in the fraction of electrons in higher, magnetic, Rabi-Landau levels following suitable excitation are expected to show up when the electrons emerging from the fringing field are analyzed by time-of-flight spectrometry.

THE WASHINGTON EXPERIMENTS

Based on experiments in which a low signal/noise ratio and lack of time prevented doing an ample number of runs and controls, Walls & Stein (1973) have published preliminary measurements of g-2 in an electron gas at $80^{O}K$ by means of a bolometric technique. In this technique (Dehmelt & Walls, 1968) the temperature of the cloud is inferred from noise measurements on an LC circuit coupled to the axial motion. This noise thermometry is used to detect heating of the cloud caused by excitation of the cyclotron motion brought about by sequentially inducing spin- and g-2 transitions. The inhomogenous magnetic rf field inducing the g-2 transitions was created by opposing loop currents flowing in the especially cut and latticed ring electrode (Walls 1970). The measured distribution of this field near the center of the trap, could be approximated by a gradient $\partial H_x/\partial x = -\partial H_z/\partial z$, $\partial H_y/\partial y = 0$. An electron moving in a cyclotron orbit of radius r_c will consequently see an oscillating field $r_c(\partial H_x/\partial x)\cos\omega_c t$. Since $(\partial H_x/\partial x) \propto \cos\omega_d t$ this field will consist of two side bands at $\omega_c \pm \omega_d$ and will induce spin transitions when one choses $\omega_c + \omega_d = \omega_s$. Values $\partial H_x/\partial x \simeq 10$ G/cm were practical, yielding with $r_c = 5 \times 10^{-5}$ cm Rabi frequencies of about 600 Hz. Graeff & al. (1969) and Church & Mokri (1971) have used a different configuration in which the orbiting electron sees a field rotating at ω_c whose amplitude is $\propto \cos\omega_d t$. Both of these fields violate Byrne's (1963) unnecessary restriction that fields capable of inducing the g-2 transition should be independent of z. (Byrne has proposed to use the leading, z-independent term of the circular H-field present in a special coaxial cavity. However, even for $(\text{curl } H)_z \simeq 20$ mG/cm and a Rabi frequency of ~1 Hz the accompanying electric field $E_z = \lambda_d(\text{curl } H)_z$ assumes the value of ~1 kV/cm! at $\nu_d = 25$ MHz.) Subsequent experimental efforts were directed towards increasing the efficiency of the resonant spin/g-2 heating and of the bolometric detection. Computer simulation indicated that alternating adiabatic fast passage spin reversals and g-2 frequency π-pulses of a combined 2 msec duration should result in a quick drop of the spin temperature to ~8°K and an increase in the cloud temperature of ~10°K/sec (Dehmelt & Ekstrom, 1973a). The interaction of the tuned circuit with the electron cloud was analyzed in terms of an equivalent $\ell cr\nu_z$ series resonant circuit shunting a real LCR ν_z' parallel resonant circuit. The equivalent inductance for a single electron $\ell_1 = n\ell$, n the electron number, is a constant of the trap which for our parameters had the value $\ell_1 = 8000$ Hy. Drastic effects of the cloud on the noise spectrum of the tuned LC circuit demonstrated r<<R. The width of the notch observed for $\nu_z = \nu_z'$ allowed a convenient determination of n. From the viewpoint of calorimetry of special interest is the sharp noise peak observed at $\nu_z > \nu_z'$. It is due to the parallel resonance associated with the cloud and the area under the resonance is proportional to n x the cloud temperature. At 80°K the corresponding electron cloud calorimeter is characterized by a heat input integration time of ~5 s and a temperature sensitivity of about ½°K for $n \simeq 10^4$. (Dehmelt & Wineland, 1973, Wineland & Dehmelt, 1975b). Assuming with Gardner (1951), Franken & Liebes (1956) and Fischer (1959) that the electric shifts of cyclotron and axial resonance frequencies should reflect fields from identical neighbors one might feel that a narrow observed relative axial line width of ~10^{-5} should indicate a comparable ~1 Hz uniformity of the cyclotron resonance shift equal to $\nu_m \simeq 10^5$ Hz. However, this is not the case (Wineland & Dehmelt, 1975a). The equations of motion of a single particle in a Penning trap under forced cyclotron/axial excitation $f_x(t)/f_z(t)$ may be written

$$m\ddot{x} - m\omega_z^2 x/2 + m\omega_c\dot{y} = f_x(t), \qquad -m\omega_z^2/2 = e\phi_{xx}$$

$$m\ddot{y} - m\omega_z^2 y/2 - m\omega_c\dot{x} = 0, \qquad -m\omega_z^2/2 = e\phi_{yy}$$

$$m\ddot{z} + m\omega_z^2 z \qquad = f_z(t), \qquad m\omega_z^2 = e\phi_{zz}.$$

Electrostatic interactions between like particles do not shift or broaden the cyclotron resonance at $\omega_c - \omega_m$ or the axial resonance at ω_z, $\omega_c \equiv eH_o/mc$, $\omega_c\omega_m - \omega_m^2 = \omega_z^2/2 = -e\phi_{xx}/m$. Rather, from the equation of the z- motion of two interacting particles

$$m\ddot{z}_1 + m\omega_z^2 z_1 = F_{z12} + f_z(t), \quad m\ddot{z}_2 + m\omega_z^2 z_2 = F_{z21} + f_z(t)$$

it follows by addition that the center of mass coordinate $Z = (z_1 + z_2)/2$ obeys the same equation as a single particle, $m\ddot{Z} + m\omega_z^2 Z = f_z(t)$. The same argument may be extended to the x and y coordinates and to an arbitrary number of identical particles. Unfortunately, this is not so for the g-2 resonance occuring in a cloud at $\omega_s - \omega_c + \omega_{ml}$. Here the individual electron spin due to its thermal cyclotron motion at $\omega_c - \omega_{ml}$ through the applied inhomogeneous magnetic field alternating at $\omega_{g-2} + \omega_{ml}$ sees a sideband at the spin resonance frequency $\omega_s = \omega_c + \omega_{g-2}$ (Dehmelt and Walls, 1968). However, $\omega_{ml} \neq \omega_m$ now reflects the micro-environment of the electron focused upon. The presence of strong e-e interactions in usable clouds has been demonstrated ad oculos by parametric excitation at $2\nu_z$ of a mode in which one half of the cloud oscillates out-of-phase against the other half, suggesting $|\omega_{ml} - \omega_m|/\omega_m \simeq .01-0.1$. Nevertheless for the center-of-mass mode a line width $\Delta\nu_z \simeq 600$ Hz at $\nu_z \simeq 60$ MHz has been realized. The cause of the residual width is most likely anharmonicity of the trapping potential. With a trap design incorporating guard rings newly developed by Van Dyck et al. (1975) it has been possible to null out the biquadratic terms in the potential and to reduce the width of the line to ~20 Hz by applying appropriate voltages to the guard rings.

In view of the above developments a single electron contained in a harmonic well, free from complex space charge shifts began to look more and more attractive. As a first step experiments were begun to continuously observe the axial oscillation of a single electron (Wineland, Ekstrom & Dehmelt, 1973). Cloud experiments had indicated a trap anharmonicity $d\nu_z/dW_1 \simeq 50$ kHz/eV. Sidestepping associated frequency pulling problems excitation at the zero-amplitude eigenfrequency ν_{zo} of the electron was chosen. In the circuit used the drive u_d was applied to one cap which via the equivalent impedance Z_e of the series resonant electron was connected to the other cap. As u_d is increased from 0 to 2mV the eigenfrequency ν_z grows by ~7 kHz and the imaginary part of the impedance Z_e increases from 0 to $4\pi\ell_1 \cdot 7$ kHz $\simeq 7\times10^8\Omega$, an rf current $i_{d1} \simeq u_d/Z_e \simeq 3\times10^{-12}$ A flows through the LC circuit of shunt impedance $R \simeq 100$ kΩ connected to the output cap developing a signal $u_s \simeq .3$ μV and exciting an electron oscillation of energy $W_1 = \ell_1 i_{d1}^2/2 \simeq .2$ eV. The experimental problems associated with the unfavorable ratio $u_s/u_d \simeq 10^{-4}$ were solved by slightly modulating ν_z at $\nu_{mod} = 1$ MHz, driving the electron oscillation on the weak $\nu_{zo} - \nu_{mod}$ side band but synchronously detecting the strong ν_{zo} carrier. Drawing upon the cloud techniques sketched earlier about 10 electrons were injected. When u_d was raised to a critical level of ~2 mV the signal decreased roughly exponentially but in equal steps. We attribute the steps to loss of single electrons associated with an instability due to sign reversal of dv_z/dW_1 occurring at $W_1 \gtrsim .2$ eV. At a lower u_d and $u_s = .2$ μV the last plateau corresponding to a single electron has been observed continuously for days. As a first application the cyclotron resonance near 22 GHz was observed on this "monoelectron oscillator" (MEO). A trigger technique based on energy transfer from the excited cyclotron motion via gas collision to the axial motion parametrically driven at $2v_z$ was used. For the purpose of monitoring cyclotron and spin quantum numbers n, m of the MEO, Dehmelt & Ekstrom (1973b) have proposed to superimpose a magnetic bottle, $H_B = 100$ G deep, over the Penning trap fields. For our standard trap this would cause a shift $\delta v_z(n,m) \simeq (2n + 1 + 2m)(3 \text{ Hz})$ of $v_z \approx 60$ MHz due to the contribution $(2n + 1 + 2m)\mu_o H_B$, $\mu_o H_B \simeq .5$ μeV, to the ~6 eV deep trapping potential. Thereby it should be possible to literally watch the MEO jump from one n,m level to another, a MEO linewidth $\Delta v_z = 8$ Hz having been demonstrated recently (Wineland & al., 1975). A problem remaining is the coupling between the cyclotron motion and the "free space" radiation field. A solution might be to enclose the trap in a small, tight high-Q microwave cavity and to choose v_c thus as to decouple the MEO as much as possible from the few resonant modes the cavity can support. The current approach is to cool the MEO to $4°$K. Thereby a dwell time in the n = 0 levels of ~1 sec may be realized at $v_c \simeq 22$ GHz during which the shifts $\delta v_z(0,-\frac{1}{2}) = 0$ and $\delta v_z(0,+\frac{1}{2}) \approx 6$ Hz might be observed (Dehmelt et al. 1974). For $v_c \simeq 200$ GHz the dwell times become so short that it becomes practical to measure the average shifts $<\delta v_z(n,\pm\frac{1}{2})>$ which again differ by 6 Hz. Hereupon and on the very slow spin relaxation the detection of g-2 transitions induced when n>0 may be based. The small shifts δv_z due to the relativistic mass changes $\delta m \simeq (n + \frac{1}{2} + m) hv_c/c^2$, may become of interest for the detection of the g-2 resonance (Dehmelt et al., 1974). Rabi (1928) has solved the Dirac equation for an electron moving in a magnetic field. In accordance with this the spin state energy has to be regarded as kinetic as far as δm is concerned. For the MEO experiments quite modest g-2 transition rates suffice and it is of interest to look for possibly more convenient alternate ways to realize them. E. g. a <u>static</u> magnetic point dipole field is inhomogeneous enough for an electron carrying out its cyclotron motion at v_c plus a forced axial motion at v_d to see a usable magnetic rf field at the combination frequency $v_c + v_d$ (Dehmelt, 1969 p. 151). Even the relativistic magnetic field $B = Ev/c \simeq .1$ μG seen by an 1 meV electron for E = 1 V/cm at v_d may find application (Dehmelt & Ekstrom, 1973b). I wish to thank Dr. D. Wineland and Dr. R. Van Dyck for reading the manuscript, and Lyn Maddox for typing it.

BLOCH, F. (1953). Physica $\underline{19}$, 821.
BYRNE, J. (1963). Can. J. Phys. $\underline{41}$, 1571.
CHURCH, D. A. & MOKRI, B. (1971). Z. Physik $\underline{244}$, 6.
DEHMELT, H. (1961). Progress Report NSF-G5955, May 1961, "Spin Resonance of Free Electrons".
DEHMELT, H. (1962), Bull. A.P.S. $\underline{7}$, 470.
DEHMELT, H. (1967 & 1969). Adv. Atom. & Mol. Physics, $\underline{3}$ & $\underline{5}$, (Academic Press).
DEHMELT, H. & MAJOR, F. G. (1962). Phys. Rev. Letters $\underline{8}$, 213.
DEHMELT, H. & WALLS, F. (1968). Phys. Rev. Letters $\underline{21}$, 127.
DEHMELT, H. & EKSTROM, P. (1973a,b). Bull. A.P.S. $\underline{18}$, 408 & 786.
DEHMELT, H. & WINELAND, D. (1973). Bull. A.P.S. $\underline{18}$, 786.
DEHMELT, H. & EKSTROM, P., WINELAND, D. & VAN DYCK, R. (1974). Bull. A.P.S. $\underline{19}$, 643.
FAIRBANK, W. H., WITTEBORN, F. C., MADEY, T. M. F. & LOCKHART, M. G. (1973). Exp. Gravit. $\underline{56}$, 310. (Academic Press).
FARAGO, P. S. (1965). Adv. Electronics & Electron Phys. $\underline{21}$, 1.
FISCHER, E. (1959). Z. Physik $\underline{156}$, 1.
FORTSON, E. N., MAJOR, F. G., & DEHMELT, H. G. (1966). Phys. Rev. Letters $\underline{16}$, 221.
FORTSON, N. E., GRAEFF, G., MAJOR, F. G., ROEDER, R. W. H., & WERTH, G. (1968). ICAP Abstracts of Contributed Papers, p. 9.
FRANKEN, P. A., & LIEBES, S. JR. (1956). Phys. Rev. $\underline{104}$, 1197.
GARDNER, J. H. (1951). Phys. Rev. $\underline{83}$, 996.
GRAFF, G., KLEMPT, E. & WERTH, G. (1969). Z. Physik $\underline{222}$, 201.
GRAFF, G. HUBER, K., KALINOWSKY, H. & WOLF, H., (1972). Phys. Lett. A$\underline{41}$, 277.
GRAFF, G., HUBER, K., & KALINOWSKY, H. (1975). Winter Meet. DPG.
HUGHES, V. W. (1959). Rec. Research Mol. Beams (Academic Press).
KIENOW, E. KLEMPT, E., LANGE, F., & NEUBECKER, K. (1974). Phys. Lett. A$\underline{46}$, 441.
KLEPPNER, D., GOLDENBERG, H. M., RAMSEY, N. F. (1962). Phys. Rev. $\underline{126}$, 603.
KNIGHT, L. V. (1965). Ph.D. Thesis, Stanford Univ..
LAND, S. E., & RAITH, W. (1974). Phys. Rev. A$\underline{9}$, 1592.
PENNING, F. M. (1937). Physica $\underline{4}$, 71.
PIERCE, T. R. (1949). "Theory and Design of Electron Beams," Chap. 4, Van Nostrand, Princeton, New Jersey.
RABI, I. I. (1928). Z. Physik $\underline{49}$, 507.
RICH, A., & WESLEY, T. C. (1972). Rev. Mod. Phys. $\underline{44}$, 250.
VAN DYCK, R. S., EKSTROM, P. A. & DEHMELT, H. G. (1975). Bull. A.P.S. $\underline{20}$, 492.
WALLS, F. (1970). Thesis, Univ. of Washington.
WALLS, F. & STEIN, T. (1973). Phys. Rev. Lett. $\underline{31}$, 975.
WINELAND, D., EKSTROM, P., & DEHMELT, H. (1973). Phys. Rev. Lett. $\underline{31}$, 1279.
WINELAND, D., & DEHMELT, H. (1975a). Int. Journ. Mass Spectroscopy & Ion Phys. $\underline{16}$, 338.
WINELAND, D., & DEHMELT, H. (1975b). Journal Appl. Phys. $\underline{46}$, 919.
WINELAND, D., VAN DYCK, R. & DEHMELT, H. (1975). Bull. A.P.S. $\underline{20}$.

NEW ELECTRON AND POSITRON g-2 EXPERIMENTS AT THE UNIVERSITY OF MICHIGAN

David Newman, Eric Sweetman and Arthur Rich

The University of Michigan

Ann Arbor, Michigan, U.S.A.

We are commencing a new series of experiments to measure the electron and positron anomalous magnetic moments. The experiments will be performed in the ten kilogauss field of a precision superconducting solenoid which is now under construction.

The technique which we plan to use in the electron work is similar to that used in previous experiments (1) though with many technical modifications. The positron measurement will utilize a new method based on resonant spin rotation (2) and spin analysis in the main magnetic field of the solenoid.

It is hoped that order of magnitude improvements in accuracy will be forthcoming in both experiments.

DEFINITIONS

The proportionality constant "g" relates the spin and the magnetic moment of a particle:

$$\vec{\mu} = (g/2)(e/m_o c)\ \vec{s}$$

Dirac's relativistic formulation of quantum mechanics predicts g=2, however the radiative corrections of quantum field theory predict that g=2(1+a). The anomaly "a" is expressed as a power series in the fine structure constant i.e.

$$a = \frac{g-2}{2} = A(\frac{\alpha}{\pi}) + B(\frac{\alpha}{\pi})^2 + C(\frac{\alpha}{\pi})^3 + \dots \qquad (\alpha = e^2/\hbar c)$$

Previous g-2 measurements at the University of Michigan have used the "precession" technique, in which polarized electrons are trapped in a magnetic well (1,3). The anomaly is measured by observing the difference frequency, ω_D, between the spin precession and the cyclotron motion of the trapped electrons in a known magnetic field B $(\omega_D \equiv \omega_s-\omega_c)$. The anomaly is related to the measured quantities by means of the relations:

$$a = \omega_D/\omega_o \qquad\qquad (\omega_o = eB/m_oc)$$

Here relativistic terms of order $(\frac{\vec{v}\cdot\hat{B}}{c})^2$ are omitted.

Other g-2 experiments (see Table 1) have used the "resonance" technique (3), in which trapped electrons undergo spin-flip transitions in an rf field. A combination of these two methods, the "resonance-precession" technique, detects the resonant spin rotation of electrons or positrons in a field oscillating with frequency ω_D (2).

RECENT HISTORY

Experimental and theoretical g-2 developments since AMCO4 are summarized in Table 1. Some pre-AMCO4 work is included for completeness. Measurements of the positron in 1969 by Gilleland and Rich (4) and of the electron in 1971 by Wesley and Rich (1), the most precise g-2 determinations of these particles to date, both used the Michigan precession technique and are in good agreement with current theoretical predictions. Other experiments include the 20 ppm measurement by Walls and Stein (5) in 1973 using the resonance method, and a test of the new resonance-precession method in 1972 using the modified Wesley-Rich apparatus (2). A calculational technique, developed in 1972 by Granger and Ford (6) to deal with the effect of magnetic inhomogeneities in spin motion, brought the revised Wilkinson-Crane measurement (7,20) into satisfactory agreement with the predicted value of g-2. In addition there have been many improvements and suggestions for new g-2 measurements based on the resonance method principally by groups at the Physikalisches Institut der Universität Mainz under Professor G. Gräff and at the University of Washington under Professor H. G. Dehmelt. These are outlined in the paper by H. Dehmelt in these conference proceedings.

The first numerical solution of the Feynman diagrams contributing to the anomaly in sixth order was completed in 1971 by Levine and Wright (8). Carroll and Yao (9), using the entirely different mass operator approach, calculated g-2 to 0.75 ppm in 1973. Subsequently various theorists (10,11) found analytical solutions to some of the diagrams. Using these solutions and improved treatment of UV and IR divergences, calculations of g-2 have been improved by Levine and Wright (12) and by Kinoshita and Cvitanovic (13,14) to the 0.3 ppm level assuming exact knowledge of α. Since a new value of α not involving QED and accurate to 0.2 ppm has been reported at this conference (P. T. Olsen and E. R. Williams, private communication) the anomaly has now been calculated to about 0.4 ppm.

PROPOSED ELECTRON g-2 EXPERIMENT

The electron g-2 experiment under construction is being designed to reach a precision up to an order of magnitude better than that reached in previous experiments. If successful it will definitively measure the sixth order coefficient, thereby permitting a test of QED to sub-ppm precision.

The standard precession technique will be used (3), with many improvements in instrumentation and an order of magnitude increase in the magnetic field (10 kG as opposed to 1.2 kG in the Wesley-Rich experiment). This will permit a ten-fold increase in the number of cycles of ω_D which can be observed in a practical running time. (The Wesley-Rich experiment required six months for data runs and systematic checks.) The field will be produced by a precision-wound superconducting solenoid 60 cm in diameter and 200 cm long, using 10,000 turns of multifilament NbTi wire carrying 120 amps. The well shape and depth will be controlled by independent superconducting shim coils.

Electrons will be produced by a 1 MeV accelerator, as compared with 110 KeV in the last experiment. They will be polarized by Mott scattering from a gold foil in the solenoid (p≈1.5%, slightly less than the 2% observed before). The polarized electrons will be trapped in a well only 10 ppm deep, one-tenth the previous trapping depth. Systematic uncertainties in averaging the precession rate over the nonuniform field in the trap will decrease by about this factor. In a further refinement, axial electric fields may be used instead of a magnetic well to contain the trapped electrons. Magnetic field inhomogeneities could then

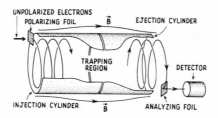

Fig. 1   The Electron Trapping Apparatus

be reduced to less than 1 ppm, and of course averaged over. The
current experiment will also be considerably less sensitive to
stray radial electric fields in the trapping region. Due to the
higher energy, smaller orbit radius, and higher magnetic field in
the current experiment, uncertainties due to radial electric
fields will be two orders of magnitude smaller than in the
Wesley-Rich experiment.

These improvements are expected to allow a measurement of
g-2 to sub-ppm precision so that a confrontation of QED theory
and experiment at this level will be possible.

CURRENT POSITRON EXPERIMENT

The new positron g-2 experiment (figure 2) will differ significantly from the last such experiment (4,17,18) in several
respects. In combination, these changes should allow at least an
order of magnitude improvement over the previous result (a =
$(11603 \pm 12) \times 10^{-7}$). Since CPT invariance requires that particle
and antiparticle have the same charge, mass and magnetic moment,
a measurement of a^+ to 100 ppm would test the equality of g^+ and
g^- at a level of 0.1 ppm and thus constitute a test of CPT invariance to the same precision, a precision exceeding the mass or
charge comparison by at least two orders of magnitude.

As in previous positron experiments, the positrons are produced by beta decay in a radioactive source. A large source is
needed (about 1 Curie of Co^{58} in the last experiment) to insure
that sufficient positrons meet the energy and angle requirements
for trapping. The small fraction of positrons meeting trapping
requirements (between 10^{-4} and 10^{-5} in the new experiment) is
partially compensated for by the fact that, due to parity non-
conservation in beta decay, positrons are emitted with polari-
zation $\vec{P} = \langle \vec{v} \rangle / c$, eliminating the need for Mott scattering before
trapping. Recent improvements in the fabrication of Ge^{68}-Ga^{68}
sources will allow us to use this source in place of the Co^{58}

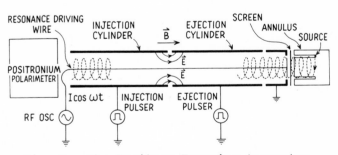

Fig. 2   The Positron Trapping Apparatus

source used in the last positron experiment. The advantages are a 280 day half-life (vs. 71 days for Co^{58}), maximum positron energy of 1.9 MeV (vs. 0.485 MeV), and a beta efficiency of 86% (vs. 15%). Selecting an energy range of 1.0-1.2 MeV will result in P=0.95 (vs. 0.76) and allow a cyclotron radius of 0.5 cm in the 10 kG field.

In all Michigan positron g-2 experiments, the spin polarization has been measured by a polarimeter which relies on the nature of positronium formation and decay in a magnetic field (3,4). The output of the polarimeter depends on $\vec{P}\cdot\vec{B}$ and the polarimeter is most efficient at a field of about 10 kG. Since in the previous precession experiments $\vec{S}$ was always perpendicular to $\vec{B}_O$ and $\vec{B}_O$ is only a few hundred gauss, an auxiliary analyzing field perpendicular to the main field was needed. Many of the problems of the previous positron experiments were caused by this requirement. A method which avoids these problems was suggested by V. L. Telegdi and tested in a modification of the Wesley-Rich electron g-2 experiment (2). An axial wire is introduced and driven by an rf current at approximately ω_D. This results in an azimuthal field $B_\theta \cos \omega_{rf} t$ (fig. 3). If ω_{rf} is equal to ω_D, the time-averaged torque $\langle \vec{\mu} x \vec{B}_\theta \rangle$ on the positron magnetic moment will tend to rotate $\vec{S}$ towards alignment with $\vec{B}_O$. The polarimeter, sensitive to $\vec{P}\cdot\vec{B}$ will then detect the component of spin along the solenoid axis.

The polarimeter (see references 4, 17, 18 for a detailed description) functions as follows. In a magnetic field, the ground states of positronium are perturbed through a mixing of the singlet and m=0 triplet spin states. This results in a reduced triplet lifetime component (τ_T'=5 ns at 10 kG), referred to as the "perturbed triplet" state. This is to be compared with the free annihilation lifetime τ_T for the triplet state of 138 ns. It can be shown (18) that the formation rate of the m=0 triplet (singlet) state is, to first order, proportional to $1-x\vec{P}\cdot\vec{B}$ ($1+x\vec{P}\cdot\vec{B}$) where $x \approx 0.0275B$ (kG). Since $\tau_s' \ll \tau_T' \ll \tau_T$, the decay γ's within a time

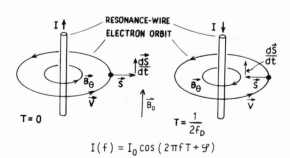

$$I(f) = I_0 \cos(2\pi f T + \varphi)$$

Fig. 3   Rotation of the Spin into the z Direction by the θ Field

window of about 1-12 ns should show an asymmetry proportional to $\vec{P} \cdot \hat{B}$. Measurement of this asymmetry as a function of the applied rf frequency on the wire should give a resonance curve whose peak will occur when $\omega_{rf} = \omega_D$.

The resolution of the polarimeter depends upon the field strength, timing circuit resolution, and the material in which the positronium is formed. In the previous two experiments, positronium formation occurred in a small piece of plastic scintillator with the positron producing a t=0 signal in the process of stopping. The γ's from positronium decay occurring between 3 and 12 ns were detected by a large plastic scintillator. The lifetimes τ_T and τ_T' were reduced to only about 2 ns and 1.4 ns respectively, due to electron pick-off in the solid, and the observed asymmetry in count rate was only about 2%.

In the new experiment, the positronium formation will occur in MgO or SiO_2 powder (19). The t=0 signal will be produced in a thin plastic scintillator through which the positrons must pass before stopping in the powder. In such powders, the observed lifetimes are the vacuum lifetimes (τ_T = 138 ns and τ_T' = 5 ns) and the positronium formation rate is 10%-30% depending upon the type of powder, size of individual grains, and handling. A polarimeter using a powder should approach an asymmetry of perhaps 14% in a 10 kG field. We are currently studying a prototype of such a device to optimize the various elements involved.

Finally we note the experiments proposed by H. Dehmelt at the University of Washington and W. Fairbank at Stanford to measure the positron anomaly at kinetic energies of 1 eV and 10^{-9} eV respectively. Since these groups plan to measure (ω_D^+/ω_O^+) and the Michigan technique measures (ω_D^+/ω_O^-), a comparison of the results could produce an indirect measurement of the ratio $(e/m)^+/(e/m)^-$ which is better than the current best direct measurement (70 ppm, ref. 16). If, as appears possible, our measurement of ω_D^+ exceeds an accuracy of 70 ppm, comparison of a^+ and a^- will be limited by the uncertainty in $(e/m_o c)^+$. A separate measurement of this ratio might therefore be in order at this time.

CONCLUSIONS

The confrontation between theoretical and experimental g-2 work may shortly be pushed to sub-ppm precision. With the completion of calculations of the sixth order coefficient, the improved measurement of α will allow the results of a successful electron g-2 experiment to confront QED theory more precisely than previous experiments by an order of magnitude. The positron g-2 experiment under construction will test CPT invariance to the 0.1 ppm level.

Table 1   Recent g-2 Developments

1969 Gilleland & Rich	Precession experiment	$(\delta a/a)_{e+}$=1000 ppm
1971 Wesley & Rich	Precession experiment	$(\delta a/a)_{e-}$=3 ppm, C=1.53
1971 Levine & Wright	Feynman Integrals	C=1.68$\pm$.33
1972 Kinoshita & Cvitanovic	Feynman Integrals	C=1.29$\pm$.06
1972 Granger & Ford	Electron Spin Motion in Nonuniform Field	
1972 Ford, Luxon, Rich, Wesley, Telegdi	Resonance-Precession	$(\delta a/a)_{e-}$=100 ppm
1973 Levine & Wright	Feynman Integrals	C=1.29$\pm$.06
1973 Walls & Stein	Resonance experiment	$(\delta a/a)_{e-}$=20 ppm
1974 Carroll & Yao	Mass Operator	C=1.01$\pm$.06
1974 Kinoshita & Cvitanovic	Feynman Integrals	C=1.195$\pm$.026

REFERENCES

(1)   J. Wesley and A. Rich, Phys. Rev. A4, 4 (1971).
(2)   G. Ford, J. Luxon, A. Rich, J. Wesley, V. L. Telegdi, Phys. Rev. Lett. 29, 25 (1972).
(3)   A. Rich and J. Wesley, Rev. Mod. Phys. 44, 250 (1972).
(4)   J. Gilleland and A. Rich, Phys. Rev. A5, 38 (1972).
(5)   F. Walls and T. Stein, Phys. Rev. Lett. 31, 975 (1972).
(6)   S. Granger and G. Ford, Phys. Rev. Lett. 28, 1479 (1972).
(7)   D. Wilkinson and H. Crane, Phys. Rev. 130, 852 (1963).
(8)   M. J. Levine and J. Wright, Phys. Rev. Lett. 26, 1351 (1973).
(9)   R. Carroll and Y.-P. Yao, Phys. Lett. 48B, 125 (1973).
(10) M. J. Levine and R. Roskies, Phys. Rev. Lett. 30, 772 (1973).
(11) Milton, Tsai, and DeRadd, Phys. Rev. D9, 1809 (1974).
(12) M. Levine and J. Wright, Phys. Rev. D8, 3171 (1973).
(13) T. Kinoshita and P. Cvitanovic, Phys. Rev. Lett. 29, 1534 (1972)
(14) P. Cvitanovic and J. Kinoshita, Phys. Rev. D10, 4007 (1974).
(15) B. Taylor, W. Parker and D. Langenberg, Rev. Mod. Phys. 41, 375 (1969).
(16) L. A. Page, P. Stehle, and S. B. Gunst, Phys. Rev. 89, 1273 (1953).
(17) A. Rich and H. R. Crane, Phys. Rev. Lett. 17, 271 (1966).
(18) J. R. Gilleland, Ph.D. Thesis (Univ. of Mich., 1970, unpublished)
(19) R. Paulin and G. Ambrosino, Le Journal de Physique 29, 263 (1968)
(20) A. Rich, Phys. Rev. Lett. 20, 967; 21, 1221 (1968).

NEW MEASUREMENT OF $(g-2)$ OF THE MUON

J. Bailey, K. Borer, F. Combley, H. Drumm, C. Eck,
F.J.M. Farley, J.H. Field, W. Flegel, P.M. Hattersley,
F. Krienen, F. Lange, G. Petrucci, E. Picasso, H.I. Pizer,
O. Runolfsson, R.W. Williams, and S. Wojcicki

(CERN Muon Storage Ring Collaboration)

The electron and the muon are the only charged particles which do not participate in the strong interaction, at present not fully understood nor amenable to rigorous calculation. Therefore measurements on the electron or muon in isolation, and on the μ^+e^- system (muonium) can give information on the fine structure constant α, free of doubts about the underlying theory. For the electron the anomalous magnetic moment $a_e \equiv (g-2)/2$ is a known function of α, and in principle provides the purest determination of this quantity. The corresponding quantity a_μ for the muon cannot yet be measured so accurately, and the theory is subject to corrections due to the strong and weak interactions which are not accurately known. Nevertheless a_μ is important in the discussion of the fundamental constants for three reasons:

(a) $g_\mu \equiv 2(1 + a_\mu)$ is needed for calculating α from the hyperfine splitting in muonium;

(b) a_μ verifies quantum electrodynamics (QED) at high $q^2 \sim (20 \text{ GeV/c})^2$ making it very unlikely that the theory will be at fault in the low q^2 region appropriate for other measurements of fundamental constants;

(c) corrections to a_e and a_μ due to modifications of QED, or to couplings to new fields in most cases satisfy[1] the rule $\Delta a_e = (m_e/m_\mu)^2 \Delta a_\mu$. So the verification of a_μ to 25 ppm validates a_e to a much higher accuracy, thus ensuring that a_e gives a good measure of α.

I report here a preliminary measurement of a_μ (12 times more accurate than before) from the new muon storage ring at CERN[2]. In

brief, polarized muons are injected into a uniform magnetic field
in a ring magnet 14 m in diameter. Vertical focusing is achieved
by means of electric quadrupoles, rather than the magnetic gradient
used before.

The various contributions to the theoretical value are given
in Table 1.

Table 1

QED terms		
α/π x	0.5	
$(\alpha/\pi)^2$ x	0.76578	
$(\alpha/\pi)^3$ x	22.92	(16)
gives 10^{-9} x	1 165 829	(2)
Strong	73	(10)
Weak	3	(3)
Total	1 165 902	(10)
Experiment	1 165 895	(27)
Experiment - theory	- 7	(29)

The coefficient of the 6th order QED term (α^3) is due to
Calmet and Peterman (in the press) and updates the figure quoted
in ref. 2, the error arising from uncertainties in a numerical
integration. The effect of strong interactions is due to modification of the photon propagator at high q^2 by virtual hadron
loops. It is obtained by dispersion theory using the experimental
cross-sections for hadron production in e^+e^- colliding beams, the
error arising from our limited knowledge of these processes.
Overall the theoretical value of a_μ is known to 9 ppm.

As in the previous experiments, the principle is to trap
longitudinally polarized muons in a magnetic field B and measure
the frequency f_a with which the muon spin rotates relative to the
momentum vector. The initial polarization is obtained by selecting only forward-going muons from pion decay, and the anisotropy
of the muon decay allows the spin motion to be followed.

$$2\pi f_a = a_\mu (e/mc)B \tag{1}$$

To convert a measurement of f_a into the value of the anomaly a_μ we need one additional frequency, which in principle could be the precession frequency of muons at rest in the same magnetic field $2\pi f_\mu = (1 + a_\mu)(e/mc)B$. In practice, we measure the proton magnetic resonance frequency f_p in the magnetic field (corrected for molecular and bulk diamagnetic shielding) and make use of the ratio of the muon and proton magnetic moments $\lambda = \mu_\mu/\mu_p = (f_\mu/f_p)$, which is known experimentally to 3 ppm. If the proton frequency is suitably averaged over the muon orbits, the anomaly can be obtained from

$$a_\mu = f_a/(\lambda \bar{f}_p - f_a). \tag{2}$$

The value of $\bar{f}_p$ is obtained by combining information from the mapping and monitoring of the magnetic field with knowledge of the distribution of muons and their motion. However, the uniformity of the field is such that the difference between the fully corrected average and the simple mean value on the central orbit is less than 2 ppm.

As the muon lifetime is only 2.2 µs the number of cycles of f_a that can be observed is limited. To extend the time available and thus enhance the accuracy we use relativistic muons with $\gamma \equiv E/m_0c^2 = 29$, thus dilating the lifetime to 64 µsec. The particles are stored in a ring magnet 14 m diameter with B = 1.47T. Each of the 40 magnet blocks was stabilized by its own NMR system. The field was measured at 250 000 points throughout the storage volume, before and after the series of runs. The azimuthally averaged map was reproducible to 1.5 ppm after switching on and cycling, but the average field then rose some 7 ppm over two days, thereafter being stable to ± 1 ppm. During a run the field was sampled, once or twice a day, by driving 37 NMR magnetometers into the storage region. As we know the relation between the readings of these monitoring magnetometers and the field map, we used these measurements to correct the data taken at different times after switching on the magnetic field.

The field is made as uniform as possible so that there is no need to study the distribution of muon orbits within the aperture. Vertical focusing is provided by electric quadrupoles. The electric field components would in general perturb the orbit and the spin motion making eqn.(1) invalid; but these effects cancel exactly at the chosen energy. It is the existence of this so called "magic" value, $\gamma = (1 + a^{-1})^{\frac{1}{2}} = 29.304$ that makes it possible to use a uniform magnetic field and so achieve a new level of accuracy.

The stored muons revolve around the 14 m diameter ring, fig.1, with a period of 0.147 μsec, while their spin precesses with respect to their momentum with a period of 4.3 μsec. An electron from the decay of a stored muon is generally of lower momentum than the muon and tends to emerge on the inside of the ring where it hits one of 20 energy-measuring shower detectors, consisting of alternate layers of lead and plastic scintillator. The resulting signal is used both to stop one of the timing scalers and to provide pulse-height information. By requiring the electron to have high energy in the laboratory we select electrons which have been emitted forwards in the muon rest frame. Since the asymmetric decay distribution rotates with the muon spin, the counting rate $N(t)$ of high-energy electrons is modulated by the spin rotation at frequency f_a as shown in fig. 2.

To extract the frequency f_a, a maximum-likelihood fit is made to the time distribution for each energy band, using the function

$$N(t) = N_o \, e^{-t/\tau}\{1 - A \cos (2\pi f_a t + \phi)\}. \qquad (3)$$

Although there is some prior knowledge of the lifetime τ and the initial phase ϕ, we allow all five parameters (N_o, τ, A, f_a, ϕ) to vary. Small modifications are made to the fitting function because of gain variations in the detection system and long-term muon losses ($\sim 1\%$). We have also allowed a background term which turns out to be $N_B = 2 \times 10^{-5}N_o$. The stability of the fit, with respect to different starting times and with respect to different corrections for system gain and muon loss, has been studied. None of these changes affects f_a by more than a small fraction of a standard deviation. Detailed checks of the analysis procedure have been carried out, and the magnitude of errors from time slewing, dead-time, or malfunction of the digitron have been found to be below 1 ppm. The quoted error on f_a is therefore purely statistical.

The lifetime is a useful check on the experiment. The time dilation predicted by relativity for an object travelling in a circle with velocity $\beta = 0.9994$ is confirmed to $\sim 1\%$.

The weighted average frequency for five groups of runs is $f_a = 0.2327449(52)$ MHz. From the map of the magnetic field the average value of f_p for the muon orbits is calculated. Corrections are applied for the diamagnetic shielding 25.64 ppm, effects of the vacuum tank and other surrounding material (measured), slow time variations of the field, vertical and horizontal oscillations of the orbit (0.75 ppm) and departures from the magic value of γ (1.6 ppm). The overall correction is 25.8(3.5) ppm, and the final value of f_p with error is 62.78310(22)MHz.

Applying eqn.(2) with[3] λ = 3.1833467(82) the experimental result for positive muons is a_μ = 10^{-9} x 1 165 895 (27) (23 ppm)

So $a_\mu^{expt} - a_\mu^{th}$ = -6 ± 25 ppm

Further details of the experiment and a discussion of the result are given in reference 2.

REFERENCES

1)   B. E. Lautrup, A. Peterman and E. de Rafael, Physics Reports 3C, 239 (1972).

2)   J. Bailey, K. Borer, F. Combley, H. Drumm, C. Eck, F. J. M. Farley, J. H. Field, W. Flegel, P. M. Hattersley, F. Krienen, F. Lange, G. Petrucci, E. Picasso, H. I. Pizer, O. Runolfsson, R. W. Williams and S. Wojcicki, Physics Letters 55B, 420 (1975).

3)   K. M. Crowe, J. F. Hague, J. E. Rothberg, A. Schenck, D. L. Williams, R. W. Williams and K. K. Young, Phys. Rev. D5, 2145 (1972).

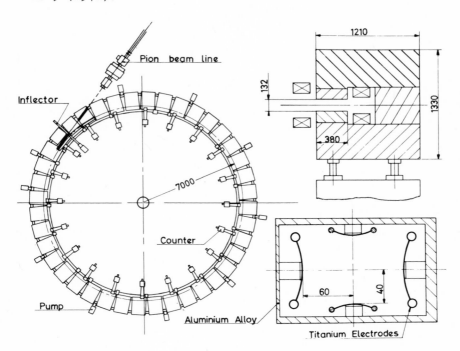

Figure 1

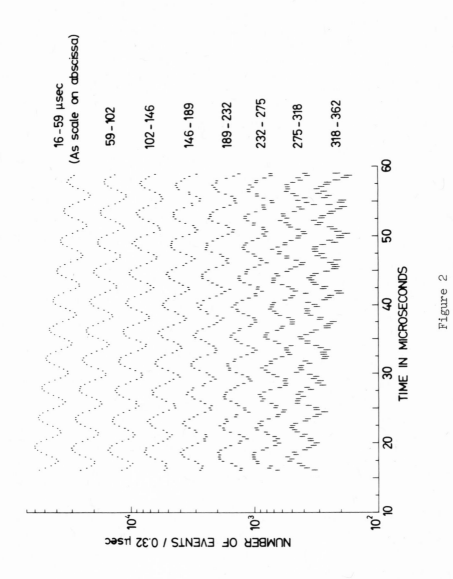

Figure 2

STATUS OF ANOMALOUS MAGNETIC MOMENT CALCULATIONS FOR ELECTRON AND MUON

R. BARBIERI   and   E. REMIDDI[+]

CERN Geneva   LPTENS 24,r.Lhomond PARIS

ABSTRACT : The present theoretical knowledge of electron and muon anomalous magnetic moments is reviewed, discussing in particular recent progresses in the analytic evaluation of sixth order quantum electrodynamical corrections.

1°) ELECTRON ANOMALY

Quantum Electro Dynamics predicts the anomalous gyromagnetic ratio (anomaly) of the electron and muon as a formal power series in the fine structure constant [1]. For the electron anomaly a_e we write

$$a_e = a_2 \left(\frac{\alpha}{\pi}\right) + a_4 \left(\frac{\alpha}{\pi}\right)^2 + a_6 \left(\frac{\alpha}{\pi}\right)^3 + \ldots \ldots \tag{1}$$

The coefficients of the expansion are given as sums of Feynman graphs integrals of rapidly increasing complexity. The first two coefficients have been evaluated a long time ago [2]

$$a_2 = \frac{1}{2} \tag{2}$$

$$a_4 = \frac{197}{144} + \frac{1}{2}\zeta(2) + \frac{3}{4}\zeta(3) - 3\zeta(2)\log 2 = -0.328479.$$

Despite the considerable efforts made in order to evaluate it, a_6 is only known approximately, as the result of multiple (up to seven dimensions) numerical integrations base on extensive use of Feynman Dyson rules in parametric space. Among the various calculations, the most

detailed one is due to M. Levine and J.Wright[3] , who
give results for all the graphs separately, and thus
provide a very useful basis for comparisons ; the most
recent and accurate one has been performed by P.Cvita-
novic and T. Kinoshita[4], who obtain

(3) $a_6 = 1.195 \pm 0.026$.

By using, for instance, the new value of the fine structure coupling constant[5]presented to this conference,
α^{-1}= 137.035987(28) , one has

$$a_e(th) = 1159652.51 \quad X \ 10^{-9}$$

with the following errors (in units 10^{-9}= 1ppm)

0.24   from the experimental error in α

0.33   from the theoretical error in a_6

0.03(?) from the unknown eight order term in the expansion
        (1), if supposed to be of the order of 1 times
        (α/π)^4.

Apart from them, there are other theoretical terms of
order $(\alpha/\pi)^2$, the so-called "QED mass dependent" and
"hadronic" contributions. They are known to be smaller
than the corresponding contributions to the μ-anomaly
(see next section and Ref. [6]) by a factor $(m_e/m_\mu)^2$,
and therefore still negligible.

The present experimental value is :

$a_e(exp) =(1159656.7 \pm 3.5) \ 10^{-9}$

The next Michigan University experiment[7] is planned to
reduce the experimental error to 0.3 ppm, while even
higher precisions can be hoped to be attained in future
resonance experiments[8].
From the above discussion we see that a measure of a_e is
a measure of α within 0.03 ppm, provided the theoretical error in a_6 is further reduced by at least a factor
ten. This is to be achieved by means of a complete analytic calculation, as the further improvement of numerical methods seems unlikely.

Apart from the practical implications, the analytic evaluation of a_6 is to be regarded as a computational challenge, which can eventually give some insight into the
structure of QED.

     The Feynman graphs contributing to a_6 can be classi-
fied into three groups : i) light-light graphs, ii) vacuum polarization insertions, iii) "3photons exchange".

The contribution of i) is known[9] , numerically, to be

0.366(10). We have thoroughly investigated the feasibility of the analytic evaluation. No insurmountable difficulties appeared, but the amount of work necessary to go from survey to results looks fearfully large.

Group ii) is the simplest one, and it is actually known analytically. Its contribution[10] is -0.09474.

Group iii) is the largest. Its contributions are depicted in fig 1).

Each graph (actually equivalent to 5 Feynman vertex graphs) can be regarded as describing an electron (full line) which propagates in an external weak magnetic field, emitting and re-absorbing three virtual photons (wavy lines) in all possible ways (mirror graphs are omitted for the sake of simplicity). The various graphs have been drawn in order of (presumably) increasing difficulty. As a matter of fact, a) b) have been evaluated, analytically, by M. Levine and R. Roskies [11]. Then, the calculation for a) has been repeated by De Raad, Milton and Tsai[12] . More recently we have developed a technique which enables us to evaluate the contributions of class c) fig 1).

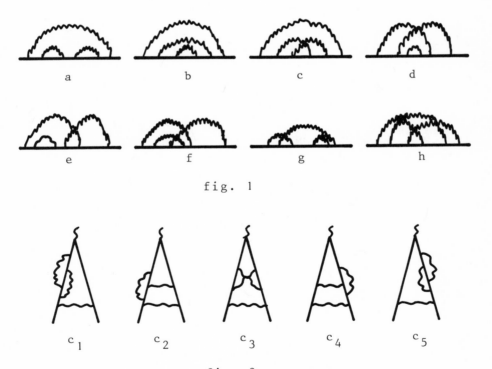

a   b   c   d

e   f   g   h

fig. 1

c_1 c_2 c_3 c_4 c_5

fig. 2

More exactly, this class contains the 5 Feynman graphs
of fig 2).
We find [13]

$$a \ (c_1 + c_2 + c_3 + c_4) = -\log^2\lambda + \left[-\frac{85}{24} + \frac{13}{6}\zeta(2) + \frac{3}{4}\zeta(3) - 3\zeta(2)\log 2\right]\log\lambda$$
$$-\frac{149}{144} + \frac{163}{18}\zeta(2) + \frac{63}{48}\zeta(3) - \frac{23}{2}\zeta(2)\log 2 + \frac{23}{6}\zeta(2)\log^2 2 - 3\zeta^2(2) +$$
$$+\frac{1}{9}\log^4 2 + \frac{8}{3}a_4 = -1.35561 + (\log\lambda's) , \tag{3}$$

in full agreement with $- 1.3592(71)$, obtained by
combining the results of Tables II, III of Ref. [4] suitably.

For the last graph, we have the preliminary result

$$a \ (c_5) = -\frac{1}{2}\log^2\lambda + \left[-\frac{23}{24} + \frac{1}{3}\zeta(2)\right]\log\lambda - \frac{5}{4} + \frac{25}{24}\zeta(2) + \frac{71}{48}\zeta(3) - \frac{3}{2}\zeta(2)\log 2$$
$$-\frac{5}{8}\zeta^2(2) - \frac{5}{6}\zeta(2)\log^2 2 + \frac{5}{36}\log^4 2 + \frac{10}{3}a_4 = -0.06149 + (\log\lambda's) \tag{4}$$

The graphs of fig 2) are being worked out also by Levine
and Roskies who obtained for the graph c_3 in semi analytic form [14]

$a(c_3) = -1.286972$, to be compared with $- 1.289(13)$ of
Ref. [3]

As in previous works, our method is based on dispersion
relations, while algebra and book-keeping of results is
achieved by means of the schoonschip program by M. Veltman.
The method is being extended to class d) graphs[15] and is expected to work for e) too. It is hoped that the experience
acquired in working out the first 5 classes could eventually provide with some hint on how to attack the remaining
classes of graphs.

Let us comment briefly the quantities appearing in
Eq.s (3) (4). The $\log\lambda's$ remind us of the infrared
divergences problem. In fact, separate graphs contributing to the anomaly are not well defined for zero mass
photons. It is customary to give a small mass λ to the
photon ; in the $\lambda \to 0$ limit, infrared divergences show
up as $\log\lambda's$. They cancel, of course, in the final
physical result (and also in judiciously chosen subsets
of contributions), but their presence is highly disturbing, especially in the dispersive approach we use. An
alternative way of parametrizing the infrared divergences
is given by the dimensional regularization. In this approach, the photon always keeps a zero mass (which simplifies all the phase space factors enormously), whereas

the number of dimensions of the loop integration momenta
is brought by analytic continuation from 4 (energy + 3
spatial momenta) to $4 + \eta > 4$. Then, infrared divergences
show up as poles in η . The physical result corresponds
to $\eta \rightarrow 0$. This method has been used in Ref. [13] to handle
"spurious" infrared divergences i.e. divergences which
appear in the various dispersive cuts, but cancel already
within each graph. Beside the $\log \lambda's$, Eq.s(3) (4) consist of rational fractions times a few transcendental
constants (numerically known with absolute precision) :
the Rieman ζ-function of integer argument p, $\zeta(p) \equiv \sum_{n=1}^{\infty} n^{-p}$
(remind $\zeta(2) = \pi^2/6$, $\zeta(4) = 2\zeta^2(2)/5$), $\log 2 = \sum_{n=1}^{\infty} (2^n n)^{-1}$, $a_p \equiv \sum_{n=1}^{\infty} (2^n n^p)^{-1}$.

A look at the not yet analytically known graphs, suggests
that their final expression is likely to involve, also,
transcendental constants of "degree 5" such as $\zeta(5)$, a_5.
Those transcendental constants correspond to particular
values of Nielsen generalized polylogarithms $S_{n,p}(z)$
($S_{1,1}(z)$ is the Euler dilogarithm, also referred to as
Spence function), which are in turn related to the hy-
pergeometric function $_2F_1(\alpha, \beta; \gamma; z)$) by

$$_2F_1(\alpha, \beta; 1+\beta; z) = 1 + \sum_{r=1}^{\infty} \sum_{n=1}^{\infty} \alpha^r \beta^n (-1)^{n-1} S_{n,p}(z) .$$

One is naturally lead to suspect the existence of a deep
connection between Feynmann graphs amplitudes and hyper-
geometric series, but the bridge between general properties
or conjectures and actual results is far from being built.

2°) μ- ANOMALY

Let us turn now to the μ-anomaly, whose measurement
was highly improved by the recent CERN experiment.

Beside the graphs just discussed for the electron
-which are independent of the lepton mass and therefore
equal for the electron and the muon- other contributions
are to be considered : the so-called "mass dependent"
terms. Corresponding contributions exist for the electron
too, but they are suppressed by a factor $(m_e/m_\mu)^2$
and are negligible in practice, as already said in the
previous section.

Mass dependent contributions can be classified into
the following classes : i) Q.E.D. vacuum polarization ;
ii) Q.E.D. light-light contributions ; iii) hadronic.

i) It corresponds to vacuum polarization insertions
with internal electron loops. In the limit $m_e/m_\mu \rightarrow 0$
the coefficients in the usual (α/π) expansion are found

to contain powers of $\log(m_\mu/m_e)\simeq 5.13$, and can therefore become large. All the contributions of this class up to 6th order are, by now, known analytically [16]

$$a_\mu^{6th}(\text{mass dep. vac. pol.}) = \frac{1075}{216} - \frac{25}{3}\zeta(2) - 3\zeta(3) + 10\zeta(2)\log 2 + \frac{11}{6}\zeta^2(2) - \frac{4}{3}\zeta(2)$$
$$- \frac{1}{9}\log^4 2 - \frac{8}{3}a_4 + \left(\frac{31}{27} + \frac{2}{3}\zeta(2) + \zeta(3) - 4\zeta(2)\log 2\right)\log\frac{m_\mu}{m_e} + \frac{2}{9}\log^2\frac{m_\mu}{m_e} = 1.94404.$$

It was shown long ago, that the leading terms in $\log(m_\mu/m_e)$ are connected to the asymptotic behavior of vac. pol., and can be evaluated by means of renormalization group techniques [17]. The simplest way of stating the results is, perhaps, given in Ref. [18]. Up to terms vanishing for $m_e/m_\mu \to 0$, the mass independent term (a_e) and the electron loop vac. pol. corrections can be accounted for by writing

$$a_\mu(\text{vac. pol.}) = \frac{1}{2}\left(\frac{\alpha_\mu}{\pi}\right) + b_4\left(\frac{\alpha_\mu}{\pi}\right)^2 + b_6\left(\frac{\alpha_\mu}{\pi}\right)^3 + \ldots$$

where α_μ is the "effective fine structure constant" for the muon, given by

$$\alpha_\mu^{-1} \equiv \alpha^{-1}\left[1 + \Pi(-m_\mu^2/m_e^2, \alpha)\right] \simeq 136.07875$$

with $b_4 = -0.74512,$ $b_6 = 1.282 \pm 0.026$

and the dependence on $\log\frac{m_\mu}{m_e}$ is absorbed in α_μ.

   ii) Q.E.D. light-light contributions due to virtual electron. They are known to be the largest [19]. They are known numerically only. The most recent result [20] is

$$(19.79 \pm 0.16)\left(\frac{\alpha}{\pi}\right)^3 = (247 \pm 2) \times 10^{-9}$$

The electron vac. pol. corrections to the above quantity have also been evaluated recently[21]. They are found to be $(111.1 \pm 8.1)(\alpha/\pi)^4 = (3.2 \pm .2) \times 10^{-9}$. This number is relatively very large, but in agreement with previous guesses. However, at variance with the vac. pol. contributions discussed above, the general structure of higher order light-light graphs is not known.

   iii) Hadronic contributions : the discussion on this important class of contributions, as well as on the implications of the measurement of the μ-anomaly for strong interactions is in [6] .
Let us just note that in the present value of hadronic contributions $(73 \pm_- 10) \times 10^{-9}$, the error is not theoretical, but of experimental origin.

Footnotes and References

+ On leave from Istituto di Fisica dell' Universita,
Bologna, Italy

[1] Detailed references to (g-2) literature   can be found
in the excellent reviews papers :
B.E. Lautrup, A. Peterman and E. de Rafael, Phys. Re-
ports $\underline{3}$, 4 (1972)
A. Rich and J.C. Wesley, Revs of Modern Physics $\underline{44}$,
250 (1972)
J. Calmet, preprint TH.1761-CERN (Geneva) october 1973
[2] J. Schwinger, Phys. Rev. $\underline{73}$, 416 (1948)
                      $\underline{76}$, 790 (1949) ;
R. Karplus and N.M. Kroll, Phys. Rev. $\underline{77}$, 536 (1950);
A. Peterman, Helv. Phys. Acta $\underline{30}$, 407 (1957)
C.M. Sommerfield, Ann.Phys. (N.Y.) $\underline{5}$, 26 (1958)
[3] M.J. Levine and J. Wright, Phys. Rev. $\underline{8}$, 3171 (1973)
[4] P. Cvitanovic and T. Kinoshita, Phys. Rev. D10, 4007
(1974)
[5] P.T. Olsen and E.R. Williams, this conference.
[6] F.J.M. Farley, this conference
[7] A. Rich, this conference
[8] H.G. Dehmelt, this conference
[9] J. Calmet and A. Peterman, Phys. Letters $\underline{47B}$, 369
(1973)
[10] R. Barbieri and E. Remiddi, Nuclear Phys. B$\underline{90}$, 233
(1975) and references therein.
[11] M.J. Levine and R. Roskies, Phys. Rev. D$\underline{9}$, 421 (1974)
[12] K.A. Milton, W. Y. Tsai and L. De Raad, Jr., Phys. Rev.
D9, 1809 (1974)
[13] R. Barbieri, M. Caffo and E. Remiddi, preprint PTENS
75/4, Ecole Normale Sup., Paris (april 1975)
[14] M.J. Levine and R. Roskies, private communication
[15] with D. Oury, in progress.
[16] T. Kinoshita, Nuovo Cimento $\underline{51B}$, 140 (1967)
[17] B. Lautrup and E. de Rafael, Nuclear Physics B$\underline{70}$,
317 (1974)
[18] R. Barbieri and E. Remiddi, preprint TH 2012-CERN
(Geneva) april 1975
[19] J. Aldins, S.J. Brodsky, A. Dufner and T. Kinoshita,
Phys. Rev. D$\underline{1}$, 2378 (1970)
[20] J. Calmet and A. Peterman, preprint TH 1978-CERN
(Geneva) february 1975
[21] J. Calmet and A. Peterman, Phys. Letters $\underline{56 B}$, 383
(1975)

A NEW MASS-SPECTROMETRIC METHOD FOR

THE MEASUREMENT OF γ_p'

B.A.Mamyrin, S.A.Alekseyenko, N.N.Aruyev
and N.A.Ogurtsova

Academy of Sciences of USSR
A.F.Ioffe Physical-Technical Institute
Leningrad

The accuracy of the fundamental constants set is determined now to the considerable extent by the accuracy of the γ_p' measurements. The existing methods of the γ_p' measurements (low field and high field) are distinguished by the method of a magnetic field B measurement. The uncertainties for the γ_p' input data in the 1973 least squares adjustment were 2 + 16 ppm, ref.(I).

The new method allows to decrease essentially the error of measurements and to get a new value for γ_p' The systematic errors in this new method must be quite different from these in the traditional methods. The new method allows to avoid the measurements which restrict an accuracy of the γ_p' measurements in existing methods.

The basic equation for γ_p' in the new method can be deduced from the formulaes for the cyclotron frequency of the ion

$$\omega_{c.i.} = q_i B / m_i \qquad (1)$$

and for the radius of trajectory of the ion

$$\rho = \frac{1}{B} \sqrt{2 U m_i / q_i} \qquad (2)$$

in the same magnetic field B, then

$$B = 2U/\omega_{c.i.}\rho^2, \tag{3}$$

where U – the potential difference corresponding to the ion energy, m_i – ion mass, q_i – ion charge. From the formula for magnetic moment of the proton in nuclear magnetons

$$\mu'_p/\mu_n = \omega'_n/\omega_{c.p.}, \tag{4}$$

where ω'_n – nuclear magnetic resonance (n.m.r.) frequency, $\omega_{c.p.}$ – proton cyclotron frequency, and from the ratio

$$\omega_{c.i.}/\omega_{c.p.} = M_p/M_i, \tag{5}$$

here M_p – proton mass, we obtain

$$B = \frac{2U}{\rho^2} \cdot \frac{(\mu'_p/\mu_n)}{\omega'_n} \cdot \frac{M_i}{M_p}. \tag{6}$$

By substituting eq.(6) in the equation for γ'_p

$$\gamma'_p = \omega'_n/B \tag{7}$$

we obtain

$$\gamma'_p = \frac{1}{2} \cdot \frac{(\omega'_n)^2}{(\mu'_p/\mu_n)} \cdot \frac{\rho^2}{U} \cdot \frac{M_p}{M_i}. \tag{8}$$

In this equation auxiliary coefficients are used. μ'_p/μ_n has the error 0.38 ppm (1,2) and M_p/M_i has the error less than 0.1 ppm (3). The quantities directly measured are ω'_n, ρ and U.

As it is seen from eq.(8) the unit of current enters the basic equation as in the low field method.

The measurement of the ion trajectory radius ρ can be done with high precision if the ion beam location will be fixed with narrow slits. The accurate measurement of the potential difference corresponding to the energy of ions generated in the ion source is much more difficult.

We have examined some designs of the energy

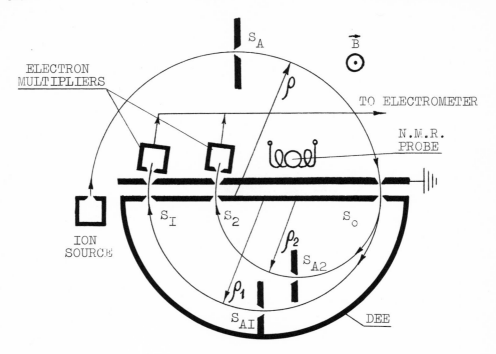

Fig.1. Scheme of the two-stage mass-spectrometric
       analyzer.

analyzers. The design of the analyzer in the form of a
two-stage static mass-spectrometer was accomplished
experimentally. It is presented schematically at the
Fig.1.

     The ions are created at the electron impact ion
source and move in the magnetic field B with the radius
of trajectory ρ . After the $180°$ deflection the ion
beam changes its energy in the gap between the dee and
the grounded plate. The ion beam enters the dee through
the slit S_o. It moves along the trajectory with the
radius ρ_1 or ρ_2 depending on the dee's potential.
Passing the dee the ion beams are accelerated and hit
the entrances of the crossed-fields electron-beam
multipliers. From the output of the multipliers the
signal is registered by electrometer and oscillograph.

The aperture slits S_A, S_{A1} and S_{A2} are used for restriction of the angular divergence and for precise fixing of the ion beam.

The equation (8) in this case is:

$$\gamma'_p = \frac{(\omega'_n)^2}{(\mu'_p/\mu_n)} \cdot \frac{(D_1^2 - D_2^2)}{8\Delta U} \cdot \frac{M_p}{M_i}, \tag{9}$$

where $D_1 = 2\rho_1$, $D_2 = 2\rho_2$, $\Delta U = U_2 - U_1$, U_1 and U_2 – dee potentials corresponding to the ions moving along the orbits with radii ρ_1 and ρ_2 respectively.

With such method of measurements one have no need to know the energy of ions produced in the ion source. It simplifies the measurements greatly. Contact potentials in the gap does not influence the result of the measurements provided they are constant when the radius of the ion trajectory is changed.

The resolution of the analyzer is approximately 10^4 with the slits S_0, S_1, S_2 about 4 mu.

The ion current signal is swept by the additional sinusoidal voltage applied to the dee. The method of the oscillographic superposition of the ion current signal and the n.m.r. signal is used as in ref.(2).

The Hartree correction is introduced with the error 0.2 ppm as in ref.(2). The correction for the certain width of the decelerating gap was calculated. It was about 3 ppm.

The extrapolation technique for the eliminating of the stray electric fields influence was used. For this reason the measurements for ions with different M/q were used.

As it is follows from our experiments the main difficulty now is the measurement of the dee potential about several kilovolts with sufficient accuracy. We have no standard apparatus of such kind and therefore we are forced to construct it ourselves. For this reason we are unable to give the numerical results of our measurements. The detailed analysis of the mass-spectrometric method shows that the measurement of γ'_p with an error of 1 ppm may be fulfilled.

When the result of the mass-spectrometric measurement of δ_p' to be used in the adjustment of the fundamental constants it is convenient to introduce the result as:

$$\delta_p' \cdot \mu_p'/\mu_n = \Gamma'',\qquad(10)$$

where

$$\Gamma'' = (\omega_n')^2 \cdot \frac{(D_1^2 - D_2^2)}{8_\Delta U} \cdot \frac{M_P}{M_i}\qquad(11)$$

does not depend on the adopted value of μ_p'/μ_n and is measured directly in the experiment. The use of eq.(10) makes it possible to introduce the one more independent observational equation for Γ'' in the adjustment.

REFERENCES

1.   COHEN,E.R., TAYLOR,B.N., J. Phys. Chem. Ref. Data, 2, No.4, (1973).

2.   MAMYRIN,B.A., ARUYEV,N.N., ALEKSEYENKO,S.A., Zh. Eksp. Teor. Fiz. 63, 3, (1972),/English transl. Sov. Phys. - JETP 36, 1, (1973)/.

3.   WAPSTRA,A.H., GOVE,N.B., Nucl. Data Tables, 9, 265, (1971).

EXPERIMENT WITH MAGNETICALLY ISOLATED CALCULABLE SOLENOID

FOR γ'_p DETERMINATION

Hisao Nakamura   and Akira Nakamura

Electrotechnical Laboratory

5-4-1, Mukodai-machi, Tanashi, Tokyo, JAPAN

1. INTRODUCTION

In the determination of the proton gyromagnetic ratio γp in a low magnetic field, there are two conditions required. The first is the precise determination of the magnetic field by calculation from the known current which passes through a solenoid. Here, very accurate information is necessary for the dimension of the solenoid, such as diameter, pitch and so on. With this condition fulfilled, the system should be placed in nonmagnetic environment. This second condition means that no unpredictable magnetic field is permissible at the sample position.

A trial experiment has been performed for achieving the above two conditions with a system which we call Magnetically Isolated Calculable Solenoid (MICS) (1,2). MICS consists of, essentially, a single layered solenoid enclosed with ferromagnetic walls of high permeability. When permeability is infinitely large, not only the second condition, but also the first condition would be satisfied. Two slabs of ferromagnet with infinite area produce images of the solenoid, when the slabs are placed closely at the both ends of the solenoid, and perpendicularly to its axis. The magnetic field at the center of the solenoid behaves as if it were produced by a solenoid with infinite length. In this situation, not so high precision would be required for the solenoid dimension.

2. STRUCTURE OF MICS

In Fig.1, the structure is shown on the MICS used in the experiment(2). Single layered principal solenoid with the length of

480 mm, the diameter of 60 mm, and the pitch of 1 mm is enclosed in
a parallelepiped of high permeability alloy, which has inside length
of 480 mm, cross-section of 250x150 mm^2 and wall thickness of 20 mm.
The solenoid fits well along its axis to the parallelepiped.

    In order to compensate the leakage flux out of the parallele-
piped, outer shield is provided. The outer shield has four walls,
that is two sides open, and its wall thickness is 2 mm. In between
the inner and the outer shields, two compensation solenoids fit well
in the gap which has the length of 60 mm. They are axially aligned,
and have the same diameter and pitch with the principal solenoid.

    Original windings are replaced by substitutional solenoids at
both ends of the principal solenoid in the length of 30 mm each.
The substitutional solenoids have the same number of turns with the
removed part of the original winding, but it is wound more densely
with thinner wire in a length of 12 mm approximately. This replacement provides for inlet spaces of the lead wire to the NMR coil
which should be placed at the center of the principal solenoid.

    Magnetic field is calculated on this structure with the follow-

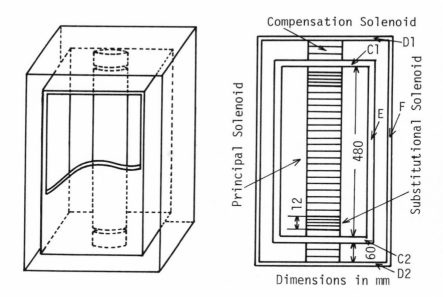

Fig.1   Structure of the MICS (Magnetically Isolated Calculable
         Solenoid). The single layered principal solenoid is
         enclosed doubly with high permeability magnetic walls.

ing assumptions; magnetic properties in the wall material is uniform, magnetic response is linear, and the effect of the side walls is negligibly small. Obviously, flux returns through the side walls, but this return passes are assumed to be the same with the case where two upper and two lower plates extend infinitely, and all side walls are removed.

Calculated magnetic field is given as

$$H = \frac{I}{g} (1 - 0.21 \times 10^{-6}) \qquad A/m \qquad , \ \ldots\ldots\ldots (1)$$

at the center of the principal solenoid, where I is current through the all solenoids in series, and g is pitch of the windings of the principal and compensation solenoids. In this calculation, relative permeability of the wall material μ / μ_0 is taken as 10,000 which is the approximate value of the observed permeability. Effect of the substitutional solenoids is estimated to be small. Deviation from the value for the infinite length solenoid amounts to 0.21 ppm.

3. EXPERIMENTS AND RESULTS

The MICS is placed in a temperature regulated compartment (20 + 0.5°C) in an ordinary laboratory surrounded by many iron works. Spherical sample of pure water with diameter of 20 mm is mounted on the center of the principal solenoid. Absorption signal is observed by a transformer type bridge at frequency around 54 kHz with the current of 1 A through the solenoids. The frequency is swept in a small span such as several tens of ppm or several Hz for the observation.

Uncertainty caused by dispersion mixing is estimated as low as 0.1 ppm from the resolution of the phase angle in the bridge. The current is stabilized against a standard cell in an air bath at 30 ±0.01°C. Standard resistor is also temperature controlled within 0.1 °C at 20 °C.

NMR signal has signal to noise ratio 10, and full width of 1 Hz(20 ppm), although original estimation of the width is one order of magnitude smaller. This broadening of the width seems to arise from the inhomogeneity of the magnetic field at the sample position. The width reduces to one third by introduction of a field gradient produced by a pair of extra coils. Two thirds of the broadening would originate from some kind of asymmetrical factor of the MICS.

The broadening of this order can be explained neither by mis-alignment of the principal solenoid (38 μ m) nor by asymmetric loca-tion of the substitutional solenoids (10 μ m). Other sources of the broadening are also checked: no effect is observed by replacing NMR coil with another coil made of high purity copper wire (99.9999%), no trace of impurities are detected in the water sample, and temperature gradient along the principal solenoid (2°C) would not explain this much broadening.

Line broadening itself would not give the serious problem as far as the center of the resonance line is properly given by the calculated magnetic field. However, above result may suggest that there will be serious sources of errors, which affect both the center and the width. Assumptions in the calculation may be violated in the real MICS.

Assumption on the uniformity would not hold, because the broadening depends on the applied magnetic field. Linewidth decreases from 75 ppm at the resonant field of 1.2 mT to 27 ppm at 0.4 mT, when larger sample is used in a toroidal form with the outer and the inner diameter of 36 mm and 14 mm respectively, and with the height of 42 mm. Inhomogeneity of the field would decrease as the applied field decreases, when some leakage flux out of the wall is distributed asymmetrically.

After removal of the applied field, residual field is detected along the axis of the principal solenoid. The residual field has rather large asymmetry. This fact would suggest again the presence of nonuniformity in the wall material.

Nonlinear response is clearly present in the wall materials, because there is residual field. Linewidth of the larger sample decreases from 60 ppm to 40 ppm by demagnetization of the MICS system with slowly decreasing 50 Hz current in two extra coils wound around two walls of E through small holes on the remaining two E walls. Line distortion is also observed before demagnetization.

However, in the smaller sample, demagnetization has little effect. Distribution of the residual field decreases clearly from 10 ppm to 1 ppm at the sample position by demagnetization, but the linewidth of the sample does not show any change under the resonant field of 1.2 mT. The broadening in this case would arise from the nonuniformity of the wall materials. Nonlinearlity would not play a principal role on the broadening and the shift.

Side walls such as E seems to have large effect on the calculation of the magnetic field. The field obtained from the experiment decreases about 20 ppm when current flows only through the principal and the substitutional solenoids but the compensation solenoids. Calculation gives the decrease of 70 ppm. This discrepancy may arise from the insertion of the side wall E, when permeability of the wall has finite value.The side wall E might compensate the above difference of 50 ppm even without the outer shield and the compensation solenoids.

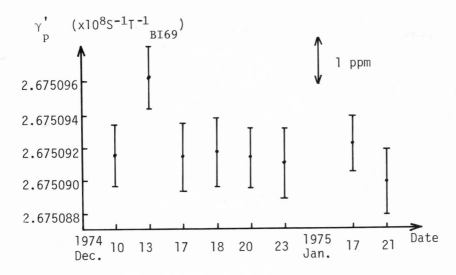

Fig.2   Observed values of $\gamma p'$ with the MICS between
         December, 1974 and January, 1975. Each points
         represents an average of fifty measurements.
         Error bar indicates random error in one stan-
         dard deviation.

        The calculation by the magnetic circuit approximation reveals
that the flux can pass through air gap such as from D1 to C1 or
from C1 to C2 in Fig.1 in the normal operation of the system. The
airgap flux through D1-C1 would make the field value lower than the
value calculated with eq.(1) in the order of 10 ppm. If the airgap
D1-C1 were directly connected by magnetic walls, this deviation
would diminish to the order of 1 ppm. Thus, side walls and their
configuration have serious effect on the calculation of the magnetic
field.

        In spite of these deficiencies, the determination of $\gamma p'$ was
tried to check the working condition of the full system. In Fig.2,
each point represents an average of fifty measurements with ten times
changes in the current direction. There are constant field in the
order of 10 nT superposed by fluctuating field of about 1 nT at the
sample position. Constant part would arise from the earth's field,
which is about 50 μT, and decreases by a factor of 1/5000 after passing through the walls of the MICS. The fluctuating part is probably
due to nonlinearity of the wall material. The constant part is eliminated by taking an average of data in the both directions of the

applied field. Long term stability of $\gamma p'$ is well within 2 ppm, and the system may be used as a secondary standard of current.

Overall average of the observed value is

$$\gamma_p' = 2.675\ 097\ 9\ (\ \pm\ 20\ \text{ppm}\ )\ \text{x}\ 10^8\ \text{rad. s}^{-1}\text{T}^{-1}\ \text{(ETL unit)}$$

$$\quad = 2.675\ 092\ 0\ (\ \pm\ 20\ \text{ppm}\ )\ \text{x}\ 10^8\ \text{rad. s}^{-1}\text{T}^{-1}\ \text{(BI unit).}$$

Conversion of the unit is made by using the results of the 1973 triennial comparison reported by BIPM. The result is not too far from the recommended value of $\gamma p'$ by Cohen and Taylor in 1973(3), that is

$$\gamma_p' = 2.675\ 130\ 1\ (\ \pm\ 2.8\ \text{ppm}\ )\ \text{x}\ 10^8\ \text{rad. s}^{-1}\text{T}^{-1}\ \text{(BI unit).}$$

Sources of uncertainty are listed in Table 1, where the largest contribution comes from the field calculation. It seems not so unfair to assign 20 ppm, which is the same as the line broadening, to the uncertainty of the field.

Finally, we conclude that the MICS used in the experiment violates the assumptions for the field calculation in a few tens of ppm. However, the original idea of MICS does work as far as the assumptions hold, for bare principal solenoid produces magnetic field of 10,000 ppm lower than the observed field in the MICS.

The authors are grateful to K.Hara for his original idea of MICS and kind discussions.

REFERENCES

1) K.Hara and H.Nakamura: NBS Spec. Publ. 343 (1971) p.123,
   Proc. Inter. Conf. Precise Measurement and Fundamental
   Constants.

2) K.Hara and H.Nakamura: Atomic Masses and Fundamental Constants
   4, ed. by J.H.Sanders and A.H.Wapstra (Plenum, 1972) p.462.

3) E.R.Cohen and B.N.Taylor: J. Phys. Chem. Ref. Data, vol.2 (1973)
   p.663.

Table 1

Sources of uncertainty in γ'_p determination
by the MICS

Sources	Relative uncertainty (one standard deviation)
1. Magnetic field	
Misalignment of MICS	0.4 ppm
Measurements of diameter and pitch	0.1
Substitutional solenoids	0.9
Temperature of MICS	0.4
Permeability of inner shield	0.5
Nonlinearity	1.0
Validity of the field calculation	20
2. Current	
Calibration of standard cell	0.3
Temperature of standard cell	0.6
Annual changes in the working standard cell	0.6
Current determination	0.4
Calibration of standard resistance	0.1
Annual change in the working standard resistance	0.6
Temperature of standard resistance	0.4
3. Frequency	
Calibration of frequency	0.2
Mixing of dispersion component	0.1
Random uncertainty	0.8
Total (RSS)	20 ppm

DETERMINATION OF THE GYROMAGNETIC RATIO OF THE PROTON γ_p'*

P. T. Olsen and E. R. Williams

Electricity Division, National Bureau of Standards

Washington, D.C.   20234

INTRODUCTION

We report a new value for the low field gyromagnetic ratio of the proton, γ_p'. Since AMCO-4, when we last gave a value of γ_p' [1], we have developed some new techniques [2,3,4] which are primarily responsible for the present reduced uncertainty. Although some further refinements in the experiment are still expected, we feel that the present improved value should be reported because of its impact on the adjusted values of the constants, especially on the fine structure constant [5,6]. In particular, α can be calculated from the equation [5]:

$$\alpha^{-2} = [(1/4R_\infty)(c\Omega/\Omega_{NBS})(\mu_p'/\mu_B)(2e/h)]/(\gamma_p') = C_o/\gamma_p'. \qquad (1)$$

All of the constants in the brackets have been measured to 5 parts in 10^8 or better [6]. The 0.42 ppm value of γ_p' that we report here will therefore provide a 0.21 ppm value of α. This provides a so-called WQED (without quantum electrodynamics) value for α that can be used to compare QED experiments with theory.

In the NBS low field γ_p' experiment, one measures the dimensions of a precision solenoid from which one then calculates the magnetic field B at the center when a current, known in NBS units, is passed through that solenoid. A spherical sample of H_2O is placed at the center of the solenoid, and the proton precession frequency ω_p'

*Contribution of the National Bureau of Standards. Not subject to copyright.

(prime denotes spherical H_2O sample) is measured; γ_p' is then given by the ratio, $\gamma_p' = \omega_p'/B$.

THE EXPERIMENT

The key to an improved γ_p' experiment lies in the measurement of the solenoid dimensions. We have developed three new methods which are used to measure 1) the axial position of each turn of wire (pitch); 2) the changes in the diameter of each turn of wire; and 3) the average diameter of selected turns of the solenoid. All three techniques use electrical methods to locate the position of the current by placing the current in selected segments of the solenoid and using the resultant magnetic field to give the desired information. The first two methods have been described [2,3] and we plan to publish a complete description of the diameter techniques later. A brief description of all three is included so that the reader can better understand the results.

Pitch and Diameter Variations

An alternating current is injected into a set of turns whose axial position is to be measured. This is easily accomplished because the single layer solenoid is wound with bare copper wire (oxygen free). These activated loops produce an ac magnetic field which in turn induces a signal in the coils A, A', B, B', and C [see Fig. 1]. The coils are fixed together on a fused silica tube, T, which can be pushed back and forth along the center axis of the solenoid. Coils A and A' are wired in such a way that the EMF's induced in them cancel when they are centered on the activated turns. The spacing between A and A' is chosen to give a maximum rate of change of voltage for axial displacements. The reader may recognize this system as a large linear differential transformer. The output of this linear differential transformer can be used in a servo system to "lock" the probe to the center of the activated wires with a resolution of better than 0.05 μm.

Coils B, B' and C are wired together such that the EMF's of B and B' add together, and the number of turns in coil C is adjusted so that its EMF as nearly as possible cancels the sum of B and B'. The voltage induced in coil C is inversely proportional to the average diameter of the activated loops D on the solenoid while the voltage induced in B and B' increases with an increase in diameter. The resultant voltage from these three coils B, B', and C is very sensitive to changes in diameter (0.1 μm resolution) and yet not sensitive to axial positioning. We calibrate this voltage change by using extra sets of windings on the ends of the solenoid--one set is 60 μm larger and the other 60 μm smaller. Because the diameter

variations of our solenoid are small, a calibration accurate to
a few percent is more than sufficient for present accuracies of
0.15 ppm in the calculated field.

One can assemble a single measurement probe out of the two
systems just described and make the axial (pitch) and diameter
variation measurements simultaneously. While the pitch system is
"locked" on the energized turns of the solenoid, a laser interferometer is used to record the axial position. The voltage output
of the coils B, B', and C is a measure of the diameter variation.
This measurement system is automated and computer-controlled.
The solenoid is measured in place with a one ampere current in
all 1000 turns. The information thus obtained about the solenoid
dimensions corresponds to the condition under which the solenoid is
to be used when measuring the precession frequency.

Diameter

We emphasize the importance of the diameter variation
measurement described in the previous section. If our solenoid were
infinitely long, we would not need to know the diameter; but the
variations in diameter would still be as important as with our
present system. Our solenoid is, of course, not infinite, so one
needs a method to measure the diameter of at least some portion of
the solenoid. With our particular solenoid a 1 ppm measure of the
diameter produces a 0.0724 ppm uncertainty in the calculated field.
Our mechanical measure of the diameter is estimated to be good
to 3.5 ppm, resulting in a 0.25 ppm error in the calculated field
which is one of the largest contributions to the total uncertainty
in γ_p'. To reduce this uncertainty we are developing a simple
electrical technique in which two segments of the solenoid are
used as a Helmholtz pair. We compare the magnetic field from this
pair to the field of the solenoid alone using either a proton or
Rb^{87} magnetometer. We know the spacing of the Helmholtz coils
(using the pitch measuring technique described above) so we can
calculate their effective diameter. This approach is feasible
because the field of a Helmholtz pair is more sensitive to the
diameter than is the field of the long solenoid.

Using this technique we hope to be able to measure the solenoid
diameter with about 1 ppm accuracy; however, an accurate measurement
has not been made in time for this conference.

Electrical Standards and Precession Frequency

The current that is passed through the solenoid is determined
by making the voltage drop across a precision 1-Ω resistor equal
to the EMF of a standard cell. For this experiment the EMF of

the cell and resistance of the resistor need to be known in NBS units to determine γ_p' in NBS units. In our present experimental arrangement, we directly transfer current by placing the solenoid in series with a 1 1/2 km long cable and the 1-Ω standard resistor which is located in the 2e/h laboratory. This arrangement does not noticeably affect the quality of the NMR experiment. We only need to make sure that the leakage current in the cable is sufficiently small. This arrangement allows us to use the same standard cells that are directly monitored via the ac Josephson effect, thus eliminating the uncertainty introduced by a voltage transfer. [4]

The 1-Ω standard resistor is periodically calibrated against the NBS maintained resistors whose values have been determined by the absolute farad and ohm experiment [7].

The method we use to measure the precession frequency (nuclear induction) has been described in a previous paper [8]. The quality environment provided by our non-magnetic facility allows us to achieve a precision on the order of 0.05 ppm in this frequency measurement and our accuracy is limited to about 0.15 ppm. Below the 0.1 ppm level there are a number of possible systematic effects which will require further attention in order to achieve even higher accuracies.

In order to achieve a uniform field we have developed a new technique in which the solenoid itself is used as the compensation coil. Two constant current sources are used to inject additional current in appropriate turns of the solenoid so that the 2nd, 4th, 6th and 8th order terms of the Legendre polynomial expansion of the field are nearly zero. A major advantage of this compensation system is that the compensation coils are nearly perfectly aligned with the solenoid. After installing this new system we discovered that we had an unexplained gradient of 0.2 ppm/cm along the axis across the center region of the solenoid. After an extensive re-examination of our compensation system failed to explain this gradient, we found that it was actually caused by the diameter and pitch variations. In fact, on careful examination of the data, one finds that this gradient is primarily caused by the diameter variations of the center 150 turns of the solenoid which have a very slight conical shape. This interesting effect serves to re-emphasize the importance of a good dimensional measurement system as the key to improved γ_p' measurements; and a well constructed solenoid is also required.

RESULTS AND CONCLUSIONS

In table 1 we list the separate contributions to the uncertainty assigned to this γ_p' measurement. The uncertainties quoted are meant to represent standard deviation or 68% confidence

level estimates. The axial position and radial variation
uncertainties are grouped as one error, and it represents both
the statistical scatter and the estimates of possible systematic
errors. The susceptibility corrections were all small (<.1 ppm),
and one such correction for the pick-up coil was measured empirically
as well as calculated. Some additional susceptibility corrections
need to be made, but they are expected to be small (~0.04 ppm).
A possible systematic effect (the NMR frequency may depend on the
sample size) has been observed; therefore, the frequency uncertainty
has been expanded to cover this possible problem. Further data
should quickly resolve this small effect (~0.1 ppm).

The final uncertainty assignment is the square root of the
sum of the squares or 0.42 ppm. Our value of γ'_{pNBS} for a spherical
sample of water corrected to 25° C is:

$$\gamma'_{pNBS} = 2.6751314(11) \times 10^8 \text{ Rad } s^{-1} \cdot T_{NBS}^{-1}(0.42 \text{ ppm}),$$

in terms of the electric units as-maintained at NBS in the spring
of 1975.

In order to arrive at a value for α we have chosen the values
listed in table 2 for the constants required to calculate C_0 in
equation 1. (See the paper by Taylor and Cohen, this conference,
for information concerning these constants [6].) The value for α
is

$$\alpha^{-1} = 137.035987(29) \quad (0.21 \text{ ppm}),$$

which is 0.39 ppm less than the 1973 recommended value that had an
assigned uncertainty of 0.82 ppm [5].

This new value for α is of sufficient accuracy to allow
theorists to treat α as a known constant when comparing their QED
theoretical calculations with experiment. We can now especially
look forward to new results in the theory of the muonium hyperfine
splitting and the experimental determination of g-2.

This value of γ'_p(low) can be combined with the value of γ'_p(high)
described in the next paper by B. Kibble to give a value of K, the
ampere conversion factor [5].

Acknowledgment

The authors would like to thank William Trimmer and Kurt Weyand
for their valuable assistance. Dr. Trimmer is a contributor to the
technique for measuring the diameter. Mr. Weyand, a guest worker
from PTB, helped with the solenoid compensation system. Our
experiment has required much assistance from other Electricity
Division personnel. In particular, T. E. Wells has made the

necessary resistance measurements and B. F. Field has helped in
the cell calibration and current transfer technique. The assistance
and advice of R. D. Cutkosky, R. L. Driscoll and B. N. Taylor is
greatly appreciated. J. E. Faller of J.I.L.A. contributed
important suggestions for the pitch measurement scheme.

Table I

Summary of Uncertainties

Item	Uncertainty (ppm)
Axial position and radial variation	0.27
Average diameter	0.25
Susceptibilities	0.08
Precession frequency	0.15
Standard resistor	0.08
Return lead alignment	0.05
Temperature coefficient of solenoid	0.03
Compensating coil	0.008
Earth's field	0.002
RSS Total	0.42 ppm

Table II

Values   assumed for constants of equation 1

Constant	Value	Uncert. ppm	Source
R_∞	10973731.43 m^{-1}	0.009	Hänsch, et al.[a]
c	299792458 ms^{-1}	0.004	CCDM[b]
Ω_{NBS}/Ω	$1 - 0.819 \times 10^{-6}$	0.054	Cutkosky[a]
μ'_p/μ_B	$1.520993136 \times 10^{-3}$	0.014	Phillips, et al.[a]
$2e/h$	$4.83593420 \times 10^{14}$ Hz$\cdot$V$^{-1}_{NBS}$	0.030	Field, et al.[c]
γ'_p(low)	2.6751314×10^{8} s$^{-1}\cdot$T$^{-1}_{NBS}$	0.42	This work
α^{-1}	137.035987	0.21	From Eq. 1

a These values are described in ref. [6].
b Value recommended by the Comite Consultatif pour la Definition
  de Metre (CCDM).
c Value used to maintain NBS Volt (ref. [9].)

REFERENCES

[1] OLSEN, P. T., and DRISCOLL, R. L., Atomic Masses and
    Fundamental Constants 4, 471 (1972).
[2] WILLIAMS, E. R., and OLSEN, P. T., IEEE Trans. Instrum. Meas.
    IM-21, 376 (1972).
[3] OLSEN, P. T., and WILLIAMS, E. R., IEEE Trans. Instrum. Meas.
    IM-23, 302 (1974).
[4] WILLIAMS, E. R., OLSEN, P. T., and FIELD, B. F., IEEE Trans.
    Instrum. Meas. IM-23, 299 (1974).
[5] COHEN, E. R., and TAYLOR, B. N., J. Phys. Chem. Ref. Data 2,
    633 (1973).
[6] TAYLOR, B. N., and COHEN, E. R., this conference.
[7] [4.1] CUTKOSKY, R. D., IEEE Trans. Instrum. Meas. IM-23,
    305 (1974).
[8] DRISCOLL, R. L., and OLSEN, P. T., Proc. Int. Conf. on Prec.
    Meas. and Fund. Constants (NBS Spec. Pub. 343), 117 (1971).
[9] FIELD, B. F., FINNEGAN, T. F., and TOOTS, J., Metrologia 9,
    155 (1973).

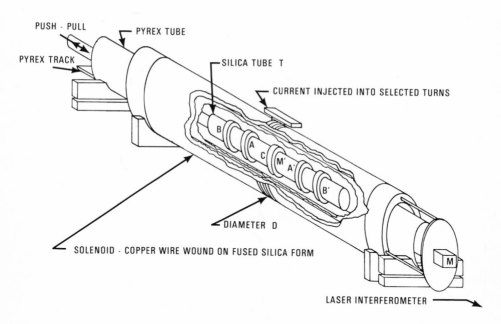

Figure 1. Dimensional Measuring System. A pick-up probe,
consisting of five coils A, A', B, B', and C wound on a glass
tube T, is used to locate the axial position and radial variation
of the turns of wire on the solenoid. Mirrors (retroreflectors)
located at M and M' are used in a Michelson type interferometer
to locate the axial position.

A MEASUREMENT OF THE GYROMAGNETIC RATIO OF THE

PROTON BY THE STRONG FIELD METHOD

B. P. Kibble

Division of Electrical Science

National Physical Laboratory, Teddington, England

1 INTRODUCTION

The gyromagnetic ratio of the proton in water, γ_p' , is a physical constant whose value is extensively used to measure magnetic fields precisely by observing precession frequencies. By determining the value of γ_p' by the weak and strong field methods we can both express in SI units magnetic fields measured in this way and realise the ampere according to its SI definition with an uncertainty of about a part in a million.

In the strong field method a current, whose value i ▪ V/R is precisely determined in terms of the laboratory's maintained units of voltage and resistance, flows in a rectangular coil of accurately known width which is suspended from one arm of a precision balance so that its lower end is in the uniform field of a permanent magnet. The precession frequency of protons in this field is observed, whilst at the same time the change in force on the coil when the current is reversed is obtained by weighing. If m is the mass which counterbalances the force change, g the local gravitational acceleration, l the mean width of the coil and ω the precession frequency, then

$$\gamma_p' = \frac{\omega}{B} = \frac{\omega \, l \, V}{m \, g \, R} \, .$$

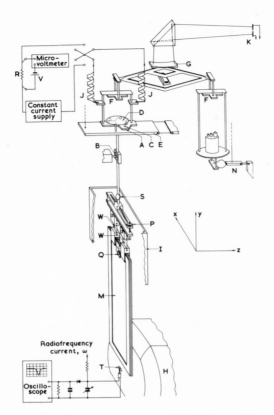

Figure 1. The Apparatus. The coil M is hung from one end of a
balance beam via supplementary knife—edges F, suspension rods,
levelling gear P and strip hinges W. The lower end of M lies in
the magnetic field B_z of a permanent magnet, one poleface, H, of
which is shown. The upper end is in the near—zero field within
the mu—metal shields I. The mechanism N acts on the
counterweight to bring planes in the suspension rod into gentle
contact with the ball B so that, without jarring the balance, the
current supplied to the coil through the strips J can be reversed.
The resulting force change is counteracted by lowering masses
placed on the pan D onto the pillars C with the mechanism E. By
open circuiting the coil and shorting the conductors leading
current to it with the switch Q a separate measurement of the
small unwanted force contributed by these conductors can be made.
A beam of light from the mirror G on the balance beam falls on
the photocells K and the amplified photocurrent flows in a
supplementary turn on the coil (not shown) to servo—control the
balance for fast and sensitive weighing. A similar system is
described by Kibble (1975). The lower circuit measures the
precession frequency of the protons in the sample T and the
upper circuit measures the current supplied to the coil in terms
of the resistor R and the standard cell V.

In practice, the measurement involves considerable attention
to detail and very many corrections are necessary to achieve an
accurate result.   It is not possible to give a complete description
in a short paper, but such a description will soon be available
(Kibble and Hunt 1975).   Some experimental details are given in
figure 1 and the accompanying legend, and in the following
paragraphs we highlight those difficulties which have not been
given explicitly by ourselves or other workers in previous
publications, and which we feel need special attention if an
accurate result is to be obtained.

2 SOME EXPERIMENTAL DETAILS

We have produced the magnetic field with a permanent magnet
which has been permanently shimmed with thin sheets of magnetic
material shaped to nullify the contours of the unshimmed
inhomogeneous field.   Thus we avoid perturbation of the balance by
hot air arising from an electrically energised magnet, and the small
remaining inhomogeneity of the field does not change with time.
The orientation of the field in the pole-gap, which was established
by methods described previously (Kibble and Hunt 1971), remained
constant over the whole duration of the measurements.

We have conducted experiments with the aim of ensuring that
the balance is unaffected by small forces and torques other than
the vertical force of interaction of current and field which we
seek to measure.   In particular we test that the important
perturbing force F_z and torque Γ_y have been made sufficiently
small by checking that there is no movement in the intermediate
suspension rod when the current in the coil is reversed.   The
necessary flexibility at either end of this rod is ensured by the
auxiliary knife-edges F at the top and the strip hinges W at the
bottom (figure 1) and by making the rod of thin walled tube so
that it can twist easily under the torque Γ_y.   The current to the
coil also flows through W.   We observed with an autocollimator
angular displacements about the x and y axes of the small mirror S
(figure 1) attached to this intermediate rod.

Measurements have been made using each of the three turns of
the coil individually and in series.   A second set was carried out
with the coil inverted, which implies use of a different significant
part of the coil.   Hence the agreement, within experimental error,
of the γ'_p results offers reassurance that the measurement of coil
width and the calculation of a weighted mean coil width has been
consistently carried out.   (The weighting function is the magnetic
field gradient $\partial B_z/\partial y$ up the length of the coil).   More
importantly, it also offers reassurance that the mean current path
accurately follows the geometrical centre of the conductors, since
if this were not so, perhaps because of non-isotropic strain or

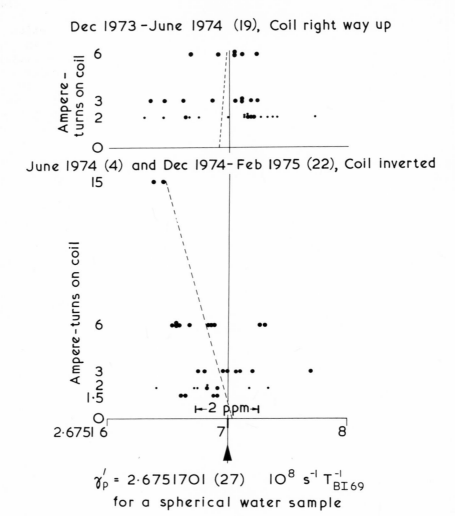

Figure 2.   The results of two series of measurements, plotted as a function of the ampere turns used on the coil.   The small points are the results of measurements made with a current of two amperes using the individual turns on the coil;   they have been averaged and are shown as large points on the same horizontal line so as to be comparable with the results shown on the other horizontal lines which were obtained with all three turns connected in series.   The broken lines are a least squares fit to the data.

impurity distribution, there would be a random difference between
the six conductors on either side of the coil forming the three
turns, and a spread of results would have been obtained. Further
reassurance on this important point is being obtained by measuring
the mean coil widths with an inductive sensing technique similar
to that of Williams and Olsen (1972).   This technique senses
directly the mean current path in the conductors.

It is necessary in the strong field method to obtain the force
change when the current is reversed, because this eliminates
unwanted forces caused by convection currents from the heated coil
and interactions between the current and the flux it induces in
the magnet.   Such forces should depend only on the square of the
current, but, for example, non-linearity of the induced field in
the magnet and the mu-metal which screens the top of the coil
would cause higher order dependence.   We have ascertained that
this particular non-linearity is small directly, by measuring the
induced fields involved, and a general test has been made by
performing measurements over as wide a range as possible of
ampere-turns on the coil.   The results are plotted in figure 2.
We have taken the intercepts of the least-squares fitted lines
(shown broken), with the appropriate standard error, to represent
our result.

Finally, the magnetic field in the vicinity of the coil
conductors is altered by the presence of the magnetically
permeable material of the coil former and conductors.   A
correction of one ppm, determined experimentally, was made for
this effect and the validity of this correction was demonstrated
by experiments in which material having a much larger permeability
was temporarily attached to the coil.

3 ACCURACY OF THE RESULT

Since space does not permit a complete discussion of errors,
we merely remark that most of the error ascribed arises from the
measurement of coil width, and since it is possible that this
error may be reduced by the inductive sensing technique the result
quoted in figure 2, although believed correct within the standard
error in parenthesis, should not necessarily be taken as final.

4 A SUGGESTION FOR A DIFFERENT WAY OF REALISING THE AMPERE

A major aim of these measurements is to determine the ratio
of the maintained ampere to the SI ampere, denoted by K, by
combining the result with that of the weak field method (Cohen and
Taylor 1973).   We take this opportunity to draw attention to a
possible way of determining K directly which needs only minor
modifications to the strong field apparatus described above.

Consider a coil carrying a current i which encloses a magnetic flux Φ Then the energy of interaction is $W = -i\Phi$ and if Φ is a function of displacement of the coil, y, then a component of force exists

$$F_y = \frac{\partial W}{\partial y} = -i\frac{\partial \Phi(y)}{\partial y}.$$

which may be determined in SI units by opposing it with a mass M in a gravitational acceleration g

$$Mg = -i\frac{\partial \Phi}{\partial y} . \qquad (1)$$

Suppose that the coil, in a separate measurement, moves with velocity dy/dt in the same flux Φ.   Then an e.m.f. V' is generated,

$$V' = -\frac{\partial \Phi}{\partial t} = -\frac{\partial \Phi}{\partial y}\frac{dy}{dt}. \qquad (2)$$

Eliminating $\partial\Phi/\partial y$ between (1) and (2), we have

$$Mg\frac{dy}{dt} = iV' = \frac{VV'}{R} \qquad (3)$$

where i is known in terms of the potential drop V it produces across a resistor R.   V, V' and R would be measured in the maintained units of the laboratory;   these are related to the SI units of (3) by

$$\frac{VV'}{R} = K^2\frac{\Omega_{MAINTAINED}}{\Omega_{SI}}\left(\frac{VV'}{R}\right)_{MAINTAINED}$$

and Ω maintained/ Ω SI can be considered as determined by the calculable capacitor with an uncertainty of about 1 in 10^7. Hence from (3)

$$K^2 = Mg\frac{dy}{dt}\left(\frac{R}{VV'}\right)_{MAINTAINED}\frac{\Omega_{SI}}{\Omega_{MAINTAINED}}$$

In practice we cannot measure an instantaneous velocity dy/dt or potential V' but an average over a well-defined time interval of both quantities, $\Delta y/ \Delta t$ and $\bar{V}' = 1/ \Delta t\int_0^{\Delta t}V'dt$ is equally exact.   Of course, precise measurement is simplified if dy/dt and V' are as constant as possible – perhaps to 1 in 10^4 or 1 in 10^5.   The mean velocity $\Delta y/ \Delta t$ could be measured with very great accuracy with a laser interferometer. The detailed distribution of Φ is not required, nor any dimensions of the coil, provided that both are stable for the short time between the weighing and moving with constant velocity parts of the experiment.

REFERENCES

Cohen, E R and Taylor, B N. Journal of Physical and Chemical Reference Data 2, 663-734 (1973)

Kibble, B P. Metrologia 11, 1 - 5 (1975)

Kibble, B P and Hunt, G J. Electrical Science Divisional Report, National Physical Laboratory. (To be published)

Kibble, B P and Hunt, G J. National Physical Laboratory Divisional Report Qu15 (1971)

Williams, E R and Olsen, P T. IEEE Trans. on Instrumentation and Measurement IM21 376-379 (1972)

RECENT ESTIMATES OF THE AVOGADRO CONSTANT

Richard D. Deslattes

National Bureau of Standards

Washington, D.C. 20234 U.S.A.

INTRODUCTION

The first microscopic information on the Avogadro constant came in 1910 (1) from the oil drop e which could be combined with the relatively well known electrochemical Faraday to obtain a value for N_A. The claimed accuracy for this route ultimately reached 1000 ppm in 1916 (2).

The second route came into use following the development of ruled grating procedures for X-ray wavelength measurements (3). Expressed in current language, the result is a modification of the expression first used by W.H. Bragg (4) for estimating the size of cell edge dimensions in crystals:

$$N_A \Lambda^3 = \frac{nA}{\rho a_o^3} \qquad [1]$$

Equation [1] is written for the case of a crystal having n atoms of atomic mass A occupying a unit cell volume, a_o^3. The macroscopic density is ρ and Λ is the conversion factor connecting some presumably unambiguous X-ray wavelength scale with the scale of optical wavelengths. (There were many problems with the X-ray scale itself which are, however, not of importance for this report.)

A long period of controversy began with the 1928 thesis of Erik Bäklin (5) where evidence was presented that the value of e (this is equivalent to N_A in these routes) obtained in X-ray measurements differed from that of the oil drop experiment. This controversy was ultimately resolved in favor of the X-ray route and the discrepancy traced an error in the rotating cylinder viscosimeter end effect corrections for air viscosity. The mid

50's summary by Cohen, Crowe and DuMond attributed to the X-ray route a 120 ppm standard deviation while that for the (corrected) oil drop route was quoted at 1500 ppm (6). The dominant contribution to the uncertainty in the X-ray route was and is due to the uncertainty in Λ which was about 20 ppm (60 ppm in Λ^3) in the late 50's and so remained until reduced to perhaps 10 ppm in the work reported by Henins in 1970 (7).

The X-ray experiments were overtaken by increased accuracy which became available in other methodologies. In fact, in adjustments completed after 1963, X-ray measurements carried zero weight except insofar as the experiments on $N_A\Lambda^3$ could be used to infer Λ.

More recently, a controversy between two groups of measurements of μ_p/μ_N has been resolved in a fashion which deletes the electrochemical Faraday. One group of μ_p/μ_N measurements was earlier preferred because it maintained consistency with the Faraday (8). Subsequently two new μ_p/μ_N measurements having substantially reduced uncertainty appear consistent with the previously excluded group. This led in the most recent adjustment (9) to rejection of the Faraday.

The effect of excision of both X-ray data and coulometric data was to leave a relatively unique (indirect) path to N_A. This has been described by Taylor (10) with the result indicated in equation [2].

$$N_A = \mu_o \frac{\left(\frac{c\Omega}{\Omega_{NBS}}\right)^{3/2}\left(\frac{\mu'_p}{\mu_B}\right)^{1/2}\gamma'\, p(low)_{NBS}^{1/2}\left(\frac{2e}{h}\right)_{NBS}^{3/2} M_p^*}{8\,R_\infty^{1/2}\,\frac{\mu'_p}{\mu_N}\,K^2} \quad [2]$$

The subscript "NBS" means that the measurement is made in terms of "as maintained" units at NBS while (*) means that the masses are expressed on the ^{12}C atomic mass unit scale. The remainder of the symbols have their conventional meaning with K expressing the conversion to the absolute ampere. According to Taylor, estimated 1 σ is near 6 ppm with the largest contribution coming from K. He also indicates an expectation that this uncertainty can be significantly reduced by work underway or contemplated (10).

In the work which my colleagues and I have reported last year, we returned to equation [1] with elimination of the factor Λ (11). This was made possible by simultaneous optical and X-ray interferometry of a common baseline. As it appeared that such an approach could lead to unit cell volume estimates with 1 ppm uncertainties, it became desirable to reexamine procedures used to establish

density and atomic mass. Whereas previous density measurements
were made with respect to an assigned value for water density, we
used highly regular solid objects whose mass and volume were
determined (12). In the case of atomic mass, isotopic abundance
measurements were made on the samples actually used in the density
measurements. These abundances were obtained with respect to
synthetic mixtures of separated isotopes in a well characterized
mass spectrometer (13). All measurements were carried out on
dislocation free silicon crystals. It may be appropriate to briefly
describe each of these steps and make some comments on limitations
which we see at present.

The principle of our X-ray optical interferometry is indicated
in Fig. 1. X-ray interferometry is of the symmetric Laue case
type first demonstrated by U. Bonse and M. Hart (14). As is well
known by now, these interferometers are achromatic and act as
linear Moire encoders whose signal has the period of the lattice
repeat distance. In our work, optical interferometry has been of
the high finesse resonant cavity Fabry-Perot type. Finesse values
in the range 1000-1500 have been routinely obtained yielding
transparency curves of comparable width to that of the X-ray
fringes. Illumination of the X-ray interferometer is by a conventional molybdenum anode diffraction tube. Illumination of the
optical interferometer is by means of an iodine stabilized HeNe
laser (15) both mode-matched and decoupled from the optical cavity.

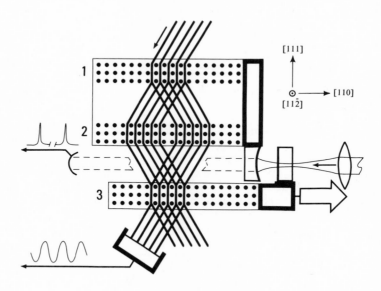

Fig. 1.   The essential features of our combined X-ray and optical
          interferometer, XROI, are shown here.

Output from such a measurement is the ratio of the optical period (nominally $\lambda/2$ but this must be corrected for diffraction phase shifts) to the lattice repeat period ($a_o/\sqrt{h^2 + k^2 + \ell^2}$) where h = k = 2, ℓ = 0 in our work). This ratio is an integer, 1648, plus a fraction, f. The fraction is obtained from a plot of the cumulative phase in the X-ray channel, modulo 1648 versus optical order number. Efficient measurement follows from a careful choice of the pattern of observation. We have used a procedure of successive approximations which in the end leads to approximating f by a rational fraction, a/b; observations at intervals of b optical orders would be degenerate in X-ray phase if f = a/b. What is measured is thus f – a/b. The results from last year's study are shown in Fig. 2 where one can see evidence of an available precision better than 0.1 ppm. A principal limitation of this result is that rotations in the plane of Fig. 1 (yaw errors) contribute to phase change in the X-ray channel without signalling their presence by means of loss of X-ray fringe contrast or intensity. Pitch errors (rotations about 111 in Fig. 1) signal their presence by loss of contrast in the X-ray channel in a very sensitive manner. Roll motion (rotations about the direction of motion, 110) does not introduce first order errors.

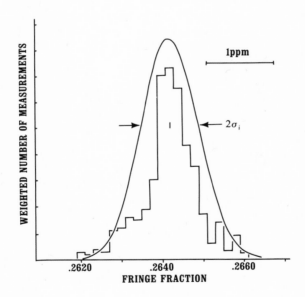

Fig. 2.   A histogram of the data reported in Ref. 11 is indicated. Progress has recently been made in understanding the source of outlying values at greater than the expected frequency.

In a search for a means of detecting the presence of yaw errors, we have operated crystals 2 and 3 (Fig. 1) in a conventional 2 crystal diffraction mode. Superimposed on the 440 $\overline{4}40$ reflection curve we have found a strikingly narrow feature, apparently due to a 6 beam interaction (16). In the case of $AgK\alpha_1$ radiation, this feature is about 0.02 arc sec (full width at half maximum). By adjustment to the flank of this curve, one can study yaw motion down to the order of 10^{-4} arc sec. The error of the motion so far evaluated amounts to 1.7 ± .06 millisec for a typical scan of 200 optical fringes. This requires a correction of 0.18 ppm to the value of a_o previously reported (11). The revised result is a_o = 543.10651 pm (0.15 ppm).

In the case of crystal density determinations carried out previously (in $N_A\Lambda^3$ measurements), density was established by hydrostatic weighing relative to an assigned value for the density of suitably prepared water (17). The two first principle measurements of water density (18, 19) unfortunately differ by about 6 ppm at the reference temperature. Also, water prepared according to the recommended procedures is unstable against redissolution of atmospheric gases and highly corrosive as well. In our work, water was replaced by fluorocarbon liquids chosen for their lack of corrosive tendencies, thermal coefficient of density and for their ability to dissolve atmospheric gases. Our procedures made use of highly spherical steel artifacts which were measured interferometrically (20) to determine their volume and then compared as to mass with kilogram replica No. 20. These became local and temporary standards of density. They were used to determine the densities of a group of 4 silicon crystals whose masses summed to near 1 kg by classical hydrostatic weighing procedures. Subsequently, the densities of a group of 3 rather good Si specimens were determined with respect to the 4 reference objects. The entire procedure resulted in uncertainties somewhat better than 1 ppm for the specimen crystals. Details may be found in reference 12.

In the case of atomic weight assignments, these were made for each specimen crystal by absolute abundance measurements. First, a uniform bar of silicon material was wafered and tested for uniformity. Samples were quantitatively converted to gas phase SiF_4 and compared synthetic mixtures SiF_4 prepared from separated isotopes. In this way, the effect of bias in the mass spectrometer is considerably reduced. These wafers form now a publicly available standard reference material (SRM 990) (21). Subsequently, fragments from each of the three specimens were compared with SRM 990 and their abundance distributions determined. In each case, the measurements were carried out in such a way that the assigned atomic mass was known to somewhat better than 0.5 ppm. These procedures will be described in detail elsewhere (13).

It is of considerable significance that, while densities and atomic masses each varied over 6 ppm for the three specimens, the formally invariant ratios A/ρ had a total spread of data of 0.3 ppm. We thus formed a weighted mean of these values. The uncertainty had to be increased to account for those parts of the A and ρ determinations common to all measurements. This together with the X-ray/optical interferometry result permit us to report $N_A = 6.0220976 \times 10^{23} \, mol^{-1}$ (0.9 ppm).

Recently (22) problems have been noted in mass dissemination procedures which, in principle, affect the present Avogadro result, the electrochemical Faraday, ampere realization and proton-gyromagnetic ratio measurements. The problem seems to be one in which traditional corrections to weighings for air buoyancy are incorrect. The extent of the magnitude of the problem is, in each mass comparison, proportional to the volume difference between the pair of objects being compared. In the case of the Si density measurement, one problem shows up at the transfer from the platinum-iridium replica to steel and a second at the transfer from steel to silicon. The problem is not yet fully understood but it could in the future require correction of the above value by about 1 ppm.

References

1. R.A. Millikan, Phil. Mag. 19, 209 (1910).

2. R.A. Millikan, Electrons + and -, Univ. of Chicago Pres (1947), Chapter 5.

3. A.H. Compton and R.L. Doan, Proc. Natl. Acad. Sci. (U.S.) 11, 598 (1926).

4. W.H. Bragg, Proc. Roy. Soc. (London) 88A, 428 (1913); 89A, 246 (1914); 89A, 430 (1914).

5. E. Bäklin, Dissertation, Uppsala (1928).

6. E.R. Cohen, K.M. Crowe and J.W.M. DuMond, The Fundamental Constants of Physics, Interscience, New York (1957), p. 129.

7. A. Henins, Precision Measurement and Fundamental Constants, D.N. Langenberg and B.N. Taylor, Editors, NBS Special Publication 343, U.S. Government Printing Office (1971), p. 255.

8. B.N. Taylor, W.H. Parker and D.N. Langenberg, Rev. Mod. Phys. 41, #3, 375-496 (1969) (see p. 429 and ff.).

9. E.R. Cohen and B.N. Taylor, J. and Chem. Ref. Data 2, No. 4, 663-734 (1973) (see p. 684, p. 704 and ff.).

10. B.N. Taylor, Metrologia 9, 21-23 (1973).

11. R.D. Deslattes, A. Henins, H.A. Bowman, R.M. Schoonover, C.L.
     Carroll, I.L. Barnes, L.A. Machlan, L.J. Moore and W.R. Shields,
     Phys. Rev. Lett. 33, 463 (1974).

12. H.A. Bowman, R.M. Schoonover and C.L. Carroll, J. Res. Nat.
     Bur. Stand., Sect. A78, 13 (1974).

13. I.L. Barnes, L.J. Moore, L.A. Machlan and W.R. Shields, to be
     published.

14. U. Bonse and M. Hart, Appl. Phys. Lett. 7, 99 (1965).

15. W.G. Schweitzer, E.G. Kessler, R.D. Deslattes, H.P. Layer and
     J.R. Whetstone, Appl. Opt. 12, 2927 (1973).

16. T. Huang, M.H. Tillinger and Ben Post, Zeits. für Naturforsh.
     28a, 600-603 (1973).

17. I. Henins and J.A. Bearden, Phys. Rev. 135, A890 (1964).

18. M. Thiessen, K. Schell and H. Disselhorst, Wiss. Abh. Phys.-
     Tech. Reichsanst. 3, 1 (1900).

19. P. Chappuis, Trav. Mem. BIPM 13, D1 (1907) and 14, B1, D1
     (1910).

20. J.B. Saunders, J. Res. Nat. Bur. Stand., Sect. C76, 11 (1972).

21. Standard Reference Materials are available through the Office
     of Standard Reference Materials, Room B306, Bldg. 222,
     National Bureau of Standards, Washington, D.C. 20234, USA.

22. Paul E. Pontius, Science, to be published.

RE-EVALUATION OF THE RYDBERG CONSTANT

T. Masui*, S. Asami** and A. Sasaki*

*Faculty of Engineering, University of
    Shizuoka, Hamamatsu

**National Research Laboratory of Metro-
    logy, Tokyo

We present here only a brief outline of our
re-evaluation of the Rydberg constant, since the
result of this evaluation has already been referred to in the latest analysis of fundamental
constants reported by Cohen and Taylor[1].

Raw materials used in the present paper
(see tables following the text) are much the same
as those reported by one of us at the Gaithersberg
Conference in 1970[2]. Two important improvements,
however, have been made on the process of reduction of data as compared with our previous report.

In the first place, the integral orders of
interference in the two series of observations
at the path difference approximately equal to
1.80 cm have been corrected by -194 with respect
to the primary standard krypton 86 line[3] and
consequently by -179 with respect to $H\alpha$.

In the second place, the intensity anomaly
has been ascribed to an enhancement in the two 2S
states instead of a suppression in a single
$2P_{3/2}-3D_{5/2}$ state. The intensity distributions
for $2S_{1/2}-3P_{1/2}$ and $-3P_{3/2}$ then assume the values
$0.056\ 557+r^{(j)}$ and $0.113\ 114+2r^{(j)}$ corresponding to the total intensity $1+3r^{(j)}$, those of
all the other components being unchanged (assump-

tion 1). The superscript (j) here and henceforth
indicates the numerical value j of the discharge
current intensity in milliampere supplied
to the hydrogen source in use.

Equation of observation of the order of interference $N^{(j)}$ as a function of path difference D can
theoretically be set up as

$$2N^{(j)} = (2D/7.2) R_H^{(j)} + \pi^{-1} argF^{(j)}, \quad \ldots\ldots(1)$$

where $R_H^{(j)}$ is the apparent Rydberg for hydrogen
and $F^{(j)}$ the Fourier transform of the fs
pattern of $H\alpha$, both at the current intensity j.
It is noteworthy to mention that the line-broaden-
ing of each component does not affect the value of
$argF^{(j)}$ nor $R_H^{(j)}$, insofar as the broadening is
symmetrical and its magnitude is common for all
the components (assumption 2). 18 series of observations at j=10 and at various values of D yield
the least-squares solution

$$R_H^{(10)} = (109\ 677.609\ 612 \pm 0.004\ 175)\ cm^{-1}$$
$$\ldots\ldots(2)$$

and

$$r^{(10)} = 0.025\ 793 \pm 0.001\ 267, \quad \ldots\ldots(3)$$

A representation of the Rydberg for an unperturbed hydrogen atom

$$R_H^{(0)} = R_H^{(10)} + correction, \quad \ldots\ldots(4)$$

$$correction = (7.2/2D) \left[(\pi^{-1} argF^{(10)} - \pi^{-1} argF^{(0)}) \right.$$
$$\left. - (2N^{(10)} - 2N^{(0)}) \right]$$
$$\ldots\ldots(5)$$

can be obtained by substituting j=10 and 0 respectively into eq. (1), if we admit that the wavelength shift due to collisions is negligibly small
(assumption 3).

Further determination of $2N^{(10)} - 2N^{(0)}$ at a
certain value of D is indispensable, while all the
other quantities in the right-hand side of eq. (4)
can be derived from eqs. (2) and (3). Considering
the trend in results of 22 series of observations
$2N^{(j)}_2 - 2N^{(j)}_1$ at an estimated path difference

$$D = (3.240\ 0 \pm 0.007\ 5)\ cm \qquad \ldots \ldots (6)$$

and at various combinations of j_2 and j_1, we set up empirically an equation of observation of the form

$$2N^{(j_2)} - 2N^{(j_1)} = a \left[(j_2/10)^b - (j_1/10)^b \right] \ \ldots (7)$$

(assumption 4), and obtain the least-squares solution

$$a = 0.009\ 669 \pm 0.001\ 405 \qquad \ldots \ldots (8)$$

and

$$b = 3.604 \pm 0.348 .$$

In addition, we have

$$2N^{(10)} - 2N^{(0)} = a \qquad \ldots \ldots (9)$$

by substituting $j_2 = 10$ and $j_1 = 0$ into eq. (7).

The correction term represented by eq. (5) can now be derived from eqs. (3), (6), (8) and (9). The result is

$$correction = (-0.023\ 157 \pm 0.001\ 678)\ cm^{-1} .$$
$$\ldots \ldots (10)$$

Eqs. (2), (4) and (10) furnish us the final result

$$R_H^{(0)} = 109\ 677.586\ 5\ cm^{-1} ,$$

in conjunction with the statistical one-standard-deviation error of $\pm 0.004\ 5\ cm^{-1}$. We tentatively estimate that either of the four "assumptions" mentioned in the text may give rise to an additional systematic error of the order of magnitude of $0.001\ 0\ cm^{-1}$, and conclude that

$$R_H = (109\ 677.586\ 5 \pm 0.008\ 5)\ cm^{-1} .$$

References

1) E. R. Cohen & B. N. Taylor: J. Phys. Chem. Ref. Data 2 (1973) 663.
2) T. Masui: NBS Spec. Pub. 343 (1970) 83.
3) CCDM 4 (1970) M36.

Table 1
Fine structure of Hα

component	separation/cm^{-1}	intensity
$2P_{3/2}-3S_{1/2}$	-0.133 862	0.010 604
Balmer line	-0.072 891	0
$2P_{3/2}-3D_{3/2}$	-0.036 134	0.054 295
$2P_{3/2}-3D_{5/2}$	0	0.488 654
$2S_{1/2}-3P_{1/2}$	+0.186 231	0.056 557+r$^{(j)}$
$2P_{1/2}-3S_{1/2}$	+0.232 011	0.005 302
$2S_{1/2}-3P_{3/2}$	+0.294 637	0.113 114+2r$^{(j)}$
$2P_{1/2}-3D_{3/2}$	+0.329 739	0.271 474
	total	1.000 000+3r$^{(j)}$

Table 2

Observed $2N^{(10)}$ in terms of 2D

2D/cm	$2N^{(10)}$
0.715 700 40	10 902.412 0
0.715 376 44	10 897.479 8
1.431 044 91	21 799.420 8
1.441 594 21	21 960.113 6
2.149 934 02	32 750.380 2
2.148 740 68	32 732.206 4
2.882 369 98	43 907.658 2
2.869 425 75	43 710.490 1
3.563 991 01	54 290.245 1
3.573 201 37	54 430.550 8
4.320 544 44	65 814.869 4
4.318 411 38	65 782.370 7
5.021 585 99	76 493.984 0
5.013 528 00	76 371.247 2
5.760 746 61	87 753.770 6
5.748 965 52	87 574.320 1
6.466 061 20	98 497.984 4
6.466 221 52	98 500.432 2

Table 3

Observed $2N^{(j_2)} - 2N^{(j_1)}$ at D=3.24 cm

j_2	j_1	$(2N^{(j_2)} - 2N^{(j_1)}) \cdot 10^4$
10	5	117
		96
11	6	43
		97
12	7	210
		197
13	8	209
		170
14	9	257
		332
15	10	277
		291
11	5	92
		147
12	6	205
		170
13	7	205
		191
14	8	325
		253
15	9	402
		330

RYDBERG CONSTANT MEASUREMENT USING cw DYE LASER AND H* ATOMIC BEAM

R. L. Barger, T. C. English and J. B. West

National Bureau of Standards
Boulder, Colorado USA

We would like to describe a new experiment which we have in progress for obtaining a greatly improved value for the Rydberg[1]. The accuracy of measurements using classical techniques[1] have been limited to about 2×10^{7} by Doppler broadening and discharge effects[2]. A recent improvement of accuracy to about 1×10^{-8} was obtained[2] by using a pulsed dye laser to obtain laser saturated-absorption of the Balmer α line in a hydrogen discharge, thereby eliminating Doppler broadening. We hope to achieve an accuracy between 1×10^{-9} and 1×10^{-10} through use of a recently developed extremely stable cw dye laser[3] to produce saturated absorption of the Balmer α line in an atomic beam of 2 s hydrogen atoms. Simultaneous excitation with RF[4] should yield a double-resonance linewidth of about 1 MHz. Use of an atomic beam should eliminate systematic errors due to collisional effects and extraneous electromagnetic fields such as are encountered in discharges. For measuring the Balmer α wavelength to obtain the Rydberg, we shall use the frequency-controlled Fabry-Perot interferometer which was used for the wavelength measurement of the 3.39 μ CH_4 line[5]. This 3.39 μ CH_4 line will be used as the length standard for the measurement. We expect the accuracy of our wavelength measurement to be limited to a few $\times 10^{-10}$ by already investigated systematic errors inherant in our interferometer, mainly diffraction effects.

With the beam of n = 2, $^1S_{\frac{1}{2}}$ hydrogen atoms, the transitions available for observation are shown in Fig. 1. The approximate level splittings are indicated. Selection rules allow the six hyperfine components shown for the optical transition, which has saturated absorption linewidth of about 30 MHz corresponding to

the 3P lifetime. Thus the four transitions
$2^2S_1(F=1)$ to $3^2P_1(F=0,1)$ and to $3^2P_{3/2}(F=0,1)$ form overlapping
doublets, but the two transitions
$2^2S_{1/2}(F=0)$ to $3^2P_1(F=1)$ and to $3^2P_{3/2}(F=1)$ are well isolated
singlets. For these singlets, the addition of RF radiation
to mix the 3S and 3P levels should result in sharp lines with a
width of about 1 MHz for the four double resonance transitions.
The sum frequencies $\nu_{opt} \pm \nu_{rf}$ give the two transition frequencies of the 2^2S_1 $(F=0)$ to 3^2S_1 $(F=0,1)$ doublet. Setting the laser
and RF on the 1 MHz-wide peak to within only 1/20 of the line
width corresponds to obtaining the transition frequency to a
part in 10^{10}.

    The experimental arrangement of the hydrogen beam apparatus
is indicated in Fig. 2. Hydrogen atoms are produced with an RF
discharge in a cell having a multi-channel slit. The atoms then
pass through an electron excitation region to convert a few per
cent to the 2^2S_1 state. Downstream, the laser beam is crossed at
right angle with the hydrogen beam and retroreflected to produce
the optical saturated absorption peak. RF in the same region
produces the narrow double resonance peak. The peak is detected
by observation of the number of 2s hydrogen atoms remaining in
the beam downstream from the optical-RF excitation region. At
the detection region, the beam passes through a quenching electric field and the resulting Lyman α photons are detected with a
solar blind photomultiplier. Modulating the laser frequency over
the peak gives the first derivative signal used to servo the laser
frequency to the center of the peak, with the RF frequency fixed.
The predicted shot-noise limited signal-to-noise ratio at line
center is about 10^3 in one second, corresponding to a pointing
precision on the lines of approximately 2×10^{-12}. This is about
2 orders of magnitude better than the expected limit of accuracy
for our interferometer.

    The stabilized cw dye laser[3] has a frequency stability of
about 6×10^{-13} for an integration time of 300 sec, better than
the shot noise limit for the hydrogen beam discussed above. The
Allan variance of the frequency noise is shown in Fig. 3**A**.The
short term stability should allow observation of lines as narrow
as a few kHz, and the long term stability indicates the measurement accuracy which can be achieved. Although such good stability
is not needed for this Rydberg measurement, it will be useful in
our other experiments[6] such as saturated absorption of a state-
selected calcium beam.

    The technique for stabilizing the dye laser is indicated in
Fig. 4. We use a jet of cresyl-violet dye dissolved in ethyl-
ene glycol in order to obtain sufficient power at the Balmer α
wavelength. Since cresyl-violet absorption is very low at the

Ar$^+$ laser wavelengths and high at 6000Å, we use the Ar$^+$ laser
to pump an auxilliary untuned rhodamine 6G dyelaser, and use the
resulting 6000Å output to pump the cresyl-violet. In this way we
have obtained about 100 mW single mode at 6563Å. The dye laser
wavelength is locked to the servo cavity fringe, and the length of
the servo cavity is in turn stabilized to either the 3.39 μ fringe,
for tunable stability, or to the Lyman α signal for locking to the
Balmer α line.   The entire dye laser is enclosed in a pressure
box, to which the servo cavity is connected, to provide the
capability of pressure scanning the stabilized wavelength. With
the frequency locked to Balmer α , part of the dye laser output is
passed through the interferometer for the wavelength measurement.

The flat-plate frequency-controlled interferometer is described elsewhere[2]. The systematic errors inherant in the interferometry should limit the accuracy of our Balmer α wavelength
measurement. These errors together with their estimated accuracy
limits are included in Table 1.

Table 1. Predicted accuracy limits, δR/R

Dye laser stability	6×10^{-13}
Hydrogen beam stability (shot noise limit)	2×10^{-12}
Interferometer systematic errors	$\sim 4 \times 10^{-10}$
1. Fringe pointing	10^{-10}
2. Diffraction correction	$\sim 4 \times 10^{-10}$
3. Alignment of laser beam and interferometer axis	10^{-10}
Uncertainty in fundamental constants $(\frac{m_e}{m_p})$	2×10^{-10}

Also given in Table I is the main uncertainty in the Rydberg
contributed by uncertainties in the fundamental constants. The
theory[7], relating the measured transition energy to the Rydberg,
involves functions of the fine structure constant and the ratio
of the electron mass to the proton mass. All terms appear to be
negligible to the order of accuracy of this experiment except
the factor $(1 + \frac{m_e}{m_p})$ used in converting the Rydberg for hydrogen

to the Rydberg for infinite mass, R_∞. This factor contributes an uncertainty to R_∞ of about 2×10^{-10}, and is probably less than the systematic errors which will arise in the interferometery.

The present status of our experiment is as follows. The stabilized laser works satisfactorily at the Balmer α wavelength, having more than the required stability and power output. Also, the interferometer system is complete. The hydrogen atomic beam apparatus is constructed; however, on our first try we were unable to detect any metastable atoms. We are now in the process of trying to locate and correct the problem.

References

1. G. W. Series, Contemp. Phys. 14, 49 (1974).
2. T. W. Hansch, M. H. Nayfeh, S. A. Lee, S. M. Curry and I. S. Shahin, Phys. Rev. Lett. 32, 1336 (1974).
3. R. L. Barger, J. B. West and T. C. English, Appl. Phys. Lett. (July 1975).
4. D. E. Roberts and E. N. Fortson, Phys. Rev. Lett. 31, 1539 (1973).
5. R. L. Barger and J. L. Hall, Appl. Phys. Lett. 22, 196 (1973).
6. R. L. Barger, Laser Spectroscopy, Ed. by R. G. Brewer and A. Mooradian, Plenum Press, New York (1974).
7. J. D. Garcia and J. E. Mack, J. Opt. Soc. Am. 55, 654 (1965).

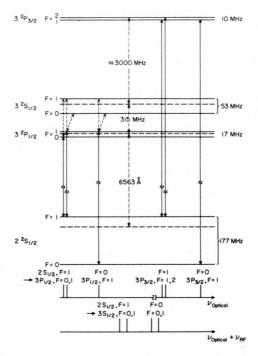

Fig. 1.   Level diagram for the n=2→n=3 double resonance transition.

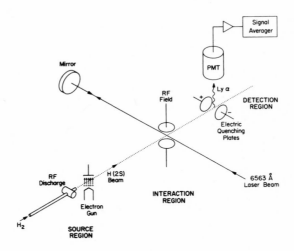

Fig. 2.   Diagram of the hydrogen beam.

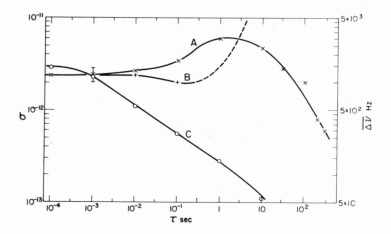

Fig. 3. Allan variance plot for cw dye laser, showing frequency
         stability vrs. integration time A.

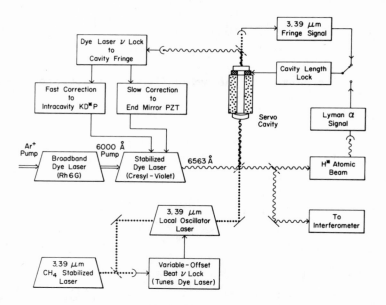

Fig. 4. Stabilization technique for cw dye laser.

FAST BEAM MEASUREMENT OF HYDROGEN FINE STRUCTURE

S.R. Lundeen

Lyman Laboratory of Physics
Harvard University
Cambridge, Massachusetts 02138

Hydrogenic atoms play a special role in physics and metrology since they are the only atomic systems whose properties can presently be calculated with confidence and high precision. This dual role is well illustrated by the history of fine structure measurements in the hydrogen atom. Figure 1 shows the fine structure of the n=2 state of hydrogen as deduced from the pioneering measurements of Lamb and co-workers in the '50's.[1] The two independent intervals shown each have their own special significance. The $2^2S_{1/2}$ and $2^2P_{1/2}$ states are separated by the Lamb shift interval $\mathcal{S}$ which is due almost entirely to the radiative corrections of Quantum Electrodynamics. Precision measurements of this interval are one of the best tests of this fundamental theory of physics.[2] On the other hand, the fine structure interval ΔE between the $2^2P_{1/2}$ and $2^2P_{3/2}$ states depends only weakly on QED and can therefore be calculated unambiguously in terms of the fine structure constant α. Experimental determinations of ΔE can be used by metrologists to improve the precision with which α and other fundamental constants are known.[3]

The best measurements of these fine structure intervals have all been made with atomic beam techniques similar to those originally used by Lamb. Unfortunately, the precision of the more recent measurements is not significantly better than that obtained by Lamb in the 50's. The basic reason for this is simple; the 1.6 ns radiative lifetime of the 2P states gives rise to a 100 MHz natural linewidth for the allowed fine structure transitions ($^2S_{1/2}-^2P_{1/2}$ and $^2S_{1/2}-^2P_{3/2}$). Consequently precision measurements have only been obtained by splitting a resonance line to a small fraction (usually about 1/1000) of its width. This is a very tricky business and it is difficult to do much better. Furthermore,

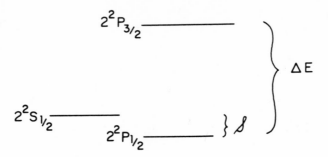

Fig. 1.   Fine structure of the n-2 state of Hydrogen.

this difficulty persists independent of Z, n, and L in all other
allowed transitions within hydrogenic fine structures.   Improvements in the precision of these measurements have therefore awaited
the development of new experimental techniques capable of observing
narrower fine structure transitions with high precision.   I will
be describing here two such techniques, utilizing a fast atomic
beam, which are currently under development at Harvard.

The first of these is the use of Ramsey's method of "separated
oscillatory fields"[4] to observe the conventional allowed transitions
with experimental widths which are significantly less than their
natural widths.[5]   The feasibility of this approach was first
demonstrated by Fabjan and Pipkin in 1971.[6]   Recently, this technique has been used to remeasure the Lamb shift in the n=2 state of
hydrogen.[7]   Fig. 2 shows the fast atomic beam apparatus used for
that measurement.   A beam of 50-100 KeV protons is converted by
charge exchange into a fast excited hydrogen beam.   Subsequently
a drift region allows the 2P state to decay away, leaving the beam
predominantly in the metastable $2S_{\frac{1}{2}}$ state.   Normally, these atoms
survive the entire length of the apparatus until they enter the
detection region where a continuous electric field quenches them
to the ground state.   The resulting Lyman α-decay radiation is used
as a monitor of the metastable population in the beam.   Before the
metastable atoms reach the detection region, however, they pass
through a spectroscopy region where rf electric fields, nearly
resonant with the Lamb shift transition frequency, are applied.
These fields have the effect of partially quenching the metastable
beam and thus reducing the level of photocurrent in the detection
region.   The quenching resonance is plotted out by simply varying
the frequency of the spectroscopic rf fields.   At no time is an
external magnetic field applied.

At zero magnetic field, the Lamb shift resonance in hydrogen
is normally a superposition of three overlapping components produc-
ed by the hyperfine structure of the $S_{\frac{1}{2}}$ and $P_{\frac{1}{2}}$ states.   This
complication can be eliminated, however, through the use of rf

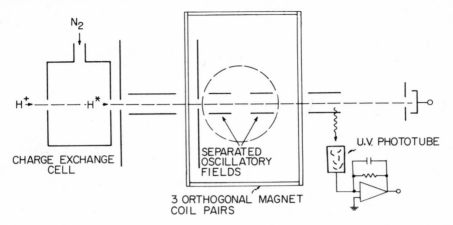

Fig. 2.   Fast atomic beam, separated oscillatory field apparatus.

hyperfine state selection techniques which have the effect of preparing the beam uniquely in the $2^2S_{\frac{1}{2}}$ (F=0) state.[6] From this state,
electric dipole selection rules allow transitions only to the F=1
state of the $2^2P_{\frac{1}{2}}$ level so that a single component quenching
signal is observed.

      In order to narrow such a resonance below its natural line-
width, the spectroscopic rf fields are applied coherently in two
separated regions, where the time separation between the regions
is comparable to the lifetime of the P-state. An atom entering
the first of these rf regions in the S-state may emerge from the
second as an S-state by two distinct paths. The first requires
transmission as an S-state through each rf field separately, while
the second requires a transition to the P-state in the first region,
survival as a P-state between the regions and a transition back
to the S-state in the second. The total transmission coefficient
for the S-state contains an interference term between these two
alternative paths which is used to effect the line narrowing. The
portion of the total quenching signal arising from this interference term can be conveniently isolated by subtracting the quenching
signals observed when the relative phase of the two rf regions is
fixed at 0 and 180 degrees. Typical results of this procedure are
shown in Fig. 3.   The plotted curves show the average and difference of the experimentally observed quenching signals in the two
cases. The difference, or "interference signal" is, of course,
nothing but the separated oscillatory field "Ramsey pattern" which
has been observed for years with conventional atomic beams. Its
width is determined not by the natural linewidth for the transition, but by the time separation between the two rf fields. By
increasing this separation, the interference signal may be made
arbitrarily narrow. The price of this linear decrease in
linewidth is an exponential decrease in the size of the interference

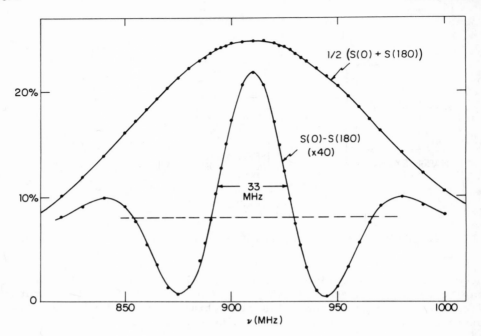

Fig. 3.   Separated oscillatory field resonance profiles for
$2^2S_{\frac{1}{2}}$(F=0) to $2^2P_{\frac{1}{2}}$(F=1) transition in Hydrogen.

signal, since only the exponentially small fraction of P-states
which survive passage between the two fields can contribute to it.
Nevertheless, with the high metastable intensities available in
a fast atomic beam ($\sim 10^{13}$/sec) it is possible to retain a high
signal-to-noise ratio even in the narrowed signal.   In Fig. 3,
for example, the width of the interference curve is about 1/3 of
the natural linewidth for the transition while the statistical
errors are indicated by the size of the plotted points.

     Precision determination of the centers of resonance lines
such as these yield a measurement of the Lamb shift.   Successful
use of this technique, however, required a careful consideration
of the details of the separated oscillatory field lineshape as
well as a complete understanding of the operation of the tunable
rf system.   These problems appear now to be well understood and
to pose no fundamental barrier to further progress with the
technique.[8]

     The result of the recent fast beam measurement is compared
in Fig. 4 with previous measurements and theory.   The solid

circles represent direct Lamb shift measurements, while the
open circles represent measurements of the $2^2S_{\frac{1}{2}}-2^2P_{3/2}$ fine
structure interval from which the Lamb shift can be determined
if a value for the fine structure constant is assumed. The
shaded vertical bars show the two most recent theoretical results
with an estimated ± one standard deviation error. The new
measurement is in good agreement with the theory and about a
factor of three more precise than past measurements. Unfortunately, however, the two calculations shown are themselves inconsistent. Hopefully this will soon be resolved, as present
indications are that additional improvements in the experimental
precision of at least a factor of three can be expected from the
separated oscillatory field technique. Finally, it is worth
noting that almost half of the quoted theoretical uncertainty
for this interval is due to uncertainty in the rms charge radius
of the proton, which is known experimentally to only about 2%.
The situation in other hydrogen isotopes and in other low Z hydrogenic systems is more severe in this respect. The experimental
uncertainties in those nuclear radii are such as to prevent a
test of the most precise QED calculations. Future improvements
in nuclear radii measurements can therefore play an important
role in improving the Lamb shift as a test of QED.

It should also be possible to apply the separated oscillatory
field-fast beam technique to a remeasurement of the $2^2S_{\frac{1}{2}}-2^2P_{3/2}$
fine structure interval in hydrogen. If such a measurement could
obtain the precision already achieved for the Lamb shift transition, it would yield a value for the fine structure constant good
to about 1 ppm. In view of the critical role of the fine
structure constant in recent adjustments of the fundamental

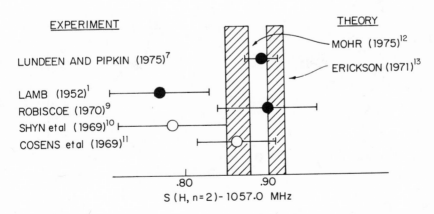

Fig. 4. Experiment and theory for the Lamb shift in H, n=2.

constants, this would be a useful development.[14]   Such an experiment is now in progress at Harvard.

Another fast beam technique that is currently emerging is the study of multiple quantum fine structure transitions. While the aim of the separated oscillatory field technique is to narrow allowed transitions below their natural linewidth, the multiple quantum technique makes it possible to observe lines which are inherently narrower. The Q, or ratio of transition frequency to natural linewidth, for allowed electric dipole transitions in hydrogenic atoms is about 100 or less. Multiple quantum transitions, however, can have much higher Q's, and thus offer an attractive tool for the high precision study of these systems.

A preliminary study of the $3^2S_{1/2}-3^2D_{5/2}$ double quantum transition in hydrogen has recently been reported.[15] The experimental technique again utilizes a fast atomic beam and is basically similar to that used in the fast beam Lamb shift experiments. In this case, a beam of $3^2S_{1/2}$ atoms is prepared and the depopulation of that state by rf transitions is measured. Of course there is no direct coupling between the $S_{1/2}$ and $D_{5/2}$ states in an rf field, but they are coupled in second order through the intermediate state $3^2P_{3/2}$.   Thus in a strong rf field

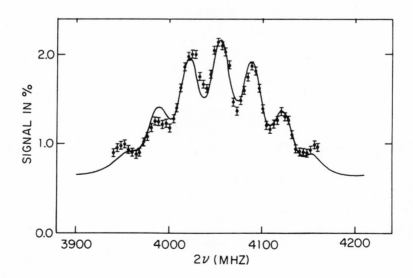

Fig. 5.   Resonance profile for $3^2S_{1/2}$(F=0) - $3^2D_{5/2}$(F=2) double quantum transition in Hydrogen.

transitions between the two states can occur and will be
resonant when the energy of two rf photons is equal to the energy
difference of the two states. Since both the $3^2S_{1/2}$ and $3^2D_{5/2}$
states are relatively long-lived, the natural linewidth for the
two photon transition is less than for any allowed transition,
giving a natural Q of 400 for this transition. Fig. 5 shows a
scan of the $3^2S_{1/2}$ (F=0) to $3^2D_{5/2}$ (F=2) transition as observed in
zero magnetic field. The resonance was obtained in a separated
oscillatory field geometry, but with the relative phase of the
two fields fixed at zero degrees. The smooth curve is a fit of
the experimental points to a lineshape of the form expected for
this geometry. The 16 MHz width (in 2ν) of the central peak of
the resonance is due to the experimental geometry and is still
somewhat larger than the 10 MHz natural width for the transition.
An effort is presently underway at Harvard to narrow this
transition even further with the separated oscillatory field
technique and obtain a Q of 1000. Measurement of the line center
to 1 part in 1000 could then give the fine structure constant to
0.5 ppm. The severity of the systematic problems associated
with this technique remains to be determined, but in the long
term, it may prove to be a very productive approach since more
highly forbidden transitions can yield progressively higher Q's.

1. E. S. Dayhoff, S. Triebwasser, and W. E. Lamb, Jr. Phys. Rev.
    89, 98 (1953).
2. S. J. Brodsky and S. D. Drell. Ann. Rev. Nuclear Science,
    20, 147 (1970).
3. B. N. Taylor, W. H. Parker, and D. N. Langenberg, Rev. Mod.
    Phys. 41, 375 (1969).
4. N. F. Ramsey, Molecular Beams, (Oxford Univ. Press, London,)
    (1956), p. 119.
5. V. W. Hughes in Quantum Electronics, (C. H. Townes, ed.
    Columbia Univ., New York, 1960), p. 582.
6. C. W. Fabjan and F. M. Pipkin, Phys. Rev. A6, 556 (1972).
7. S. R. Lundeen and F. M. Pipkin, Phys. Rev. Lett. 34, 1362 (1975)
8. S. R. Lundeen, thesis, Harvard University (unpublished).
9. R. T. Robiscoe and T. W. Shyn, Phys. Rev. Lett. 24, 559 (1970)
10. T. W. Shyn, W. L. Williams, R. T. Robiscoe, T. Rebane,
    Phys. Rev. Lett. 22, 1273 (1969).
11. T. V. Vorbunger and B. L. Cosens, Phys. Rev. Lett. 23, 1273
    (1969).
12. P. J. Mohr, Lawrence Berkeley Laboratory, Preprint A 3641.
13. G. W. Erickson, Phys. Rev. Lett. 27, 780 (1971).
14. E. R. Cohen, B. N. Taylor, J. Phys. Chem. Ref. Data 2, 741
    (1973).
15. P. B. Kramer, S. R. Lundeen, B. O. Clark and F. M. Pipkin,
    Phys. Rev. Lett. 32, 635 (1974).

A NEW DETERMINATION OF THE FARADAY

BY MEANS OF THE SILVER COULOMETER

V. E. Bower and R. S. Davis

Electricity Division, National Bureau of Standards

Washington, D.C.

INTRODUCTION

Of the several electrochemical determinations of the Faraday constant, most of them done at the National Bureau of Standards (1, 2), the experiment usually cited is that of Craig and his collaborators (1). This experiment exhibited a model concern for detail and moreover showed a somewhat higher precision than the measurements of oxalic and benzoic acids (2). Craig's measurement employed the electrochemical dissolution of silver into perchloric acid as its principle. He dissolved a highly purified sample of silver anodically into an acid which did not attack silver unless current flowed. Such obstacles to a reliable measurement as electrolyte inclusions in the silver and segregation of the silver isotopes by electrolysis were thus avoided. By separation of the cathode and anode compartments he assured the retention of all debris which inevitably falls from the ingot of silver during the electrolysis.

The idea of Craig's experiment was simple. A highly purified sample of silver was weighed carefully. The sample was placed in the perchloric-acid coulometer and was dissolved anodically at constant current. The sample was washed, dried, and weighed again. The debris in the anode compartment was washed out into a glass filter crucible and was, of course, counted as undissolved sample. From the weight loss of silver m_{ag}, the atomic weight of silver, $M^*(Ag)$, the current, i, and the duration of the experiment, t, he could calculate:

$$F = \frac{it}{m_{Ag}/M^*(Ag)} \cdot \qquad (1)$$

There were of course corrections for impurities and for the
isotopic abundance ratio for silver. Independent critics (3)
assigned a 6 - 7 ppm error to the determination.

Confronting Craig's result is the indirect value of the
Faraday which may be calculated a variety of ways from a number of
physical measurements. An example is (3)

$$F = \frac{M_p \gamma_p'(\text{low})}{\mu_p'/\mu_N \, K^2}$$

Here M_p is the proton atomic mass, γ_p' is the gyromagnetic ratio of
the proton and μ_p'/μ_N is the proton magnetic moment in nuclear
magnetons, K is the ratio of the as-maintained ampere to the
absolute ampere. The primes indicate measurements on protons in
water samples. A high accuracy can be assigned to the Faraday thus
obtained, but it is discrepant with respect to the less accurate
electrochemical value to the extent of about 26 ppm (4).

A direct electrochemical measurement may make some contribution
toward the resolution of this difficulty if the measurement can be
made with a precision and accuracy of the same order as the derived
physical value. To this end, we report here the general method and
the preliminary result of our redetermination of the Faraday.

METHOD

The iodine coulometer was at first attempted and our progress
was described at the last conference AMCO-4 (5). The precision
reported to the conference at that time was never improved and a
better behaved system was sought. Consideration of the various
methods of measuring the Faraday led to the conclusion that the
silver-silver ion coulometer would provide the best way of obtaining
a precise value of the Faraday. The anodic dissolution of silver
proceeds smoothly, is devoid of interfering reactions, and results
in a single oxidation state of the metal. The metal is dense enough
that the usual bouyancy corrections in the weighing may be applied
with confidence. Furthermore, silver spontaneously dissolves in a
solution 20% in perchloric acid and 0.5% in silver perchlorate
at the rate of only a few micrograms over months (6).

The major contribution to the scatter in the Craig experiment
was, in our opinion, the transfer of the silver debris from the
anode compartment of the coulometer to another vessel for weighing.
In our experiment, the debris is analysed within the compartment
and for the transfer error is substituted the much smaller analytical
error of controlled potential coulometric titration. This method,
along with modern improvements in balances, current controls,
detectors, and purification and analysis of materials, should permit

a measurement that will yield higher precision than was possible
at the time of the earlier silver experiment 20 years ago.

The scheme of the experiment is as simple as Craig's. A first
weighing is performed on the high-grade silver sample. The sample
is suspended in the coulometer and accurately controlled constant
current is switch on for an accurately known time. The sample
is washed and dried and the second weighing is made. Some of the
silver thus lost goes directly into solution. The charge on this
portion is known immediately; it is simply i x t. The rest, a
small amount of silver (usually below 1 mg in our experiment) falls
as debris in the anode compartment. The coulometer electrolyte
is aspirated off. The silver debris is dissolved, and its equivalent
charge (q) is measured by controlled-potential coulometery. The
electrochemical equivalent is then calculated:

$$e.e. = \frac{m_{Ag}}{it + q(controlled\ potential)}$$

where all electrical units are NBS as-maintained. The division of
this quantity into the atomic weight of silver yields the Faraday
(eq. 1) in NBS units.

The coulometer used in this experiment may be briefly described
(Fig. 1). The anode compartment consists of a beaker with an

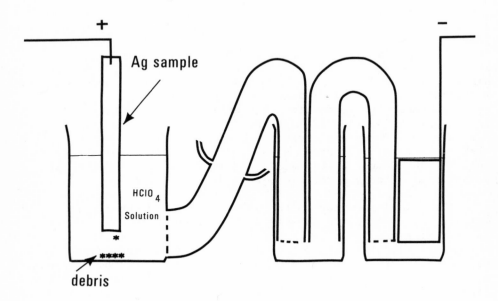

Fig. 1. Silver coulometer used in this experiment.

arm attached to the side near the bottom. At the junction of the
side-arm and the beaker is a fritted glass disk of fine porosity.
At the end of the side-arm is a spherical glass joint. The glass
joint attaches to other separating compartments and the cathode
compartment which do not participate in the measurement except to
conduct current and complete the cell. The spherical joint allows
removal of the anode compartment from the whole cell. It also
allows removal by the anolyte through the fritted disc by suction.
The silver debris is thus left in the anode compartment and (after
some washing with water) is entirely separated from silver ion.
The silver debris may be converted to silver nitrate by nitric
acid and its equivalent charge determined by controlled potential
coulometry (7,8).

The silver samples were purchased from a commercial source of
high-purity metals by the National Bureau of Standards for issue
as a standard reference material. Preliminary mass spectrographic
analyses were made on the sample at the time of purchase and
acceptance. A more searching analysis of the sample is required
and is now being undertaken. We cannot report to this conference
a detailed statement of purity but an estimated error of 2 ppm
is placed on the purity of the sample in the treatment of our data.

Other chemicals were perchloric and nitric acids which were
reagent-grade material distilled in vycor, and silver oxide which
was reagent grade.

The current control, of the photocell-galvanometer type, held
the current constant through the course of the experiment to within
about 0.2 ppm. The coulometer current also passes through a
standard resistor and the resulting IR drop was balanced against a
standard cell. The resistor and standard cell are assigned an
error of about 0.2 ppm each. A ganged switch transfers the current
to the coulometer from a dummy load at the same time it starts
the counter counting the standard 10 KHz frequency. The simultaneity
of closure of the switches is within 1 millisecond or less than
0.2 ppm on all runs 1 1/2 hour or more long.

In the analysis of the debris we used a potentiostat which is
a modification of that used by Jones and his colleagues for his
analyses of uranium (7,8). The uncertainty on the analytical
determinations was about 3 μg of silver which in our experiment
amounted to an error of about 0.6 ppm.

The balance used in this experiment was of the single-pan
constant-load type and all weighings were by the method of double
substitution (9). Weights were of the single piece stainless-
steel NBS class "M" type (10). The estimate of errors arising from
the weighing process is about 0.5 ppm.

RESULTS

Table 1 lists the electrochemical equivalent values we obtained in 8 runs on 7 samples.

Table 1

Lot no.	Sample no.	Electrochemical Equivalent of Silver Anode	mg/coul$_{NBS}$
55	A-7	1.11796109	
	B-7	6668	
	B-7	6239	
	D-7	5880	
	E-7	6214	
	F-7	6350	
	G-Y2	6625	
	G-Y3	6091	
	Average	1.11796272	
	σ	0.00000268	2.40 ppm
	σ_m	0.00000095	0.85 ppm

It will be seen that the experiment scatter is low for a determination of this type, 2.4 ppm. The list of estimates of systematic error at the 1 σ level is given in table 2.

Table 2

Uncertainty on debris recovery	0.6 ppm
Weighing	0.5 ppm
Volt	0.2 ppm
Resistance	0.2 ppm
Time	0.2 ppm
Atomic Weight of Silver (3,10)	2.1 ppm
Estimate in error of composition	2.0 ppm
Estimated total systematic error, 1 σ	3.02 ppm
Random experimental error, 1 σ	0.85 ppm
Total estimate of error	3.1 ppm

Of the errors listed in table 2 there is some hope of reducing at least one of the two large ones. The isotopic abundance ratio of silver will be redetermined on our samples this summer. A three or four-fold reduction in error is hoped for (12). The error in estimating the impurities is, at the present time, admittedly a guess. Two laboratories are independently assessing the impurities in these samples and their reports are not expected until well after this conference.

With the cautions imposed by the above considerations, we advance the tentative value for the electrochemical Faraday as

$$F = 96,486.53 \pm 0.31 \ A_{NBS} \cdot s \cdot mole^{-1}$$

where A_{NBS} signifies the ampere as at present maintained at the National Bureau of Standards and where the atomic weight of silver used in the calculation is that given in Taylor, Parker and Langenberg,

$$M*(Ag) = 107.86834 \text{ amu} \tag{3}$$

Another determination of the Faraday by Diehl through the titration of 4-aminopyridine has come to our attention (13). The agreement of Diehl's result and our own appears good.

The possible impact of this experiment on the other fundamental constants will be taken up later in the conference (14).

REFERENCES

[1] D. N. CRAIG, J. E. HOFFMAN, C. A. LAW, and W. J. HAMER, J. Res. National Bureau of Standards 64A, 381 (1960).
[2] G. MARINENKO and J. K. TAYLOR, Anal. Chem. 39, 1568 (1967).
[3] B. N. TAYLOR, W. H. PARKER, and D. N. LANGENBERG, Rev Mod. Phys., 41, 477 (1969).
[4] E. R. COHEN and B. N. TAYLOR, J. Phys. Chem. Ref. Data, 2, 663 (1973).
[5] V. E. BOWER, Proceedings of the Fourth International Conference on Atomic Masses and Fundamental Constants, Teddington, England, 1971, Plenum Press, New York, 1972.
[6] D. N. CRAIG, C. A. LAW, and W. J. HAMER, J. Res., Nat. Bur. Standards 64A, 127 (1960).
[7] R. S. DAVIS and V. E. BOWER, J. Res., Nat. Bur. Standards, to be published.
[8] H. C. JONES, W. D. SHULTZ and J. M. DALE, Anal. Chem. 680 (1965)
[9] P. E. PONTIUS and J. M. CAMERON, Nat. Bur. Standards Monograph 103 USGPO Washington 1967.
[10] T. W. LASHOF and L. B. MACURDY, NBS Circular 547 USGPO, Washington, 1954.
[11] W. R. SHIELDS, D. N. CRAIG and V. H. DIBELER, J. Am. Chem. Soc. 82, 5033 (1960).
[12] I. L. BARNES, NBS, Personal communication.
[13] H. DIEHL, this conference.
[14] B. N. TAYLOR, this conference.

A VALUE FOR THE FARADAY BASED ON A HIGH-PRECISION COULOMETRIC TITRATION OF 4-AMINOPYRIDINE

Harvey Diehl

Department of Chemistry, Iowa State University

Ames, Iowa, U. S. A. 50010

In the two most recent least-squares adjustments of the fundamental physical constants, B. N. Taylor and his co-workers, 1969[1] and 1973[2], call repeatedly for a further experimental determination of this constant. The existing value, 96,486.67 1972 NBS coulombs per mole is based on the silver dissolution work of Craig, Hoffman, Law and Hamer[3], and although closely supported by coulometric titrations of benzoic acid[4] and of oxalic acid[4], it is greater by some 21 ppm than a value calculated from other physical measurements.

The present evaluation of the Faraday is based on the coulometric titration of a highly-purified specimen of the organic base 4-aminopyridine, $C_5H_6N_2$. The titrations were carried out in two essentially different methods, one in which the working electrode was a cathode, the other in which the working electrode was an anode. The average of the two results obtained, 96,486.69 1972 NBS coulombs per mole, is identical with the silver dissolution value.

4-Aminopyridine was selected for the present work because the chemical and physical characteristics of it meet unambiguously several severe requirements. 4-Aminopyridine can be purified by sublimation, thus assuring the absence of occluded solvent. The melting point, 159.09°, lies in a convenient range to obtain a freezing curve for an estimate of the total impurity present. One titratable group is present, $pK_B = 4.63$, sufficiently strong to provide a good endpoint. The molecule is made up only of carbon, hydrogen, and nitrogen, elements for which variation in the isotope abundance ratios, natural or as a result of processing, is probably negligible, but elements for which, if necessary, the abundance ratios can be determined without too much difficulty. Thus a calculation may be made

from titration data of a value for the Faraday based directly on
carbon-twelve.

The 4-aminopyridine used was purified by repeated sublimation
in an atmosphere of dry nitrogen. The purity of this preparation
was determined by measuring the depression of the freezing point.

The freezing curve of the highly-purified 4-aminopyridine was
obtained with an instrument of new design[5] employing a gold crucible
without stirring, a platinum resistance thermometer, electrical heating elements on crucible and heat shield, and accessary electrical
control devices such that the temperature difference between crucible and shield could be maintained constant within 0.005K while the
temperature of the 4-aminopyridine and crucible dropped through the
freezing range. Methods were devised for obtaining and handling the
data by a differential method and of correcting the freezing curve
for the heat capacity of the instrument and charge. Although interpretation of the freezing curves obtained on the 4-aminopyridine was
confused by attack on the gold by the molten 4-aminopyridine, the
initial impurity in the 4-aminopyridine was probably less than the
detection limit, 0.001 mole per cent.

The coulometric titration apparatus used, Figure 1, was that
manufactured by the Leeds & Northrup Company, Philadelphia, as described by Eckfeldt and Shaffer[6], the specific standard resistance
and details of techniques used being those of Knoeck and Diehl[7,8].

The electrolyte used in all three chambers of this cell was
1 M sodium perchlorate. The level of electrolyte around the counter
electrode was kept just above the top of the electrode thus insuring
continual flow of electrolyte from the intermediate chamber into the
electrode compartments.

The titration was followed by measuring the pH with a glass
electrode and an Expanded Range pH Meter manufactured by the Hach
Chemical Company, Ames, Iowa, readable to 0.001 pH.

Current was measured by measuring the potential drop over a 20-
ohm resistance with a Leeds & Northrup Type K-5 potentiometer with
electronic null-point detector. The 20-ohm resistor was calibrated
at the National Bureau of Standards and found to have the value
19.999,703 ohms at 25.00° with an uncertainty of 0.2 ppm. The working standard cell was an unsaturated Weston cell calibrated periodically against a bank of saturated Weston cells (Eppley Model 121).
The latter in turn were calibrated using the travelling volt standard
of the National Bureau of Standards. The potentials are thus on the
basis of the 1972 NBS volt based on the Josephson junction. The
error in the potentiometer was estimated to be less than 3 ppm.

Time was measured with an electronic counter calibrated using

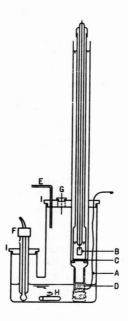

Figure 1.   Coulometric titration cell.   A, Platinum working electrode.   B, Platinum counter electrode.   C, Glass frit on bottom of inner shield tube.   D, Unfired Vycor on bottom of intermediate chamber.   E, Nitrogen inlet.   F, Combination pH electrode.   G, Sample inlet.   H, Magnetic stirring bar.   I, Plexiglass covers.

the time signals of Radio Station WWV of the National Bureau of Standards; the error was less than 0.2 ppm.

The mass of 4-aminopyridine taken for the individual titration was 3 g., weighed to the nearest 3 μg on an equal-arm, micro balance using stainless steel weights calibrated at the National Bureau of Standards.   The weighings were made by the substitution method and corrected for the buoyancy of air using for the density of 4-aminopyridine 1.2695 (found by the pynometer method using mineral oil) and for the density of air a value calculated from the prevailing barometric pressure, temperature and humidity.

In the individual titrations, nitrogen was passed through the electrolyte to remove carbon dioxide and preliminary titrations were made forward and backward through the equivalence-point region, between pH 3 and 6, by electrolyzing with the working electrode alternately anodic and cathodic.   During the third electrolysis data were taken, pH versus time, at a current of approximately 6.4 mA.   The weighed sample of 4-aminopyridine was introduced by lowering the weighing boat plus the amine into the electrolyte by a platinum wire.

The solution was then electrolyzed, the current being approximately 63.4 mA. The major part of the titration required about 10.5 hours, for a 3-g sample. At 30-minute intervals the potential across the standard resistor, the temperature of the resistor (in a constant temperature oil bath), and the pH of the solution were recorded. When the pH of the main solution reached 5.5 the titration was interrupted and the walls of the titration chamber were rinsed with triply distilled water. The titration was continued at a current of approximately 6.4 mA delivered in increments of twenty seconds, pH, time, and potential drop across the resistor being measured. Between additions sufficient time was allowed for the solution to equilibrate before the pH was recorded.

In each analysis, two equivalence points were determined, that in the pre-titration and that in the actual titration. The point of inflection was found by fitting the data obtained through the entire equivalence point region empirically by computer[9] to a cubic equation, setting the second derivative of the equation to zero, and solving for the point of inflection. Given then the total number of coulombs passed and the weight of the 4-aminopyridine, a value could be calculated for the Faraday.

Basically, the titration proposed for the 4-aminopyridine was a neutralization titration with hydrogen ion generated at the anode (1 M sodium perchlorate supporting electrolyte): $2H_2O = 4 H^+ + O_2 + 4e^-$. The titration, failed, however, some two to four per cent more electricity being required than expected. The cause of the trouble was traced to a side reaction occurring at the anode, the formation of a small amount of a peroxyperchlorate, presumably Cl_2O_8 This difficulty was circumvented two ways: 1. By the invention[10] of a new procedure, designated as the "hydrazine-platinum anode", by which hydrogen ion can be generated with 100 per cent current efficiency. 2. By adding to the 4-aminopyridine a weighed amount of perchloric acid in slight excess, and back titrating the excess of the latter with hydroxyl ion generated at the cathode; $H_2O + e^- = \frac{1}{2}H_2 + OH^-$. In turn the perchloric acid solution was standardized coulometrically in the same way.

The "hydrazine-platinum anode" is based on the reaction: $N_2H_5^+ = 5H^+ + N_2 + 4e^-$. Five hydrogen ions are formed for each four electrons passed. The hydrazinium ion, $N_2H_5^+$, is formed when any hydrazine compound is brought to pH 4.5. This is done during a preliminary titration; the 4-aminopyridine was then added and the 4-aminopyridine then titrated with acid generated at a bright platinum anode. By separate studies[10] the method was shown to be 100 per efficient in the generation of hydrogen ion.

In the back-titration method, the working electrode was a cathode rather than an anode. The perchloric acid, delivered from a weight buret, was neutralized by base generated at the negative

electrode and the procedure for titrating the perchloric acid in
excess of that needed to neutralize the 4-aminopyridine and when
alone (standardization of the perchloric acid solution) were identical. The quantities involved are again mass, resistance, potential
drop, and time; the molecular weight of the perchloric acid cancels
out in the calculation of the value of the Faraday. Inherently the
perchloric acid-back-titration method is more accurate than the hydrazine-platinum anode method because the end-points are sharper, and
in practice the standard deviation of the individual result was about
one-third that of the hydrazine-platinum anode method.

Six titrations of 4-aminopyridine were made by each method;
six additional titrations were made to standardize the solution of
perchloric acid used in the second method. The results obtained
were: by the hydrazine-platinum anode method, F = 96,486.40 1972
NBS coulombs per mole, σ_i = 1.53 (five degrees of freedom, σ_m = 0.62;
by the back titration method (platinum anode), F = 96,486.78 1972
NBS coulombs per mole, σ_i = 0.57 (five degrees of freedom), σ_m =
0.23, in which σ_i and σ_m are the standard deviations of the individual observations and of the mean, respectively, random errors only.

We advance for the value of the Faraday a weighted average of
these values: F = 96,486.69(0.81)(8.40 ppm) 1972 NBS coulombs per
mole, in which the numbers in parentheses are the uncertainty resulting from combining random and systematic errors and expressed as the
standard deviation of the mean first in coulombs per mole and second
in parts per million. The results are based on a purity of 100.000
per cent. The systematic errors are our best estimate at the one
standard deviation (70 per cent confidence) level rather than the
maximum possible error.

Fifteen or so data points of pH at time at which the low current
(0.064 A) flowed are involved in arriving at each end-point. From
the slope of the curve through the end-point region, the combined
error from the two end-points is placed at 7 ppm in the titrations
with the hydrazine-platinum anode and at 3 ppm in the back titrations.

The error in the measurement of the potential drop over the
standard 2-ohm resistor with the Type K-5 potentiometer was estimated to be 3 ppm. Variation and drift in the current supply were
averaged out by making potential measurements at intervals not exceeding 30 minutes over the 8-hour runs.

It is unfortunate that the freezing-point study[5] placed only
an upper limit, 10 ppm, on the impurity possibly present in the 4-
aminopyridine. It is the current state of the art that organic impurities at this level in organic compounds (metals having been
shown to be absent) cannot be identified and determined. To the extent that the impurities in the 4-aminopyridine are the possible

isomeric materials 2-aminopyridine and 3-aminopyridine, the impurities are without effect; the equivalent weights are identical and
both isomers are weaker bases and would be titrated together with
the 4-aminopyridine in both of the procedures used in this work,
that is, by the hydrazine-platinum anode method and by the back titration of excess perchloric acid method. Lacking any definite evidence as to the presence or nature of impurities better than the
10-ppm set by the sensitivity in the freezing point experiment, re-
liance must be placed on our feel for the chemistry involved. We
believe that impurities if present are 2-aminopyridine, 3-amino-
pyridine and the homologous aminomethylpyridines, the effects of
which are either zero or negligible. At the 70 per cent confidence
level we place the error at 3 ppm.

Both nitrogen atoms, that of the ring and that of the amino
group, are derived from ammonia. Thus, the nitrogen in the 4-amino-
pyridine used in this work was derived from nitrogen of the atmosphere of recent geologic time.

The question of the alteration by the chemical processing during purification of the ratios of the abundances of the various species of 4-aminopyridine which exist as a result of the existence of
isotopes of the three elements making up the compound is not so easily answered. In the repeated sublimation steps by which the 4-amino-
pyridine was purified probably not more than twenty successive sub-
limations were made but the number is uncertain owing to the complexity of what happened during sublimation. It seems unlikely, however, that the successive steps were anywhere near sufficient in number to disturb the abundance ratios by more than the 10 ppm aimed for
in this work.

For calculating the molecular weight of 4-aminopyridine, the
following values for the atomic weights were used: C = 12.011,
15±0.000,05, H = 1.007,97±0.000,01, N = 14.006,72±0.000,01. The values for carbon and hydrogen are those given in the 1961 Table of
Atomic Weights and were used rather than those of the 1971 Table in
which the 1961 numbers have been rounded off for general chemical
use. The value for the atomic weight of nitrogen was calculated
using the abundance ratio $^{14}N/^{15}N$ = 272±0.3 and the recent values for
the absolute masses of the isotopes of nitrogen of Wapstra and
Gove[12]: ^{14}N = 14.003,074,40±0.000,000,13 and ^{15}N = 15.000,109,3±
0.000,000,5. The value for the abundance ratio is that of Junk and
Svec[13], based on mass spectrographic analysis of air of different but
recent geographical origin. In these numbers the ± sign represents
the maximum variation reported in various studies resulting from
natural variation in isotopic composition. In calculating the molecular weight of 4-aminopyridine these numbers were handled by taking
the square root of the sum of the squares and dividing by 13, the
number of atoms in the molecule. The molecular weight of

4-aminopyridine is $C_5H_6N_2$ = 94.117,02±0.000,03. Thus the natural variation in the abundance of the isotopes of carbon, hydrogen and nitrogen which have been reported, when transferred to 4-aminopyridine, is 0.3 ppm and thus 1.5 orders of magnitude smaller than the error in the coulometric titrations reported in this work.

The present work was conceived as a three-part project: the preparation and titration of the 4-aminopyridine, the proof of the purity of the 4-aminopyridine by the freezing point method, and the determination of the abundance ratios of the isotopes of the carbon, hydrogen and nitrogen atoms making up the particular lot of 4-aminopyridine titration. Only the first two parts have been carried ou The freezing point method set an upper limit of 1 in 10^5 on the possible impurity present but it is thought that the compound was purer than this and that the titrations have been carried out with an error less than 1 in 10^5. The close agreement of the value found with that from the silver dissolutions method of Craig, Hoffman, Law and Hamer is rather astonishing in view of the difference in the chemistries involved.

The coulometric titration method involving 4-aminopyridine presented here has the merit that it is based on a material prepared by sublimation and therefore free of occluded solvent and that no recovery and filtration of a solid is involved. Offsetting this is the necessity of locating an end-point with the attendant problems. The unique feature of the present work is that it involves the titration of the basic substance, the 4-aminopyridine, in two directions, that is, by electrolysis at the anode and then again at the cathode.

ACKNOWLEDGEMENTS

The author hastens to record the part played by others in this work on the Faraday: Mr. William F. Koch who carried out the high-precision titrations; Dr. William C. Hoyle who prepared the 4-aminopyridine, uncovered the discrepancy in the generation of hydrogen ions at a platinum anode and checked out the new hydrazine-platinum anode for the generation of hydrogen ions, and who with Dr. Fred Kroeger and Professor C. A. Swenson built the freezing point apparatus. The contributions of these men are acknowledged by co-authorship on the more detailed papers being published. The author is particularly indebted to Mr. Wayne Rhinehart, to Professor C. A. Swenson, to Dr. Barry N. Taylor and to Dr. Vincent Bower for their efforts in the calibration of the various standards of mass, time, resistance and potential used in this work.

BIBLIOGRAPHY

1. B. N. Taylor, W. H. Parker and D. N. Langenberg. Rev. Modern Phys., 41, 477 (1969).

2. E. R. Cohen and B. N. Taylor. J. Phys. Chem. Reference Data, 2, 663 (1973).

3. D. N. Craig, J. I. Hoffman, C. A. Law and W. J. Hamer. J. Res. Natl. Bur. Stand., 64A, 381 (1960).

4. G. Marinenko and J. K. Taylor. Anal. Chem., 39, 1568 (1967).

5. F. R. Kroeger, W. C. Hoyle, C. A. Swenson and H. Diehl. To be published in Talanta during 1975.

6. E. L. Eckfeldt and E. W. Shaffer, Jr., Anal. Chem., 37, 1534 (1965).

7. J. Knoeck and H. Diehl. Talanta, 16, 181 (1969).

8. J. Knoeck and H. Diehl. Talanta, 16, 569 (1969).

9. W. F. Koch, D. P. Poe and H. Diehl. To be published in Talanta during 1975.

10. W. C. Hoyle, W. F. Koch and H. Diehl. To be published in Talanta during 1975.

11. W. F. Koch, W. C. Hoyle and H. Diehl. To be published in Talanta during 1975.

12. A. H. Wapstra and M. B. Gove. Nucl. Data, Sect. A., 9, 265 (1971).

13. G. Junk and H. J. Svec. Geochim. Cosmochim. Acta, 14, 234 (1958).

INITIAL RESULTS FROM A NEW MEASUREMENT OF THE NEWTONIAN GRAVITATIONAL CONSTANT

G.G. Luther, W.R. Towler,[*] R.D. Deslattes, R. Lowry,[*]
J. Beams[*]

National Bureau of Standards and [*]University of Virginia
Washington, D.C. 20234 and Charlottesville, Virginia 22901

DESCRIPTION OF EXPERIMENT

The Newtonian gravitational constant, G, remains the least well-known of the fundamental constants of physics in spite of several recent and current efforts. One of these is underway at the National Bureau of Standards in cooperation with people from the University of Virginia. Our experiment is of the type originated by Beams et al.[1] In such an experiment gravitational force is balanced against the force of inertial reaction in an accelerated rotating reference frame. It is this acceleration which is measured to yield a value for G. In its present incarnation, we feel that the measurement has a potential accuracy of about 10 ppm. The principle of the measurement is recalled in Fig. 1. The small mass system (first a cylindrical bob and later a dumbbell), the large masses, the auto-collimator, the fiber which supports the small mass and an encoder disc are mounted on the rotating table. The auto-collimator senses the tendency of the small mass system to rotate toward alignment with a line joining centers of the large masses in response to the gravitational attraction. The torquer accelerates the rotating table so as to cancel the gravitational attraction, thus keeping the small mass stationary in the rotating frame. As seen in a laboratory-based frame the table accelerates uniformly over some (tolerable) number of revolutions. Clock readings at various table positions indicated by the disc encoder are the raw data of the measurement. The entire apparatus is enclosed in an acoustic chamber about 2.5 meters on an edge and mounted on a reinforced concrete slab of about 5000 kilograms.

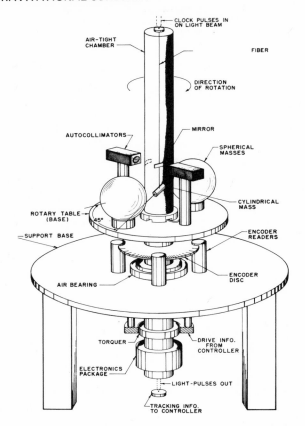

Fig. 1   Schematic of the Apparatus.

Accelerations are measured with the large masses in place and with the large masses removed. The difference between these two accelerations is due to the gravitational torque on the small mass introduced by the large masses. This acceleration may be converted to a value for G by means of Eq. [1].

$$G = \frac{R^3 \, I_T}{Mm} \, \frac{[\alpha_{on} - \alpha_{off}]}{K} \qquad\qquad Eq. \ [1]$$

where

   R = 1/2 distance between the centers of large masses
   I_T = Total moment of inertia of the small mass system
   α = Acceleration of the table
   M = Average mass of the large masses
   m = Mass of the dumbbell
   K = Geometric factors.

CONTRIBUTIONS TO THE ERROR

In order to measure G to 1 part in 10^5, the R in Eq. [1] must be known to 3 parts in 10^6. The uncertainty in the center of mass of the large masses limits the precision of the experiment to about 1 part in 10^5. These masses are the same ones used in the experiment of Beams et al.[1] They were machined from sintered tungsten by the Y-12 plant of Union Carbide. Their characteristics are listed in Table 1.

Table 1

Physical Characteristics of the Finished Tungsten Spheres

Size (inches)	Balance Period (sec)	Mass-Center Dislocation from the Geometrical Center (microinches)	Weight (kgs ± 0.00007)	Average Density (gms/cc)
4.001997	30	181.5	10.489980	19.073914
4.002011	24	298.0	10.490250	19.074866
4.000936	15	726.2	10.445982	19.009009

The time is recorded by counting 10 microsecond pulses from an NBS in-house frequency standard.

The positions of the large masses on the rotating table are determined by comparison with two sapphire rods epoxied to the rotating table, which are in turn compared to gauge blocks whose length determination has been carried out to better than 3 parts in 10^7.* The rotating table on which the large masses are mounted is a plate whose coefficient of thermal expansion is $1.5 \times 10^{-7}/°C$ and thus, the change in position of the large masses due to temperature is insignificant.

The auto-collimator signal which reflects the small mass system's angular position is converted to a digital signal by battery operated electronics mounted on the rotating system (there is no mechanical connection between the rotating system and the laboratory, all signals being transmitted to and from the rotating system via modulated light beams). A dedicated mini-computer converts this digital signal after amplification, integration and differentiation into an appropriate analog control signal as required to govern the torque.

*We are indebted to C. Tucker for this measurement and for valuable discussions.

The auto-collimator has a 2.5 cm aperture and consequently a diffraction pattern of a few seconds of arc. This diffraction pattern is divided into 10 bits (1024 parts) by the auto-collimator and electronics. The stability of the system is such that in operation the deviation of the small mass system is never more than a few bits with a standard deviation of about 1/3 of a bit under quiet conditions.

Twenty-four times per revolution as determined by the disc encoder, an elapsed time value accumulated from the house standard (as noted above) is stored in computer memory.

DATA REDUCTION

A run usually consists of 8 revolutions of the system (196 recordings of the time) with large tungsten masses on the table, 8 revolutions with the masses removed, and finally, 8 revolutions with the masses back on. Typically, the first revolution takes about half an hour and the 8th revolution is completed after about two hours. The times for each run are converted to an acceleration by a least-squares-fit to an equation of the form of Eq. [2].

$$\theta = C_o + C_1 t + C_2 t^2 + C_3 t^3 \qquad\qquad \text{Eq. } [2]$$

A typical output is shown in Fig. 2.

The meanings of the coefficients are as follows: C_o = the initial position of the table in sectors (1 sector = $2\pi/24$ radians C_1 = the initial velocity of the table in sectors/second. C_2 = the acceleration of the table π_o/sectors/2/second. C_3 = the rate of change (instability) of the acceleration. The residuals shown in column 4, Fig. 2 are the residuals of the position, that is, the difference between the actual position of the table when the mark signal occurs and that position which it would have if the acceleration were uniform. Note that the residuals have a predominantly second-harmonic component, i.e., a period of 12 sectors and that their amplitude is inversely proportional to the number of revolutions, precisely what one would expect for a gravitational gradient horizontally across the room. The acceleration for the two "balls-on" runs are averaged and algebraically subtracted from the acceleration with the balls off to obtain the gravitational acceleration.

In our first measurement we used a small mass system whose weight was about 16 gm and whose moment of inertia was about 5 gm-cm^2. It was supported by a fused quartz fiber and had a torsion period of about 140 seconds. A table of the accelerations is given in Table 2.

BIG 6 4 11 75 BALLS ON RUN 3 5.00

THIS IS TRY NUMBER 1 WITH N = 4 SET NUMBER 1 THE TOLERANCE FOR THIS FIT IS .0053

X	LAB.	CALC.	INT	RESID	WT
1878.23	21.0000	20.9989		.0011	1.00
1927.29	22.0000	21.9976		.0024	1.00
1974.93	23.0000	22.9972		.0028	1.00
2021.25	24.0000	23.9979		.0021	1.00
2066.37	25.0000	24.9993		.0007	1.00
2110.34	26.0000	26.0012		-.0012	1.00
2153.24	27.0000	27.0029		-.0029	1.00
2195.12	28.0000	28.0040		-.0040	1.00
2236.04	29.0000	29.0044		-.0044	1.00
2276.06	30.0000	30.0039		-.0039	1.00
2315.24	31.0000	31.0028		-.0028	1.00
2353.64	32.0000	32.0012		-.0012	1.00
2391.31	33.0000	32.9995		.0005	1.00
2428.32	34.0000	33.9982		.0018	1.00
2464.70	35.0000	34.9975		.0025	1.00
2500.49	36.0000	35.9974		.0026	1.00
2535.71	37.0000	36.9979		.0021	1.00
2570.39	38.0000	37.9988		.0012	1.00
2604.54	39.0000	38.9997		.0003	1.00
2638.17	40.0000	40.0004		-.0004	1.00
2671.31	41.0000	41.0008		-.0008	1.00
2703.96	42.0000	42.0008		-.0008	1.00
2736.16	43.0000	43.0004		-.0004	1.00
2767.92	44.0000	43.9997		.0003	1.00
2799.28	45.0000	44.9991		.0009	1.00
2830.24	46.0000	45.9986		.0014	1.00
2860.83	47.0000	46.9986		.0014	1.00
2891.07	48.0000	47.9990		.0010	1.00
2920.96	49.0000	48.9997		.0003	1.00
2950.51	50.0000	50.0005		-.0005	1.00
2979.73	51.0000	51.0013		-.0013	1.00
3008.63	52.0000	52.0019		-.0019	1.00
3037.20	53.0000	53.0020		-.0020	1.00
3065.46	54.0000	54.0017		-.0017	1.00
3093.41	55.0000	55.0011		-.0011	1.00
3121.08	56.0000	56.0003		-.0003	1.00
3148.48	57.0000	56.9993		.0007	1.00
3175.62	58.0000	57.9986		.0014	1.00
3202.50	59.0000	58.9982		.0018	1.00
3229.15	60.0000	59.9982		.0018	1.00
3255.56	61.0000	60.9985		.0015	1.00
3281.74	62.0000	61.9990		.0010	1.00
3307.69	63.0000	62.9996		.0004	1.00
3333.41	64.0000	64.0001		-.0001	1.00
3358.90	65.0000	65.0003		-.0003	1.00
3384.18	66.0000	66.0003		-.0003	1.00
3409.24	67.0000	67.0001		-.0001	1.00
3434.09	68.0000	67.9997		.0003	1.00
3458.75	69.0000	68.9994		.0006	1.00
3507.50	71.0000	70.9991		.0009	1.00
3531.60	72.0000	71.9994		.0006	1.00
3555.53	73.0000	72.9999		.0001	1.00
3579.28	74.0000	74.0005		-.0005	1.00
3602.86	75.0000	75.0010		-.0010	1.00
3626.26	76.0000	76.0013		-.0013	1.00
3649.50	77.0000	77.0014		-.0014	1.00
3672.56	78.0000	78.0012		-.0012	1.00
3695.46	79.0000	79.0007		-.0007	1.00

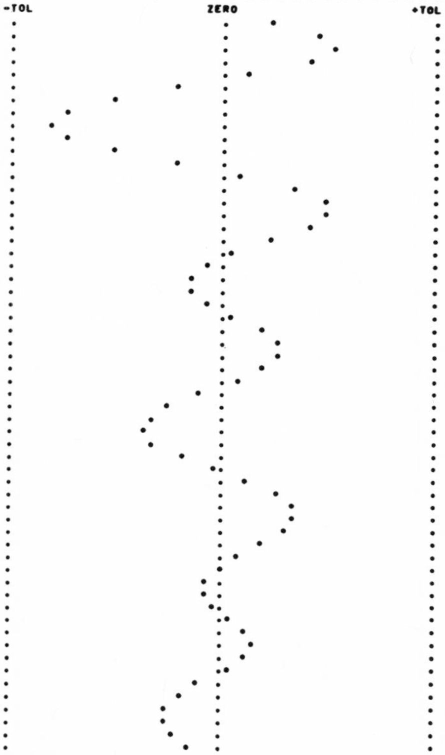

THE C(I) ARE AS FOLLOWS

6.4607668995+000 -4.5562015805-003 6.5491361459-006 -1.1991433063-012

Fig. 2. Computer Print-Out and Plot of a Typical Measurement.

Table 2

Acceleration Values Using the First Fiber and Small Mass System
from Equation 2 with C_3 Set Equal to Zero

Date	Acceleration in Radians/Second^2
3-25-75	4.9684
3-27-75	4.9605
3-31-75	4.9486
4-01-75	4.9594
4-02-75	4.9608
4-09-75	4.9660
4-11-75	4.9476
Average	4.9586

$$\sigma_m = .0042$$

Table 3

Acceleration Values Using the Second Fiber and Small Mass System
from Equation 2 with C_3 Set Equal to Zero

Date	Acceleration in Radians/Second^2
5-06-75	5.69408
5-07-75	5.69414
5-08-75	5.68927
5-13-75	5.69649
5-14-75	5.69335
5-15-75	5.69419
Average	5.6936

$$\sigma_m = .0012$$

The standard deviation of the mean of the results in Table 2 is about 9 parts in 10^4. Most of this error is believed to be caused by instabilities in the fiber's null position. A second small mass system was built with a mass of approximately 6gm and a moment of inertia the same as the original one: 7 gm-cm^2. This reduction in mass allowed a smaller fiber to be used whose torsion period is 360 seconds. The reduction in the mass while preserving moment of inertia was accomplished by making the bob in the shape of a dumbbell instead of a cylinder. The increase in the torsion period was due to the combined effect of the mass reduction and the use of extremely pure quartz for the torsion fiber. A typical set of accelerations for this new system are shown in Table 3.

The gravitational constant as calculated from the above acceleration is

$$6.6699 \times 10^{-11} \quad N.m^2kg^{-2}$$
$$\pm .0014$$

where the error includes only the statistical fluction in the acceleration. This value of G should be viewed with extreme care since it will be re-evaluated when the series of measurements with this fiber is finished and the metrology on the small mass system is completed.

The best fiber so far which supports the small mass system has a period of 1800 seconds which is expected to improve this result significantly. This fiber has not been installed to date.

Further work on this experiment is being planned using various small masses and other fibers, perhaps filamentary carbon fibers.

References

1. R.D. Rose, H.M. Parker, R.A. Lowry, A.R. Kuhlthau, and J.W. Beams, "Determination of the Gravitational Constant, G," Phys. Rev. Lett. <u>23</u>, 655 22 Sept. 1969 .

CONSTANTS OF ELECTRIC AND

MAGNETIC POLARIZABILITIES OF PROTON

V.I.Goldanskii (Institute of Chemi-
cal Physics) and V.A.Petrun'kin
(P.N.Lebedev Physical Institute)
II7334, Vorobjevskoje Shausse, 2-b
Academy of Sciences of the USSR,
Moscow

I. Introduction

One of the main problems of elementary particles physics is the study of the structure of such particles, e.g. of the electromagnetic structure of strong interacting particles - so called hadrons.

The experimental study of electromagnetic structure of hadrons was started by well-known investigation of elastic electron-proton scattering (Hofstadter [I]).

The measurements of differential cross-sections of (ep) scattering over a wide interval of angles and energies can lead to the determination of electromagnetic form-factors of proton, i.e. to the finding of spatial distribution of protons electric charge and magnetic dipole moment.

In such a way it was determined the size of proton, i.e. the mean square radius of electric charge distribution in proton $\langle r_e \rangle^2_p \approx 0.8 . 10^{-13}$ cm.

However, the (ep) scattering alone can't give a comprehensive information on the proton structure. For instance, the electromagnetic form-factors of proton by themselves give no information concerning the mobility of charge "components" within the proton, concerning the rigidity of bonds between such "components".

Recent investigations have manifested that the sizes of such particles as proton and pion are quite close to each other ($\langle r_e \rangle^2_\pi \approx 0,8 . 10^{-13}$ cm), although the theo-

retical ideas on their structure differ considerably
E.g., the proton is assumed to consist of three
quarks, pion-of only two, with a strong difference
in binding energies.

On the base of classical physics one can guess
that the elastic (γP) scattering can give necessary information, supplementary to what is obtained from (ep) data.

Indeed, the study of Rayleigh scattering of
light by atoms and molecules is a way to the determination of such structural characteristics, e.g. the
constants of polarizability.

It can be strictly proved, that at the scattering of low-energy photons on protons, the structure
of targets manifests itself starting from the quadratic (in frequency) terms of scattering amplitude.

Frequency-independent and linear terms depend
only on the charge, mass and anomalous magnetic moment of proton[2, 3].

Quadratic terms depend on two new structural
constants, which were called (by analogy with corresponding non-relativistic expressions) correspondingly the electric and the magnetic polarizabilities
- α_p and β_p [4,10]. First crude theoretical estimates of α_p and β_p -constants for proton were made
already ca. fifteen years ago [5,6]: $\alpha_p \approx 10^{-42}$ cm^3,
$\beta_p \approx 10^{-43}$ cm^3.

Let's note here, that such estimates are quite
different from what could be expected from the analogies with atomic physics. For instance, the α_p estimate is by at least two orders of magnitude less,
than its expected at the first glance value of
($\langle r_e^2 \rangle_p^{1/2}$)^3 . Furthermore, the β_p and α_p estimates are close enough, although in atomic and nu-
clear systems magnetic polarizabilities are less
than electric by several orders of magnitude.

II. The description of experiment.

The experimental determination of constants of
electric and magnetic polarizabilities of proton is
an important but complicated problem. The expected
contribution of electric polarizability of proton
to the differential cross-section of (γP) -
scattering at 60 and 100 MeV is $\sim 10\%$ and $\sim 20\%$
respectively. The contribution of magnetic polarizability is considerably weaker. Therefore one needs
to measure the (γP) - scattering cross sections
with the precision of several per cent. Because of
very small value of measured cross-section

($\sim 10^{-32}$ cm^2), the performance of such precise
measurements in the presence of strong background
is connected with considerable experimental difficulties.

The first experimental data for the cross-sec-
tion of (γp) and α_p, β_p, were obtained as early as
fifteen years ago, at the synchrotron of Lebedev
Physical Institute in Moscow [7].

Recently such measurements were repeated at the
same accelerator, at the average γ -energy ca. 95
MeV [8].

In these recent experiments the 265 MeV synchrotron was used with the maximum energy of bremsstrahlung beams equal to 127 and 148 MeV, and the scattered γ -rays were detected at the angles of 90° and
150°.

The observed effect was larger in the recent
experiments because of the higher energy of γ -rays,
but that also lead to a larger theoretical uncertainty in determination of α_p and β_p -constants.

In both earliest and recent experiments the study of (γp) scattering was performed by the detection of only one particle -the scattered photon.

Direct ways of absolutization of obtained cross-
sections within the precision of several per cent are
quite difficult because of various sources of errors,
typical for works with the bremsstrahlung beams. We
can mention in this connection the problems of determination of the shape of bremsstrahlung spectrum, of
its absolute intensity, of the efficiency of photons
detectors etc.

Most of such errors could be eliminated by using
the well-known Compton scattering of γ -rays by electrons as a monitoring process. In earliest experiments
[7] such process was used for the calibration of a
telescope of counters, detecting the photons scattered by protons. In recent experiments [8] the same
device was used for consequent observations of the
studied (γp) and monitoring (γe) processes.
The differential cross-section of monitoring process
was measured at the angle 146°, which corresponded
to the best coincidence of the energies of γ -rays
scattered by electrons and - at the angle of 90°-
by protons. The details of experiments can be found
elsewhere [7,8] . These experiments included the measurements of γ -counting rates from the empty evacuated target, the target filled with liquid hydrogen and
the background.

Knowing the ratios of counting rates for the studied (B_γ) and monitoring (B_γ^o) process and the cross-section of the latter process, one can determine the cross-section of (γp) - scattering:

$$\frac{d\sigma}{d\Omega} = \left(\frac{d\sigma}{d\Omega}\right)^o \frac{B_\gamma J_\gamma^o}{B_\gamma^o J_\gamma} \tag{I}$$

where J_γ and J_γ^o are the phase integrals over the target volume, solid angle of γ -detection and the energy of primary photons, with the account to the spectrum of bremsstrahlung and energy dependence of response of γ - detecting telescope. J_γ and J_γ^o values were calculated by computer using the method of statistical probes. The $(d\sigma/d\Omega)^o$ cross-section was taken from Klein-Nishina-Tamm formulae, with the corresponding radiation corrections (both eigencorrections and the corrections for the radiation of an additional high-energy photon- these latter corrections were necessary because ca. 40 MeV γ -rays were detected in the experiments). The absolutization of data based on their comparison with the monitoring process was supplemented by a direct absolutization. The cross-sections obtained on this latter way contained an additional systematic uncertainty of ca. 8% because of the errors in the maximum energy of electron beam (2%), in the bremsstrahlung flux (3%) and in the efficiency of γ -detecting telescope (7%).

However, the cross-sections, obtained by both ways, were in agreement within the limits of errors.

Table II represents the experimental values of (γp) - cross-sections, absolutized by corresponding (γe) - cross-sections. The given errors take into account both random errors of measurements and treatment and the small ($\sim$ I%) systematic error of absolutization of (γp) -cross-sections of monitoring process. Numbers I and II in the tables mark different variants of complectation of γ -detecting telescope for maximum γ -energies of I27 MeV and I48MeV respectively (an additional absorber of 8,I2 g.cm^{-2} was introduced in the second case).

The values of (γp) -cross-sections obtained in Ref. 8 are in good agreement with the much earlier results of Ref. 7, and other results (see the survey [9]) having at the same time much lesser total errors.

III. The determination of constants of electric and magnetic polarizability.

The theoretical expression for the differential cross-section of elastic (γP) - scattering in laboratory system can be represented as an expansion by primary photon energy ω ($\hbar = c = 1$) [I0 - I2] :

$$\frac{d\sigma}{d\Omega} = \left(\frac{d\sigma}{d\Omega}\right)^P - \frac{e^2}{M}\omega^2\left[\alpha_P\left(1+\cos^2\theta\right)+2\beta_P\cos\theta\right]\left[1-3\frac{\omega}{M}\left(1-\cos\theta\right)\right]+0.\omega^4 \quad (2)$$

The first term in (4) is the so-called Powell cross-section of the elastic scattering of γ -ray on a point particle with $S = \frac{1}{2}$ -spin. It looks like [I3]:

$$\left(\frac{d\sigma}{d\Omega}\right)^P = \frac{1}{2}\left(\frac{e^2}{M}\right)^2\left\{\left[1-2\frac{\omega}{M}\left(1-\cos\theta\right)+3\left(\frac{\omega}{M}\right)^2\left(1-\cos\theta\right)^2 - \right. \quad (3)$$

$$\left. -4\left(\frac{\omega}{M}\right)^2\left(1-\cos\theta\right)^3\right]\left(1+\cos^2\theta\right)+\left(\frac{\omega}{M}\right)^2\left[\left(1-\cos\theta\right)^2+f(\theta)\right]\left[1-3\frac{\omega}{M}\left(1-\cos\theta\right)\right]\right\}$$

where
$$f(\theta) = a_0 + a_1\cos\theta + a_2\cos^2\theta$$
$$\omega_0 = 2\lambda + \frac{9}{2}\lambda^2 + 3\lambda^3 + \frac{3}{4}\lambda^4 \quad (4)$$
$$a_1 = -4\lambda - 5\lambda^2 - 2\lambda^3$$
$$a_2 = 2\lambda + \frac{1}{2}\lambda^2 - \lambda^3 - \frac{1}{4}\lambda^4$$

Above, in (2), (3), (4), θ is the scattering angle, λ -the anomalous magnetic moment of proton (in nuclear magnetons), e, M -the charge and the mass of proton respectively.

The second term in (2) depends on two new constants - α_P and β_P , which characterize the structure of proton.

The estimate of the amplitude of forward γP -scattering, based on the dispersion Kramers-Kroenig relation [I4] has shown, that the contribution of terms neglected in (2) is less than 2% for $\omega \sim$ I00 MeV and $\theta = 0^\circ$. Due to the contribution of pole diagram with the exchange by π° -meson in the t -channel, it can be larger for other angles but strict estimates for such case are still absent. The higher is the energy of primary γ -rays, the larger is the contribution of terms neglected in (2).

The equation (2) was used for the approximation of experimental data from the Table I. As a result, the following values of α_p and β_p -constants were obtained [8]:

$$\alpha_p^{exp} = (10,7 \pm 1,1) \cdot 10^{-43} \text{ cm}^3 \qquad (5)$$
$$\beta_p^{exp} = (-0,7 \pm 1,6) \cdot 10^{-43} \text{ cm}^3$$

and, correspondingly:

$$(\alpha_p + \beta_p)^{exp} = (10,0 \pm 2,3) \cdot 10^{-43} \text{ cm}^3 \qquad (6)$$

The errors in (5) and (6) are determined by the full errors in experimental values of (γp) cross-sect-ions and don't include the uncertainty (ca. $\pm 1 \cdot 10^{-43}$ cm^3) connected to the contribution of terms neglect-ed in (2).

The error in (6) is given with the account to the correlation of α_p^{exp} and β_p^{exp} -values.

Let us compare the sum of α_p^{exp} and β_p^{exp} with the theoretical value, given by the sum rule [5, 15]:

$$\alpha_p + \beta_p = \frac{1}{2\pi^2} \int_{\omega \, thresh.}^{\infty} \frac{\sigma(\omega)}{\omega^2} \, d\omega \qquad (7)$$

where $\sigma(\omega)$ is the total cross-section of hadronic photoabsorption on proton.

Using the experimental data on the $\sigma(\omega)$ up to $\omega = 30$ GeV and the usual Regge-type extrapolation above this energy [16,17], one gets

$$(\alpha_p + \beta_p)^{theor.} = (14,1 \pm 0,3) \cdot 10^{-43} \text{ cm}^3 \qquad (8)$$

In such a way, there is a small (ca.two standard errors) exceeding of theoretical sum ($\alpha_p + \beta_p$) over the experimental one (the difference equals to $(4,1 \pm 2,3) \cdot 10^{-43}$ cm^3). If we take this difference as really exis-ting, we should keep in mind, that it is larger than the contribution to the ($\alpha_p + \beta_p$)$^{theor.}$ from the possible fixed pole with $J = 2$, which is limited from above by the value of $0,25 \cdot 10^{-43}$ cm^3 [18,19].

The experimental value of α_p^{exp}, obtained in Ref. 8, is in agreement with the first experimental value of $\alpha_p^{exp} = (9 \pm 2) \cdot 10^{-43}$ cm^3 found in Ref. 7.

Unfortunately, there are no completely definite theoretical predictions for α_p and β_p values.

Results of various dispersion-type calculations are given below:

α_p^{theor} x 10^{43} cm^3 10,4   4   12,8   16,5

β_p^{theor} x 10^{43} cm^3 0,7   10   1,3   -2, 5

Reference   20   21   22   22
                                    (I variant)(2 variant)

In the Ref.20 there were used dispersion relations for S at fixed t[*], for six different invariant amplitudes of γp -scattering $A_i(S, t)$, and the unknown substraction functions in dispersion relations were replaced by corresponding Born expressions. It can be shown, that from its very beginning such approximation assumes, that β_p is small in compare to α_p .

In the Ref.21 only two dispersion relations were used - correspondingly for averaged by spin forward and backward amplitudes of scattering, but the calculation of α_p and β_p was performed under the complete neglect of the contribution of t -channel.

In the Ref. 22 the substraction functions in dispersion relations for S at fixed t were calculated with the use of several assumptions, some of them seem to be quite dubious.

Therefore more consistent calculations of the constants of electric and magnetic polarizabilities of proton are necessary. There are also quite desirable more precise investigations of the elastic γp-scattering below $\omega \approx 70$ MeV.

In conclusion there should be recalled some recent theoretical estimates of the constants of electric and magnetic polarizabilities of pions and kaons [23, 24].

Their absolute values are expected to be several times larger, than for proton. The improvement of the precision in the determination of γ -energies for the transitions between high excited levels of π - and K - mesoatoms should open the possibility of an experimental determination of polarizabilities of mesons[25].

The problem of an experimental investigation and theoretical treatment of the constants of electromagnetic polarizabilities of other elementary particles of various classes seems to be one of interesting and urgent problems of fundamental constants in the elementary particles physics.

---

[*]S - the square of the total energy in CMS
t - the square of transferred momentum

References

I. R.Hofstadter. Ann. Rev. Nucl. Sci, 7, 231(1957)
2. F.Low.Phys. Rev. 96, 1428(1954)
3. M.Gell-Mann, M.Goldberger, Phys. Rev. 96, 1433(1954)
4. A.Klein.Phys. Rev. 99, 988 (1955)
5. A.M.Baldin.Nucl. Phys. 18, 310(1960)
6. V.S.Barashenkov et al.Nuovo Cim., 20, 593(1961)
7. V.I.Goldanskii et al. Nucl. Phys. 18, 473(1960)
8. P.S.Baranov et al., Phys. Lett (B), 52, 122(1974)
9. P.S.Baranov et al., Fortschr. d.Phys. 16, 595(1968)
10. V.A.Petrun'kin. ZhETF 40, 1148 (1961)
11. V.Barashenkov et al.Phys. Lett.2,33(1962)
12. V.A.Petrun'kin. Trudy FIAN 41, 165 (1968)
13. J.Powell.Phys. Rev.75, 32(1949)
14. M.Gell-Mann et al.Phys.Rev. 95, 1612(1954)
15. L.I.Lapidus. ZhETF 43, 1358 (1962)
16. P.Joos.Preprint DESY-HERA, 70-I, Sept. 1970
17. A.Belousov et al. Doklady AN SSSR, 215, 76(1974)

18. S.Drell.Comments Nucl. and Part. Phys. I,196(1967)
19. J.Walker. Phys. Rev. Lett.21, 1618(1968)
20. V.K.Fedyanin. ZhETF 44, 633 (1963)
21. J.Bernaben et al. Phys. Lett.(B), 49, 381 (1974)
22. D.I.Akhmedov, I.V.Filkov. Kratkiye soobshcheniya
    po fizike, FIAN,NI, 13(1975).
23. M.V.Terent'ev. Pis'ma ZhETF 16, 628(1972)
24. M.K.Volkov, V.N.Pervushin. Preprint. OI.Ya.I,
    Dubna, P2-8165 (1974)
25. H.Iachello, A.Laude. Phys. Lett.(B) 35, 205(1971)

TABLE I

The monitor process (γe) theoretical cross-section at the angle $147°$

The different gamma-ray counter sets	$\left(\dfrac{d\sigma}{d\Omega}\right)$ 10^{-26}cm^2/st	Radiative and double Compton: Scattering Corrections %	$\left(\dfrac{d\sigma}{d\Omega}\right)$ 10^{-26}cm^2/st (the corrections are included)
I	6.89	+ 3.7	7.15
II	6, 92	+ 4.0	7.19

TABLE II

The experimental results on the photon elastic scattering by protons

The different gamma-ray counter sets	θ Lab.Syst. (degree)	ω (Mev) Lab. Syst.	The ratio of the principal to the monitor process cross-sections 10^{-7}	$\left(\dfrac{d\sigma}{d\Omega}\right)$ 10^{-32}cm^2/st Lab.Syst.
I	90	85.4	1.52 ± 0.05	1.09 ± 0.04
II	90	80.9	1.60 ± 0.09	1.15 ± 0.05
I	150	86.3	1.92 ± 0.14	1.37 ± 0.10
II	150	81.9	2.02 ± 0.17	1.44 ± 0.12
I	90	109.9	1.44 ± 0.07	1.03 ± 0.06
I	150	111.1	2.02 ± 0.06	1.44 ± 0.06
II	150	106.7	2.22 ± 0.09	1.60 ± 0.08

A NEW DETERMINATION OF THE GAS CONSTANT BY AN ACOUSTIC TECHNIQUE

T.J. Quinn, A.R. Colclough, and T.R.D. Chandler

National Physical Laboratory

Teddington, Middlesex, England

ABSTRACT

A new determination of the gas constant has been made using a method in which the velocity of sound in argon was measured by means of an acoustic interferometer operated close to the triple point of water. Ninety eight independent velocity measurements were made over a pressure range from 30 to 200 kPa to enable the velocity at zero pressure to be obtained by extrapolation. Using this experimental value, a new value for the gas constant of 8316.00 ± 0.17 J K^{-1} kmole^{-1} was deduced. This is higher than the old value for R, obtained from measurements of the density of oxygen, by 191 ppm.

Prior to the present work our knowledge of the gas constant, R, was derived from several determinations based on the measurement of limiting density.[1-3] Reliable estimates of systematic and statistical uncertainty for these results are difficult to make in retrospect[4] though they are all likely to suffer from the same principal sources of experimental error — those associated with the problems of measuring extensive properties of gas samples, notably mass and volume. It was felt, therefore, that an entirely independent evaluation would be useful where every effort was made to assess systematic errors.

An acoustic technique was chosen where the velocity of sound, c, in a gas was measured at ever lower pressures so that an acoustic isotherm of c^2 versus pressure, p, could be plotted.

An ideal gas value, c_0^2, of c^2 could be obtained by extrapolation to zero pressure and the gas constant could then be calculated from the relation

$$R = Mc_0^2/\gamma T \tag{1}$$

where T represents the thermodynamic temperature of the gas, γ the ratio of its principal specific heats and M its molecular weight. This method is free of the systematic errors associated with the direct measurement of density, all quantities appearing in equation 1 being intensive ones. For a monatomic gas γ takes the theoretical value 5/3 while the value of M may, for an isotopically analysable gas, be obtained to an accuracy far in excess of that to which c_0 may be measured. In order to avoid problems of absolute temperature determination, we chose to carry out the velocity measurements close to the triple point of water where the thermodynamic temperature is, by definition, 273.16 K.

For the measurement of c a low frequency variable path acoustic interferometer was constructed as shown in Reference 4. Ambiguities as to exactly what velocity was being measured were avoided by working at a frequency below the cut-off frequency of the first higher mode of propagation in the cavity so that only the plane wave could propagate. It was mainly for this reason, that ultrasonic techniques were avoided. The acoustic wavelength was obtained from the separations of the positions of the moving reflector which brought the cavity into resonance. These separations were measured by an optical (laser) interferometer located isothermally in the acoustic interferometer unit. The velocity of sound was then calculated from the frequency (5.6 kHz) of the frequency standard used to drive the transducer which excited the cavity. The latter was a moving coil driven diaphragm whose mechanical admittance (i.e. reciprocal impedance) was directly obtained in arbitrary units in terms of the signal from a small piezo electric accelerometer attached to its rear face. The variation in admittance, which was a measure of the variation in loading of the transducer by the gas in the cavity, was plotted in the complex plane with the aid of phase sensitive detectors. A series of resonance circles resulted which enabled acoustic absorption coefficients and the velocity of sound to be determined very precisely.[5]

One of the principal systematic errors in this technique is due to the acoustic boundary layer which arises from the effects of viscosity and thermal conductivity upon the boundary conditions at the wall of the cylindrical cavity. It leads to a decrease in the observed velocity of sound of 0.1% to 0.3% at the pressures and frequency used in this work. Associated with this velocity decrease, there is a boundary layer absorption coefficient which is higher than the normal "classical" absorption coefficient by

TABLE I. The Procedure for Calculating the Value of the Gas Constant and Correcting for Systematic Effects in the Measurements

Measured Quantity	Correction or Source of Intrinsic Uncertainty	Uncertainty in Correction	Resulting Uncertainty in c_0^2 and R		Intrinsic Systematic Uncertainty in Quantity
			Systematic	Random	
Absorption Coefficient α	Calculate and subtract classical absorption, α_o, to give boundary layer absorption, α_{KH} $\alpha_{KH} = \alpha - \alpha_o$, $\alpha_o/\alpha = 0.25\% - 1\%$	< 10% α_o	< 0.5 ppm		
	Instrument Correction 1.71% α	0.18% α	12 ppm		
	Fit α_{KH}^{-1} vs. $p^{\frac{1}{2}}$ and calculate smoothed boundary layer correction $\Delta c/c = \alpha_{KH}/k$, $k = 2\pi/\lambda$ Standard Error in α_{KH} = .064% at all pressures				
Velocity c	Correct for temperature drift during meast. (typically 1 mK) and for error in triple point realisation (typically 7 mK)	1 mK	4 ppm		
	Correct for boundary layer using α_{KH}^{-1} fit. $\Delta c/c = 1.1\times10^{-3}$ at 200 kPa and 2.7×10^{-3} at 30 kPa			4 ppm*	
	Correct boundary layer theory $\Delta c/c = 3.4 (\Delta c/c)^2$	0.34 $(\Delta c/c)^2$	3 ppm		
	Correct optical meast. for refractive index of Ar	1 ppm	2 ppm		

TABLE I (Contd)

Measured Quantity	Correction or Source of Intrinsic Uncertainty	Uncertainty in Correction	Resulting Uncertainty in c_o^2 and R		Intrinsic Systematic Uncertainty in Quantity
			Systematic	Random	
Fit acoustic isotherm with quadratic: $c^2 = c_o^2 + A_1 p + A_2 p^2$ (95 degrees of freedom) $c_o^2 = (94772.2 \pm 2.0) m^2 s^{-2}$ (4 ppm error * above included) $A_1 = (0.46 \pm 3.69) \times 10^{-5}\ m^2 s^{-2} Pa^{-1}$ $A_2 = (7.95 \pm 1.55) \times 10^{-10}\ m^2 s^{-2} Pa^{-2}$ (standard errors quoted)					
Velocity c_o	Raise value to allow for error in 1st order theory of interferometer $\Delta c/c = (\alpha/k)^2$ $\Delta c_o/c_o = 20$ ppm				
	Result $c_o^2 = (94774.1 \pm 2.0)\ m^2 s^{-2}$ (standard error quoted)				
Molecular Weight, M	Uncertainty due to uncertainty in isotopic constitution and atomic weights of individual isotopes (2×10^{-4} g/mole)				5 ppm
	Impurities < 10 ppm Nitrogen				2 ppm
	Result $R = 8316.00 \pm 0.17\ JK^{-1} kmole^{-1}$ (standard statistical error quoted)				
	If systematic uncertainties are combined in quadrature an overall systematic uncertainty of 14 ppm results.				

corresponding factors of 100 to 400. A first order theory of the effect due to Kirchoff and Helmholtz[6,7] has been further investigated by a number of workers[8,9,10] and, in particular, by Lee, Shields and Wiley[11] who have assessed in detail the effect of its approximations and recommended further comparatively minute corrections at the 10 ppm level. Detailed investigations were carried out on the boundary layer absorption coefficient and velocity change with regard to their dependence on frequency and pressure. They were found to conform to theoretical expectations to within the limits of observational accuracy (which was not, however, quite equal to the accuracy to which we needed to use the boundary layer relations). The theory allows both the velocity change and the absorption coefficient to be calculated from such values of gas viscosity and thermal diffusivity as are available. However, we chose not to rely on the accuracy of these values and the detailed theoretical relations in which they appear. Instead the velocity correction at a given pressure was calculated directly in terms of a measured value of absorption coefficient to which it is intimately related at a more fundamental level in the theory. Nevertheless, calculated values were found to agree with measured values. The corrections were themselves corrected by the minute amounts recommended by Lee, Shields and Wiley. (Table I)

In choosing a gas for the measurement we had to take into account ease of isotopic analysis, interferometer sensitivity ($\propto$ density, ρ, through the specific acoustic impedance, ρc) and the smallness of the boundary layer velocity change ($\propto \rho^{-\frac{1}{2}}$). The first criterion indicated a light gas whilst the latter two indicated a dense one. After considering a number of monatomic gases, it became apparent that argon was the only really suitable one. Purity was always monitored with a mass spectrometer.

The final acoustic isotherm is presented in Figure I and consists of 98 points taken between 30 kPa and 200 kPa pressure. Some absorption coefficients were measured at even lower pressures, but velocities could not be obtained with a useful accuracy below 30 kPa. Those taken between 30 and 65 kPa were given weightings of $w^2 = 0.25$ to allow for their increased scatter. Smoothed boundary layer velocity corrections were made using values of absorption coefficients fitted as an appropriate function of pressure. Theoretically one expects the pressure dependence of c^2 to be given by an expansion of pressure terms of the form

$$c^2 = \frac{\gamma RT}{M} + A_1(T)p + A_2(T)p^2 + \ldots \qquad (2)$$

where $A_1(T)$, $A_2(T)$, $\ldots$ are referred to as the second, third, $\ldots$ acoustic virial coefficients respectively by analogy with the virial coefficients proper from which they arise. It was found that no improvement in the quality of the fit could be obtained by raising its order above the second and so c_o^2 was obtained as the intercept of a least squares quadratic representation. The statistical uncertainty quoted in the gas constant is that associated with this fit combined in quadrature with that arising from the fitting of the absorption coefficients. A list of corrections made to the measured velocities are presented in Table I together with our estimation of systematic and statistical uncertainty. In Table II we quote our value of the gas constant together with the previously accepted values from which it differs by 191 ppm — considerably more than can be explained by combining our random and systematic uncertainties with the random uncertainty normally quoted for the old value. No estimate of systematic uncertainty exists for the old value. Also shown are values of Boltzmann's constant, the second radiation constant of Planck's law and Stefan's constant as calculated from our new value of R.

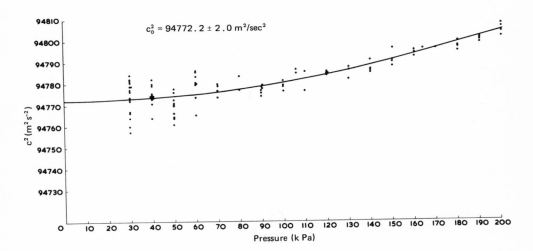

FIGURE 1.   The Isotherm at the Triple Point of Water

REFERENCES

1.   BATUECAS, T. and GARCIA MALDE, G., 1950, An. Soc. Esp. Fis. Quim., 46, 517.
     BATUECAS, T., 1952, ibid., 48, 4.
     BATUECAS, T., 1972, "Atomic Masses and Fundamental Constants-4", 534. Ed. Sanders, J.H. and Wapstra, A.H., London-New York: Plenum.

2.   BAXTER, G.P. and STARKWEATHER, H.W., 1924, Proc. Nat. Acad. Sci., 10, 479.
     BAXTER, G.P. and STARKWEATHER, H.W., 1926, ibid., 12, 699.

3.   MULES, E., 1938, "Collections Scientifique d'Institute Internationale Cooperative Intellectual", 1-75, Paris.

4.   QUINN, T.J., 1972, "Atomic Masses and Fundamental Constants-4", 529, Ed. Sanders, J.H. and Wapstra, A.H., London-New York: Plenum.

5.   QUINN, T.J., COLCLOUGH, A.R. and CHANDLER, T.R.D., To be published.

6.   HELMHOLTZ, H., 1863, Verhandl. naturhist. med. Uer. Heidelberg, 3, 16.

7.   KIRCHOFF, G., 1868, Ann. Physik, 134, 177.

8.   HENRY, P.S.J., 1931, Proc. Phys. Soc., 43, 340.

9.   WESTON, D.E., 1973, Proc. Phys. Soc. B66, 695.

10. FRITCHE, L., 1960, Acustica, 10, 199.

11. SHIELDS, F.D., LEE, K.P. and WILEY, W.J., 1965, J. Acoust. Soc. Amer., 37, 724.

TABLE II

The New Value of R and Related Quantities

Quantity	Value	Uncertainty	
		Statistical	Systematic
New Value of R	8316.00 $JK^{-1}kmole^{-1}$	20 ppm	9 ppm
Old Value of R	8314.41 $JK^{-1}kmole^{-1}$	31 ppm	unknown
Boltzmann's Constant	$1.38093 \pm 0.00003 \times 10^{-23}$ JK^{-1}	20 ppm	10 ppm
Second Radiation Constant of Planck's Equation $c_2=hc/k$	$1.43851 \pm 0.00003 \times 10^{-2}$ mK	20 ppm	11 ppm
Stefan-Boltzmann Constant $\sigma=2\pi^5k^4/15h^3c^3$	$5.6747 \pm 0.0005 \times 10^{-8}$ $Wm^{-2}k^{-4}$	80 ppm	44 ppm

Systematic errors have been calculated by combining in quadrature errors of 9 ppm in R, 5 ppm in Avogadro's number and 5 ppm in Planck's constant.

A DETERMINATION OF THE DENSITY AND DILATATION OF PURE WATER OF KNOWN ISOTOPIC COMPOSITION

G.A. Bell and A.L. Clarke

National Measurement Laboratory, CSIRO

University Grounds, Chippendale, NSW, Australia 2008

INTRODUCTION

For over a century water has been the most commonly used working standard of density and its history has been recently reviewed by Menache and Girard (1). The basic measurements of the maximum density of water were made at the BIPM (2) and the dilatation measurements were made by Chappuis (3) and by Thiesen et al. (4). Despite the discrepancies between their results it is remarkable that no fewer than four sets of tables based on this work have been published between 1937 and 1973. A fifth table has recently been published by Kell (5).

The only new experimental work since 1907 was by Steckel and Szapiro (6) whose main interest was in waters which were enriched with the heavy isotopes ^2D and ^{18}O and whose work was not of sufficient accuracy to be of real significance in resolving the differences between the earlier measurements. Their work is important as they showed that variations in isotopic composition of natural water have no significant effects on its dilatation.

EXPERIMENTAL METHODS

In the establishment of water as a standard of density there are two distinct measurements to be undertaken.

(1) The absolute density (in kg/m^3) of a sample of pure water of known isotopic composition at its temperature of maximum density.

(2)   The thermal expansion (dilatation) over a suitable
temperature range (e.g. $0°C$ to $40°C$) of naturally occurring
water.

In (1) the requirement is to measure the mass of an accurately
known volume of water at a temperature of approximately $4°C$.

The dilatation may be measured by making absolute measurements
over a range of temperatures or by some differential technique
such as a magnetic or pressure controlled float in which only
changes in density are measured.

Largely because of availability of equipment and familiarity
with the techniques it was decided to use the method of hydrostatic
weighing in which the buoyant force of the water on an object of
accurately known volume is measured by suspending the body in the
liquid from one arm of a balance.

THE STANDARD VOLUME

This must be a body of such shape and perfection of form that
its volume can be computed from its measured linear dimensions
with an uncertainty less than 1 in 10^6.

The form chosen was a hollow sphere made of Corning ULE low
expansion glass. It was made from a cube of glass, 100 mm edge,
which was ground approximately spherical then parted through a
diameter. Each half was then internally ground and polished and
the two annular surfaces were optically polished flat within
200 nm, coated with a special epoxy cement and brought together in
a vacuum at $80°C$. As soon as contact had been properly
established air was admitted to the system which pressed the two
halves of the sphere together with a force of approximately 500 N.
Experiments with the cement indicate that the thickness of the
film is of the order of 2 μm. It is essential that the gas
pressure within the sphere should be very low in order to avoid
expansion of the sphere due to increasing internal pressure when
it is heated.

After the cement had hardened for several days the external
surface was ground and optically polished to an accurately
spherical shape. The external diameter of the finished sphere is
nominally 76.8 mm, its volume at $20°C$ is 228.516 cm^3 and its mass
is 329.616 g. Its weight in water is therefore just over 100 g.

MEASUREMENT OF THE SPHERE

Departures from roundness in a number of diametral planes

were measured using a "Talyrond" roundness measuring instrument
and showed that departures from true sphericity do not exceed
0.1 μm. Preliminary interferometric measurements of the diameter
of the sphere confirm this figure.

The thermal expansion data for the glass was supplied by the
manufacturers. In addition, independent measurements of the
expansion have been made on a block cut from one face of the blank
from which the sphere was made.

The mass of the sphere has been measured on a number of
occasions over a period of a year and it was found that after
months of use in which the sphere was immersed for hundreds of
hours in water its mass had changed by less than 0.1 mg.

THE HYDROSTATIC BALANCE AND BATH

This is a three knife, equal arm balance of capacity 500 g in
each pan. The deflection of the beam is read on a scale 5 m from
the balance and the sensitivity is 0.5 mg for a 20 mm division.

The water sample is contained in a glass U-tube with arms of
unequal diameter. The U-tube is supported from a flat perspex
plate which rests on the upper rim of a large cylindrical perspex
vessel through which is circulated water from a thermostatically
controlled bath. Thus the vessel containing the water sample is
surrounded on all sides except the top by water at constant
temperature. The water sample is stirred by a pyrex glass helical
stirrer located in the smaller arm of the U-tube which is stopped
when a steady temperature has been reached. The balance and
thermostatic bath are shown in figure 1.

TEMPERATURE MEASUREMENT

The basis of the temperature measurement is a standard
platinum resistance thermometer and an ac resistance bridge
(Thompson, 1972) (7). A small platinum resistance thermometer with
a curved sheath is immersed in the water just below the equatorial
plane of the sphere. At each temperature at which measurements
were made, this thermometer was calibrated against the standard
platinum resistance thermometer. Variations in the temperature of
the water are shown on a recorder connected to the bridge output.

MEASUREMENT PROCEDURE

The sphere was held in a cage of stainless steel wire
suspended from the balance by a 50 μm tungsten wire. The level of

Fig. 1. Hydrostatic balance and bath (inset) sphere, cage,
         support and thermometer

the water relative to the sphere was adjusted until the centre of
the sphere was 120 mm below the surface of the water. When the
temperature appeared to be satisfactorily steady the balance was
poised and a rest point taken. As soon as this had been done the
sphere was lifted from the suspension cage and the cage alone was
weighed.

     The sphere was resuspended and a further set of weighings
made. At the time of the weighing readings were made of the
thermometer bridge, and of the temperature, pressure and relative
humidity of the air.

PREPARATION OF WATER SAMPLES

     The measurements of dilatation were made on Sydney tap water
distilled once in a continuous flow still. The conductivity of
the water was measured immediately after collection and was
usually very close to 1 μ siemens. Measurements made at 3.98°C
over a period of several months showed no variation in density as

great as 2 in 10^7 which would indicate that the isotopic composition of the water has remained substantially constant.

RESULTS

Measurements have been made at seven temperatures in the range 3.98° C to 40.1° C. All measurements were made on water in equilibrium with ambient air and at each temperature at least two independently prepared water samples were used.

The sphere was located with its centre 120 mm below the surface of the water and all results have been corrected to a reference atmospheric pressure of 101 325 Pa (760 mm Hg). The density ρ_T at temperature T° C was calculated using the equation

$$\rho_T = \frac{M - W(1 - \frac{\sigma}{D})}{V_T} \{1 + \alpha_T(t - T)\} \times \{1 + \beta_T(760 - p)\}$$

where
M   is the mass of the sphere

V_T is the volume of the sphere at temperature T

W   is the weight of the sphere immersed in water

σ is the density of the air

D   is the density of the weights

α_T is the approximate dilatation of water at T

t   is actual measured temperature

β_T is the compressibility of water at temperature T

p   is atmospheric pressure

In view of the small temperature differences involved in correcting each measurement in a set to a fixed temperature the use of approximate values of α will not lead to errors greater than 1 in 10^7.

The values used for the compressibility β are those given by Kell and Whalley (8).

Since there is a considerable uncertainty (of the order of 5×10^{-6}) in the actual volume of the sphere the results for the dilatation of water are expressed as values of ρ_T/ρ_{MAX} and these are given in Table 1.

The values for air free water were obtained from the measured values using the data of Wagenbreth (9) for the effect of dissolved air on the density of water.

Each of the sets of values has been fitted to a Thiesen type equation

$$\rho_T/\rho_{MAX} = 1 - \frac{(T - T_{MAX})^2}{A} \times \frac{T + B}{T + C} \; .$$

The constants in the equations and calculated values are given with the observed values in Table 1.

TABLE   1

	Water in Equilibrium with ambient air			Air free water		
$T°$ C T_M = 3.985	A = 519 985 B = 299.666 C = 69.223			A = 503 375 B = 283.263 C = 67.335		
	Observed	Calculated	O-C ×10⁶	Observed	Calculated	O-C ×10⁶
3.985	1 000 000 0	1 000 000 0	0	1 000 000 0	1 000 000 0	0
10.4	.999 691 4	691 8	-.4	.999 691 5	691 2	+.3
15.4	.999 066 2	067 0	-.8	.999 065 1	065 5	-.4
20.0	.998 233 4	232 9	+.6	.998 230 9	230 7	+.2
30.0	.995 675 7	675 8	- 0	.995 672 6	672 9	-.3
34.8	.994 128 6	128 6	+.2	.994 125 5	125 5	0
40.1	.992 204 4	204 5	+.1	.992 201 3	201 2	+.1

The calculated values for air free water are compared graphically in figure 2 with those Chappuis and Thiesen and the various reworkings of their data. The results of Steckel and Szapiro are also included. No attempt has been made to adjust any of these results for changes in the temperature scale as it is considered that the link between the temperature scales used by Chappuis and by Thiesen, and IPTS-68 is not sufficiently precise to justify it.

DISCUSSION

An analysis of the uncertainties in the various measurement parameters leads to an estimate of uncertainty for ρ_T/ρ_{MAX} ranging from 0.4×10^{-6} up to $20°$ C to approximately 0.7×10^{-6} at $40°$ C and these estimates are confirmed by a statistical analysis of the

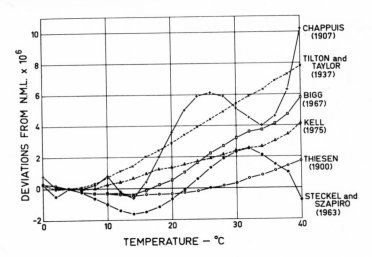

Fig. 2. Deviations of earlier tables from NML values.

measured values. These will be discussed in detail in a paper now in course of preparation

FUTURE WORK

The construction of the equipment for measurement of the diameter of the sphere is well advanced and arrangements have been made for isotopic analysis of the water samples used.

Preparations are also being made to measure the effect of dissolved atmospheric gases on the water density.

REFERENCES

1. Menaché, M. and Girard, G., Metrologia 9 2 (1973).
2. Benoit, J.R., Bur.Int.des Poids et Mesure.Trav. et Mém.
    14 1 (1910)
3. Chappuis, P., Bur.Int.des Poids et Mesure.Trav. et Mém.
    13 D1 (1907)
4. Thiesen, M., Physik Tech.Reichs.Wiss.Abhandl. 3.1;4.1 (1900)
5. Kell, G.S., J.Chem.Eng.Data 20 97 (1975)
6. Steckel, F. and Szapiro, S., Trans.Faraday Soc. 54 331 (1963)
7. Thompson, A.M. and Small, G.W., Proc.IEE 118 11 November (1971)
8. Kell, G.S. and Whalley, E., Phil.Trans.Roy.Soc. London,
    Ser.A 258 565 (1965)
9. Wagenbreth, H., and Blanke, W., PTB-Mitteilungen 6 412 (1971)

RECENT SOLAR OBLATENESS OBSERVATIONS: DATA, INTERPRETATION, AND SIGNIFICANCE FOR EXPERIMENTAL RELATIVITY*

H. A. Hill

University of Arizona

Tucson, Arizona 85721

R. T. Stebbins

High Altitude Observatory

Boulder, Colorado 80303

T. M. Brown

University of Colorado

Boulder, Colorado 80303

I.   INTRODUCTION

The shape of the sun has been under study for almost two and a half centuries, during which time the accuracy of the measurements performed has improved considerably. However, it is only within the last decade or so that these measurements have attained sufficient precision to have a significant impact on solar physics and experimental relativity. At SCLERA,[†] contributions to this improved accuracy have included the use of a daytime astrometric telescope (Oleson et al. 1974) and the development of an explicit definition of the sun's edge (Hill, Stebbins and Oleson 1975). This Finite Fourier Transform Definition (FFTD) of an edge of the solar disk offers extreme sensitivity to changes in the limb darkening function while minimizing the effects of atmospheric and instrumental seeing.

Observations at SCLERA utilizing these techniques reveal a time varying excess equatorial brightness (Hill et al. 1974). That is, a difference between the polar and equatorial limb darkening functions.

This excess equatorial brightness lies at the root of much of the controversy currently surrounding solar oblateness. Its presence vindicates many alternate explanations of previous oblateness measurements, while its time varying character precludes the accurate interpretation of these observations. Thus the properties of the excess brightness are of great interest, and are discussed with reference to the available observations in the following section.

Changes in the solar shape with time have recently become observable, and may also contain significant information concerning solar structure. In particular, normal mode oscillations of the sun would provide a useful probe of the interior. Oscillations with many of the characteristics of normal modes have been observed at SCLERA, first in the 1973 data of Hill and Stebbins (1975a, 1975b) and more recently in observations which are summarized below.

II. EXCESS EQUATORIAL BRIGHTNESS

When the excess equatorial brightness discussed above is present, the SCLERA observations indicate a large apparent oblateness. However, during a period when the excess brightness was small or absent, the observed oblateness, the difference between equatorial and polar diameters, was 18.4 ± 12.5 arc msec (Hill and Stebbins 1975b), in good agreement with the 15.7 arc msec expected from a sun rotating uniformly at the surface rate. Thus some observations indicate, and several excess brightness models which are discussed below suggest that the large oblateness observed by Dicke and Goldenberg (1967, 1974) may also be a manifestation of the excess brightness. The important question is then whether all of the available data can be explained in a unified way.

The data considered here are the Dicke-Goldenberg oblateness results, obtained primarily at 700 nm, and the SCLERA observations of excess brightness and oblateness, both obtained at 550 nm. The Dicke-Goldenberg results appear as oblatenesses observed with three different amounts of limb exposed, while the SCLERA data is best discussed in terms of the quantities $E(a_1,a_2)$ and $R(a_1,a_2)$, which are defined below. In this context a_1 and a_2 represent particular values (in arc sec) of the variable scan amplitude used in the edge definition, and $\Delta(a_i)$ is the observed oblateness at the ith scan amplitude. Finally, we consider a false oblateness $\Delta_E(a_i)$ which is due entirely to the effects of the excess brightness, and which is easily identified because the excess brightness varies with time. Then

$$E(a_1,a_2) \equiv \Delta(a_2) - \Delta(a_1) \tag{1}$$

and

$$R(a_1,a_2) \equiv \Delta_E(a_1)/E(a_1,a_2) \ . \tag{2}$$

A thorough discussion of the properties of these quantities can be
found in Hill, Stebbins and Oleson (1975) and Hill and Stebbins
(1975a). However, for the purposes of this discussion $E(a_1,a_2)$
may be considered a measure of the magnitude of the excess
brightness, while $R(a_1,a_2)$ reflects its intensity variations as
a function of radial position.

    A comparison between the observations and models of excess
brightness can be obtained by considering the ratio $R(a_1,a_2)$, the
oblateness data of Dicke and Goldenberg and the color dependence of
the phenomenon. Two types of models will be considered, differing
in their dependence on u , the radial distance between the point
in question and the nominal solar limb (i.e., $u = r_\odot - r$). In the
first of these, suggested by Ingersoll and Spiegel (1971) and later
based on faculae (Chapman and Ingersoll 1972), the excess brightness
varies as $u^{-\frac{1}{2}}$. The Chapman-Ingersoll facular model will be taken
as representative of this type. The second type includes the two-
layer models (Durney and Roxburgh 1969, Durney and Werner 1971,
Durney 1973, Dicke 1973), which postulate that the equator is hotter
than the poles at small optical depths and cooler at large depths.
The radial dependence of this model is more complicated than that
of the facular model, and was computed using Dicke's (1973) equation (11) and the Bilderberg Continuum Atmosphere (Gingerich and
de Jager 1958). The temperature structure in this model was adjusted
to fit the Dicke-Goldenberg observations at 700 nm, assuming an
intrinsic oblateness of 16 arc msec. Many temperature structures
meet this requirement, for instance

$$\Delta T \ = \begin{cases} +25°K \ , & \tau < 0.02 \\ \\ -3°K \ , & \tau > 0.02 \ , \end{cases} \tag{3}$$

where ΔT is the temperature excess of the equator over the pole
and τ is the normal optical depth at 500 nm. The resulting
excess brightness function $\Delta g(u)$ is shown in figure 1 for two
wavelengths. The fit with the Dicke-Goldenberg data is shown in
figure 2, both at 700 nm, where the fit is quite good, and at
500 nm, where no reliable data is available.

    Using the FFTD formalism (Hill, Stebbins and Oleson 1975) and
the excess brightness functions just discussed, values of $R(a_2,a_1)$

may be computed for both types of models. The results of both
these calculations of R(6.8, a) are plotted in figure 3, along
with the empirical points R(6.8, 27.2) and R(6.8, 54.4). Though
the Chapman-Ingersoll model does not fit the data very well, the
two layer model does. Thus, if the excess brightness changes in
magnitude but not in form, the Dicke-Goldenberg data at 700 nm and
that of SCLERA at 550 nm can all be explained with a two-layer model.

The above calculations provide a modicum of assurance regarding the color dependence of the two-layer model. Unfortunately
this is the best that can be done, since, as noted by Hill and
Stebbins (1975a), virtually all of the published Dicke-Goldenberg
data was taken at 700 nm and the SCLERA observations were all performed at 550 nm.

The above arguments lend considerable weight to the earlier
interpretation of the SCLERA observations. In particular they
strengthen the argument that the oblateness 18.4 ± 12.5 arc msec is
the intrinsic oblateness, uncontaminated by systematic errors. This
value indicates that Mercury's perihelion advance does stand in
strong support of Einstein's General Theory of Relativity. Further
such a small oblateness is incompatible with solar models which
invoke a rapidly rotating core (Demarque et al. 1973, Roxburgh 1974)
to explain the low neutrino flux reported by Davis (1972).

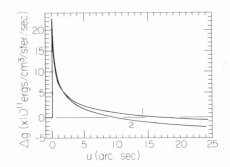 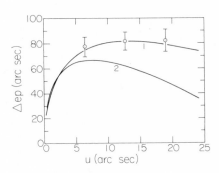

Fig. 1. (left) The radial dependence of the excess brightness
at 700 nm (curve 1) and 500 nm (curve 2) for the two layer excess
brightness model with $\Delta T = +25°K$, $\tau < 0.02$, and $\Delta T = -3°K$ for
$\tau > 0.02$.

Fig. 2. (right) The oblateness data of Dicke and Goldenberg
taken mostly at 700 nm and the predictions of the two layer excess
brightness model at 700 nm (curve 1) and 500 nm (curve 2).

III. NORMAL MODES OF THE SUN

Oscillations have been noticed in the SCLERA data for several years. In the September 1973 oblateness data the oblateness signal was observed to be beating with the sampling frequency at an ampli- tude of ~ 40 arc msec and a period of about an hour. Further, the phase difference between oscillations on consecutive days was observed to be constant for four days. These observations suggested normal mode oscillations, which would be expected to have good spatial coherence and long damping times. The period indicated either a Y_0^0 mode with a period of ~ 26 min. or a quadrupole Y_2^m mode with a period of ~ 13 min., or possibly some combination of the two.

In November 1973, Brailey (1974) undertook a further study of the oscillations by measuring the equatorial diameter every 5 sec for $3\frac{1}{2}$ hours. The power spectra of these data (Brailey 1974, SCLERA 1974) contained several lines which were reasonable candi- dates for the oscillations observed in the September 1973 oblate-

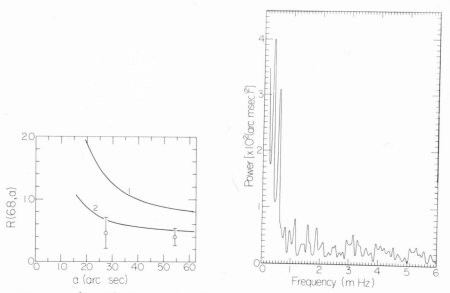

Fig. 3. (left) Excess brightness data obtained at SCLERA in 1972 and 1973, the prediction of the facular model (curve 1), and the prediction of the two layer excess brightness model at 500 nm (curve 2).

Fig. 4. (right) Power spectrum of polar diameter measurements obtained in the spring of 1975.

Table 1

K	$T_{obs}(eq)$ minutes	$T_{obs}(pole)$ minutes	T_{cal} minutes
0	52	47.9	64.04
1	33	30.3	41.08
2	23.8	21.0	31.31
3	16.7	17.1	24.49
4	13.3	14.6	20.20

ness data. The periods of the first five lines are listed in the $T_{obs}(eq)$ column of table 1.

The most recent diameter measurements were made in the spring of 1975. Some twenty hours of data in which the diameter was sampled every 8 sec have been taken and analyzed. This data was broken into five four-hour runs and the power spectrum computed for each run. The average of these five spectra is shown in figure 4 and the periods of its first five lines are listed under $T_{obs}(pole)$ in table 1.

It can be seen that, within the resolution of a $3\frac{1}{2}$-hour data set, the 1973 periods are in good agreement with those obtained $1\frac{1}{2}$ years later on a different part of the sun. Further, both are in broad agreement with the periods of low-order radial oscillations in a standard solar model as calculated by Hansen (1975), and tabulated in table 1 under T_{cal}. If, as seems quite possible, the longest-period radial mode has been overlooked, then the agreement with theory becomes much better. Remaining discrepancies might be attributable to the approximations used in calculating the periods, to intruders in the form of non-radial oscillations, or to both.

From figure 4 an estimate of the lower limit to the Q of the oscillations can be made. This is ≥ 20, which is consistent with both the September 1973 oblateness observation and the theoretical estimates of Wolff (1972).

It is important to note that since the FFTD was used to define the solar edge in these data, the observed amplitudes depend on changes in the limb darkening function as well as on actual mass motions. Consequently, the actual displacement of the surface may be only a small fraction of the observed effect. Presently, this is thought to be the case.

IV. SUMMARY

A more detailed analysis of available oblateness observations has revealed that all results are consistent with a single time-varying model of the excess equatorial brightness. This result (1) strengthens the conclusion that the excess brightness contributes an irremediable systematic error in previous oblateness observations, and (2) places on a firmer basis the oblateness results of Hill and Stebbins (1975b).

Observations of solar oscillations have been taken over the past two year at different regions on the sun, and under different observing conditions. These observations are internally consistent and agree reasonably well with calculated periods for normal modes of the sun. Consequently, the normal mode interpretation is suggested. Should this identification prove to be correct, these modes could have a significant impact on the energy and momentum transport in the sun (Hill et al. 1975), and would in any case prove extremely useful as a probe of the solar interior.

*Work supported by the National Science Foundation
†SCLERA is a research facility jointly operated by the University of Arizona and Wesleyan University.

Brailey, A. C. 1974, M.S. Thesis, University of Arizona.
Chapman, G. A., and A. P. Ingersoll 1972, Astrophys. J., 175, 819.
Davis, R., Jr. 1972, Bull. Am. Phys. Soc. II, 17, 527.
Demarque, P., J. G. Mengel, and A. V. Sweigert 1973, Astrophys. J., 183, 997.
Dicke, R. H. 1973, Astrophys. J., 180, 293.
Dicke, R. H., and H. M. Goldenberg 1967, Phys. Rev. Lett., 18, 313.
_____ 1974, Astrophys. J. Suppl., 27, 131.
Durney, B. R. 1973, Astrophys. J., 183, 665.
Durney, B. R., and I. W. Roxburgh 1969, Nature (Lond.), 221, 646.
Durney, B. R., and N. W. Werner 1971, Sol. Phys., 21, 21.
Gingerich, O., and C. de Jager 1958, Sol. Phys., 3, 5.
Hansen, C. 1975, (Private Communication).
Hill, H. A., P. D. Clayton, D. L. Patz, A. W. Healy, R. T. Stebbins, J. R. Oleson, and C. A. Zanoni 1974, Phys. Rev. Lett., 33, 1497.
Hill, H. A., J. D. McCullen, T. M. Brown, and R. T. Stebbins 1975, submitted to Phys. Rev. Lett.
Hill, H. A., and R. T. Stebbins 1975a, Ann. N.Y. Acad. Sci., in press.
_____ 1975b (Astrophys. J., 200, in press).
Hill, H. A., R. T. Stebbins and J. R. Oleson 1975, Astrophys. J., 200, in press.
Ingersoll, A. P., and E. A. Spiegel 1971, Astrophys. J., 163, 375.
Oleson, J. R., C. A. Zanoni, H. A. Hill, A. W. Healy, P. D. Clayton, and D. L. Patz 1974, Appl. Opt., 13, 206.
Roxburgh, I. W. 1974, Nature (Lond.), 213, 1077.
SCLERA Progress Report, December 1974, unpublished.
Wolff, C. L. 1972, Astrophys. J., 176, 833.

A Laboratory Experiment to Measure the Time Variation of Newton's Gravitational Constant

R. C. Ritter, J. W. Beams, and R. A. Lowry

Department of Physics, University of Virginia

Charlottesville, Virginia   22901

1.   Introduction

The possibility that Newton's gravitational constant, G, varies in time was suggested theoretically by Dirac on numerological grounds.[1)]   Brans and Dicke[2)] have proposed a specific scalar-tensor theory which predicts a variation, with time, in G; and it is also true that most metric theories of gravity[3)] make such a prediction. As evaluated thus far the most serious of these theories predict that $\dot{G}/G \sim 10^{-10}$ to 10^{-11} per year.

Because of the small value predicted for $\dot{G}/G$ experiment has not had any effective bearing on this problem until recently. In fact, no direct laboratory experiment has reported on this, and only a few indirect results from astronomical observations are available.   In particular, the most recent implications from analysis of radar ranging to inner planets[4)] has set $\dot{G}/G = (4 \pm 8) \times 10^{-11}$ per year.   Lunar occultation experiments[5)] have given a value $(-7.5 \pm 2.7) \times 10^{-11}$ per year.   Another result,[6)] the rate of slowdown of the period of pulsar JP 1953, infers an upper limit of 7×10^{-10} per year for $\dot{G}/G$.   The improvement, in time, of the radar-ranging precision is automatic and, thus, offers hope of a serious test at the level of 10^{-11} in a few years.

It is true, regardless of the astronomical results, that a laboratory test of $\dot{G}/G$ at or near the level of 10^{-11} per year would be scientifically of high interest.   Such an experiment is what we will describe.

A fedback Cavendish balance configuration, optimized for sensitivity and stability rather than absolute accuracy, can, in principle, provide the requisite precision.   It would be critically damped, cooled to $\lesssim 1°$ K and would rotate at a constant angular velocity of a few rpm.   The previous experiment which most nearly approaches this

one, in terms of properties and problems, is the Eotvos-type of
experiment by Dicke.[7]

2.   General Features

Large and small dumbells (Fig. 1) provide a gravitational mass
quadrupole–quadrupole interaction. The torsional restraint for
such a system will most likely not be a torsion fiber, thus avoiding questions of solid state stability of the fiber. It would be
desirable, in fact, to use centrifugal forces for the restraining
torque, but we have not found satisfactory configuration for that.
Present plans are to use a magnetic support and an eddy current drag
for the main part of the restraint. This, admittedly, puts electromagnetic forces into the question of any observed variation, but it
seems our best present option. Calculations indicate that special
high-resistance alloys can be made which will have resistence with
the needed stability and temperature independence.

Scaling of the experiment is important in a number of the
aspects of design. The gravitational torque experienced by the small
mass system, in the rotating coordinate frame, is factored into a
scaling factor f and an angle function $F(x,\theta)$:

$$T_g = f\, F(x,\theta); \quad f = 2Gm\, m\, \frac{s_1^2}{\ell_1^3} = \frac{1}{8}\frac{GQq}{\ell_1^5},$$

$$F(x,\theta) = \frac{\sin\theta}{x}\left\{\frac{1}{(1 + x^2 - 2x\cos\theta)^{3/2}} - \frac{1}{(1 + x^2 + 2x\cos\theta)^{3/2}}\right\} \tag{1}$$

where Q = the quadrupole moment of the large mass, q = the quadrupole

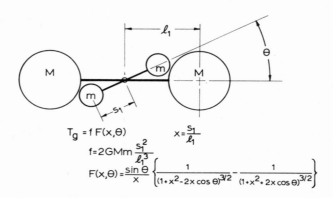

$$T_g = f\, F(x,\Theta) \qquad x = \frac{s_1}{\ell_1}$$

$$f = 2GMm\,\frac{s_1^2}{\ell_1^3}$$

$$F(x,\Theta) = \frac{\sin\Theta}{x}\left\{\frac{1}{(1+x^2-2x\cos\Theta)^{3/2}} - \frac{1}{(1+x^2+2x\cos\Theta)^{3/2}}\right\}$$

Fig. 1.   Torsion balance configuration showing the factorization of
the torque into a scaling part f and an angle-dependent.
part $F(x,\theta)$

moment of the small mass, and other symbols are shown in Fig. 1.
If ℓ_1 = 10 cm, m = 0.5 Kg, M = 5 Kg, and x = 0.5,
then f $\sim$ 0.01 dyne-cm. Tungsten balls with these masses will fit
into these dimensions, but very much less dense materials will not.

Our primary use of scaling is in connection with the signal-to-
noise ratio. The signal due to a change in G is given by

$$d\theta = \frac{\delta T_g}{C} \qquad (2)$$

where $\delta = \dfrac{dT_g}{T_g}$ = the fractional change in gravitational torque during
the measuring period and C is the restoring torque coefficient.

For an angle-independent restraining torque, such as an eddy
current drag or a fiber with many turns, the restoring torque coefficient C is purely gravitational, as can be seen from integrating
equation (1) with respect to angle (Fig. 2). $V(x,\theta)$ provides the

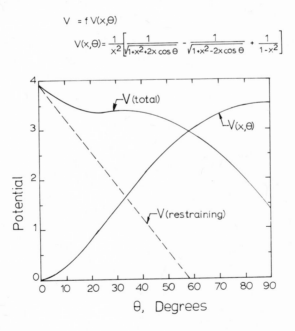

$$V = f\, V(x,\theta)$$

$$V(x,\theta) = \frac{1}{x^2}\left[\frac{1}{\sqrt{1+x^2+2x\cos\theta}} - \frac{1}{\sqrt{1+x^2-2x\cos\theta}} + \frac{1}{1-x^2}\right]$$

Fig. 2. Potential as a function of angle showing the gravitational
and restraining parts.

curvature in the total potential function of angle and hence the restoration to an equilibrium angle.   Thus, $C = \dfrac{\partial^2 V}{\partial\theta^2} = \dfrac{\partial T_g}{\partial\theta}$.   This can be determined by a combination of the scaling factor f and the setting of the equilibrium angle θ_0 by adjustment of the slope of the drag potential V(restraint).   Once a factor f is chosen, varying θ_0 changes C radically with small change in the equilibrium T_g.
    The rms value of the thermal noise is given by the usual relation

$$\Delta\theta_n = \sqrt{\frac{kT}{C}}, \tag{3}$$

where k is Boltzmann's constant and T is the temperature.
    The signal-to-noise ratio is therefore given by

$$S/N = \frac{d\theta}{\Delta\theta_n}\, W = \frac{\delta T_g}{\sqrt{CkT}}\, W = \sqrt{\frac{\delta T_g\, d\theta}{kT}}\, W = \frac{\delta\, T_g}{2\pi}\frac{\tau}{\sqrt{IkT}}\, W \tag{4}$$

where W is a signal-averaging factor related to the way the final data is treated, τ is the natural period of the balance and I is the moment of inertia of the small mass system.

3.   Design of a Cavendish Balance Experiment

    Since W is independent of f, Eq. (4) shows that $S/N \sim \sqrt{f}$.   That is, a larger experiment should improve the signal-to-noise ratio. Other factors such as the effect of external masses will favor a smaller experiment.
    To use Eq. (4) to evaluate parameters of the experiment, we need to determine the signal-averaging factor W.   An expression

$$W = \sqrt{\frac{S}{\tau_e}}, \tag{5}$$

is derived from elementary statistical considerations, assuming independent samples of the signal at intervals τ_e, the effective response time of the experiment.   In the equation S is the total observation period.
    This expression considers the fact that the rms value of fluctuations cannot be reduced below Eq. (3) by varying the damping or the moment of intertia, or by feedback.   Feedback can, however, reduce the effective period τ_e and thereby increase the number of samples in time S.
    If a reasonable experimental strategy is to observe changes in G of 10^{-11} per year in a two month period to an accuracy (S/N) of 10, this puts a great burden of precision on the experiment.   Operating at 10 mK°, using positional and derivative feedback to greatly reduce the value of τ_e, and using masses and lengths as previously given, we arrive at the following parameters: $\tau = 2.44 \times 10^6$ sec (28 days), $C = 1.7 \times 10^{-7}$ dyne-cm/radian and $d\theta = 3.9 \times 10^{-7}$ radians. The accomplishment of this requires that certain feedback criteria and angle sensing sensitivity be met.

Predicted disturbances or drifts in the system are listed in Table I.

Table I: Sources of Disturbances or Drifts

A. Internal
1. Mass Stability
2. Dimensional Stability
3. Angle-sensor Stability
4. Thermal Noise
5. Electrostatic Variations
6. Thermal Variations
7. Pressure Variations
8. Stability of Angular Velocity

B. External
1. Magnetic Field Variations
2. Surrounding Mass Vibrations
3. Ground Vibrations

Mass, dimensional, and angle-sensing stabilities to better than 10^{-12} per year are within present technology, since the system is cooled to $\lesssim 1°$ K. This cooling also establishes an adequate level of thermal noise and is expected to control thermal and electrostatic variations. At the present time, little is known about the variation of small electrostatic forces in a supercooled system, but superconductivity is obviously an important feature in this respect. A superconducting magnetic shield, essential in this experiment, will be a concommittant aspect of the cooled design.

Changes in gravitational gradients, such as tidal effects or extraneous masses, disturb the equilibrium point of the balance. In analogy with the torque calculation of Eq. (1) we can determine the anomalous torque T_a of some extraneous mass M at a distance r. Its relative magnitude is

$$\frac{T_a}{T_g} = \frac{M}{2M} \cdot \frac{\ell_1^3}{r^3}. \tag{6}$$

The instantaneous value of this is 10^{-12} for a 100 Kg man at 2100 meters with a system scaled as above. For this reason Dicke[7] used an octopole configuration to gain a higher inverse power of dependence on distance. In our experiment, however, rotation of the entire system with a period much shorter than the natural period of the balance provides an averaging and filtering function which is the equivalent of an off-resonance driving torque. Thus, the actual amplitude is reduced by a factor of approximately $\frac{\omega_o^2}{\omega^2 Q}$, where ω_o is the resonant frequency of the balance and ω is twice the rotation frequency. The expected parameters are such that $\omega \sim 10^5 \omega_o$ and Q should be $\sim$ unity (critical damping).

Vibrations are expected to be one of the most serious of the disturbances. The nature of the problem can be seen in Fig. 3,

which is an electrical circuit analog. Due to the high sensitivity
(small value of C) needed in the balance, the coupling between mass
systems is extremely weak, represented by a high impedance Z_2. Isolation from ground (the driving point) and the large mass system
will be as good as possible, but cannot be expected to lead to
$Z_1 \gg Z_2$. Methods used in gravitational wave detection will be employed to some extent, but the fact that our signal is nearly dc
as compared with $\sim 10^3$ Hz adds complications. A feedback-corrected
vibrational, isolating support table with response down to 0 Hz will
be used and can be expected to give improvement of a factor of 30.
Fortunately, the same filtering effect of rotation which applied to
external mass effects will also apply to vibration. Judicious design is needed to assure the appropriate directional properties for
the vibrational isolation.

Feedback in the balance will be used to help in many of these
problems, for example, in the reduction of the effect of vibrations
as in Fig. 3. If the impedances Z_1 and Z_2 are equal, the improvement is $\sim 1/2\ g_1 g_2$.

The system differential equation is

$$I\ddot{\theta} + K\dot{\theta} + C\theta = F(t), \tag{7}$$

where the driving function F(t) includes all disturbances, as well
as the signal. Feedback, in addition, is incorporated in F(t).
Position-sensitive feedback, in effect, adds to the stiffness C,
while derivative feedback can increase the damping to optimum without
increasing noise.[8] In order to perform this with appropriately
long time constants, a digital computer will be needed.

Angle sensing must be sensitive and stable, and must not contribute appreciable noise. Two methods are still being considered:
laser interferometry and position-sensing based on the SQUID magnetometer. Either are technically capable of meeting the requirements,
but the laser interferometry is perhaps simpler.

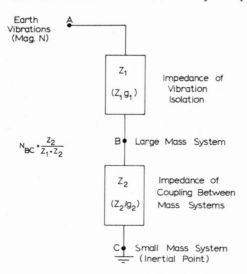

Fig. 3. Electrical circuit
equivalent to the main vibration properties of the experiment.

A concept of design envisaged now is shown in Fig. 4. Four
small mass systems in a symmetrical arrangement can provide intrinsic cancellation of vibrational effects, averaging of noise and internal testing of drift. The signals 1-4 can be summed electronically, mechanically (in part) or optically. Electronic summation of
the four separate sets of data offers the best internal testing of
drift but is the most complex.

Caution must be applied to any attemt to measure to a part in
10^{11}, but the intrinsic aspects of this design do not imply a limitation to our ability to achieve this.

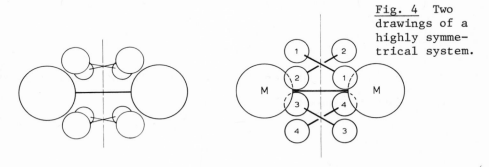

Fig. 4   Two
drawings of a
highly symme-
trical system.

References

1. DIRAC, P.A.M., Nature <u>139</u>, 323 (1937); Proc. Roy. Soc. <u>A165</u>,
    199 (1938).

2. BRANS, C. and DICKE, R.H., Phys. Rev. <u>124</u>, 925 (1961).

3. MISNER, CHARLES W., THORNE, KIP S., and WHEELER, JOHN ARCHIBALD,
    <u>Gravitation</u> (W. H. Freeman and Company, San Francisco, 1973),
    p. 1122.

4. COUNSELMAN, C.C. and SHAPIRO, IRWIN I., Private Communication

5. VAN FLANDERN, T.C., B.A.P.S. II, <u>20</u>, 543 (1975).

6. COUNSELMAN, C.C. and SHAPIRO, I.I., Science <u>162</u>, 352 (1968);
    RICHARDS, D.W., RANKIN, J.M., and ZESSIG, G.A., Nature <u>251</u>,
    37 (1974).

7. ROLL, P.G., KROTKOV, R. and DICKE, R.H., Annals of Physics <u>26</u>,
    442 (1964).

8. MCCOMBIE, C.W., Repts. Prog. in Phys. <u>16</u>, 266 (1953).

Research supported by the National Science Foundation

AN EXPERIMENTAL LIMIT ON THE TIME VARIATION

OF THE FINE STRUCTURE CONSTANT[*]

J. P. Turneaure and S. R. Stein[†]

Department of Physics and High Energy Physics Laboratory

Stanford University, Stanford, California , U.S.A.

INTRODUCTION

The possible time variation of the time structure constant has been investigated using both astronomical[1] and geophysical[2] data. The time variation of the fine structure constant is characterized by R_α which is defined as

$$R_\alpha = \frac{1}{\alpha} \frac{d\alpha}{dt} \quad .$$

The lowest upper limit on $|R_\alpha|$ has been determined using geophysical data[2] and is $|R_\alpha| \leq 5 \times 10^{-15}$ year^{-1} for a measurement time of 10^9 years. The measurement time of 10^9 years is also common to upper limits placed on $|R_\alpha|$ by astronomical data.[1]

Similar upper limits on $|R_\alpha|$ appear to be potentially possible but with much shorter measurement times on the order of one year. This is possible if two types of frequency sources based on appropriately different physical phenomena exist with adequate frequency stability. Frequency sources based on atomic hyperfine transitions, including the hydrogen MASER and Cs frequency standards, have already demonstrated fractional frequency drift rates of 10^{-13} year^{-1} or less.[3] At Stanford, we have been developing a very-stable frequency source[4-7] based on a different physical phenomena which has demonstrated the capability of very small frequency drift rates. This frequency source is called the superconducting-cavity

*   This work supported in part by the U.S. Office of Naval Research.
†   Present Address: National Bureau of Standards, Boulder, Colorado, U.S.A.

stabilized oscillator (SCSO), and it derives its frequency from the resonant frequency of a superconducting cavity. The resonant frequency of the superconducting cavity depends on the cavity size, which is proportional to the interatomic spacing, and on the velocity of light. A calculation[7] of the ratio of the frequency of a hyperfine transition, for example ν_{Cs}, to the resonant frequency of the superconducting cavity results in:

$$\frac{\nu_{Cs}}{\nu_{SCSO}} \cong \text{constant} \times g \left(\frac{m}{M}\right) \alpha^3 \, ,$$

where m is the mass of an electron, g is the gyromagnetic ratio and M is the nuclear mass of the Cs nucleus, and the constant is a pure mathematical number. If one assumes that g and m/M are pure mathematical numbers and uses the fact that $\nu_{SCSO} \simeq \nu_{Cs}$, then

$$|R_\alpha| \cong \frac{1}{3 \, \nu_{SCSO}} \left| \frac{d}{dt} < \nu_{Cs} - \nu_{SCSO} > \right| \, .$$

Comparison of the frequencies of an ensemble of three SCSO's with the frequency of an ensemble of Cs frequency standards has already resulted in a measurement of an upper limit on $|R_\alpha|$ of 3×10^{-11} year^{-1} for a measurement time of 70 days (Ref. 7). This article reports a reduction in the upper limit on $|R_\alpha|$ for a measurement time of 12 days by comparing the frequencies of an ensemble of SCSO's and of an ensemble of Cs frequency standards, and it suggests that further improvements may be made.

EXPERIMENTAL TECHNIQUES

The frequency of an SCSO is derived from the resonant frequency of a superconducting cavity. The resonant frequency of a superconducting cavity is very stable because of its low temperature operation which reduces the dependence of frequency on temperature and reduces frequency drift resulting from thermal activation of stress relaxation and creep. Further, a superconducting cavity can achieve quality factors (Q's) of up to 10^{11} (Ref. 8) which is of great importance to frequency stabilization since the required degree of phase control for a given frequency stability is inversely proportional to Q .

A variety of frequency stabilization methods are possible. We have used a method[6,7] which contains all of the mechanical adjustments and electronics at room temperature, reduces dispersive effects and line length changes of the transmission lines, provides a very large dc open loop frequency gain (~ 220 dB), and provides

a modest useful power output of 25 mW. The stabilization technique
utilizes a voltage tunable Gunn-Effect oscillator of about 8.6 GHz
as the power source. A portion of its output is phase modulated
(PM) at 1 MHz, and this PM signal is transmitted to the cavity.
If the carrier frequency of the PM signal is at the cavity resonant
frequency, a pure PM signal is reflected from the cavity. However,
if the carrier frequency is just above or below the resonant frequency, a 1 MHz amplitude modulated (AM) signal component is reflected from the cavity whose amplitude is proportional to the
deviation from the cavity resonant frequency and whose sign depends
on whether the deviation is above or below the resonant frequency.
The 1 MHz AM signal component is detected, amplified and finally
demodulated with a 1 MHz reference. The demodulated signal is then
dc amplified with appropriate frequency response shaping, dc biased
and applied to the voltage tunable Gunn-Effect oscillator.

TABLE I

SCSO GENERAL OPERATING CONDITIONS

Cavity Temperature	1.1 K
Cavity Fractional Frequency Offset Due to Stored Energy	$(-2 \pm 1) \times 10^{-12}$
Cavity Fractional Frequency Temperature Coefficient at 1.1 K	10^{-10} K^{-1}
Cavity Fractional Frequency Offset Due to Strain in Gravitational Field	$\sim 10^{-9}$
Effective dc Open Loop Frequency Gain	$\sim$ 220 dB

An ensemble of three "essentially" independent SCSO's were
operated in these experiments. Each SCSO consists of a superconducting niobium cavity, an independent temperature servo loop to
control the cavity temperature, a Gunn-Effect oscillator, and a
system of electronics for the frequency stabilization circuit. The
superconducting niobium cavities are cylindrical and are operated
in the TM_{010} mode at about 8.6 GHz. The inside diameter of the
cavities is about 25 mm, and the cavities have massive walls of
about 15 mm to reduce elastic strain in the gravitational field.
The three superconducting cavities are placed in a single copper
vacuum can so each cavity can be independently temperature regulated at some temperature above that of the surrounding helium
bath. The temperature of each cavity can be maintained within
$\pm$ 10 μ K of the set temperature. The nitrogen jacketed helium
dewar can maintain a temperature of 1.1 K for 12 days without

refilling with liquid helium; however, the liquid nitrogen jacket requires refilling twice daily. The helium dewar is located on a vibration isolation pad with about a 10 Hz resonant frequency, and the floor of the room has a relatively low vibrational level. The entire SCSO system is located in an acoustically shielded room. The SCSO general operating conditions are given in Table I, and the individual SCSO cavity characteristics and incident power levels are given in Table II.

TABLE II

SCSO CAVITY CHARACTERISTICS AND INCIDENT POWER

SCSO #	Q_L	Q_o	Incident Power (μW)	Resonant Frequency (GHz)
1	4.8×10^9	39.0×10^9	0.23	8.648
2	12.0×10^9	47.0×10^9	0.26	8.621
3	1.1×10^9	2.5×10^9	2.5	8.605

The frequencies of the ensemble of three SCSO's at Stanford were compared with an ensemble of four Cs frequency standards at the Hewlett-Packard Measurement Standards Laboratory which is located about 20 km from Stanford. At Stanford, a phase lock loop is used to lock a voltage tunable 5 MHz (f_{xo}) quartz oscillator to the frequency of SCSO # 1 (f_1) according to the relationship:

$$f_{xo} = f_1 /(1728 + x) ,$$

where x is some rational fraction. A commercial digital clock driven by the 5 MHz quartz oscillator completes the implementation of a clock with time T_{SCSO1} based on the frequency of SCSO # 1. By similar means, a clock with time T_{Cs} is implemented at Hewlett-Packard based on a Cs frequency standard. Frequency measurements are then accomplished by periodically comparing the time difference $T_{SCSO1} - T_{Cs}$. The time difference is measured using the passive TV line-10 system.[9] For this measurement, it is necessary that both laboratories be located within the line-of-sight of the same television transmitter so that the time delay of arriving information does not change significantly with time. The time transfer is accomplished by measuring the time of arrival of the same distinguishable portion of the TV signal according to the clocks in both laboratories and by communicating this information at a later time. The stability of such time transfers between Hewlett-Packard and Stanford is about 5 ns. At Stanford, the time difference between SCSO # 1, SCSO # 2 and SCSO # 3 are recorded about every 400 s, and thus $T_{SCSO2} - T_{Cs}$ and $T_{SCSO3} - T_{Cs}$ can also be calculated.

MEASUREMENT RESULTS

An experiment utilizing an ensemble of three SCSO's was made from 9-May-75 to 21-May-75 during which nine time transfers were made with a clock based on one of the Cs frequency standards at Hewlett-Packard. During the 12 day measurement period, the ensemble of four Cs frequency standards did not deviate significantly from each other, and thus time transfers were made with respect to only one of the Hewlett-Packard clocks. Unfortunately, on the tenth day of the experiment a dc power failure stopped the Stanford clock for about 1 hour, and thus uninterrupted data only exists for the first 7 days and the last 2 days of the 12 day measurement period.

Two models for extracting the linear fractional frequency drift rate of the SCSO's relative to the Cs frequency standard are used: a time model and a frequency model. The time model assumes that the time difference is modeled by the following equation:

$$T_{SCSO} (t + t_p) - T_{Cs} (t) =$$

$$T_a + F_a \times (t - t_o) + \frac{1}{2} D_a \times (t - t_o)^2 + y(t) ,$$

where t_p is the propagation delay time between the two laboratories, t_o is the initial measurement time, T_a is the initial time difference, F_a is the initial fractional frequency offset, D_a is the relative linear fractional frequency drift rate, and $y(t)$ represents fluctuations. The frequency model assumes that the frequency difference is modeled by the following equation:

$$\frac{\left[T_{SCSO}(t_{i+1} + t_p) - T_{Cs}(t_{i+1})\right] - \left[T_{SCSO}(t_i + t_p) - T_{Cs}(t_i)\right]}{t_{i+1} - t_i}$$

$$= F_b + D_b \times \left(\frac{t_{i+1} + t_i}{2} - \frac{t_2 + t_1}{2}\right) + z(t_i) ,$$

where F_b is the initial fractional frequency offset at $(t_2 + t_1)/2$, D_b is the relative linear fractional frequency drift rate, the subscript i refers to the i'th time transfer, and $z(t_i)$ represents fluctuations.

The results of a least square fit to the frequency model for the entire 12 days of data is given in Table III. The linear fractional frequency drift rates of the SCSO's relative to a Cs frequency standard varies between $+5.6 \times 10^{-14}$ and -6.0×10^{-14}

TABLE III

LINEAR FREQUENCY DRIFT RATE OF SCSO's
RELATIVE TO Cs FREQUENCY STANDARD FOR 12 DAY EXPERIMENT

SCSO #	Fractional Drift (Day^{-1})
1	$- 0.8 \times 10^{-14}$
2	$+ 5.6 \times 10^{-14}$
3	$- 6.0 \times 10^{-14}$

day^{-1}. The estimate of the standard deviation of the population of three measurements is 5.8×10^{-14} day^{-1} ; while the ensemble average is $- 0.4 \times 10^{-14}$ day^{-1} which has a standard deviation equal to 3.4×10^{-14} day^{-1} . It was not possible to analyze the data for the entire 12 days using the time model because of the clock interruption. However, the data for the 7 uninterrupted days of operation were analyzed using both the time model and the frequency model, and those results are consistent with the results for the 12 days of operation. The results for the entire 12 days can be used to calculate an upper limit on $|R_{\alpha}|$. This calculation gives the result that $|R_{\alpha}|$ has a 68% probability of being less than 4.1×10^{-12} $year^{-1}$ for a measurement time of 12 days.

CONCLUSIONS

The result that $|R_{\alpha}|$ has a 68% probability of being less than 4.1×10^{-12} $year^{-1}$ for a 12 day measurement time is only preliminary and is based on a single experiment with an ensemble of three SCSO's. Additional experiments will not be done, however, without a number of modifications. Important correlations of frequency shift with both nitrogen jacket refilling and ambient temperature changes have been observed. Nitrogen jacket refilling will be eliminated by providing continous gas cooling to the nitrogen jacket. There is strong evidence that the correlation of frequency shift with ambient temperature is the result of a temperature dependence of the phase modulator or of dispersion in the transmission lines which gives rise to a temperature dependent AM signal offset. This problem will be reduced by improving the microwave design, controlling the ambient temperature, as well as increasing the Q's of the superconducting cavities. Finally, a tilt meter will be mounted on the apparatus to monitor any correlations of frequency shift with tilt. A considerable reduction in

the upper limit on $|R_\alpha|$ is anticipated; however, with the current
measurement time limited to 12 days and the utilization of Cs
frequency standards for the comparison, upper limits on $|R_\alpha|$ will be
limited to about 5×10^{-13} year^{-1}. To make further reductions,
either the measurement time will have to be increased or the utilization of high performance hydrogen MASER's located in the same
laboratory as the SCSO's will be required.

REFERENCES

1.   BAHCALL, J.N. and SCHMIDT, M., Phys. Rev. Lett. 19, 1294 (1967).
2.   DYSON, F.J., in Aspects of Quantum Theory, A. Salam and E. P.
      Wigner, eds. (Cambridge Univ. Press, Cambridge, 1972) pp. 213-
      236.
3.   HELLWIG, H., in Proc. 28th Annual Symposium on Frequency Control
      (Electronic Industries Association, Wash. D.C., 1974) pp. 315-
      339.
4.   STEIN, S.R. and TURNEAURE, J. P., Electron. Lett. 8, 321 (1972).
5.   STEIN, S. R. and TURNEAURE, J. P., in Proc. 27th Annual Sym-
      posium on Frequency Control (Electronic Industries Association,
      Wash., D.C., 1973) pp. 414-420.
6.   STEIN, S. R. and TURNEAURE, J. P., in Low Temperature Physics-
      LT13 (Plenum Press, N. Y., 1974) pp. 535-541.
7.   STEIN, S. R., Ph.D. Dissertation, Stanford University (1974).
8.   TURNEAURE, J. P. and VIET, NGUYEN TUONG, Appl. Phys. Lett.
      16, 333 (1970).
9.   PARCELIER, P., IEEE Trans. Instrum. Meas. IM-19, 233 (1970).

BOUND ON THE SECULAR VARIATION OF THE GRAVITATIONAL INTERACTION

R. D. Reasenberg and I. I. Shapiro

Massachusetts Institute of Technology

Cambridge, Massachusetts   02139

Abstract

Analysis of several thousand radar observations of the inner planets, spanning the period from 1966 through 1975, discloses no significant change with (atomic) time of the gravitational constant, G. Consideration of this result in combination with those of lunar and other planetary data leads to the (one-standard-deviation) bound

$$|\frac{\dot{G}}{G}| < 10^{-10} \text{ yr}^{-1}.$$

Continued radar observations appear to offer the best hope for a significant improvement in this bound.

I.   Introduction

Speculation about a possible time variation of the gravitational constant has persisted for 50 years and centers mainly on a large, dimensionless number. This number, of the order of 10^{39}, can be formed approximately, either from the present age of the universe, measured in atomic-time units, or from the ratio of the electric to the gravitational attraction between a proton and an electron. If one assumes, as does Dirac (1938), that the near equality of these two dimensionless combinations of physical quantities is no mere coincidence, then one is forced to consider the possibility that the second combination is a function of time such as to maintain its current relation to the first, as the universe ages. This possibility may be viewed as a weakening of the gravitational interaction, or, equivalently, as a decrease of the gravitational constant with time, measured in atomic units.

Under such an hypothesis, Dirac noted that the time kept by a gravitating system, i.e. the period of a two-body orbit, would slow with respect to the time kept by a non-gravitational (atomic) clock. Other theories of gravitation, including the scalar-tensor theory of Brans and Dicke (see, for example, Weinberg, 1972), also predict a time-dependent gravitational interaction. In general, such theories imply that the fractional yearly change in the gravitational constant, G, is

$$|\dot{G}/G| \lesssim 10^{-10}/\text{yr}.$$

In this paper, we describe primarily the use of measurements of the round-trip (atomic time) delays of radar signals reflected from the various inner planets to place an experimental bound on $\dot{G}/G$. We also discuss briefly the determination of $\dot{G}/G$ from optical observations of the moon and planets. All such measurements are potentially very sensitive to $\dot{G}/G$ because the mean longitude, L, of the moon or a planet will show an extra change, quadratic with (atomic) time, if $\dot{G} \neq 0$. More precisely, if two test particles in orbit about the same massive primary were to have identical initial conditions but with the first controlled by $G = G_O + \dot{G}t$, $\dot{G} < 0$, and the second by $G = G_O$, then the mean longitude of the first would lag behind that of the second according to the relation

$$\Delta L(t) \approx \frac{2\pi}{P} \frac{\dot{G}}{G} t^2, \tag{1}$$

easily derived for $|(\dot{G}/G)| \ll P^{-1}$, where P is the orbital period and t the elapsed time.

II. Bound on $\dot{G}/G$ from Radar Observations of the Inner Planets

About 1000 time-delay measurements from radar observations of each of Mercury and Venus were available for analysis as were over 5000 from similar observations of Mars. The data spanned the period from 1966 to 1974. Most of the Mercury and Venus measurements were made at the Haystack (Massachusetts) and Arecibo (Puerto Rico) observatories with some having been made at the Goldstone Tracking Station (California); the Mars measurements consisted of almost equal contributions from these three radar facilities. In addition, we used about 100 values of earth-Mars delay inferred from Mariner 9 measurements which spanned the period from late 1971 to late 1972.

The MIT Planetary Ephemeris Program (PEP) was used to analyze these data. In particular, the linearized likelihood equations were used to obtain estimates of $\dot{G}/G$ and the other relevant parameters. Iteration of the procedure, usually needed for convergence to the maximum-likelihood estimates, was unnecessary in our analysis since the ephemerides used for the relevant solar-system bodies (the nine planets, the moon, the four most massive asteroids, and a ring representation of the remainder) were sufficiently accurate to insure the linearity of the dependence of the computed time delays on each parameter. We also numerically integrated the partial derivatives of the relevant orbital initial conditions with respect to the necessary parameters for use in the linearized likelihood equa-

tions. A more detailed discussion of our approach is given in Ash
et al., 1967 and 1971; a description of the computer program is
given in Ash, 1972.

Our analysis included over 100 numerical experiments in which
the time-delay data were used to estimate, simultaneously with $\dot{G}$,
parameters drawn from the following set: the initial conditions for
the orbits of the inner planets; the initial mean anomaly (orbital
position) of the moon and Jupiter; the masses of the moon, the major
asteroids, and the planets (through Saturn); the coefficients of
two-dimensional Fourier series representations of the topography in
the equatorial region* of each of the inner planets; and parameters
that represent measurement biases as well as those that differentiate
between various metric theories of gravitation. The numerical experiments differed primarily in the choices of the data to be included and of the subset of parameters to be estimated. The purpose
of such an extended study was to expose as clearly as feasible the
effects of possible systematic errors on our results for $\dot{G}$. The most
significant source of errors is undoubtedly the irregularity of the
topography of the target planets. The "peak-to-valley" variations
in topography reach values of the order of 10 kilometers, corresponding to an effect on the (round-trip) time delays of the order of
70 μsec; this value may be contrasted with the delay measurement
accuracy which, for some of the data, has been as low as 0.1 μsec.
Even after removal of the bulk of the lower-frequency effects through
the 123-term Fourier series available to model the topography of each
target planet, a substantial "noise" remains with a possibly important corresponding systematic contribution to the estimate of $\dot{G}$.
The second, and much less serious, source of possible systematic
error is introduced by the high variability of the ionosphere and
the interplanetary medium. The corresponding effects on delay vary
with the inverse square of the operating frequency, f, of the radar
and are negligible for the Haystack (f = 7840 MHz) and Goldstone
(f = 2388 MHz) data, but reach values up to several microseconds
for the Arecibo (f = 430 MHz) data.

The effect of systematic errors on the estimate of $\dot{G}$ is exa-
cerbated by high correlations between this estimate and those of
some of the other parameters. Consider as the prime example of the
"masking" phenomenon, the following situation: For $\dot{G}/G = -10^{-11}$/yr, the
accumulated effect of the orbital phase "lag" ΔL (discussed earlier)
on the earth-Mercury round-trip time-delay measurements would amount,
when Mercury is near elongation, to about 8 μsec over the 9-year
time span of the radar data. However, since one of the independent
combinations of the orbital initial conditions corresponds to the

---

*The radar measurements of time delay are made with respect to the
so-called subradar point on the target planet -- the intersection
with the surface of the line from the radar site to the planet's
center of mass. This point is restricted by the orbital inclinations and spin-axis orientations of the planets to always lie in
the equatorial regions.

initial orbital position and another to the initial average orbital angular velocity, the observable effect of the change ΔL is reduced to a maximum of under 2 μsec: the maximum "excursion" of a parabola $\Delta L(t) = at^2$, $0 \leq t \leq T$, is reduced sixfold after subtraction of the "best-fitting" straight line, $-a/6T^2 + aTt$. The simultaneous estimation of over 100 additional parameters reduces further the uncorrelated part of the effect and increases further the sensitivity to systematic errors. Thus it should be clear that the determination of $\dot{G}/G$, if it is of the order of 10^{-11} per year or smaller, is a non-trivial task.

What, then, have we been able to conclude from our numerical experiments? Some representative results are shown in Table 1. Equation (1) implies that our sensitivity to $\dot{G}$ should be greatest for observations of the innermost planet and least for those of the outermost if ΔL is determined with the same accuracy in both cases. (Recall that the ratio of the orbital periods of the earth and Mercury is about 4 and that of the periods of Mars and the earth about 2.) The result for $\dot{G}/G$ based on the earth-Mars time-delays is relatively more precise because of the higher accuracy of the measurements; however, because the intrinsic effect (ΔL) is smaller for Mars, the result for $\dot{G}/G$ is relatively more prone to the effects of systematic errors. For such reasons we deem it quite unwarranted to conclude that $\dot{G}$ is positive; rather we conclude that our work allows us only to place the following approximate bound on $\dot{G}$:

$$-0.5 \times 10^{-10} \text{ yr}^{-1} < \frac{\dot{G}}{G} < 1.5 \times 10^{-10} \text{ yr}^{-1}. \qquad (2)$$

Nevertheless, this bound represents about a fourfold improvement over our previous result, based on nearly five years of radar data (Shapiro et al., 1971).

III. Bounds on $\dot{G}/G$ from Other Observations

The only other relevant estimates of $\dot{G}/G$ involve analyses of observations of the moon in combination with those of the inner planets. Lunar observations, by themselves, are insufficient because the tidal effects on the moon's orbit completely mask any that might be due to $\dot{G} \neq 0$, and they cannot yet be estimated accurately by any independent means. The first such analysis was carried out by Slade (1971). He considered a combination of optical, spacecraft, and radar observations of the moon's motion, and optical and radar observations of the inner planets. The radar and spacecraft observations of time delays and Doppler shifts were all made with respect to atomic time and spanned a time interval of under 10 years; the optical observations extended over a period of 220 years and, for the most part, were based on the time scale defined by the earth's rotation with only a small fraction ($\approx 7\%$) being based directly on an atomic time scale. The only relevant data omitted were laser observations of the retroreflectors on the moon (too few such observations were then available), and observations of stellar occultations by the moon (insufficient time was available to complete

Table 1

Preliminary Results for $\dot{G}/G$ from Analysis
of Radar Interplanetary Time-Delay Data

Data Set	Estimate of $\dot{G}/G$* (units of 10^{-11} yr^{-1})
Earth-Mercury	6±4
Earth-Venus	6±6
Earth-Mars	25±33

*In an attempt to account for unmodelled systematic effects, we show as uncertainties for the individual results numbers which are threefold larger than the standard errors determined from scaling the rms of the postfit residuals to unity.

---

the necessary modifications to PEP).

Slade analyzed all of his data simultaneously, obtaining the maximum likelihood estimates of $\dot{G}$, a parameter representing the scale of the tidal interaction, and a large number ($\gtrsim 200$) of other relevant parameters. This analysis, carried out with PEP, utilized the same approach as was described in Section II. Slade's preliminary conclusion was given as

$$\left|\frac{\dot{G}}{G}\right| < 6 \times 10^{-11} \text{ yr}^{-1} . \tag{3}$$

This result was not published since it was felt that more sensitivity studies in regard to the possible effects of systematic errors were required before reliable conclusions could be drawn.

More recently, Van Flandern has analyzed those lunar-occultation observations that have been based on atomic time. This analysis yielded a value for the rate of change, $\dot{n}$, of the moon's mean motion which presumably contains, additively, contributions from the tidal interaction and from any possible time variation of G. Van Flandern (1975a) then subtracted the mean of the most "reasonable" of the estimates for $\dot{n}$ obtained by others and based on observations keyed to a "gravitational" clock. The resultant (difference) value of $\dot{n}$ was assumed to represent only the effect of G. The latest published value for $\dot{G}$ obtained using this approach (Van Flandern, 1975b) is

$$\frac{\dot{G}}{G} = (-7.5\pm2.7) \times 10^{-11} \text{ yr}^{-1} . \tag{4}$$

It is our opinion, however, based on our familiarity with the methods used and the various results obtained for both the minuend and the subtrahend, that the standard error given by Van Flandern is considerably underestimated and that his negative value for $\dot{G}/G$ is actually not significantly different from zero.

Other attempts (see, for example, Dicke, 1966 and Van Flandern, 1975a) to infer $\dot{G}$ through use of the earth's rotation to provide a sufficiently stable "non-gravitational" clock (see Weinberg, 1972) are rendered suspect by ill-understood geophysical processes, such as possible core growth, core-mantle coupling, and crustal rebound from glacial melting, whose effects on the earth's rotation could well destroy its usefulness as a clock for this purpose. Alternatively, suggestions (Counselman and Shapiro, 1968) for the use of (atomic-time-based) observations of any changes in the periods of distant, massive, rapid rotators to infer $\dot{G}$ are severely hampered by a lack of sufficient knowledge of other effects that might have greater influence on their rotation.

For the present, then, we can conclude reliably only that

$$\left| \frac{\dot{G}}{G} \right| < 10^{-10} \text{ yr}^{-1}. \tag{5}$$

IV. Possible Improvements in the Bound on $\dot{G}/G$

What are the prospects for improvement? Unless a sufficiently accurate and independent model for the effects of tides on the moon's motion becomes available, continued observations of lunar occultations cannot yield a stricter bound on any possible time variation of G. The reason is simple. The occultation data, as stated above, cannot be used alone to separate tidal from $\dot{G}$ effects accurately*; other observations, such as the older optical observations of the moon and planets, must be used in addition to provide the estimate of the tidal contributions to the moon's orbit. But these older observations cannot be improved substantially by re-analysis. On the other hand, there is no point in making new observations for this indirect purpose of disentangling the tidal and $\dot{G}$ effects. One can do better by using the new observations of the planets to estimate G directly since these observations will be made with respect to an atomic-time scale and since the planetary orbits are virtually free from tidal effects. Radar and spacecraft interplanetary time-delay measurements are by far the most accurate available, and will continue to be so for some time to come. They thus offer the best prospects for improvement in the bound on $\dot{G}$ or for a measurement of it. Even if no further advances are made in modelling planetary topography and no improvements are forthcoming in interplanetary delay measurement accuracy, the bound on $\dot{G}/G$ will decrease approximately as $T^{-5/2}$, where T is the time span of the measurements (Shapiro et al., 1971). Measurement of $\dot{G}/G$ in the laboratory with an accuracy competitive with that presently achieved by astronomical methods may be feasible (Ritter et al., 1975). However, the uncertainty in the bound placed on $\dot{G}/G$ from laboratory measurements can be expected to decrease at best as $T^{-3/2}$, and perhaps no better than $T^{-1/2}$, depending upon the coherence time achieved with the

---

*The lunar laser ranging data suffer from the same problem.

measurement system.

    To conclude, we deem the prospects excellent for obtaining a substantial, nearly tenfold, improvement in the bound on $\dot{G}/G$ by 1985. Should no variation be discerned at this level, it is nonetheless doubtful that theoreticians will cease speculation on the possible causes and consequences of a variation in the gravitational constant. Theoreticians are far too resourceful to be discouraged by a limit so palpably weak as $|\dot{G}/G| < 10^{-11} \text{ yr}^{-1}$.

V. Acknowledgement

We thank D. B. Campbell, R. B. Dyce, R. M. Goldstein, R. R. Green, R. P. Ingalls, G. A. Morris, G. H. Pettengill, and A. E. E. Rogers for the observations of the interplanetary time delays upon which this paper is primarily based. We also thank T. C. Van Flandern for his very helpful comments on his analyses of the lunar occultation data, and R. C. Ritter for information on the proposed laboratory experiment. Our work was supported by the National Science Foundation, Grant No. MPS72-05104 A02 (formerly GP-37107).

VI. References

Dirac, P.A.M., Proc. Roy. Soc. A, 165, 199 (1938).

Weinberg, S., Gravitation and Cosmology (Wiley, New York, 1972).

Ash, M.E., Shapiro, I.I., and Smith, W.B., Astron. J., 72, 338 (1967).

___, Science, 174, 551 (1971).

Ash, M.E., Massachusetts Institute of Technology Lincoln Laboratory Technical Note 1972-5 (1972).

Shapiro, I.I., Smith, W.B., Ash, M.E., Ingalls, R.P., and Pettengill, G.H., Phys. Rev. Lett., 26, 27 (1971).

Slade, M.A., Ph.D. Thesis, Massachusetts Institute of Technology (1971).

Van Flandern, T.C., Mon. Not. Roy. Astron. Soc., 170, 1 (1975a).

Van Flandern, T.C., Bull. Amer. Phys. Soc. Ser. II, 20, 543 (1975b).

Dicke, R.H., in The Earth-Moon System, eds. B.G. Marsden and A.G.W. Cameron (Plenum, New York, 1966), p. 98.

Counselman, C. C., and Shapiro, I.I., Science, 162, 352 (1968).

Ritter, R.C., Beams, J.W., and Lowry, R.A., in these Proceedings (1975).

UNDERSTANDING THE FUNDAMENTAL CONSTANTS

Brandon CARTER

Groupe d'Astrophysique Relativiste

Observatoire de Paris   92190 MEUDON

In order to avoid confusion I wish to make clear at the outset that in this discussion the term "fundamental constants" is not intended to refer to what would be better described as "standard dimensional coefficients", such as Newton's gravitational constant G whose value is well defined only with respect to some(more or less arbitrarily chosen) system of dimensional standards. I intend that the term fundamental constant should be interpreted as refering to quantities that play a genuinely fundamental role in physical theory. At the present stage of development of physics the most important examples are coupling constants such as the electromagnetic("fine structure") constant $e^2/\hbar c$ and the proton gravitational coupling constant $Gm_p^2/\hbar c$ whose values(which are roughly 1/137 and of the order of 10^{-39} respectively) are independent of the choice of measuring standards. Since I do not wish to prejudge the question of whether such quantities can vary in time or space it would perhaps be better to refer to them as fundamental "parameters" rather than "constants".

Before proceeding I would like to express my regret as a theoretician that experimenters do not take more trouble to express their results in terms of genuinely fundamental quantities such as these instead of in terms of quantities such as "G" whose significance depends entirely on sophisticated technicalities such as whether it is specified with respect to an atomic or an astronomical system of measurement. I mention this particular example because the present session of this conference includes two papers describing interesting experiments whose results are unfortunately expressed in terms of "variations of G". To a theoretician, who may have chosen to work with units defined so that G = 1, a result expressed in this form is quite meaningless. What in fact is being measured is variation of

some composite function of fundamental constants, doubtless including the fine structure constant and the proton/electron mass ratio
m_p/m_e as well as the gravitational coupling constant, presumably
in a product of powers of the form $\left(\dfrac{Gm_p^2}{\hbar c}\right)^P \left(\dfrac{e^2}{\hbar c}\right)^q \left(\dfrac{m_p}{\hbar c}\right)^r$ ... .

A so called "varying G" experiment could produce a positive result
that had nothing to do with gravity as a result e.g. of a variation
in the fine structure constant $e^2/\hbar c$,(assuming the index q is
non zero). In order for their results to be theoretically interpreta-
ble, I therefore plead that experimenters should always work out and
clearly state the powers p, q, r etc.. of the various genuinely fundamental constants appearing in the formula for whatever is actually
measured.

Of course the question of what is actually fundamental depends
on the theory one is using. It may be hoped that as our understanding
advances, the number of independent fundamental constants may be able
to be progressively reduced. For example the interelation of what
had previously been independent strong interaction coupling constants
by the SU(3) symmetry scheme was a significant stride in this direction. It may be that we shall ultimately arrive at a theory containing no fundamental constants at all, i.e. a theory in which the fine
structure constant and the rest can all be given by purely mathematical formulae, so that the need for experimental measurement (except
as a check on the theory) will be obviated.

I wish to consider what may be said if this is not the case,
i.e. if there are some coupling constants or other fundamental parameters that really are ultimately fundamental after all. A priori
there is of course not much that can be said if the values are not
determined mathematically. However what most scientists since rennaissance times have tended to overlook is the fact that as far as comparison with observation is concerned it is not the a priori probability of finding a particular parameter value that is relevant, but
rather the a posteriori probability, i.e. the probability subject to
the condition that the measurement can be and has been carried out.

Although most scientists are accustomed – in the name of scientific objectivity – to allow for bias in their apparatus and in themselves as individuals, there has been hardly any serious attempt to
allow for the collective bias of humanity as a whole. A very simple
example of the kind of error into which one may fall (and into which
many people have fallen) is to deduce that because our sun – which
has all the characteristic properties of a typical main sequence
star – has a planetary system, therefore planetary systems must be
a commonly occuring feature of main sequence stars. This argument
overlooks the fact that even if there were only one star in the whole
galaxy(or for that matter the whole visible universe) posessing a
planetary system, the conditions necessary for our existence would
ensure that we could not be anywhere else. Thus in fact our observation of a planetary system here tells us virtually nothing about what
occurs on a typical randomly chosen star. The case for believing in

planetary systems elsewhere must depend entirely on the (at present very meagre) direct observational evidence, and on the (as yet unreliable) detailed theory of stellar formation from diffuse gas clouds.

Errors of this kind result from the fact that post rennaissance scientists have learnt perhaps too well the "Copernican Principle" according to which we must avoid the medieval error of assuming that mankind occupies a priviledged central position in the universe. It is in fact no less important to bear in mind what may be described as the "anthropic principle" according to which our situation in the universe must certainly be untypical, at least to the extent required by the conditions necessary for our existence.

In its "weak" form the anthropic principle may be interpreted as applying only to our situation in space and time. It is this weak anthropic principle that is relevant to the foregoing example concerning planetary systems. Another example, most directly relevant to the discussion of fundamental constants, is the explanation given by Dicke(1961) of thefact that the gravitational coupling constant $Gm_p^2 / \hbar c$ is roughly equal in order of magnitude to the ratio of the proton Compton wavelength (about one fermi) to the Hubble radius of the universe as observed at the present epoch (about 10^{10} light-years).

In the context of the present meeting I wish to draw attention to the pertinence of the more general "strong" anthropic principle which is to be interpreted as applying to our situation in the sense not only of our space-time position but also with respect to all other physically observable features of our environment, including the measured values of fundamental parameters.

A good illustration of the application of the principle is the case of the "strong interaction coupling constant", meaning more precisely the pseudscalar pion-nucleon coupling constant $f^2/ \hbar c$ (whose value is ~ 15). It is possible to argue from the anthropic principle that this coupling constant could not possibly have been observed to differ by too large a factor from the nucleon/pion mass ratio m_p/m_π (with which it agrees in fact within a factor of two) since this would have produced such a drastic modification of nuclear physics as to make life as we know it impossible. It is well knownthat this "strong" coupling constant is in fact only just strong enough for bound states of nucleons to exist at all, and that if it were significantly weaker there would be no other elements than hydrogen, which could hardly provide a chemistry rich enough to allow life. On the other hand if the coupling constant were significantly stronger there could exist pure neutron bound states(i.e. states with zero nuclear electric charge number) with any atomic mass number. Having no Coulomb barriers, such nuclei would combine easily to form compound nuclei of ever inceasing size, so that even quite small bodies would tend to condense to form what would in effect be mini neutron stars. It would require a detailed examination of thermonuclear evolution in such a high strong coupling constant universe to determine how far this tendency would go. However as a guess, it does not seem unlikely that under any conditions where small nuclei could fuse, the interaction might go all the way to the formation of ultra large nuclei instead without leaving any of the medium weight elements(sulphur, iron etc..) on which life as we know it depends.

I would like to see a serious research programme aimed at exploring fully what would happen in this and in analogous eventualities. It seems to me that in order to have a correct understanding of the significance of the actually observed values of fundamental parameters we must always consider how the appearance of the universe, and in particular our own existence would be affected if they were varied.

Having myself worked mainly on gravitation theory, I have been particularly interested in the gravitational coupling constants $Gm_p^2 / \hbar c$, $Gm_e^2 / \hbar c$ etc, and in the question of whether there would be any interesting qualitative effects if their values were altered by a rescaling factor (so that the mass ratios m_p/m_e etc remained unaffected). Since even the electric coupling constant($e^2/\hbar c \sim 1/137$) is so weak that it only affects "fine structure", my first impression was that the gravitational coupling constants($\sim 10^{40}$) would be vastly too weak for there to be any noticable qualitative effects at all, even under rescalling by a factor of ten or more. All that would happen would be that the overall sizes of the various standard kinds of gravitationally bound systems(main sequence stars, white dwarfs, neutron stars etc..) would be correspondingly rescaled, but without any perceptible change in the qualitative characters. However after a prolonged search I finally came accross an interesting phenomenon that <u>would</u> after all be significantly modified, namely the mechanism of energy transport in main sequence stars (Carter 1974). If the gravitational coupling were significantly weaker than its actual value, which is related to the fine structure constant and the electron/proton mass ratio by the rough order of magnitude equality

$$\left(\frac{Gm_p^2}{\hbar c} \right) \sim \left(\frac{e^2}{\hbar c} \right)^{12} \left(\frac{m_e}{m_p} \right)^4$$

then the energy transfer in the bulk of the star (I am not thinking of the extreme central core or surface skin) would be mainly by radiative conduction.If the gravitational coupling constant were significantly stronger, the transfer would be mainly by convection. As it actually is, both cases occur, radiative conduction being predominant in the larger(blue) stars, and convection being predominant in the smaller(red) stars. Convection is also predominant during the Hyashi stage of formation during which the stars are contracting towards their main sequence states. If gravitation were significantly stronger, the convection would be suppressed in this formative stage of stellar evolution, which is also the stage when planetary systems are formed. It is may plausibly be guessed that when the mechanisms of planetary formation are better understood, it may turn out that the convection and the associated turbulence play a vital role.

If correct, this would mean that a stronger gravitational coupling
constant would be incompatible with the very existence of planetary
systems. Thus, using the "strong" anthropic principle, we would have
an explanation of why the gravitational coupling is in fact observed
to be so extremely weak.

References: R.H. DICKE(1961) Nature <u>192</u>, 440.

             B. CARTER(1974) in Confrontations of Cosmological Theo-
ries with Observational Data. P-P.291-298. Proceedings of 1963 IAU.
Symposium, ed. M.Longair).

PRESENT STATUS OF ATOMIC MASSES

A.H. Wapstra

Instituut voor Kernphysisch Onderzoek, Amsterdam
and University of Technology, Delft

The author keeps up to date a list of information
on atomic masses. Regularly (at least once a year) this
list is used as input for a least squares calculation
of "best" values for these masses[x]. Both the list and
the results of the calculation are made available to
interested parties. It has been found necessary, though,
to give some warning about the status of these data.
Their function is definitely not the same as that of
published tables [1]. Just in order to test the validity
of some new data, or that of some old data thrown in
doubt by newer ones, some items are not used in the calculations, or trial data are added, both in a way that
could certainly not be tolerated in a publication. Also,
the intermediate tables are not checked as carefully for
completeness and consistency as published ones.

My method of working is as follows. I personally
collect new data published in the most important journals
in the field (about a dozen), or sent to me by the
authors. This yields considerably over 90 % of the new
data. Almost complete coverage is then obtained by the
help of the Nuclear Data Group, in two ways. They send
me regularly lists of all new reference key numbers of
papers reporting Q-values. And in prepairing their A-
chains, they check my latest list for completeness and

---

[x]   At present these calculations are carried out in
Amsterdam and in Delft by K.Bos.

correctness and give me message. The last task is more
important than it may seem. Information on decay schemes,
which I can not possibly collect completely myself, may
well necessitate revision of reported reaction energies.
Most recently this happened for the $^{74}Se(n,\gamma)^{75}Se$
reaction where dr.Horen, evaluating the data on the level
scheme of ^{75}Se, found that the reported reaction energy
8021.3 ± 1.0 keV had in fact to be increased by 6.1 keV.
As an added precaution - and one that I found not at all
unnecessary - I check the collected data for each mass
number group every time that a new compilation of the
corresponding A-chain appears in the Nuclear Data Sheets.
It is therefore quite important for my work that the
NIRA-scheme recently helped taking care of a revision
of all A-chains, soon to be completed.

The new data collected this way are compared with
the existing files and with various kinds of systematics[2]
They are then either accepted for use in subsequent mass
calculations, or not; in the latter case they are marked
by flags as : probably false, unimportant, or judgment
suspended. Of course, the new data may also lead to
changes in flags attached to earlier data. Especially
the last case is interesting. An example form the mass
values for ^{31}Na and ^{32}Na found in the mass spectroscopic
work of Klapisch' group, which gave a very unexpected
deviation in the systematics of two-neutron separation
energies. In this case, I marked them "judgment suspended"
and added values as would follow from systematics (marked
again with a proper flag; all "SYST" values are treated
in the calculations as if their errors are 1000 keV, for
analytical reasons; the resulting errors can be changed
back again in the printouts [3] to the indication "SYST").
Of course the number of such cases is made as small as
possible in published tables!

But even other values may be added for try-out purposes. A case in point are the ground state masses of
several light $T_z = -2$ nuclides. Using a quadratic isobaric mass equation, one can calculate their mass from
masses of isobaric T = 2 states. The error to be assigned
to these calculated values should contain an allowance
for a possible cubic term; we therefore added the variation caused by a term $+5T_z^3$ keV to the error calculated
from the quadratic IME (root sum squares).

Different opinions are possible about the question
whether such values should be admitted as "experimental"
input data. Values could also be derived from the other
ground state masses of the same isobaric quintet assuming

charge symmetry (these values agree quite well with those
discussed above); but I would certainly not use these.
The important difference is in my opinion that the values
derived in the last way do not add independent experimental material, but the first ones do : they contain the
excitation energies of these T = 2 states. (Such adopted
energies are of course explained in notes attached to the
table of input values.) I feel yet not sure, though, that
these values will remain in next published version of the
mass table.

What is then new in this field, since 1971 ? The
mass energy conversion factor is now [4] 931501.6 ± 2.6
keV/u, not too far from the rather unofficial value
931504 ± 10 keV/u adopted in last mass table calculation;
we hear at the present conference that the new value may
be about two error values off. The decreased error has
the effect that its systematic contribution to the calculated mass excesses or (especially) total binding
energies has been reduced to nearly negligible.

As to mass spectroscopy, an essential contribution
for the lightest masses (A < 20) has been made by L.G.
Smith. They not only yield better mass values, but also
calibration points for (n,γ) reactions that promise a
order of magnitude more precise reaction energies in
that field. Also, they are so precise that in order to
reduce them to atomic mass differences, corrections have
to be made for atomic electron binding energies. This
is only necessary too for a few nuclear reaction measurements: the α-decay of ^{8}Be, a few (p,n) reaction energies
and the β-decay of ^{3}H. The matter of the difference between the last result [5] and Smith's measurement has not
been solved, but it is now known [6] that the first series
of measurements of Smith contain a systematic deviation.
This is a pity since these measurements contain probably
the best available information on the masses of ^{1}H, ^{35}Cl
and ^{37}Cl (the last ones important for interpretation of
mass spectroscopic doublets of the type ^{A}Z ^{37}Cl -
$_{A+2}$Z' ^{35}Cl). How important, can be seen from table 1,
where the last column gives the values obtained without
the use of these data. The mass excess value for ^{1}H has
been given in table 1 both in μu and in keV, in order to
show the influence of the decreased error in the mass
energy conversion constant.

At the opposite end of the periodical system, mass
spectrometric measurements in Minnesota on Th and U
isotopes improved precision of their masses by a factor

Table 1.   Some important atomic mass excess values, in
μu, as determined now and in the 1971 compilation. In parenthesis : same value in keV

	1972		1975		1975[+)]	
^{1}H	7825.22	.04	7825.040	.011	7825.063	.020
	(7289.22	.09)	(7289.037	.023)	(7289.058	.027)
^{4}He	2603.04	.29	2603.266	.047	2603.360	.080
^{14}N	3074.42	.13	3074.015	.023	3074.070	.040
^{16}O	-5084.98	.20	-5085.354	.048	-5085.252	.081
^{35}Cl	-31146.41	.33	-31147.26	.07	-31146.98	.26
^{37}Cl	-34096.96	.27	-34097.37	.11	-34097.15	.26

[+)]   without Smith 1972 values

three and their dependability even more (earlier values
were obtained from a rather longish chain of data). They
now report new values for Sm - and Winnipeg does so for
Lu - that are quite important for determining the absolute mass values in the region of the heavier rare
earths.

The impressive list of new measurments of ^A_Z ^{37}Cl -
$_{A+2}Z'$ ^{35}Cl chlorine doublets of these groups in the same
region determines relative masses with quite high accuracy.

Perhaps the most exciting development in the field
of mass spectroscopy, though,is the success of the Saclay
group in directly measuring the masses of highly unstable
nuclei. They have at once been rewarded by unexpected
results, as discussed above. I hope that this will en-
courage them to go on, and others to share this very
important new branch of research.

As to nuclear reactions, a wealth of data is now
available on (n,γ) reactions. In a few cases it has been
somewhat difficult to encertain which levels in the final
nucleus are fed by which capture gamma rays. Misassign-
ments in this field can lead to serious errors, just
because the assigned errors are so small. A relatively
minor case has been mentioned above and ^{133}Cs(n,γ) has
been another rather serious case. Just for this reason,
independent checks by the chlorine doublets mentioned
above and by parallel (d,p) and (d,t) reactions are important. Considerable progress in the precision with
which such - and other - charged particle reactions can
be determined have been reported in Notre Dame and in

Munich. The first group has, among others, succeeded in solving the matter of the error suspected in last mass adjustment around ^{57}Fe.

The other group, together with the Chalk River one, has also made an important contribution to the measurement of mass differences between isobars by (^{3}He,^{3}H) reactions. This effort is parallel to that of the Harwell group, who unexpectedly found that earlier precision measurements of (p,n) reaction thresholds may have been influenced by near-threshold resonances. Other reactions in this field are (p,γ) and (^{3}He,d). A contribution to the precision of the first ones has now been reported by Rolfs et al. Some important (p,γ) reactions in the neighbourhood of ^{89}Y presented to this conference by the Frankfurt group essentially solve another puzzle noticed in the previous mass table calculation. Still, the total number of available (p,γ) reaction energies is limited. A very useful series of (^{3}He,d) reaction energies, determining some seventeen proton binding energies in odd Z rare earth isotopes, has just been published [7]; such measurements in critical places are very welcome.

Taking these points together, both the dependability and the precision of known masses along the line of beta-stability is in an unprecedented shape. This is demonstrated by table 2 (evidently the most accurately determined masses occur along the line of stability). There are only a few places left where better data on mass differences along the line are urgently needed. About the values of the masses themselves I feel considerably less certain, even though their calculated errors are now all below 10 keV, with as sole exception A = 193-196. Some more "absolute" mass doublet measurement in critical places would be welcome indeed. Fortunately, the real

Table 2   Comparison of number of masses determined with given accuracy (first line, in keV) in 1975(1971)

A	0-10	11-30	31-100	) 100
0 - 60	198(185)	33(28)	29(13)	22(10)
61 - 130	218(197	76(66)	53(57)	52(35)
131 - 200	180(28)	85(218)	71(58)	38(29)
201 - 262	149(86)	64(109)	30(35)	7(5)
0 - 262	745(496)	258(422)	183(163)	119(79

importance of a mass table, anyhow for nuclear physics,
is not in the absolute masses, but in the mass differences.

Summing the numbers in table 2, it is seen that
some 150 new masses have been determined, alsmost all
of them for far unstable nuclides. Already for A < 30
some 18 new masses have been determined, slightly over
half of them for neutron rich nuclides. Except for ^{11}Li
and the Na-isotopes measured in the Orsay mass spectrometer, they have mainly been determined by somewhat un-
orthodox nuclear reactions. This is also true for the
new nuclides in the region 30 - 60, among them Ar isotopes up to A = 46 and new T_z = -1/2 mirror nuclides up
to ^{55}Ni.

In the region 60 - 120 most new data come from beta
desintegration work, among this the beautiful and revealing Isolde work around neutron deficient Kr isotopes.
We also meet here the first delayed proton emitter giving
mass information, ^{73}Kr(ϵ,p)^{72}Se. There is some reason to
distrust this particular item [8] , but around mass number
120 the endpoints of proton spectra in delayed proton
emitters give very welcome information on neutron deficient nuclei such as ^{111}Te and ^{118}Cs, 9 and 15 (!)
neutrons away from stability and yielding their mass
values with standard errors of 70 and 320 keV respectively. In this region two gaps occurred in the system of
mass connections between mass numbers : 81 - 82 and 106 -
107. Both have been closed now in sufficient agreement
with the 1971 mass lists, the first one with not better
than 20 keV standard error.

The next higher mass region saw measurement of
nuclides as neutron rich as ^{132}Sn (8 away from stability)
from β-decay measurements on fission isotopes. But an
even more important break-through occurred for the far
neutron deficient alpha stable isotopes between A = 150
and A = 200. In 1971 three groups were available with
gaps at A = 158 - 167 and at Z = 80 - 82. Both gaps have
been closed very efficiently, so that now complete chains
of alpha particle energies are available between, e.g.,
^{149}Dy and ^{173}Pt (20 neutrons from stability !), and between ^{193}Po and ^{177}Os. Our joy with these results is
marred by two defects : in few of these interesting
chains the mass of any member has been determined experimentally; and in many cases we do not have a real
guarantee that the reported alpha decays really occur
between ground states. We can, however, estimate what
error we make in the average by neglecting the last fact.

In the decay of even-even nuclides we feel quite sure
that the main alpha branch occurs between ground states.
In making a systematics of the reported alpha particle
energies (or even better of their sums over a few successive decays), a systematic difference would at once
show up. From such a study I estimate that this effect
would in this region be taken adequately into account
in a general way by adding 30 ± 30 keV to the reported
energies, for odd A nuclides. As to the first defect,
this can not really be mended. Yet, in the special case
of the ^{173}Pt - ^{149}Dy chain, the last members can be connected to nuclides with known masses by several two-
proton and two-neutron separation energies that can be
obtained from systematics by a very short extrapolation
that gives the feeling of being certainly dependable
within some 200 keV. Thus, a very fair estimate is obtained for the mass of ^{173}Pt, which can be used as a
milestone in estimating the masses of all the many
nuclides between this nuclide and the line of stability.

Problems with alpha decay energies also occur at
higher mass numbers. Some alpha energies being determined with a precision of a few tens of eV's , one might
ask whether corrections for atomic binding energies (as
discussed above) might be required. This would indeed be
so if the energy would be determined from the centre of
an alpha ray peak at a resolution of a few keV; the
correction would then be about 200 eV, not much dependent on the charge number. The standard alpha particle
energies have been determined, however, from an extrapolation procedure [9] which makes the above correction
unnecessary. And since most of the other energies have
been determined by comparison with these standards, no
first order corrections are necessary.

Some beautiful cases of reported alpha transitions
not originating in a ground state occur for odd mass Bi
isotopes between A = 191 and 201. For the mass numbers
197 and 199 the α-decay of the ground state is unobserved.
The data for the other ones indicate a very regular increase of the excitation energies with mass number.
Thus, the known data for the isomers allow a very fair
guess of the ground state decay energies in these two
cases. Also for the heavier nuclides, one has often to
suspect that the observed transition does not feed the
ground state. I have adopted the policy that the input
data list gives then an estimate for the ground state
decay energy - treated in the same way as the "SYST"
values discussed above - but that a remark is added

stating how much this value differs from the decay energy
of the observed transition.

A particular difficult case in this respect occurs
for the newly discovered elements 104 and 105. One could
wonder whether better estimates could be obtained here
from a mass formula. The recent Russian discovery that
the rather pronounced influence of the semi-magic number
152 is so much less important for the newest elements
than for the lower Z ones, form an indication, though,
that both approaches may be problematic here.

References

1) A.H.Wapstra and N.B.Gove, Nuclear Data Tables 9
   (1971) 357.

2) N.B.Gove and A.H.Wapstra, Nuclear Data Tables 9
   (1971) 457.

3) A.H.Wapstra and N.B.Gove, Nuclear Data Tables 9
   (1971) 1, 303.

4) E.R.Cohen and B.N.Taylor, J.Phys.Chem.Ref.Data 2
   (1973) 663.

5) K.E.Bergquist, Nuclear Physics 39 (1972) 37.

6) L.G.Smith and A.H.Wapstra, Phys.Rev. 11 (1975) 1392.

7) D.G. Burke and J.M. Balogh, Can.J.Phys. 53 (1975) 948.

8) K.Bos, N.B.Gove and A.H.Wapstra, Z.Physik 271
   (1974) 115.

9) B.Grennberg and A.Rytz, Metrology 7 (1971) 65.

PRESENT STATUS OF THE FUNDAMENTAL CONSTANTS

Barry N. Taylor*   and   E. Richard Cohen

Electricity Division   Science Center
National Bureau of Standards   Rockwell International
Washington, DC 20234   Thousand Oaks, CA 91360

I. INTRODUCTION

Since the completion of the authors' 1973 least-squares adjustment of the fundamental constants [1], and its subsequent adoption for international use by the CODATA Task Group on Fundamental Constants and the 8th CODATA General Assembly [2], several relevant experiments and theoretical calculations have been completed. It is the purpose of this paper to examine the effect of some of these results, and in as far as possible, the effect of the most important new experiments and calculations reported at this Conference, on the recommended output values of our 1973 least-squares adjustment. It should be emphasized, however, that it is not our intention here to carry out a new adjustment nor to present an updated list of recommended values. Rather, we wish only to identify and examine the areas of major change which have occurred since 1973.

Because of space limitations, we shall be very brief in our discussions and shall lean heavily on our 1973 article, and shall assume the reader is reasonably familiar with its contents.

II. REVIEW OF THE DATA

We can only briefly discuss the new data which have become available during the past two years. We shall be particularly brief in regard to those data which, although significant, will not produce large changes in the recommended values of the fundamental constants. If there has been no significant change with regard to an item, we shall generally not mention it in the paragraphs below; thus unless there is a specific statement to the contrary, the information in CT73 is to be considered to remain valid.

Speed of Light. The CCDM in 1973 recommended for general use the value

$$c = 299792458 \pm 1.2 \text{ m/s} \qquad\qquad (0.004 \text{ ppm}) \qquad (1)$$

based on a number of measurements of the wavelength of the 3.39 μm methane line [3]. This recommendation occurred after the completion of the major portion of the 1973 adjustment which used Evenson's published value, $c = 299792456.2 \pm 1.1$ m/s (0.0035 ppm), which would however be raised to 299792457.4 m/s when expressed in terms of the same definition of the center of the ^{86}Kr line. A new and completely independent experiment by Blaney et al. [4] using a CO_2 stabilized CO_2 laser gives $c = 299792459.0 \pm 0.8$ m/s (0.0027 ppm) which is also in good agreement with the CCDM recommendation.

Electrical Standards. A regular BIPM triennial international comparison of the maintained units of voltage and resistance of the various national laboratories was completed in early 1973 with a central date of 25 February [5]. During the same period measurements of $2e/h$ were also carried out by NBS, NPL, NSL and PTB. These laboratories, except for NPL, have now defined the as-maintained unit of voltage in terms of adopted values of $2e/h$. In general, the volt comparison differences and the $2e/h$ measurements are self-consistent and consistent with the linear drift model used in CT73.

The ratio Ω_{BI69}/Ω used in CT73 was derived from the calculable capacitor measurements carried out at NSL. Although a similar determination has now been completed by Cutkosky at NBS, it suffers from some as-yet unresolved resistance transfer problems [6]. Since the 1973 BIPM triennial resistance intercomparisons are consistent with the previous data, we may consider Eqs. (4.3) and (4.5) of CT73 to be adequate for reducing resistance measurements to Ω_{BI69} units. However, we should for completeness correct Ω_{BI69} in accordance with the CCDM recommendation on the speed of light and use

$$\bar{R} \equiv \Omega_{BI69}/\Omega = 1 - (0.53 \pm 0.19) \text{ ppm} \quad . \qquad (2)$$

Muon g-Factor. A measurement of g_μ which is an order of magnitude more precise than their previous measurement has been carried out by the group at CERN [7]. The new value for the muon g-factor is

$$g_\mu/2 = 1 + a_\mu = 1.001165895(27) \qquad\qquad (0.027 \text{ ppm}) \quad . \qquad (3)$$

The anomaly, a_μ, is 230 ppm smaller than their previous value used in CT73, but the assigned uncertainty of the earlier value is 270 ppm.

Proton Moment in Bohr Magnetons. Phillips et al. [8] have determined $g_j(H)/g_p(H_2O)$ using a hydrogen maser. Their best result at the time of writing is $g_j(H)/g_p(H_2O) = 658.2160103(92)$ at $t = 34.7°C$ which has a quoted uncertainty approximately 1/5 that of Lambe's value given in CT73. This is reduced to its value at 25°C using the temperature coefficient for the proton diamagnetic shielding correction given by Hindman [9], -0.0106 ppm/°C. This is then combined with the values for μ_e/μ_B and $g_j(H)/g_e$ given respectively in Eqs. (6.1) and (7.2b) of CT73, to give

$$\mu'_p/\mu_B = 1.520993136(21) \times 10^{-3} \qquad (0.014 \text{ ppm}) . \qquad (4)$$

On the other hand we may combine the measurement with $g_p(H)/g_p$ and $g_j(H)/g_p(H)$ to obtain a value for the diamagnetic shielding factor

$$\sigma(H_2O, 25°C) = 25.688(17) \times 10^{-6} . \qquad (5)$$

Rydberg Constant. Hänsch et al. [10] have used laser saturation spectroscopy to resolve and measure the fine structure of the Balmer-α line of H and D. From H_α they obtain $R_\infty = 109737.3130(6)$ cm^{-1} and from D_α, $R_\infty = 109737.3150(6)$ cm^{-1}. Neglecting any (possible) discrepancy between these two values and including an allowance for systematic uncertainties, Hänsch et al. give as a final value

$$R_\infty = 109737.3143(10) \text{ cm}^{-1} \qquad (0.009 \text{ ppm}) . \qquad (6)$$

Ampere Determinations. To the one NPL and two NBS values of the ampere included in the 1973 adjustment we can now add the VNIIM determination discussed in the "Notes Added in Proof" in CT73 and a recently reported ASMW measurement [11]. The results are

$$K \equiv A_{BI69}/A = 1 - (2.3 \pm 6.0) \text{ ppm} \qquad \text{VNIIM, 1966-69} \qquad (7.1)$$
$$= 1 + (1.7 \pm 8.0) \text{ ppm} \qquad \text{ASMW, 1974 .} \qquad (7.2)$$

Together with the CT73 data the weighted mean of the five determinations is

$$K = 1 - (0.4 \pm 3.1) \text{ ppm} \qquad (7.3)$$

with a Birge ratio, $R_B = 0.24$.

Faraday. Two new determinations of the Faraday have become available. Craig, et al.'s fifteen year old silver coulometer experiment at NBS has now been repeated with improved precision [12]. The provisional result of these measurements is

NBS: $F = 96486.41(31)\ A_{BI69}\cdot s\cdot mol^{-1}$ (3.2 ppm) .   (8.1)

We consider this new NBS measurement to replace the older measurement. The other new experiment [13] utilizes the coulometric titration of the organic compound 4-aminopyridine ($C_5H_6N_2$), which has not been previously used in a precision Faraday determination. The result, using both oxidation and reduction reactions, is

ISU: $F = 96486.41(82)\ A_{BI69}\cdot s\cdot mol^{-1}$ (8.5 ppm)   (8.2)

when recalculated using the procedure of CT73 to determine the atomic weight of $C_5H_6N_2$ [A = 94.11684(11)], a value 2 ppm less than the atomic weight used by Koch et al. [13]. Equations (8.1) and (8.2) may be combined with the average of the two determinations of Marinenko and Taylor used in the 1973 adjustment; the mean of the three is

$F = 96486.46(27)\ A_{BI69}\cdot s\cdot mol^{-1}$ (2.8 ppm)   (8.3)

with a Birge ratio, $R_B = 0.39$.

Proton Gyromagnetic Ratio. A provisional low field value of γ_p' is now available from the work of Olsen and Williams which has been underway at NBS since 1971 [14-16]. Since this new series of measurements is closely related to and significantly more accurate than the earlier value of CT73, the new value,

NBS: $\gamma_p'(low) = 2.6751340(11)\ 10^8 s^{-1}\cdot T^{-1}_{BI69}$ (0.4 ppm) ,   (9.1)

replaces the older one. As previously noted, the other three available γ_p' (low) values (ETL, NPL, VNIIM) are highly consistent among themselves, giving a weighted mean

$\gamma_p'(low) = 2.6751158(68)\ 10^8 s^{-1}\cdot T^{-1}_{BI69}$ (2.6 ppm)   (9.2)

with a Birge ratio, $R_B = 0.32$. The new NBS value, Eq. (9.1), exceeds the mean of the other three, Eq. (9.2), by 6.8 ppm or 2.6 times the standard deviation of the difference.

A preliminary value of $\gamma_p'(high)$ is also now available from the significantly improved experiment of Kibble and Hunt at NPL [17]. Their new result,

NPL: $\gamma_p'(high) = 2.6751702(54)\ 10^8 A_{BI69}\cdot s\cdot kg^{-1}$ (2.0 ppm), (9.3)

exceeds their earlier value by 36 ppm but is eight times more accurate. This new NPL value exceeds that obtained by Yagola et al. at Kharkov by 15 ppm or 1.9 standard deviations. The mean of the NPL and Kharkov high field measurements is

$$\gamma_p' = 2.6751675(51) \ 10^8 A_{BI69} \cdot s \cdot kg^{-1} \qquad (1.9 \ ppm) \qquad (9.4)$$

with $R_B = 1.96$.

Proton Moment in Nuclear Magnetons. There have been no new measurements of μ_p'/μ_N beyond those reported in CT73. The only change from the 1973 adjustment is the final uncertainty assignment of Petley and Morris reduced from 0.82 ppm to 0.72 ppm (CT73, Notes Added in Proof). The weighted mean of the Mamyrin et al. and the Petley and Morris values is

$$\mu_p'/\mu_N = 2.7927741(10) \qquad (0.37 \ ppm) \qquad (10)$$

with $R_B = 0.43$.

X-Ray Data. The very accurate combined x-ray optical inter-ferometer measurements of Λ and of N_A using high purity silicon single crystals by Deslattes and his collaborators at NBS have made all previous x-ray-dependent data obsolete with respect to the determination of fundamental constants. In comparison with the part-per-million determination of N_A by Deslattes et al. [18],

$$N_A = 6.0220943(63) \ 10^{23} \ mol^{-1} \qquad (1.05 \ ppm) \ , \qquad (11)$$

the older determinations of $N_A \Lambda^3$ which were used in CT73 will have negligible weight. Furthermore, the two measurements of the electron Compton wavelength in CT73 have uncertainties of 15 and 33 ppm and hence, with the fine-structure constant known with an uncertainty of 1 ppm, they in essence become merely determinations of Λ. But it is no longer necessary to include Λ as a variable in a least-squares adjustment of the fundamental constants since its only links to them, $\lambda_C = 10^{10} \alpha^2/2\Lambda R_\infty$ and $N_A \Lambda^3$, are now no more than determinations of Λ itself. Deslattes and Henins [19] have measured Λ to 1 ppm,

$$\Lambda = 1.0020802(10) \qquad (1.0 \ ppm) \ , \qquad (12)$$

in comparison to which even the best determination used in CT73, with an uncertainty of 9.8 ppm, may be disregarded.

Lepton Anomalous Moments. There is now an improved calculation of the third order coefficient in the theoretical expression for a_e [20], $C = 1.195(26)$, which implies a decrease in the derived value of α^{-1} of 1 ppm from that used in CT73. The new result is

$$\alpha^{-1}(a_e) = 137.03549(42) \qquad (3.0 \ ppm) \ . \qquad (13)$$

Although both experimental [7] and theoretical [21] improvements have also occurred in a_μ, the uncertainty of a deduced value of α is still too large for it to be of immediate interest.

Hyperfine Structure. Improvements have been made in both theory [22] and experiment [23–24] for the positronium hyperfine splitting, ν_{hfs}(Ps), but a value of α of sufficient accuracy to be of interest cannot yet be derived. There have been no changes of significance in either the theoretical or experimental values of ν_{hfs}(H). The theory of muonium hfs remains unchanged, but new measurements by the Yale group under V. W. Hughes working at LAMPF [25] are yielding results with a precision of 0.36 ppm. The reported value is ν_{hfs}(M) = 4463301.1(1.6) kHz, which is 2.7 ± 2.2 kHz, (0.6 ± 0.5 ppm) lower than the value given in CT73 (Notes Added in Proof). Uncertainties in the theory however are still of the order of 2 ppm.

Fine Structure. There is some indication that Erickson's theoretical calculation of the Lamb shift [26], S_H = 1057.916(10) MHz may be in error. Mohr has recently carried out an independent calculation of self-energy contributions to the Lamb shift [27] and finds S_H = 1057.864(14) MHz, which is lower by 52 ± 17 kHz, or 3 standard deviations of the difference. The measurements of the shift in the n=2 state of H by Lundeen and Pipkin [28], S_H = 1057.893(20) MHz, lies conveniently between the two. If one uses the Lundeen-Pipkin measurement, the three values of α^{-1} derived from measurements of ΔE–S in CT73 would have to be raised by 1.1 ppm, or if one uses Mohr's calculated value, by 2.4 ppm.

The theory of the fine structure of atomic helium has also improved; Lewis has now calculated the ν_{01} interval to an accuracy of 4 ppm [29] which, when combined with the experimental measurement of Kponou et al. [30] (CT73, Eq. (23.8)), gives

$$\alpha^{-1} = 137.03598(28) \qquad\qquad (2.0\ \text{ppm}) , \qquad (14)$$

and further work is expected to reduce the uncertainty to less than 1 ppm.

Gravitational Constant. Luther et al. at NBS have preliminary results from a new determination of G, to be presented at this conference [31], which are consistent with the value recommended in CT73.

Gas Constant. The determination of the gas constant using an acoustic thermometer by Quinn et al. at NPL [32] yields a result,

$$R = 8.31600(17)\ \text{J·mol}^{-1}\text{·K}^{-1} \qquad\qquad (21\ \text{ppm}) \qquad (15)$$

which is more than 190 ppm larger than that recommended in CT73 on the basis of Batuecas' measurements of the density of O_2. The cause of the apparent discrepancy is not known.

III. ANALYSIS AND CONCLUSIONS

In our 1973 adjustment we discarded four stochastic input data in order to obtain a reasonably consistent set. These were Knowles x-ray value of λ_C, the two measurements at NBS of the Faraday and the Kaufman et al. measurement of the $2P_{3/2} - 2S_{1/2}$ interval in hydrogen. Since the x-ray wavelength scale can now be by-passed in determining the fundamental constants, we can omit all of the older x-ray data from consideration. Furthermore, the changes in the QED data have been so minor since 1973 (except perhaps for the possible change in the Lamb shift mentioned above) that we need not examine them in detail - the interesting changes lie with the new stochastic data.

The discussions in Section II lead to the conclusions that the five direct measurements of $K \equiv A_{BI69}/A$, (Eq. (7.3)), and the three measurements of F, (Eq. (8.3)), are each quite self-consistent but that the four values of $\gamma_p'(low)$, (Eqs. (9.1,9.2)) and the two values of $\gamma_p'(high)$, (Eq. (9.4)) are only marginally so. However the new γ_p' measurements are significantly more accurate than the older ones and the weighted means of the new and the old differ little from the new values themselves (a shift of -0.2 ppm for $\gamma_p'(low)$ and -1.0 ppm for $\gamma_p'(high)$ - each roughly 0.5 σ of the mean and 0.1 σ of the older data). Since our purpose here is to focus on possible changes in the 1973 adjustment as a result of the availability of new data, we shall consider only the new γ_p' data here. We shall therefore restrict our discussion to the inconsistency of K, F, $\gamma_p'(low)$, $\gamma_p'(high)$, and N_A, using the data of Eqs. (7.3), (8.3), (9.1), (9.3) and (11) respectively and including also Eq. (2) for $\bar{R}$ and Eq. (10) for μ_p'/μ_N.

The overall consistency of these data may be examined in several ways. Using the various equations derived in CT73 we may compare them by means of the implied values of K, indicating in the first column the dependence of the calculated quantity on the stochastic data:

Direct, Eq. (7.3) $\qquad\qquad\qquad$ $K - 1 = - (0.4 \pm 3.1)$ ppm

$\left[\gamma_p'(low)/(\mu_p'/\mu_N)^2 N_A^2 \bar{R}^3\right]^{\frac{1}{4}}$ $\qquad = - (3.27 \pm 0.55)$ ppm

$\left[\gamma_p'(low)/\gamma_p'(high)\right]^{\frac{1}{2}}$ $\qquad\quad = - (6.8 \pm 1.0)$ ppm

$\left[\gamma_p'(low)/F(\mu_p'/\mu_N)\right]^{\frac{1}{2}}$ $\qquad = - (9.1 \pm 1.4)$ ppm

These values are to be compared with the 1973 recommended value, $K - 1 = (0.7 \pm 2.7)$ ppm.

An alternative and equivalent comparison can be presented in terms of the implied values of the Faraday:

Direct, Eq. (8.3) $F = 96486.46(27)A_{BI69} \cdot s \cdot mol^{-1}$ (2.8 ppm)

$N_A[\gamma'_p(low)\bar{R}^3]^{\frac{1}{2}}$ $= 96485.32(11)$ (1.1 ppm)

$\gamma'_p(high)/(\mu'_p/\mu_N)$ $= 96486.00(20)$ (2.0 ppm)

$\gamma'_p(low)/K^2(\mu'_p/\mu_N)$ $= 96484.76(60)$ (6.2 ppm).

The 1973 recommended value is $F = 96484.49(50)A_{BI69} \cdot s \cdot mol^{-1}$.

Finally, we compare these data using the implied value of α^{-1}. A "direct" QED value can be obtained from a weighted mean of the QED data of CT73 (Table 23.1), but with the new values of $\alpha^{-1}(a_e)$, Eq. (13), and $\alpha^{-1}(\nu_{01})$, Eq. (14), replacing the old and with minor changes from the new values of R_∞, c and μ'_p/μ_B. The QED-independent values use the defined value of $2e/h$:

QED, CT73 and Eqs. (13,14) $\alpha^{-1} - 137 = 0.03592(11)$ (0.80 ppm)

$[\bar{R}\gamma'_p(low)]^{-\frac{1}{2}}$ $= 0.035982(30)$ (0.22 ppm)

$N_A\bar{R}/F$ $= 0.03436(42)$ (3.0 ppm)

$[K^2\bar{R}\gamma'_p(high)]^{-\frac{1}{2}}$ $= 0.03510(44)$ (3.3 ppm)

$[K^2\bar{R}F(\mu'_p/\mu_N)]^{-\frac{1}{2}}$ $= 0.03478(47)$ (3.4 ppm)

The 1973 recommended value is $\alpha^{-1} = 137.03604(11)$.

It is clear from these comparisons that there are inconsistencies in the data. If we hold the three sub-ppm quantities, $\gamma'_p(low)$, $\bar{R}$, and μ'_p/μ_N fixed, we can partially solve the remaining equations to determine the corrections which are required to achieve consistency. Although we can not solve the system of equations completely, we find that in order to achieve consistency the quantity N_AK^2 must decrease 5.8 ppm, FK^2 must decrease 17.6 ppm and $\gamma'_p(high)K^2$ must decrease 12.8 ppm. Using a least squares fitting to allocate the corrections in such a way as to do minimum violence to the data yields the following approximate adjustments: N_A should increase 2.4 ppm or 2.3 σ; F, decrease 9.4 ppm or 3.3 σ; $\gamma'_p(high)$, decrease 4.6 ppm or 2.3 σ; K decrease 4.1 ppm or 1.3 σ. Of course, at the present moment there is no _experimental_ justification for expecting such changes to occur. Clearly, much work must be done in terms of additional measurements and extensions of theory to resolve these discrepancies if the provisional values used here remain unchanged. In particular the new high precision value of $\gamma'_p(low)$, which when combined with $2e/h$ leads to a value of α with an uncertainty of 0.2 ppm, makes available a new tool for testing QED. To be able to utilize this precision fully will require improvements in theory of one or two orders of magnitude for fine structure and hyperfine structure splittings, particularly for muonium and positronium.

TABLE I.   Possible changes in the 1973 recommended values

Quantity	1973 ppm Uncert.	ppm change in 1973 value, & new ppm uncertainty					
		Case A		Case B		Case C	
α^{-1}	0.82	-0.54	0.30	-0.41	0.30	-0.42	0.30
K	2.6	-4.7	0.54	-4.1	0.56	-7.8	1.1
N_A	5.1	9.7	0.95	8.4	1.0	16	2.2
e	2.9	-4.1	0.69	-3.6	0.70	-7.4	1.2
h	5.4	-8.7	1.2	-7.7	1.2	-15	2.3
m_e	5.1	-9.8	1.0	-8.6	1.1	-16	2.2
F	2.8	5.6	0.67	4.8	0.70	8.7	1.1

We conclude this survey with Table I which compares the 1973 recommended values with the results of three least-squares adjustments labeled A, B and C.   In Case A, the new data discussed here are considered together with the 1973 data except for the x-ray data, the 1972 NBS value of γ_p'(low) which is replaced by Eq. (9.1), and the 1971 NPL value of γ_p'(high) which is replaced by Eq. (9.3). To achieve consistency among sets of similar data (as in CT73) the uncertainties in γ_p'(low) were expanded by a factor of 1.53; γ_p'(high), a factor of 1.96; QED data (including the new data), a factor of 1.31.   For this adjustment the number of input items is 28 and χ^2 is 28.74 with 23 degrees of freedom (R_B = 1.12; the probability of χ^2 exceeding this value by chance is 19%).   For the 1973 adjustment the corresponding values were χ^2 = 14.50 with 21 degrees of freedom and R_B = 0.83.

Case B uses the same data as Case A except that the three Faraday measurements are omitted.   In this case there is a significant decrease in χ^2 to 14.03 with 20 degrees of freedom; the Birge ratio is 0.84, (probability = 83%).   Case C represents a similar censoring of the data; in this case N_A rather than the Faraday data is omitted.   There are now 22 degrees of freedom and χ^2 is 17.96.   The Birge ratio is 0.90 (probability = 71%). The internal consistency of all three of these adjustments, as determined by the values of χ^2, is statistically acceptable. These adjustments therefore serve to indicate the changes in the 1973 values which could be expected if the new experimental data presented at this conference are confirmed.   Of particular significance is the reduced uncertainties which arise from the new part-per-million level data and its importance as a severe test of current theoretical calculations.

IV. REFERENCES

*Supported in part by U.S. National Bureau of Standards Office of Standard Reference Data.

[1] COHEN, E.R., and TAYLOR, B.N., J. Phys. Chem. Ref. Data 2, 663 (1973). We shall refer to this paper as CT73.

[2] CODATA Bulletin No. 11 (December 1973).

[3] TERRIEN, J., Metrologia 10, 9 (1974).

[4] BLANEY, T.G., et al., Nature 251, 46 (1974).

[5] LECLERC, G., Report on the 13th Comparison of the National Standards of Electromotive Force, Bureau International des Poids et Mesures, Sèvres, France (1974); LECLERC, G., Report on the 13th Comparison of the National Standards of Electric Resistance, BIPM, Sèvres, France (1974); see also GALLOP, J.C., and PETLEY, B.W., IEEE Trans. Instrum. Meas. IM-23, 267 (1974); KOSE, V., et al., ibid, p. 271.

[6] CUTKOSKY, R.D., IEEE Trans. Instrum. Meas. IM-23, 305 (1974).

[7] BAILEY, J., et al., Phys. Lett. 558, 420 (1975), and this Conference.

[8] PHILLIPS, W.D., COOKE, W.E., and KLEPPNER, D., private communication, and this Conference.

[9] HINDMAN, J.C., J. Chem. Phys. 44, 4582 (1966).

[10] HÄNSCH, T.W., et al., Phys. Rev. Lett. 32, 1336 (1974).

[11] BENDER, D., and SCHLESOK, W., Metrologia 10, 1 (1974).

[12] BOWER, V.E., and DAVIS, R.S., private communication, and this conference.

[13] KOCH, W.F., HOYLE, W.C., and DIEHL, H., Talanta, to be published, and this Conference.

[14] OLSEN, P.T., and WILLIAMS, E.R., private communication, and this Conference.

[15] OLSEN, P.T., and WILLIAMS, E.R., IEEE Trans. Instrum. Meas. IM-23, 302 (1974).

[16] WILLIAMS, E.R., and OLSEN, P.T., IEEE Trans. Instrum. Meas. IM-21, 376 (1972).

[17] KIBBLE, B.P., private communication, and this Conference.

[18] DESLATTES, R.D., et al., Phys. Rev. Lett. 33, 463 (1974).

[19] DESLATTES, R.D., and HENINS, A., Phys. Rev. Lett. 31, 972 (1973).

[20] CVITANOVIC, P., and KINOSHITA, T., Phys. Rev. D 10, 4007 (1974).

[21] CALMET, J., and PETERMAN, A., preprint CERN/TH 1978 (1975).

[22] SAMUELS, M.A., Phys. Rev. A 10, 1450 (1974).

[23] MILLS, A.P., Jr., and BEARMAN, G.H., Phys. Rev. Lett. 34, 246 (1975).

[24] EGAN, P.O., et al., Bull. Am. Phys. Soc. 20, 703 (1975).

[25] CASPERSON, D.E., et al., Bull. Am. Phys. Soc. 20, 702 (1975).

[26] ERICKSON, G. W., Phys. Rev. Lett. 27, 780 (1971).

[27] MOHR, P.J., Phys. Rev. Lett. 34, 1050 (1975).

[28] LUNDEEN, S.R., and PIPKIN, F.M., Phys. Rev. Lett., to be published, and this Conference.

[29] LEWIS, L.M., Proceedings of the Fourth International Conference on Atomic Physics (1974), to be published.
[30] KPONOU, A. et al., Phys. Rev. Lett. 26, 1613 (1971).
[31] LUTHER, G.G., et al., private communication, and this Conference.
[32] QUINN, T.J., CHANDLER, T.R.D., and COLCLOUGH, A.R., Nature 250, 218 (1974), and this Conference.

SUMMARY OF THE CONFERENCE

E. Richard Cohen

Science Center, Rockwell International

Thousand Oaks, California 91360

I hope that this conference has been a success. Certainly if size and scope are taken as indicators it has been the most successful conference in the AMCO series. As I pointed out in my opening remarks the size of these conferences has grown from an attendance of approximately 65 at the first conference in 1960 to approximately 90 at the Teddington conference in 1971. Today the registration at this conference has passed 200.

The scope of the conference has also grown. In addition to to the topics covered in previous conferences - atomic masses, reaction energetics and Q-values, mass theories, determinations of the values of the fundamental constants - we have also had papers on the application of lasers to metrology and to precise physical measurements; the application of nuclear mass systematics to astrophysics, nucleogenesis and the process of element-building in stars; and the implications of theories of gravitation and general relativity to the question of the constancy of the physical "constants".

The conference therefore was bigger and broader in scope than previous conferences. The true measure of a successful conference however, is not size but the significance of the new material presented, and the evaluation of this is the primary purpose of a conference summary. Unfortunately, I can not do this completely. Because of the necessity to hold simultaneous sessions it was impossible for any one person to hear more than 75% of the papers. Therefore my remarks are to some extent (more so in the area of atomic masses than in the area of fundamental constants) based on a subjective evaluation of what I was attracted to or what I thought to be important or interesting.

In the area of mass determinations we have seen a significant increase in accuracy. The masses of the stable elements are in many cases now known with accuracies of better than 10 keV. Unstable nuclei of course are known with less accuracy. There has been increased emphasis on the measurement of the masses of radioactive elements. Kashy, Wollnik and others have reported on mass spectrometric measurements of reaction products or fission fragments and the direct coupling of the mass analyser and a cyclotron or nuclear reactor.

There has been increased activity reported at this conference on measurements of Q-values of unstable nuclei. These regions of the N-Z plane, the neutron deficient region, (N<Z), and the neutron-rich region, (N>2Z), are extremely interesting. Q-valued determinations here not only are severe tests of nuclear systematics and the various algorithms for predicting the masses of unmeasured (or unmeasurable) nuclei but also provide crucial tests of various mass formulae. In my summary of AMCO-4 I pointed out the two distinct uses of mass formulae: one, as an interpolation algorithm to evaluate masses of unmeasured nuclides or to resolve the experimental question whether a measured mass referred to the ground state or an excited nuclear state; and two, to gain a theoretical understanding of nuclear matter, shell structure and deformed nuclei. Until we have a complete theory, rather than simply a model, of nuclear structure there is no à priori reason why the "best" formula for each of these two very different purposes should be identical. Prof. Bleuler has this week reviewed for us these two domains of mass-formula development. In previous conferences the emphasis in mass formulae appeared to me to be in terms of global models, extensions of Weizsacker-type formulae for the nuclidic-mass surface. The papers presented here appear to concentrate more on "local" formulae, particle separation energies and shell effects.

Energetics and Q-value measurements can be combined with mass doublet measurements only if one has an accurate energy scale and an accurate conversion factor between mass and energy. The accurate determination of the conversion factor was the original link between "atomic masses" and "fundamental constants" in Commission 13 and in the AMCO conferences. In order to provide an accurate energy scale there must be an accurate intercomparison of the several gamma-ray energies which are used as standards in these measurements. A task group was established by the Commission to look at this problem. Some of its work has been reported here in the papers by Helmer et al. and van Assche et al. The inter-calibration of gamma-ray energies however still leaves the value of the mass-energy conversion factor undefined. Wapstra's least squares mass adjustment can take this factor as one of its unknowns. Does this lead to an accurate determination and if so, could it be used as an input datum to Cohen and Taylor's adjustment of the fundamental physical

constants? The 1973 adjustment gives the conversion factor as
931.5016(26) MeV with an uncertainty of 2.8 ppm. However the
calibration of most gamma-ray lines is traced ultimately to a
comparison with the wavelength of the electron annihilation radiation and hence with the energy equivalent of the electron mass.
Therefore the majority of Q-value determinations are actually mass
equivalents in terms of electron mass units and the introduction
of nuclear reaction Q-values into a mass adjustment in reality
involves only the atomic mass of the electron or the ratio
m_p/m_e = 1836.1515. As a result of the accurate determination of
the ratio of the Bohr magneton to the nuclear magneton this ratio
has an uncertainty of less than 0.4 ppm. The determining uncertainty in Q-value measurements is thus the calibration standards and
not the conversion factor itself.

In the first atomic masses conference at McMaster University
in 1960 L. G. Smith described an ingenious new mass spectrometer
design, and in AMCO-IV at Teddington in 1971 he presented his first
accurate measurements. His tragic death last year was deeply felt
by all of us who knew him. The afternoon session on Wednesday at
Orsay was dedicated to him and his work will not be lost because,
as was described by Koets, his instrument has been transferred to
Delft and should be in full operation soon. It is unfortunate
that the scheduling of the sessions was such that Prof. Wapstra's
words to the memory of Lincoln Smith could not have been heard
by the entire conference.

In mass spectrometer development Matsuda described a second-
order double-focussing machine which is under construction and
which is designed to have a resolution of a few tens of thousands,
while Ogata described a two-stage double-focussing machine now in
operation at Osaka University with a resolution of almost one
million. Another interesting new development in mass spectrometers
is the time-of-flight instrument described by Torgerson and
Macfarlane of Texas A&M University which is designed specifically
for the measurement of radioactive nuclei.

Turning to fundamental constants I find there, also, much of
interest. The equivalent of more than a full day was devoted to
the subjects of time, frequency, wavelength, and the velocity of
light. The ability to determine the wavelength and frequency of
the methane stabilized He-Ne laser at 3.39μm by a direct comparison respectively with the [86]Kr length standard and with the
[133]Cs time standard led the CCDM in 1973 to recommend the adoption
of a new value for the velocity of light, c = 299792458 m.s[-1] and
this recommendation has just this week been confirmed, if my understanding is correct, by the CGPM. It is now clear that absorption-
stabilized lasers are superior to the [86]Kr lamp as a length standard,
and a redefinition of the metre is to be expected. It is not yet
clear which particular line should be used, and it may well be that

the optimum laser transition has not yet been found. Certainly the measurements reported here on laser wavelengths, line shapes, and stability will be important in reaching that decision in the future. We are closely approaching the question as to whether independent definitions of the metre and the second are necessary or whether they should be defined by a single atomic transition.

An important development in this area is the two-photon laser, described by Roberts and Fortson, in which the doppler width is cancelled. Further work in this area should determine if the expected higher stability is achieved and demonstrate the extent to which two-photon lasers may be superior to one-photon lasers as metrological standards.

Another interesting paper was the report from MIT presented by Ezekiel on the I_2 molecular-beam-stabilized Ar laser. This paper not only demonstrated the stability of the laser but also showed that it could be used to extract some significant physics of the I_2 molecule. The hyperfine structure of the line at 514.5 nm was interpreted to provide extremely precise determinations of the nuclear electric quadrupole, spin-rotation and spin-spin interaction strengths.

Measurements of the Josephson-effect constant, 2e/h, have become so precise that they now define the realization of the volt and hence also yield information on the realization of the ampere (since the ohm is much more precisely determined). The low-field gyromagnetic ratio then also gives us the most accurate determination of the fine structure constant, and the spectroscopic and atomic beam measurements of hydrogen, helium and muonium have become essentially tests of the adequacy of quantum electrodynamics.

Lest one jump to the conclusion that the 1973 adjustment of the constants is the final word, we should remember that the conference heard from Deslattes of his direct measurement of the Avogadro constant which is somewhat higher than the 1973 recommended value. We also heard of new measurements of the Faraday constant and of measurements with increased precision of the gyromagnetic ratio of the proton – a low field measurement of NBS by Olsen and Williams and a high field measurement at NPL by Kibble. As Taylor pointed out in his lucid presentation of our review of the status of the fundamental constants these four measurements are not mutually consistent, indicating that changes of the order of 5 to 10 ppm may be necessary in one or more of these data.

I should also recall the interesting papers on the possible variations in time of the magnitudes of the physical constants. Studies of this sort are important, philosophically certainly, but also if confirmed, for practical metrology in view of the increased precision with which the fundamental standards of time and length

can be maintained and measured. I would like to thank the program
committee for adding this new area to the scope of AMCO.

Finally I must again thank the local committee, Prof. Grivet,
Prof. Audoin, Prof. Petit and all the others who contributed to
making this conference the success that it appears to be. I also
want to thank the staff which insured the smooth operation of the
conference this week. Thank you.

INDEX